T0141942

Cognitive Science and Technology

Series editor

David M.W. Powers, Adelaide, Australia

More information about this series at http://www.springer.com/series/11554

Jim makes his first step toward developing cephalopod measure for his new class in comparative cognition

James K. Peterson

Calculus for Cognitive Scientists

Partial Differential Equation Models

 Springer

James K. Peterson
Department of Mathematical Sciences
Clemson University
Clemson, SC
USA

ISSN 2195-3988   ISSN 2195-3996   (electronic)
Cognitive Science and Technology
ISBN 978-981-13-5721-3   ISBN 978-981-287-880-9   (eBook)
DOI 10.1007/978-981-287-880-9

© Springer Science+Business Media Singapore 2016
Softcover re-print of the Hardcover 1st edition 2016
This work is subject to copyright. All rights are reserved by the Publisher, whether the whole or part
of the material is concerned, specifically the rights of translation, reprinting, reuse of illustrations,
recitation, broadcasting, reproduction on microfilms or in any other physical way, and transmission
or information storage and retrieval, electronic adaptation, computer software, or by similar or dissimilar
methodology now known or hereafter developed.
The use of general descriptive names, registered names, trademarks, service marks, etc. in this
publication does not imply, even in the absence of a specific statement, that such names are exempt from
the relevant protective laws and regulations and therefore free for general use.
The publisher, the authors and the editors are safe to assume that the advice and information in this
book are believed to be true and accurate at the date of publication. Neither the publisher nor the
authors or the editors give a warranty, express or implied, with respect to the material contained herein or
for any errors or omissions that may have been made.

Printed on acid-free paper

This Springer imprint is published by SpringerNature
The registered company is Springer Science+Business Media Singapore Pte Ltd.

I dedicate this work to my students who have learned this material in its various preliminary versions, to my students who interested in working out models of various small creatures brains, have settled down and devoted months of effort to learning about excitable nerve cells in this abstract way and to the practicing scientists from biology, cognitive science, neurobiology and computational cognitive science who have helped an outsider think better in this very interdisciplinary world. As always, I also thank my family who have listened to my ideas in the living room, over dinner, in my study and on walks in the forest for many years. I hope that this new text helps to inspire all my students to consider mathematics and computer science as indispensable tools in their own work in the trying to understand cognition. This is the last preparatory text on the underlying mathematics and computation to study cognitive systems. In the next volume, the study of neural systems begins!

Acknowledgments

We would like to thank all the students who have used the various iterations of these notes as they have evolved from handwritten to the fully typed version here. We particularly appreciate your interest as it is hard to find students who want to learn how to wear so many hats! We have been pleased by the enthusiasm you have brought to this interesting combination of ideas from many disciplines. These ideas have been taught as an overload to small numbers of students over the last 10 years. We appreciate them very much!

Contents

List of Figures

List of Tables

List of Code Examples

Abstract

This book again tries to show cognitive scientists in training how mathematics, computer science, and science can be usefully and pleasurably intertwined. This is a follow-up text to our first two sets of notes on mathematics for cognitive scientists, which are our companion volumes (J. Peterson, *Calculus for Cognitive Scientists: Derivatives, Integration and Modeling* (Springer, Singapore, 2015 in press)) and (J. Peterson, *Calculus for Cognitive Scientists: Higher Order Models and Their Analysis* (Springer, Singapore, 2015 in press)). In these notes, we add the mathematics and computational tools needed to understand how to compute the terms in the Fourier Series expansions that solve the cable equation. We derive the cable equation from the first principles by going back to cellular biology and the relevant biophysics. We discuss carefully ion movement through cellular membranes and then show how the equations that govern such ion movement lead to the standard transient cable equation. Then we learn how to solve the cable model using separation of variables. We go over carefully why Fourier Series converge and implement MATLAB tools to compute the solutions. We then discuss how to solve linear PDE in general using these tools. Finally, we develop the standard Hodgkin–Huxley model of an excitable neuron and solve it using MATLAB tools. As always, our chosen models are used to illustrate how these three fields influence each other in interesting and useful ways. We also stress our underlying motto: always take the modeling results and go back to the scientists to make sure they retain relevance.

History

Based On:
Research Notes: 1992–1998
Class Notes: MTHSC 982 Spring 1997
Research Notes: 1998–2000
Class Notes: MTHSC 860 Summer Session I 2000
Class Notes: MTHSC 450, MTHSC 827 Fall 2001
Research Notes: Fall 2007 and Spring 2008
Class Notes: MTHSC 450 Fall 2008
Class Notes: MTHSC 450 Fall 2009
Class Notes: MTHSC 450 Spring 2010 and Fall 2010
Class Notes: MTHSC 450 Spring 2011 and Fall 2011
Class Notes: MTHSC 450 Spring 2012
Class Notes: MTHSC 434 Summer Session II 2010 and 2011
Class Notes: MTHSC 450 Fall 2011 and Spring 2012
Class Notes: MTHSC 434 Summer Session II 2012
Class Notes: MTHSC 434 Summer Session II 2014

Part I
Introductory Matter

Chapter 1
Introduction

In this book, we will continue to discuss how to blend mathematics, computational tools and science into a cohesive point of view. Hence, this is a follow up text to our books on mathematics for Cognitive Scientists, (Peterson 2015a, b). In these volumes, we add the new science, mathematics and computational tools to get you to a good understanding of how excitable neurons generate action potentials. Hence, we have chapters on some of the basics of cellular processing and protein transcription and a thorough discussion of how to solve the kinds of linear partial differential equation models that arise. The equations we use to model how an action potential develops are derived by making many assumptions about a great number of interactions inside the cell. We firmly believe that we can do a better job understanding how to build fast and accurate approximations for our simulations of neural systems if we know these assumptions and fully understand why each was made and the domain of operation in which assumptions are valid.

We also spend a lot more time discussing how to develop code with MatLab/Octave to solve our problems numerically. As we have done in Peterson (2015a, b), we continue to explain our coding choices in detail so that if you are a beginner in the use of programming techniques, you continue to grow in understanding so that you can better build simulations of your own. We always stress our underlying motto: always take the modeling results and go back to the scientists to make sure they retain relevance.

Our model of the excitable neuron is based on the cable equation which is a standard equation in biological information processing. The cable equation is one example of a linear model phrased in terms of partial derivatives: hence, it is a linear partial differential equation model or Linear PDE. We begin by solving the time independent version of the cable equation which uses standard tools from linear differential equations in both the abstract infinite length and the finite cable version. The time dependent solutions to this model are then found using the separation of variables technique which requires us to discuss series of functions and Fourier series carefully. Hence, we have to introduce new mathematical tools that are relevant to solving linear partial differential equation models. We also discuss other standard

© Springer Science+Business Media Singapore 2016
J.K. Peterson, *Calculus for Cognitive Scientists*, Cognitive Science
and Technology, DOI 10.1007/978-981-287-880-9_1

Linear PDE models such as the diffusion equation, wave equation and Laplace's equation to give you more experience with the separation of variables technique.

We then use the cable equation to develop a model of the input side of an excitable neuron—the dendrite and then we build a model of the excitable neuron itself. This consists of a dendrite, a cell body or soma and the axon along which the action potential moves in the nervous system. In the process, we therefore use tools at the interface between science, mathematics and computer science.

We are firm believers in trying to build models with explanatory power. Hence, we *abstract* from biological complexity relationships which then are given mathematical form. This mathematical framework is usually not amenable to direct solution using calculus and other such tools and hence part of our solution must include simulations of our model using some sort of computer language. In this text, we focus on using MatLab/Octave as it is a point of view that is easy to get started with. Even simple models that we build will usually have many parameters that must be chosen. For example, in Peterson (2015b) we develop a cancer model based on a singe tumor suppressor gene having two allele states. There are two basic types of cell populations: those with no chromosomal instability and other cells that do have some sort of chromosomal instability due to replication errors and so forth. The two sets of populations are divided into three mutually exclusive sets: cells with both alleles intact, cells having only one allele and cells that have lost both alleles. Once a cell loses both alleles, cancer is inevitable. Hence, the question is which pathway to cancer is most probable: the pathway where there is not chromosomal instability or the pathway which does have chromosomal instabilities. The model therefore has 6 variables—one for each cell type in each pathway and a linear model is built which shows how we think cell types turn into other cell types. The problem is all the two dimensional graphical tools to help us understand such as phase plane analysis are no longer useful because the full system is six dimensional. The model can be solved by hand although it is very complicated and it can be solved using computational tools for each choice of our model's parameters. If you look at this model in Peterson (2015b), you'll see there are 4 parameters of interest. If we had 10 possible levels for each parameter, we would have to solve the model computationally for the 10^4 parameter choices in order to explore this solution space. This is a formidable task and it still doesn't answer our question about which pathway to cancer is dominant. In Peterson (2015b), we make quite a few reasonable assumptions about how this model works to reduce the number of parameters to only 2 and then we approximate the solutions using first order methods to find a relationship between these two parameters which tells us when each pathway is probably dominant. This is the answer we seek: not a set of plots but a algebraic relationship between the parameters which gives insight. Many times these types of questions are the ones we want to ask.

In this book, we are going to build a model of an excitable neuron and to do this we will have to make many approximations. At this point, we will have a model which generates action potentials but we still don't know much about how to link such excitable neuron models together or how to model second messenger systems and so forth. Our questions are much larger and more ill-formed than simply generating a plot of membrane voltage versus time for one excitable neuron. Hence, this entire book is just a starting point for learning how to model networks of neurons. In Peterson (2015c), we develop our understanding further and develop MatLab/Octave code to actually build neural circuit models. But that is another story.

The book (McElreath and Boyd 2007) also caution modelers to develop models that are appropriately abstract yet simple. In McElreath and Boyd (2007, pp. 7–8), we find a nice statement of this point of view. This book looks carefully at how to build models to explain social evolution but their comments are highly relevant to us.

> Simple models never come close to capturing all the details in a situation.... Models are like maps – they are most useful when they contain details of interest and ignore others....Simple models can aid our understanding of the world in several ways. There are ... a very large number of possible accounts of any particular biological ... phenomenon. And since much of the data needed to evaluate these accounts are ... impossible to collect, it can be challenging to narrow down the field of possibilities. But models which formalize these accounts tell us which are internally consistent and when conclusions follow from premises....Formalizing our arguments helps us to understand which [models] are possible explanations. [Also] formal models are much easier to present and explain. ...They give us the ability to clearly communicate what we mean. [Further], simple models often lead to surprising results. If such models always told us what we thought they would, there would be little point in constructing them. ...They take our work in directions we would have missed had we stuck to verbal reasoning, and they help us understand features of the system of study that were previously mysterious.... Simple formal models can be used to make predictions about natural phenomena.

Further, they make the same important points about the need for abstraction and the study of the resulting formal models using abstract methods aided by simulations. In McElreath and Boyd (2007, p. 8) we have

> There is a growing number of modelers who know very little about analytic methods. Instead these researchers focus on computer simulations of complex systems. When computers were slow and memory was tight, simulation was not a realistic option. Without analytic methods, it would have taken years to simulate even moderately complex systems. With the rocketing ascent of computer speed and plummeting price of hardware, it has become increasingly easy to simulate very complex systems. This makes it tempting to give up on analytic methods, since most people find them difficult to learn and to understand.

> There are several reasons why simulations are poor substitutes for analytic models. ...Equations – given the proper training – really do speak to you. They provide intuitions about the reasons a ... system behaves as it does, and these reasons can be read from the expressions that define the dynamics and resting states of the system. Analytic results *tell us* things that we must infer, often with great difficulty, from simulation results.

This is our point with our comments about the cancer model. Simulation approaches for the cancer model do not help us find the parameter relationships we seek to estimate which pathway to cancer is the most likely for a given set of parameter choices. The analytic reasoning we use based on our dynamic models helps us do that. In the same way, simulation approaches alone will not give us insight into the workings of neural systems that can begin to model portions of the brain circuitry humans and other animals have. The purpose of our modeling is always to gain insight.

1.1 Chapter Guide

This text covers the material in bioinformation processing in the following way.

Part I: Introductory Matter   This is the material you are reading now.

Part II: Quantitative Tools   In this part, we introduce computational tools based on *MatLab* (although you can use the open source *Octave* just as easily. The students we teach are not very comfortable with programming in other languages and paradigms and so we have not used code based on *C*, *C++* and so forth although it is available. We discuss and provide code for

- Chapter 2: After introducing integration approximation routines in code, we move on to more abstract discussion of vectors which leads to the idea of vector spaces, linear independence and so forth. We finish with the Graham–Schmidt Orthogonalization process in both theory and code.
- Chapter 3: We look again at Runge–Kutta methods for solving systems of differential equations. We look closely at how the general Runge–Kutta methods are derived and finish with a discussion of an automatic step size changing method called Runge–Kutta Fehlberg.

Part III: Deriving The Cable Model   We now derive the basic cable equation so we can study it. We start with a discussion of basic cell biology and the movement of ions across cell membranes. Then, with the basic cable equation available for study, we then turn to its solution in various contexts. We begin by removing time dependence and discuss its solution in the three flavors: the infinite cable, the half-infinite cable and the finite cable. To do the nonhomogeneous infinite cable model, we introduce the method of variation of parameters and discuss what we mean by linearly independent functions.

- Chapter 4: We go over the biophysics of how ions move inside cells and discuss how proteins are made.
- Chapter 5: We discuss how ions move across biological membranes leading to the derivation of the Nernst–Planck equation and the notion of Nernst equilibrium voltages for different ions. We introduce the idea of signaling with multiple ion species crossing a membrane at the same time.
- Chapter 6: We derive the basic cable model for the change in membrane voltage with respect to space and time. We end with the transient membrane cable equation. We are now need to study how to solve linear partial differential equations.
- Chapter 7: We solve the cable equation when there is no time independence. We start with an infinite cable and finish with the half-infinite cable case.
- Chapter 8: We solve the cable equation when the cable has a finite length. This is messy technically but it rewards your study with a deeper understanding of how these solutions behave. Our method of choice is Variation of Parameters.

Part IV: Excitable Neuron Models   The standard model of an excitable neuron then glues together a dendrite model (a finite length cable) to an equipotential cell body and an axon. The dendrite and cell body have a complicated interface where they share a membrane and the resulting complicated cable equation plus boundary conditions is known as the *Ball and Stick Model*. This model is solved using the separation of variable technique and gives rise to a set of functions whose properties must be examined. The numerical solutions we build then require new tools in *MatLab* on root finding and the solution of linear equations. Once the dendrite cable models have been solved, we use their solution as the input to classical Hodgkin–Huxley models. Hence, we discuss the voltage dependent ion gates and their usual Hodgkin–Huxley model in terms of activation and interaction variables. Once the axon models are built, we then glue everything together and generate action potentials using the dendrite–axon model we have built in this part. The necessary *MatLab* scripts and code to implement these models are fully described.

- Chapter 9: We learn how to solve the cable equation using a new technique called separation of variables. To do this, we need to build what are called infinite series solutions. Hence, once we find the infinite series solution to the cable equation, we need to backtrack and describe some mathematical ideas involving series of functions. We then include a discussion of Fourier series and their convergence. The convergence of these series is a complicated topic and we include a fairly careful explanation of most of the relevant ideas. We finish with a detailed development of the way we would implement Fourier series computations in MatLab/Octave code.
- Chapter 10: We then use separation of variables to solve a variety of additional linear partial differential equations. We finish with an implementation of how to build approximations to the series solutions in MatLab/Octave.
- Chapter 11: We build a primitive, although illuminating, model of an excitable neuron which has a single dendritic cable, a body called the soma and a simplified output cable called the axon. This model is the **Ball Stick** model and it is complicated to solve. The usual method of separation of variables has some complications which we must work around. We provide a full discussion of this and the relevant MatLab/Octave code needed for approximating the solutions.
- Chapter 12: We finish this volume with a nice discussion of a classical Hodgkin–Huxley model. We introduce the Hodgkin–Huxley ion gate model and add it to the already developed Ball Stick model so that we have a full excitable nerve cell model. We include our development of relevant code also.

Our style here is to assume you the reader are willing to go on this journey with us. So we have worked hard at explaining in detail all of our steps. We develop the mathematical tools you need to a large extent in house, so to speak. We also try to present the algorithm design and the code behind our MatLab/Octave experiments in a lot of detail. The interactive nature of the MatLab integrated development environment makes it ideal for doing the exploratory exercises we have given you in these pages. If you work hard on this material, you will be ready for the later journeys which begin to explain bioinformation processing in more detail along with more advanced software techniques.

1.2   Code

All of the code we use in this book is available for download from the site Biological Information Processing (http://www.ces.clemson.edu/~petersj/CognitiveModels. html). The code samples can then be downloaded as the zipped tar ball **Cognitive-Code.tar.gz** and unpacked where you wish. If you have access to MatLab, just add this folder with its sub folders to your MatLab path. If you don't have such access, download and install **Octave** on your laptop. Now Octave is more of a command line tool, so the process of adding paths is a bit more tedious. When we start up an Octave session, we use the following trick. We write up our paths in a file we call **MyPath.m**. For us, this code looks like this

Listing 1.1: How to add paths to Octave

```
function MyPath()
%
s1 = '/home/petersj/MatLabFiles/BioInfo/:';
s2 = '/home/petersj/MatLabFiles/BioInfo/GSO:';
s3 = '/home/petersj/MatLabFiles/BioInfo/HH:';
s4 = '/home/petersj/MatLabFiles/BioInfo/Integration:';
s5 = '/home/petersj/MatLabFiles/BioInfo/Interpolation:';
s6 = '/home/petersj/MatLabFiles/BioInfo/LinearAlgebra:';
s7 = '/home/petersj/MatLabFiles/BioInfo/Nernst:';
s8 = '/home/petersj/MatLabFiles/BioInfo/ODE:';
s9 = '/home/petersj/MatLabFiles/BioInfo/RootsOpt:';
s10 = '/home/petersj/MatLabFiles/BioInfo/Letters:';
s11 = '/home/petersj/MatLabFiles/BioInfo/Graphs:';
s12 = '/home/petersj/MatLabFiles/BioInfo/PDE:';
s13 = '/home/petersj/MatLabFiles/BioInfo/FDPDE:';
s14 = '/home/petersj/MatLabFiles/BioInfo/3DCode';
s = [s1,s2,s3,s4,s5,s6,s7,s8,s9,s12];
addpath(s);
end
```

The paths we want to add are setup as strings, here called **s1** etc., and to use this, we start up Octave like so. We copy **MyPath.m** into our working directory and then do this

Listing 1.2: Set paths in octave

```
octave>> MyPath();
```

We agree it is not as nice as working in MatLab, but it is free! You still have to think a bit about how to do the paths. For example, in Peterson (2015c), we develop two different ways to handle graphs in MatLab. The first is in the directory **GraphsGlobal** and the second is in the directory **Graphs**. They are not to be used together. So if we wanted to use the setup of **Graphs** and nothing else, we would edit the **MyPath.m** file to set **s = [s11];** only. If we wanted to use the **GraphsGlobal** code, we would edit **MyPath.m** so that **s11 = '/home/petersj/MatLabFiles/BioInfo/GraphsGlobal:';** and then set **s = [s11];**. Note the directories in the **MyPath.m** are ours: the main directory is **'/home/petersj/MatLabFiles/BioInfo/** and of course, you will have to edit this file to put your directory information in there instead of ours.

All the code will work fine with **Octave**. So pull up a chair, grab a cup of coffee or tea and let's get started.

References

R. McElreath, R. Boyd, *Mathematical Models of Social Evolution: A Guide for the Perplexed* (University of Chicago Press, Chicago, 2007)

J. Peterson, *Calculus for Cognitive Scientists: Derivatives, Integration and Modeling*, Springer Series on Cognitive Science and Technology (Springer Science+Business Media Singapore Pte Ltd., Singapore, 2015a, in Press)

J. Peterson, *Calculus for Cognitive Scientists: Higher Order Models and Their Analysis*, Springer Series on Cognitive Science and Technology (Springer Science+Business Media Singapore Pte Ltd., Singapore, 2015b, in Press)

J. Peterson, *BioInformation Processing: A Primer On Computational Cognitive Science*. Springer Series on Cognitive Science and Technology (Springer Science+Business Media Singapore Pte Ltd., Singapore, 2015c, in press)

Part II
Quantitative Tools

Chapter 2
Graham–Schmidt Orthogonalization

In our work, we need to talk about vectors and functions using more advanced ideas. In this chapter, we begin by discussing how to integrate functions numerically with MatLab so that we can calculate the inner product of two functions in code. In the second text, we introduced the ideas of vector spaces, linear dependence and independence of vectors. Now we apply these ideas to vector spaces of functions and finish with the Graham–Schmidt Orthogonalization process.

2.1 Numerical Integration

We now discuss how to approximate $\int_a^b f(x)dx$ on the interval $[a, b]$. These methods generate the **Newton–Cotes Formulae**. Consider the problem of integrating the function f on the finite interval $[a, b]$. Let's assume that the interval $[a, b]$ has been subdivided into points $\{x_1, \ldots, x_m\}$ which are uniformly spaced; we let the function values at these points be given by $f_i = f(x_i)$. This gives

$$x_i = a + \frac{i-1}{m-1}(b-a)$$
$$x_1 = a$$
$$x_2 = a + 1 \times h, \quad \text{where } h = \frac{m-1}{b-a}$$
$$\cdots = \cdots$$
$$x_m = a + (m-1) \times h$$

The functions we use to build the interpolating polynomial for the m points x_1 to x_n have the form

$$p_0(x) = 1$$
$$p_j(x) = \Pi_{i=1}^{j}(x - x_i), \quad 1 \leq j \leq m.$$

© Springer Science+Business Media Singapore 2016
J.K. Peterson, *Calculus for Cognitive Scientists*, Cognitive Science
and Technology, DOI 10.1007/978-981-287-880-9_2

Thus,

$$p_0(x) = 1$$
$$p_1(x) = (x - x_1)$$
$$p_2(x) = (x - x_1)(x - x_2)$$
$$p_3(x) = (x - x_1)(x - x_2)(x - x_3)$$
$$\vdots$$
$$p_{m-1}(x) = (x - x_1)(x - x_2) \cdots (x - x_{m-1})$$

The Newton Interpolating polynomial to f for this partition is given by the polynomial of degree $m - 1$

$$P_{m-1}(x) = c_1 p_0(x) + c_2 p_1(x) + c_3 p_2(x) + \cdots + c_m p_{m-1}(x)$$
$$= 1 + c_2 p_1(x) + c_3 p_2(x) + \cdots + c_m p_{m-1}(x)$$

where the numbers c_1 through c_m are chosen so that $P_{m-1}(x_j) = f(x_j)$. For convenience, let's look at a four point uniform partition of $[a, b]$, $\{x_1, x_2, x_3, x_4\}$ and figure out how to find these numbers c_j. Hence the uniform step size here is $h = \frac{b-a}{3}$ here. The Newton interpolating polynomial in this case is given by

$$p_3(x) = c_1 + c_2(x - x_1) + c_3(x - x_1)(x - x_2) + c_4(x - x_1)(x - x_2)(x - x_3)$$

Since we want $p_3(x_j) = f(x_j) = f_j$, we have

$$p_3(x_1) = f_1 = c_1$$
$$p_3(x_2) = f_2 = c_1 + c_2(x_2 - x_1)$$
$$p_3(x_3) = f_3 = c_1 + c_2(x_3 - x_1) + c_3(x_3 - x_1)(x_3 - x_2)$$
$$p_4(x_4) = f_4 = c_1 + c_2(x_4 - x_1) + c_3(x_4 - x_1)(x_4 - x_2)$$
$$+ c_4(x_4 - x_1)(x_4 - x_2)(x_4 - x_3)$$

For convenience, let $x_{ij} = x_i - x_j$. Then we can rewrite the above as

$$f_1 = c_1$$
$$f_2 = c_1 + c_2 x_{21}$$
$$f_3 = c_1 + c_2 x_{31} + c_3 x_{31} x_{32}$$
$$f_4 = c_1 + c_2 x_{41} + c_3 x_{41} x_{42} + c_4 x_{41} x_{42} x_{43}$$

This is the system of equations which is in lower triangular form and so it is easy to solve.

$$
\begin{bmatrix}
1 & 0 & 0 & 0 \\
1 & x_{21} & 0 & 0 \\
1 & x_{31} & x_{31}x_{32} & 0 \\
1 & x_{41} & x_{41}x_{42} & x_{41}x_{42}x_{43}
\end{bmatrix}
\begin{bmatrix}
c_1 \\ c_2 \\ c_3 \\ c_4
\end{bmatrix}
=
\begin{bmatrix}
f_1 \\ f_2 \\ f_3 \\ f_4
\end{bmatrix}
$$

We find $c_1 = f_1$. Then, $c_1 + c_2 x_{21} = f_2$ and letting $f_{ij} = f_i - f_j$, we have $c_2 = f_{21}/x_{21}$. Next, we have

$$
c_1 + c_2 x_{31} + c_3 x_{31} x_{32} = f_3
$$

or

$$
\left(f_{21}/x_{21} \right) x_{31} + c_3 x_{31} x_{32} = f_{31}
$$

$$
c_3 x_{31} x_{32} = f_{31} - f_{21}/x_{21}
$$

$$
c_3 = \left(f_{31}/x_{31} - f_{21}/x_{21} \right)/x_{32}.
$$

The final calculation is the worst. Again, for convenience, let $g_{ij} = f_{ij}/x_{ij}$. Then we can rewrite c_3 as

$$
c_3 = \left(g_{31} - g_{21} \right)/x_{32}.
$$

We know

$$
c_1 + c_2 x_{41} + c_3 x_{41} x_{42} + c_4 x_{41} x_{42} x_{43} = f_4.
$$

Solving for $x_{41} x_{42} x_{43} c_4$, we obtain

$$
c_4 x_{41} x_{42} x_{43} = f_4 - c_1 - c_2 x_{41} - c_3 x_{41} x_{42}
$$

$$
= f_{41} - \left(f_{21}/x_{21} \right) x_{41} - \left(\left(g_{31} - g_{21} \right)/x_{32} \right) x_{41} x_{42}
$$

Now divide through by x_{41} to find

$$
c_4 x_{42} x_{43} = \left(f_{41}/x_{41} \right) - \left(f_{21}/x_{21} \right) - \left(\left(g_{31} - g_{21} \right)/x_{32} \right) x_{42}
$$

$$
= g_{41} - g_{21} - \left(\left(g_{31} - g_{21} \right)/x_{32} \right) x_{42}.
$$

Now divide by x_{42} to solve for $x_{43}c_4$:

$$c_4 x_{43} = \left(g_{41} - g_{21}\right)/x_{42} - \left(g_{31} - g_{21}\right)/x_{32}.$$

which tells us (finally!) that

$$c_4 = \left(\left(g_{41} - g_{21}\right)/x_{42} - \left(g_{31} - g_{21}\right)/x_{32}\right)/x_{43}.$$

Since all the differences $x_{ij} = h$, the uniform step size, these formulae can be then be rewritten.

$$c_1 = f_1$$
$$c_2 = \frac{f_2 - f_1}{h}$$
$$c_3 = \frac{\frac{f_3 - f_1}{2h} - \frac{f_2 - f_1}{h}}{h}$$
$$= \frac{f_3 - 2f_2 + f_1}{2h^2}$$
$$c_4 = \frac{f_4 - 3f_3 + 3f_2 - f_1}{6h^3}$$

Although this is messy to do by hand, it is easy to do in MatLab. First, we set **f** to be the vector of function values $f(x_1), \ldots, f(x_m)$. If **func** is the function we are using in MatLab for the function f, this is done with the lines **f = func(x)** where **x** has already been setup with a **linspace** command.

Listing 2.1: Implementing the differences in Matlab

```
m = length(x);
for k = 1:m−1;
  f(k+1:m) = (f(k+1:m) − f(k) )./(x(k+1:m) − x(k) );
end
c = f;
```

For our example, $m = 4$ and so when $k = 1$, we find **f(2:4) = (f(2:4) − f(1))./(x(2:4) − x(1))**. This looks forbidding, but remember **f(2:4)** is the vector **f(2), f(3), f(4)**. So **f(2:4) − f(1)** becomes **f(2) − f(1), f(3) − f(1), f(4) − f(1)** or the quantities f_{21}, f_{31}, f_{41}. Then **x(2:4) − x(1)** is **x(2) − x(1), x(3) − x(1), x(4) − x(1)** or x_{21}, x_{31}, x_{41}. Hence, when **k = 1**, we find **f(2) = (f(2) − f(1))/ (x(2) − x(1)** or f_{21}/x_{21} **f(3) = (f(3) − f(1))/ (x(3) − x(1)** or f_{31}/x_{31} and **f(4) = (f(4) − f(1))/ (x(4) − x(1)** or f_{41}/x_{41}. At this point the vector **f** contains the proper value of c_1 in **f(1)** and the right value of c_2 in the component $\mathbf{f(2)}$. The values for c_3 and c_4 are not yet right. The next pass through the loop will set c_3 correctly. When $\mathbf{k = 2}$, the MatLab code becomes $\mathbf{f(3:4)} = \mathbf{(f(3:4)} - \mathbf{f(2)}) ./ (\mathbf{x(3:4)} - \mathbf{x(2)})$ which is short for $\mathbf{f(3)} = \mathbf{(f(3)} - \mathbf{f(2)})/(\mathbf{x(3)}-\mathbf{x(2)})$ and $\mathbf{f(4)} = \mathbf{(f(4)} - \mathbf{f(2)})/(\mathbf{x(4)}-\mathbf{x(2)})$. Now remember what $\mathbf{f(2)}$, $\mathbf{f(3)}$ are at this point and plug them in. We find $\mathbf{f(3)}$ has become $(f_{31}/x_{31} - f_{21}/x_{21})/x_{32}$ or $(g_{31} - g_{21})/x_{32}$. This is the proper value c_3 should have. We now have $\mathbf{f(4)}$ is $(f_{41}/x_{41} - f_{21}/x_{21})/x_{42}$ or $(g_{41} - g_{21})/x_{42}$. Finally, the last pass through the loop uses $\mathbf{k = 3}$ and results in the line $\mathbf{f(4:4)} = \mathbf{(f(4:4)} - \mathbf{f(3)})./ (\mathbf{x(4:4)} - \mathbf{x(3)})$ which is just $\mathbf{f(4)} = \mathbf{(f(4)} - \mathbf{f(3)})/ (\mathbf{x(4)} - \mathbf{x(3)})$. Now plug in what we have for $\mathbf{f(3)}$ to obtain

$$f_4 = \left(\left(g_{41} - g_{21} \right)/x_{42} - \left(g_{31} - g_{21} \right)/x_{32} \right)/x_{43}$$

which is exactly the value that c_4 should be. This careful walk through the code is what we all do when we are trying to see if our ideas actually work. We usually do it on scratch paper to make sure everything is as we expect. Typing it out is much harder!

These polynomials are called **Newton interpolants** and we can use these ideas to approximate $\int_a^b f(x)dx$ as follows. We approximate $\int_a^b f(x)ds$ by replacing f by a Newton interpolant on the interval $[a, b]$. For an m point uniformly spaced partition, we will use the $m - 1$ degree polynomial P_{m-1} whose coefficients can be computed recursively like we do above. We replace the integrand f by P_{m-1} to find

$$\int_a^b f(x)dx \approx \int_a^b P_{m-1}(x)dx$$
$$= \int_a^b \left(c_1 + \left(\sum_{k=1}^{m-1} c_{k+1} \Pi_{i=1}^k (x - x_i) \right) \right) dx$$
$$= c_1(b - a) + \sum_{k=1}^{m-1} c_{k+1} \left(\int_a^b \Pi_{i=1}^k (x - x_i) \right) dx$$

To figure out this approximation, we need to evaluate

$$\int_a^b \Pi_{i=1}^k (x - x_i)dx$$

Make the change of variable s defined by $x = a + sh$. Then at $x = a$, $s = 0$ and at $x = b$, $s = (b - a)/h = m - 1$ (m points uniformly spaced means the common interval is $(b - a)/m$). Further, $x - x_i$ becomes $(s - i + 1)h$. Thus, we have

$$\int_a^b \Pi_{i=1}^k (x - x_i)dx = \int_0^{m-1} h^{k+1} \Pi_{i=1}^k (s - i + 1)ds$$

We will let

$$S_{m1} = m - 1$$

$$S_{m,k+1} = \int_0^{m-1} \Pi_{i=1}^k (s - i + 1)ds, \ 1 \leq k \leq m - 1$$

Next, note $c_1(b - a) = c_1 h((b - a)/h)$ and since $(b - a)/h = m - 1$, we have $c_1(b - a) = c_1 h(m - 1)$, which gives us the final form of our approximation:

$$\int_a^b f(x)dx \approx c_1(b - a) + \sum_{k=1}^{m-1} c_{k+1} h^{k+1} S_{m,k+1}$$

$$= c_1 h(m - 1) + \sum_{k=1}^{m-1} c_{k+1} h^{k+1} S_{m,k+1}$$

We can rewrite the sum above letting $n = k + 1$ to

$$\int_a^b f(x)dx \approx c_1 h(m - 1) + \sum_{n=2}^{m} c_n h^n S_{mn} = \sum_{k=1}^{m} c_k h^k S_{mk}.$$

2.1.1 Evaluating S_{mk}

The value of m we choose to use gives rise then to what is called an m point rule and given m we can easily evaluate the needed coefficients S_{m1} through S_{mm}. Here are some calculations:

$$S_{m2} = \int_0^{m-1} sds$$

$$= \frac{(m - 1)^2}{2}$$

$$S_{m3} = \int_0^{m-1} s(s - 1)ds$$

$$= \frac{(m - 1)^2}{6}(2m - 5)$$

$$S_{m4} = \int_0^{m-1} s(s - 1)(s - 2)ds$$

$$= \frac{(m - 1)^2(m - 3)^2}{4}$$

We will denote this Newton Polynomial approximation to this integral using m points by the symbol $Q_{NC(m)}$. Hence, for the case $m = 4$, we have

$$S_{41} = 3$$
$$S_{42} = \frac{9}{2}$$
$$S_{43} = \frac{9}{2}$$
$$S_{44} = \frac{9}{4}$$

This leads to the approximation

$$\int_a^b f(x)dx \approx c_1 h S_{41} + c_2 h^2 S_{42} + c_3 h^3 S_{43} + c_4 h^4 S_{44}$$

$$= 3f_1 h + \frac{9}{2}\frac{f_2 - f_1}{h}h^2 + \frac{9}{2}\frac{f_3 - 2f_2 + f_1}{h^2}h^3 + \frac{9}{4}\frac{f_4 - 3f_3 + 3f_2 - f_1}{6h^3}h^4$$

$$= 3f_1 h + \frac{9}{2}(f_2 - f_1)h + \frac{9}{4}(f_3 - 2f_2 + f_1)h + \frac{9}{24}(f_4 - 3f_3 + 3f_2 - f_1)h$$

$$= 3f_1 h + \frac{9}{2}(f_2 - f_1)h + \frac{9}{4}(f_3 - 2f_2 + f_1)h + \frac{3}{8}(f_4 - 3f_3 + 3f_2 - f_1)h$$

$$= \frac{3h}{8}(8f_1 + 12f_2 - 12f_1 6f_3 - 12f_2 + 6f_1 f_4 - 3f_3 + 3f_2 - f_1)$$

$$= \frac{3h}{8}(f_1 + 3f_2 + 3f_3 + f_4)$$

But h is $\frac{b-a}{3}$ here, so our final 4 point formula is

$$\int_a^b f(x)dx \approx \frac{b-a}{8}(f_1 + 3f_2 + 3f_3 + f_4)$$

2.1.2 Homework

Exercise 2.1.1 *Show that for $m = 2$, we get the* **Trapezoidal Rule**

$$\int_a^b f(x)dx \approx (b-a)\left(\frac{1}{2}f_1 + \frac{1}{2}f_2\right)$$

Exercise 2.1.2 *Show that for $m = 3$, we get the* **Simpson Rule**

$$\int_a^b f(x)dx \approx \frac{b-a}{6}(f_1 + 4f_2 + f_3)$$

2.1.3   Matlab Implementation

We store the Newton–Cotes Weight vectors in this short piece of Matlab code

Listing 2.2: Storing Newton–Cotes Weight Vectors

```
   function w = WNC(m)
   %
   % m     an integer from 2 to 5
   % w     this is the Newton Cotes Weight Vector
 5 %
   if m==2
     w = [1  1]'/2;
   elseif m==3
     w = [1  4  1]'/6;
10 elseif m==4
     w = [1  3  3  1]'/8;
   elseif m==5
     w = [7  32  12  32  7]'/90;
   elseif m==6
15   w = [19  75  50  50  75  19]'/288;
   elseif m==7
     w = [41  216  27  272  27  216  41]'/840;
   elseif m==8
     w = [751  3577  1323  2989  2989  1323  3577  751]'/17280;
20 elseif m==9
     w = [989  5888  -928  10496  -4540  10496  -928  5888  989]'/28350;
   else
     disp('You must use a value of m between 2 and 9');
     w = [0  0]';
25 end
```

The Newton–Cotes implementation is then given by

Listing 2.3: Newton–Cotes Implementation

```
   function numI = QNC(fname,a,b,m)
   %
   % fname   the name of the function which is to integrated
   % a,b     the intergation interval is [a,b]
 5 % m       the order of the Newton-Cotes Method to use
   %         this is the same as the number of points in
   %         the partition of [a,b]
   %
   % numI    the value of the Newton-Cotes approximation to
10 %         the integral
   %
   if m >= 2 & m <= 9
     w = WNC(m);
     x = linspace(a,b,m)';
15   f = feval(fname,x);
     numI = (b-a)*(w'*f);
   else
     disp('You need to use an order m between 2 and 9');
   end
```

and finally, a script to run the numerical integration routines is as follows:

Listing 2.4: Numerical Integration Script

```
1 while input ('Another Example? (1=yes, 0=no). ');
    fname = input ('Enter function name: ');
    a = input ('Enter left endpoint ');
    b = input ('Enter right endpoint: ');
    s = ['QNC(' fname sprintf(',%6.3f,%6.3f,m )',a,b)];
6   disp ([' m            ' s])
    disp (' ')
    for m = 2:9
      numI = QNC(fname,a,b,m);
      disp (sprintf(' %2.0f        %20.16f',m,numI))
11  end
   end
```

2.1.4 Run Time Output

We use the functions defined in **func1.m**, $\left(f(x) = \sin(x) \right)$, and **func2.m**, $\left(f(x) = e^{-x^2} \right)$, which have matlab codes

Listing 2.5: Integrand Function I

```
function y = func1(x)
y = sin(x);
```

Listing 2.6: Integrand Function II

```
function y = func2(u)
%
3 z = -u.*u;
y = exp(z);
```

Here is our runtime. The first example is $\int_0^\pi \sin(x)dx = 1$ and the second is $\int_0^2 e^{-x^2} dx$ which can only be evaluated numerically and has the value approximately 0.882. You can see this by using the built in MatLab numerical integration function **quad**. The following session computes the integral using **quad**: here, we use the anonymous function **h** to pass into **quad**.

Listing 2.7: Using quad

```
1 h = @(x) exp(-x.^2);
  c = quad(h,0,2)
  c = 0.8821
```

We then compare this to the Newton–Cotes methods in the session below.

Listing 2.8: A Newton–Cotes session

```
  ShowNC
2 Another  Example?  (1=yes ,  0=no ).  1
  Enter  function  name:  'func1'
  Enter  left  endpoint  0
  Enter  right  endpoint:  pi/2
     m          QNC( func1 ,  0.000 ,  1.571 ,m )
7
     2               0.7853981633974483
     3               1.0022798774922104
     4               1.0010049233142790
     5               0.9999915654729927
12   6               0.9999952613861668
     7               1.0000000258372352
     8               1.0000000158229039
     9               0.9999999999408976
  Another  Example?  (1=yes ,  0=no ).  1
17 Enter  function  name:  'func2'
  Enter  left  endpoint  0
  Enter  right  endpoint:  2.0
     m          QNC( func2 ,  0.000 ,  2.000 ,m )

22   2               1.0183156388887342
     3               0.8299444678581678
     4               0.8622241875991991
     5               0.8852702891231793
     6               0.8838030970892903
27   7               0.8819161924221999
     8               0.8819818734329694
     9               0.8820864256236417
  Another  Example?  (1=yes ,  0=no ).  0
```

2.2 Linearly Independent Functions

The ideas of linear independence and dependence are hard to grasp, so even though we went through these ideas in Peterson (2015), it is a good idea to do a repeat performance. So let's go back and think about vectors in $\Re^2$. As you know, we think of these as arrows with a tail fixed at the origin of the two dimensional coordinate system we call the x–y plane. They also have a length or magnitude and this arrow makes an angle with the positive x axis. Suppose we look at two such vectors, E and F. Each vector has an x and a y component so that we can write

$$E = \begin{bmatrix} a \\ b \end{bmatrix}, \quad F = \begin{bmatrix} c \\ d \end{bmatrix}$$

The cosine of the angle between them is proportional to the inner product $< E, F > = ac + bd$. If this angle is 0 or π, the two vectors lie along the same line. In any case, the angle associated with E is $\tan^{-1}(\frac{b}{a})$ and for F, $\tan^{-1}(\frac{d}{c})$. Hence, if the two vectors lie on the same line, E must be a multiple of F. This means there is a number β so that

$$E = \beta F.$$

We can rewrite this as

$$\begin{bmatrix} a \\ b \end{bmatrix} = \beta \begin{bmatrix} c \\ d \end{bmatrix}$$

Now let the number 1 in front of E be called $-\alpha$. Then the fact that E and F lie on the same line implies there are 2 constants α and β, both not zero, so that

$$\alpha E + \beta F = 0.$$

Note we could argue this way for vectors in $\Re^3$ and even in $\Re^n$. Of course, our ability to think of these things in terms of lying on the same line and so forth needs to be extended to situations we can no longer draw, but the idea is essentially the same. Instead of thinking of our two vectors as lying on the same line or not, we can *rethink* what is happening here and try to identify what is happening in a more abstract way. If our two vectors lie on the same line, they are not *independent* things in the sense one is a multiple of the other. As we saw above, this implies there was a linear equation connecting the two vectors which had to add up to 0. Hence, we might say the vectors were *not linearly independent* or simply, they are *linearly dependent*. Phrased this way, we are on to a way of stating this idea which can be used in many more situations. We state this as a definition.

Definition 2.2.1 (*Two Linearly Independent Objects*)
Let E and F be two mathematical objects for which addition and scalar multiplication is defined. We say E and F are **linearly dependent** if we can find non zero constants α and β so that

$$\alpha E + \beta F = 0.$$

Otherwise, we say they are **linearly independent**.

We can then easily extend this idea to any finite collection of such objects as follows.

Definition 2.2.2 (*Finitely many Linearly Independent Objects*)
Let $\{E_i : 1 \leq i \leq N\}$ be N mathematical objects for which addition and scalar multiplication is defined. We say E and F are **linearly dependent** if we can find non zero constants α_1 to α_N, not all 0, so that

$$\alpha_1 E_1 + \cdots + \alpha_N E_N = 0.$$

Note we have changed the way we define the constants a bit. When there are more than two objects involved, we can't say, in general, that *all* of the constants must be non zero.

2.2.1 Functions

We can apply these ideas to functions f and g defined on some interval I. By this we mean either

- I is all of $\Re$, i.e. $a = -\infty$ and $b = \infty$,
- I is half-infinite. This means $a = -\infty$ and b is finite with I of the form $(-\infty, b)$ or $(-\infty, b]$. Similarly, I could have the form (a, ∞) or $[a, \infty$,
- I is an interval of the form (a, b), $[a, b)$, $(a, b]$ or $[a, b]$ for finite $a < b$.

We would say f and g are linearly independent on the interval I if the equation

$$\alpha_1 f(t) + \alpha_2 g(t) = 0, \text{ for all } t \in I.$$

implies α_1 and α_2 must both be zero. Here is an example. The functions $\sin(t)$ and $\cos(t)$ are linearly independent on $\Re$ because

$$\alpha_1 \cos(t) + \alpha_2 \sin(t) = 0, \text{ for all } t,$$

also implies the above equation holds for the derivative of both sides giving

$$-\alpha_1 \sin(t) + \alpha_2 \cos(t) = 0, \text{ for all } t,$$

This can be written as the system

$$\begin{bmatrix} \cos(t) & \sin(t) \\ -\sin(t) & \cos(t) \end{bmatrix} \begin{bmatrix} \alpha_1 \\ \alpha_2 \end{bmatrix} = \begin{bmatrix} 0 \\ 0 \end{bmatrix}$$

for all t. The determinant of the matrix here is $\cos^2(t) + \sin^2(t) = 1$ and so picking any t we like, we find the unique solution is $\alpha_1 = \alpha_2 = 0$. Hence, these two functions are linearly independent on $\Re$. In fact, they are linearly independent on any interval I.

This leads to another important idea. Suppose f and g are linearly independent differentiable functions on an interval I. Then, we know the system

$$\begin{bmatrix} f(t) & g(t) \\ f'(t) & g'(t) \end{bmatrix} \begin{bmatrix} \alpha_1 \\ \alpha_2 \end{bmatrix} = \begin{bmatrix} 0 \\ 0 \end{bmatrix}$$

only has the unique solution $\alpha_1 = \alpha_2 = 0$ for all t in I. This tells us

$$\det\left(\begin{bmatrix} f(t) & g(t) \\ f'(t) & g'(t) \end{bmatrix} \right) \neq 0$$

for all t in I. This determinant comes up a lot and it is called the **Wronskian** of the two functions f and g and it is denoted by the symbol $W(f, g)$. Hence, we have the implication: if f and g are linearly independent differentiable functions, then

$W(f, g) \neq 0$ for all t in I. What about the converse? If the Wronskian is never zero on I, then the system

$$\begin{bmatrix} f(t) & g(t) \\ f'(t) & g'(t) \end{bmatrix} \begin{bmatrix} \alpha_1 \\ \alpha_2 \end{bmatrix} = \begin{bmatrix} 0 \\ 0 \end{bmatrix}$$

must have the unique solution $\alpha_1 = \alpha_2 = 0$ at each t in I also. So the converse is true: if the Wronskian is not zero on I, then the differentiable functions f and g are linearly independent on I. We can state this formally as a theorem.

Theorem 2.2.1 (Two Functions are Linearly Independent if and only if their Wronskian is not zero)
If f and g are differentiable functions on I, the **Wronskian** *of f and g is defined to be*

$$W(f, g) = \det\left(\begin{bmatrix} f(t) & g(t) \\ f'(t) & g'(t) \end{bmatrix}\right).$$

where $W(f, g)$ is the symbol for the Wronskian of f and g. Sometimes, this is just written as W, if the context is clear. Then f and g are linearly independent on I if and only if $W(f, g)$ is non zero on I.

Proof See the discussions above. ∎

If f, g and h are twice differentiable on I, the Wronskian uses a third row of second derivatives and the statement that these three functions are linearly independent on I if and only if their Wronskian is non zero on I is proved essentially the same way. The appropriate theorem is

Theorem 2.2.2 (Three Functions are Linearly Independent if and only if their Wronskian is not zero)
If f, g and h are twice differentiable functions on I, the **Wronskian** *of f, g and h is defined to be*

$$W(f, g, h) = \det\left(\begin{bmatrix} f(t) & g(t) & h(t) \\ f'(t) & g'(t) & h'(t) \\ f''(t) & g''(t) & h''(t) \end{bmatrix}\right).$$

where $W(f, g, h)$ is the symbol for the Wronskian of f and g. Then f, g and h are linearly independent on I if and only if $W(f, g, h)$ is non zero on I.

Proof The arguments are similar, although messier. ∎

For example, to show the three functions $f(t) = t$, $g(t) = \sin(t)$ and $h(t) = e^{2t}$ are linearly independent on $\Re$, we could form their Wronskian

$$W(f, g, h) = \det\left(\begin{bmatrix} t & \sin(t) & e2t \\ 1 & \cos(t) & 2e^{2t} \\ 0 & -\sin(t) & 4e^{2t} \end{bmatrix}\right) = t\begin{bmatrix} \cos(t) & 2e^{2t} \\ -\sin(t) & 4e^{2t} \end{bmatrix} - \begin{bmatrix} \sin(t) & e2t \\ -\sin(t) & 4e^{2t} \end{bmatrix}$$

$$= t\left(e^{2t}(4\cos(t) + 2\sin(t))\right) - \left(e^{2t}(4\sin(t) + \sin(t))\right)$$

$$= e^{2t}\left(4t\cos(t) + 2t\sin(t) - 5\sin(t)\right).$$

Since, e^{2t} is never zero, the question becomes is

$$4t\cos(t) + 2t\sin(t) - 5\sin(t)$$

zero for all t? If so, that would mean the functions $t\sin(t)$, $t\cos(t)$ and $\sin(t)$ are linearly dependent. We could then form another Wronskian for these functions which would be rather messy. To see these three new functions are linearly independent, it is easier to just pick *three* points t from $\Re$ and solve the resulting linearly dependence equations. Since $t = 0$ does not give any information, let's try $t = -\pi$, $t = \frac{\pi}{4}$ and $t = \frac{\pi}{2}$. This gives the system

$$\begin{bmatrix} -4\pi & 0 & 0 \\ \pi & 2\frac{\pi}{4}\frac{\sqrt{2}}{2} & -5\frac{\sqrt{2}}{2} \\ 0 & 2\frac{\pi}{2} & -5 \end{bmatrix} \begin{bmatrix} \alpha_1 \\ \alpha_2 \\ \alpha_3 \end{bmatrix} = \begin{bmatrix} 0 \\ 0 \\ 0 \end{bmatrix}$$

in the unknowns α_1, α_2 and α_3. We see immediately $\alpha - 1 = 0$ and the remaining two by two system has determinant $\frac{\sqrt{2}}{2}(-10\frac{\pi}{4}) + 10\frac{\pi}{2} \neq 0$. Hence, $\alpha_2 = \alpha_3 = 0$ too. This shows $t\sin(t)$, $t\cos(t)$ and $\sin(t)$ are linearly independent and show the line $4t\cos(t) + 2t\sin(t) - 5\sin(t)$ is not zero for all t. Hence, the functions $f(t) = t$, $g(t) = \sin(t)$ and $h(t) = e^{2t}$ are linearly independent. As you can see, these calculations become messy quickly. Usually, the Wronskian approach for more than two functions is too hard and we use the *pick three suitable points t_i, from I* approach and solve the resulting linear system. If we can show the solution is always 0, then the functions are linearly independent.

2.2.2 Homework

Exercise 2.2.1 *Prove e^t and e^{-t} are linearly independent on $\Re$.*

Exercise 2.2.2 *Prove e^t and e^{2t} are linearly independent on $\Re$.*

Exercise 2.2.3 *Prove $f(t) = 1$ and $g(t) = t^2$ are linearly independent on $\Re$.*

Exercise 2.2.4 *Prove e^t, e^{2t} and e^{3t} are linearly independent on $\Re$. Use the pick three points approach here.*

Exercise 2.2.5 *Prove* $\sin(t)$, $\sin(2t)$ *and* $\sin(3t)$ *are linearly independent on* $\Re$. *Use the pick three points approach here.*

Exercise 2.2.6 *Prove* 1, *t and* t^2 *are linearly independent on* $\Re$. *Use the pick three points approach here.*

2.3 Vector Spaces and Basis

We can make the ideas we have been talking about more formal. If we have a set of objects u with a way to add them to create new objects in the set and a way to *scale* them to make new objects, this is formally called a **Vector Space** with the set denoted by $\mathcal{V}$. For our purposes, we scale such objects with either real or complex numbers. If the scalars are real numbers, we say V is a vector space over the reals; otherwise, it is a vector space over the complex field.

Definition 2.3.1 (*Vector Space*)
Let $\mathcal{V}$ be a set of objects u with an additive operation $\oplus$ and a scaling method $\odot$. Formally, this means

1. Given any u and v, the operation of adding them together is written $u \oplus v$ and results in the creation of a new object w in the vector space. This operation is *commutative* which means the order of the operation is not important. Also, this operation is associative as we can group any two objects together first, perform this addition $\oplus$ and then do the others and the order of the grouping does not matter.
2. Given any u and any number c (either real or complex, depending on the type of vector space we have), the operation $c \odot u$ creates a new object. We call such numbers *scalars*.
3. The scaling and additive operations are compatible in the sense that they satisfy the *distributive* laws for scaling and addition.
4. There is a special object called o which functions as a *zero* so we always have $o \oplus u = u \oplus o = u$.
5. There are *additive inverses* which means to each u there is a unique object $u^\dagger$ so that $u \oplus u^\dagger = o$.

Comment 2.3.1 *These laws imply*

$$(0+0) \odot u = (0 \odot u) \oplus (0 \odot u)$$

which tells us $0 \odot u = 0$. *A little further thought then tells us that since*

$$0 = (1-1) \odot u$$
$$= (1 \odot u) \oplus (-1 \odot u)$$

we have the additive inverse $u^\dagger = -1 \odot u$.

Comment 2.3.2 *We usually say this much simpler. The set of objects $\mathscr{V}$ is a vector space over its scalar field if there are two operations which we denote by $\boldsymbol{u} + \boldsymbol{v}$ and $c\boldsymbol{u}$ which generate new objects in the vector space for any $\boldsymbol{u}$, $\boldsymbol{v}$ and scalar c. We then just add that these operations satisfy the usual commutative, associative and distributive laws and there are unique additive inverses.*

Comment 2.3.3 *The objects are often called* vectors *and sometimes we denote them by $\boldsymbol{u}$ although this notation is often too cumbersome.*

Comment 2.3.4 *To give examples of vector spaces, it is usually enough to specify how the additive and scaling operations are done.*

- *Vectors in $\mathfrak{R}^2$, $\mathfrak{R}^3$ and so forth are added and scaled by components.*
- *Matrices of the same size are added and scaled by components.*
- *A set of functions of similar characteristics uses as its additive operator, pointwise addition. The new function $(f \oplus g)$ is defined pointwise by $(f \oplus g)(t) = f(t) + g(t)$. Similarly, the new function $c \odot f$ is defined by $c \odot f$ is the function whose value at t is $(cf)(t) = cf(t)$. Classic examples are*

1. *$C[a, b]$ is the set of all functions whose domain is $[a, b]$ that are continuous on the domain.*
2. *$C^1[a, b]$ is the set of all functions whose domain is $[a, b]$ that are continuously differentiable on the domain.*
3. *$R[a, b]$ is the set of all functions whose domain is $[a, b]$ that are Riemann integrable on the domain.*

   There are many more, of course.

Vector spaces have two other important ideas associated with them. We have already talked about linearly independent objects. Clearly, the kinds of objects we were focusing on were from some vector space $\mathscr{V}$. The first idea is that of the span of a set.

Definition 2.3.2 (*The Span Of A Set Of Vectors*)
Given a finite set of vectors in a vector space $\mathscr{V}$, $\mathscr{W} = \{\boldsymbol{u}_1, \ldots, \boldsymbol{u}_N\}$ for some positive integer N, the span of $\mathscr{W}$ is the collection of all new vectors of the form $\sum_{i=1}^{N} c_i \boldsymbol{u}_i$ for any choices of scalars $c_1, \ldots, c_N$. It is easy to see $\mathscr{W}$ is a vector space itself and since it is a subset of $\mathscr{V}$, we call it a *vector subspace*. The span of the set $\mathscr{W}$ is denoted by $Sp\boldsymbol{W}$. If the set of vectors $\mathscr{W}$ is not finite, the definition is similar but we say the span of $\mathscr{W}$ is the set of all vectors which can be written as $\sum_{i=1}^{N} c_i \boldsymbol{u}_i$ for some finite set of vectors $\boldsymbol{u}_1, \ldots \boldsymbol{u}_N$ from $\mathscr{W}$.

Then there is the notion of a *basis* for a vector space. First, we need to extend the idea of linear independence to sets that are not necessarily finite.

Definition 2.3.3 (*Linear Independence For Non Finite Sets*)
Given a set of vectors in a vector space $\mathscr{V}$, $\mathscr{W}$, we say $\mathscr{W}$ is a linearly independent subset if every finite set of vectors from $\mathscr{W}$ is linearly independent in the usual manner.

Definition 2.3.4 (*A Basis For A Vector Space*)
Given a set of vectors in a vector space $\mathcal{V}$, $\mathcal{W}$, we say $\mathcal{W}$ is a *basis* for $\mathcal{V}$ if the span of $\mathcal{W}$ is all of $\mathcal{V}$ and if the vectors in $\mathcal{W}$ are linearly independent. Hence, a basis is a linearly independent spanning set for $\mathcal{V}$. The number of vectors in $\mathcal{W}$ is called the *dimension* of $\mathcal{V}$. If $\mathcal{W}$ is not finite is size, then we say $\mathcal{V}$ is an *infinite dimensional vector space*.

Comment 2.3.5 *In a vector space like $\Re^n$, the maximum size of a set of linearly independent vectors is n, the dimension of the vector space.*

Comment 2.3.6 *Let's look at the vector space $C[0, 1]$, the set of all continuous functions on $[0, 1]$. Let $\mathcal{W}$ be the set of all powers of t, $\{1, t, t^2, t^3, \ldots\}$. We can use the derivative technique to show this set is linearly independent even though it is infinite in size. Take any finite subset from $\mathcal{W}$. Label the resulting powers as $\{n_1, n_2, \ldots, n_p\}$. Write down the linear dependence equation*

$$c_1 t^{n_1} + c_2 t^{n_2} + \cdots + c_p t^{n_p} = 0.$$

Take n_p derivatives to find $c_p = 0$ and then backtrack *to find the other constants are zero also. Hence $C[0, 1]$ is an infinite dimensional vector space. It is also clear that $\mathcal{W}$ does not span $C[0, 1]$ as if this was true, every continuous function on $[0, 1]$ would be a polynomial of some finite degree. This is not true as $\sin(t)$, e^{-2t} and many others are not finite degree polynomials.*

2.4 Inner Products

Now there is an important result that we use a lot in applied work. If we have an object u in a Vector Space $\mathcal{V}$, we often want to find to *approximate* u using an element from a given subspace $\mathcal{W}$ of the vector space. To do this, we need to add another property to the vector space. This is the notion of an *inner product*. We already know what an inner product is in a simple vector space like $\Re^n$. Many vector spaces can have an inner product structure added easily. For example, in $C[a, b]$, since each object is continuous, each object is Riemann integrable. Hence, given two functions f and g from $C[a, b]$, the real number given by $\int_a^b f(s)g(s)ds$ is well-defined. It satisfies all the usual properties that the inner product for finite dimensional vectors in $\Re^n$ does also. These properties are so common we will codify them into a definition for what an inner product for a vector space $\mathcal{V}$ should behave like.

Definition 2.4.1 (*Real Inner Product*)
Let $\mathcal{V}$ be a vector space with the reals as the scalar field. Then a mapping ω which assigns a pair of objects to a real number is called an inner product on $\mathcal{V}$ if

1. $\omega(u, v) = \omega(v, u)$; that is, the order is not important for any two objects.
2. $\omega(c \odot u, v) = c\omega(u, v)$; that is, scalars in the *first slot* can be pulled out.

3. $\omega(\boldsymbol{u} \oplus \boldsymbol{w}, \boldsymbol{v}) = \omega(\boldsymbol{u}, \boldsymbol{v}) + \omega(\boldsymbol{w}, \boldsymbol{v})$, for any three objects.
4. $\omega(\boldsymbol{u}, \boldsymbol{u}) \geq 0$ and $\omega(\boldsymbol{u}, \boldsymbol{u}) = 0$ if and only if $u = 0$.

These properties imply that $\omega(\boldsymbol{u}, c \odot \boldsymbol{v}) = c\omega(\boldsymbol{u}, \boldsymbol{v})$ as well. A vector space $\mathscr{V}$ with an inner product is called an inner product space.

Comment 2.4.1 *The inner product is usually denoted with the symbol* $< , >$ *instead of* $\omega(,)$. *We will use this notation from now on.*

Comment 2.4.2 *When we have an inner product, we can measure the size or magnitude of an object, as follows. We define the analogue of the euclidean norm of an object $\boldsymbol{u}$ using the usual* $|| \ ||$ *symbol as*

$$||\boldsymbol{u}|| = \sqrt{< \boldsymbol{u}, \boldsymbol{u} >}.$$

This is called the norm induced by the inner product *of the object. In $C[a, b]$, with the inner product $< f, g > = \int_a^b f(s)g(s)ds$, the norm of a function f is thus $||f|| = \sqrt{\int_a^b f^2(s)ds}$. This is called the L_2 norm of f.*

It is possible to prove the Cauchy–Schwartz inequality in this more general setting also.

Theorem 2.4.1 (General Cauchy–Schwartz Inequality)
If $\mathscr{V}$ is an inner product space with inner product $< , >$ and induced norm $|| \ ||$, then

$$| < \boldsymbol{u}, \boldsymbol{v} > | \leq ||\boldsymbol{u}|| \, ||\boldsymbol{v}||$$

with equality occurring if and only if $\boldsymbol{u}$ and $\boldsymbol{v}$ are linearly dependent.

Proof The proof is different than the one you would see in a Calculus text for $\mathfrak{R}^2$, of course, and is covered in a typical course on beginning analysis. ∎

Comment 2.4.3 *We can use the Cauchy–Schwartz inequality to define a notion of angle between objects exactly like we would do in $\mathfrak{R}^2$. We define the angle θ between $\boldsymbol{u}$ and $\boldsymbol{v}$ via its cosine as usual.*

$$\cos(\theta) = \frac{< \boldsymbol{u}, \boldsymbol{v} >}{||\boldsymbol{u}|| \, ||\boldsymbol{v}||}.$$

Hence, objects can be perpendicular or orthogonal even if we can not interpret them as vectors in $\mathfrak{R}^2$. We see two objects are orthogonal if their inner product is 0.

Comment 2.4.4 *If $\mathscr{W}$ is a finite dimensional subspace, a basis for $\mathscr{W}$ is said to be an orthonormal basis if each object in the basis has L_2 norm 1 and all of the objects are mutually orthogonal. This means $< \boldsymbol{u}_i, \boldsymbol{u}_j >$ is 1 if $i = j$ and 0 otherwise. We typically let the Kronecker delta symbol δ_{ij} be defined by $\delta_{ij} = 1$ if $i = j$ and 0 otherwise so that we can say this more succinctly as $< \boldsymbol{u}_i, \boldsymbol{u}_j > = \delta_{ij}$.*

Now, let's return to the idea of finding the best object in a subspace $\mathcal{W}$ to approximate a given object u. This is an easy theorem to prove.

Theorem 2.4.2 (Best Finite Dimensional Approximation Theorem)
Let u be any object in the inner product space $\mathcal{V}$ with inner product $<,>$ and induced norm $\|\ \|$. Let $\mathcal{W}$ be a finite dimensional subspace with an orthonormal basis $\{w_1, \ldots w_N\}$ where N is the dimension of the subspace. Then there is an unique object p^* in $\mathcal{W}$ which satisfies

$$\|u - p^*\| = \min_{p \in \mathcal{W}} \|u - p\|$$

with

$$p^* = \sum_{i=1}^{N} < u, w_i > w_i.$$

Further, $p - p^*$ is orthogonal to the subspace $\mathcal{W}$.

Proof Any object in the subspace has the representation $\sum_{i=1}^{N} a_i w_i$ for some scalars a_i. Consider the function of N variables

$$E(a_1, \ldots, a_N) = \left\langle u - \sum_{i=1}^{N} a_i w_i, u - \sum_{j=1}^{N} a_j w_j \right\rangle$$

$$= < u, u > -2 \sum_{i=1}^{N} a_i < u, w_i > + \sum_{i=1}^{N} \sum_{j=1}^{N} a_i a_j < w_i, w_j > .$$

Simplifying using the orthonormality of the basis, we find

$$E(a_1, \ldots, a_N) = < u, u > -2 \sum_{i=1}^{N} a_i < u, w_i > + \sum_{i=1}^{N} a_i^2.$$

This is a quadratic expression and setting the gradient of E to zero, we find the critical points

$$a_j = < u, w_j > .$$

This is a global minimum for the function E. Hence, the optimal p^* has the form

$$p^* = \sum_{i=1}^{N} < u, w_i > w_i.$$

Finally, we see

$$< p - p^*, w_j > \ = \ < p, w_j > - \sum_{k=1}^{N} < p, w_k > < w_k, w_j >$$
$$= \ < p, w_j > - \ < p, w_j > = 0,$$

and hence, $p = p^*$ is orthogonal of $\mathscr{W}$.   ∎

2.5   Graham–Schmidt Orthogonalization

Let's assume we are given two linearly independent vectors in $\mathfrak{R}^2$. Fire up
MatLab/Octave and enter two such vectors.

Listing 2.9: Setting two vectors

```
V1 = [6;2];
V2 = [-1;4];
```

In Fig. 2.1 we see these two vectors.

Now *project* V_2 to the vector V_1. The unit vector pointing in the direction of V_1 is
$E_1 = V_1/||V_1||$ and the amount of V_2 that lies in the direction of V_1 is given by the

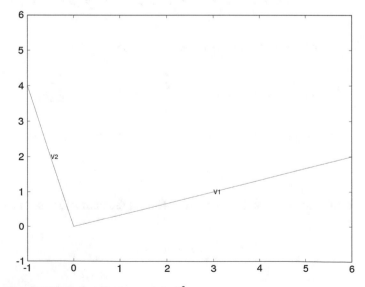

Fig. 2.1 Two linearly independent vectors in $\mathfrak{R}^2$

inner product $< V_1, E_1 >$. Hence, the vector $< V_1, E_1 > E_1$ in the vector portion of V_2 which lies in the direction of V_1. Now subtract this from V_2. Hence, define W by

$$W = V_2 - < V_2, E_1 > E_1.$$

Note by construction W is perpendicular to V_1 as

$$< W, E_1 > \ = \ < V_2 - < V_2, E_1 > E_1, E_1 > \ = \ < V_2, E_1 > \ - \ < V_2, E_1 > < E_1, E_1 >$$
$$= \ < V_2, E_1 > \ - \ < V_2, E_1 > \ = \ 0,$$

since $< E_1, E_1 > \ = \ ||E_1||^2 = 1$. Let E_2 be the unit vector given by $W/||W||$. Then from the linearly independent vectors V_1 and V_2, we have constructed two new linearly independent vectors which are mutually orthogonal, E_1 and E_2. We also see both $\{V_1, V_2\}$ and $\{E_1, E_2\}$ are bases for $\Re^2$ but the new one $\{E_1, E_2\}$ is preferred as the basis vectors are orthogonal. We can see the new vectors in Fig. 2.2.

You should be able to see that this procedure is easy to extend to three linearly vectors in $\Re^3$ and finite sets of linearly independent functions as well. This type of procedure is called **Graham–Schmidt Orthogonalization** and we can see this graphically in $\Re^2$. First we graph the vectors. This done with calls to **plot**. We start with the function **Draw2DGSO** and add the vector drawing code first. We do all of our drawing between a **hold on** and **hold off** so we keep drawing into the figure until we are done.

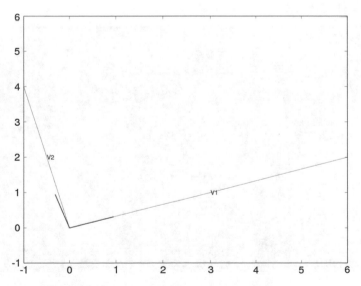

Fig. 2.2 Two linearly independent vectors in $\Re^2$

Listing 2.10: Draw2DGSO: drawing the vectors

```
    function  Draw2DGSO(V1,V2)
    %
    T = linspace(0,1,2);
    hold on
5   % find  equation  of  line  for  vector  V1
    if  V1(1) == 0
        f1 = @(t)  t;
    else
        f1 = @(t)  (V1(2)/V1(1))*t;
10  end
    T = [0:V1(1):V1(1)];
    plot(T,f1(T));
    text(V1(1)/2,V1(2)/2, 'v1');
    % find  equation  of  line  for  vector  V2
15  if  V2(1) == 0
        f2 = @(t)  t;
    else
        f2 = @(t)  (V2(2)/V2(1))*t;
    end
20  plot(T,f2(T));
    T = [0:V2(1):V2(1)];
    text(V2(1)/2,V2(2)/2, 'v2');
    hold off
    end
```

Note we also use the new **text** command to add text at desired points in the graph.
Next, we find the new basis vectors E_1 and E_2 and graph them.

Listing 2.11: Graph the new basis vectors

```
    % Now  plot  e1
    E1 = V1/norm(V1);
    % find  equation  of  line  for  vector  V1
    if  E1(1) == 0
5       e1 = @(t)  t;
    else
        e1 = @(t)  (E1(2)/E1(1))*t;
    end
    T = [0:E1(1):E1(1)];
10  plot(T,e1(T),'r','linewidth',4);
    % find  equation  of  line  for  vector  e2
    W =   V2 - dot(V2,E1)*E1
    E2 = W/norm(W);
    % find  equation  of  line  for  vector  e2
15  if  E2(1) == 0
        e2 = @(t)  t;
    else
        e2 = @(t)  (E2(2)/E2(1))*t;
    end
20  T = [0:E2(1):E2(1)];
    plot(T,e2(T),'r','linewidth',4);
```

Now if you just did this, you will probably see the E_1 and E_2 don't necessarily
look perpendicular in the figure. This is because by itself, MatLab/Octave does not
necessarily use a square picture frame for our graph. So the x and y axes need not be

the same size. If they are not this, will make us lose the appearance of orthogonality! So to fix this, we make sure the *x* and *y* axes are the same size. This requires some messy code as follows.

Listing 2.12: Setting the aspect ratio

```
   % set axis limits
   xmin = V1(1);
   if (V2(1) < xmin)
4     xmin = V2(1);
      xmax = V1(1);
   else
      xmax = V2(1);
   end
9  ymin = V1(2);
   if (V2(2) < ymin)
      ymin = V2(2);
      ymax = V2(2);
   else
14    ymax = V2(1);
   end
   %
   umin = xmin;
   if (ymin < umin)
19    umin = ymin;
   end
   umax = xmax;
   if (ymax > umax)
      umax = ymax;
24 end
   axis([umin,umax,umin,umax]);
```

This sets the axes correctly, but we still need to make sure the figure is square. We do this be setting the figure size with **figure('Position',[0,0,600,600]);** at the top of the file. The full code to see what is happening in $\Re^2$ is then given in the function **Draw2DGSO** given below.

Listing 2.13: Drawing a two dimensional Graham–Schmidt orthogonalization result: Draw2DGSO

```
   function Draw2DGSO(V1,V2)
   %
   T = linspace(0,1,2);
   figure('Position',[0,0,600,600]);
5  hold on
      % find equation of line for vector V1
      if V1(1) == 0
         f1 = @(t) t;
      else
10       f1 = @(t) (V1(2)/V1(1))*t;
      end
      T = [0:V1(1):V1(1)];
      plot(T,f1(T));
      text(V1(1)/2,V1(2)/2, 'v1');
15    % find equation of line for vector V1
      if V2(1) == 0
         f2 = @(t) t;
      else
         f2 = @(t) (V2(2)/V2(1))*t;
20    end
      plot(T,f2(T));
```

```
     T = [0:V2(1):V2(1)];
     text(V2(1)/2,V2(2)/2, 'V2');
     % Now plot e1
25   E1 = V1/norm(V1);
     % find equation of line for vector V1
     if E1(1) == 0
        e1 = @(t) t;
     else
30      e1 = @(t) (E1(2)/E1(1))*t;
     end
     T = [0:E1(1):E1(1)];
     plot(T,e1(T),'r','linewidth',4);
     % find equation of line for vector e2
35   W =  V2 - dot(V2,E1)*E1
     E2 = W/norm(W);
     % find equation of line for vector e2
     if E2(1) == 0
        e2 = @(t) t;
40   else
        e2 = @(t) (E2(2)/E2(1))*t;
     end
     T = [0:E2(1):E2(1)];
     plot(T,e2(T),'r','linewidth',4,);
45   % set axis limits
     xmin = V1(1);
     if (V2(1) < xmin)
        xmin = V2(1);
        xmax = V1(1);
50   else
        xmax = V2(1);
     end
     ymin = V1(2);
     if (V2(2) < ymin)
55      ymin = V2(2);
        ymax = V2(2);
     else
        ymax = V2(1);
     end
60   %
     umin = xmin;
     if (ymin < umin)
        umin = ymin;
     end
65   umax = xmax;
     if (ymax > umax)
        umax = ymax;
     end
     axis([umin,umax,umin,umax]);
70 hold off
   end
```

It is simple to use this code. The example we just worked out would be done this way in MatLab/Octave.

Listing 2.14: A 2DGSO sample session

```
  V1 = [6;2];
  V2 = [-1;4];
  Draw2DGSO(V1,V2);
4 print -dpng '2DGSO.png'
```

We can certainly do this procedure for three linearly independent vectors in $\Re^3$. We graph lines using the **plot3** function. For example, to draw the line between V_1 and V_2, we set up vectors X, Y and Z as follows:

Listing 2.15: Setting coordinates for a line plot

```
X = [0;V1(1)];
Y = [0;V1(2)];
Z = [0;V1(3)];
```

The way to look at this is that the first column of these three vectors specifies the start position $[0, 0, 0]^T$ and the second column is the end position coordinates V_1^T. We then plot the line and add text with

Listing 2.16: Plotting the lines

```
  plot3(X,Y,Z,'r','linewidth',2);
2 text(V1(1)/2,V1(2)/2,V1(3)/2, 'V1');
```

Note we set the width of the plotted line to be of size 2 which is thicker than size 1. Once this line is plotted, we do the others. So the code to plot the three vectors as lines starting at the origin is as follows; note, we wrap this code between **hold on** and **hold off** statements so that we draw into our figure repeatedly until we are done.

Listing 2.17: Plotting all the lines

```
   hold on
      % plot V1
      X = [0;V1(1)];
      Y - [0;V1(2)];
5     Z = [0;V1(3)];
      plot3(X,Y,Z,'r','linewidth',2);
      text(V1(1)/2,V1(2)/2,V1(3)/2,'V1');
      % plot V2
      X = [0;V2(1)];
10    Y = [0;V2(2)];
      Z = [0;V2(3)];
      plot3(X,Y,Z,'r','linewidth',2);
      text(V2(1)/2,V2(2)/2,V2(3)/2, 'V2');
      % plot V3
15    X = [0;V3(1)];
      Y = [0;V3(2)];
      Z = [0;V3(3)];
      plot3(X,Y,Z,'r','linewidth',2);
      text(V3(1)/2,V3(2)/2,V3(3)/2, 'V3');
20 hold off
```

We then do the steps of the Graham–Schmidt orthogonalization. First, we set up $E_1 = V_1/\|V_1\|$ as usual. We then project V_2 to V_1 to obtain

$$W = V_2 - \; < V_2, E_1 > E_1$$

which by construction will be perpendicular to E_1. We then set $E_2 = W/\|W\|$. Finally, we project V_3 to the plane determined by V_1 and V_2 as follows:

$$W = V_3 - \; < V_3, E_1 > E_1 - \; < V_3, E_2 > E_2$$

which by construction will be perpendicular to both E_1 and E_2. Finally, we let $E_3 = W/\|W\|$ and we have found a new mutually orthogonal basis for $\Re^3$ $\{E_1, E_2, E_3\}$. It is easy put this into code. We write

Listing 2.18: Finding and plotting the new basis

```
  % Do GSO
  E1 = V1/norm(V1);
  W = V2 - dot(V2,E1)*E1;
  E2 = W/norm(W);
5 W = V3 - dot(V3,E1)*E1 - dot(V3,E2)*E2;
  E3 = W/norm(W);
  % Plot new basis
  % plot E1
  X = [0;E1(1)];
10 Y = [0;E1(2)];
  Z = [0;E1(3)];
  plot3(X,Y,Z,'b','linewidth',4);
  % plot E2
  X = [0;E2(1)];
15 Y = [0;E2(2)];
  Z = [0;E2(3)];
  plot3(X,Y,Z,'b','linewidth',4);
  % plot E3
  X = [0;E3(1)];
20 Y = [0;E3(2)];
  Z = [0;E3(3)];
  plot3(X,Y,Z,'b','linewidth',4);
```

Then to make this show up nicely, we draw lines between V_1 and V_2 to make the plane determined by these vectors stand out.

Listing 2.19: Drawing the plane between V_1 and V_2

```
  % Draw plane determined by V1 and V2
  delt = 0.05;
3 for i = 1:10
    p = i*delt;
    X = [p*V1(1);p*V2(1)];
    Y = [p*V1(2);p*V2(2)];
    Z = [p*V1(3);p*V2(3)];
8   plot3(X,Y,Z,'r','linewidth',2);
  end
  text( (V1(1)+V2(1))/4, (V1(2)+V2(2))/4,(V1(3)+V2(3))/4, 'V1 -
      V2 Plane');
```

The full code is then given in the function **Draw3DGSO**.

Listing 2.20: Drawing a three dimensional Graham–Schmidt orthogonalization result: Draw3DGSO

```
   function Draw3DGSO(V1,V2,V3)
   %
   T = linspace(0,1,2);
   figure('Position',[0,0,600,600]);
 5 hold on
     % plot V1
     X = [0;V1(1)];
     Y = [0;V1(2)];
     Z = [0;V1(3)];
10   plot3(X,Y,Z,'r','linewidth',2);
     text(V1(1)/2,V1(2)/2,V1(3)/2, 'v1');
     % plot V2
     X = [0;V2(1)];
     Y = [0;V2(2)];
15   Z = [0;V2(3)];
     plot3(X,Y,Z,'r','linewidth',2);
     text(V2(1)/2,V2(2)/2,V2(3)/2, 'v2');
     % plot V3
     X = [0;V3(1)];
20   Y = [0;V3(2)];
     Z = [0;V3(3)];
     plot3(X,Y,Z,'r','linewidth',2);
     text(V3(1)/2,V3(2)/2,V3(3)/2, 'v3');
     %
25   % Draw plane determined by V1 and V2
     delt = 0.05;
     for i = 1:10
       p = i*delt;
       X = [p*V1(1);p*V2(1)];
30     Y = [p*V1(2);p*V2(2)];
       Z = [p*V1(3);p*V2(3)];
       plot3(X,Y,Z,'r','linewidth',2);
     end
     text( (V1(1)+V2(1))/4, (V1(2)+V2(2))/4,(V1(3)+V2(3))/4, 'V1 - V2
        Plane');
35   %
     % Do GSO
     E1 = V1/norm(V1);
     W = V2 - dot(V2,E1)*E1;
     E2 = W/norm(W);
40   W = V3 - dot(V3,E1)*E1 - dot(V3,E2)*E2;
     E3 = W/norm(W);
     % Plot new basis
     % plot E1
     X = [0;E1(1)];
45   Y = [0;E1(2)];
     Z = [0;E1(3)];
     plot3(X,Y,Z,'b','linewidth',4,);
     % plot E2
     X = [0;E2(1)];
50   Y = [0;E2(2)];
     Z = [0;E2(3)];
     plot3(X,Y,Z,'b','linewidth',4,);
     % plot E3
     X = [0;E3(1)];
55   Y = [0;E3(2)];
     Z = [0;E3(3)];
     plot3(X,Y,Z,'b','linewidth',4,);
   hold off
   end
```

It is again easy to use this code. Here is a sample session. Since we make up our vectors, we always set up a matrix A using the vectors as rows and then find the determinant of *A*. If that was 0, we would know our choice of vectors was not linearly independent and we would have to make up new vectors for our example.

Listing 2.21: A Draw3DGSO example

```
1 V1 = [6;2;1];
  V2 = [-1;4;3];
  V3 = [8;2;-1];
  A = [V1';V2';V3'];
  det(A)
6 ans = -48.000
  Draw3DGSO(V1,V2,V3);
  print -dpng '3DGSO.png'
```

We can use this code to generate a nice 3D plot of this procedure as shown in Fig. 2.3. If you do this in MatLab/Octave yourself, you can grab the graph and rotate it around to make sure you see it in the orientation that makes the best sense to you.

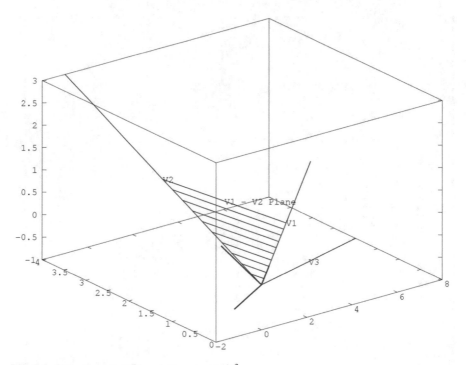

Fig. 2.3 Three linearly independent vectors in $\Re^3$

2.5.1 Making It Formal

If we had a finite number of linearly independent objects in an inner product space $\mathscr{V}$, then they form a basis for their span. There is a recursive procedure to find a new orthonormal basis from these vectors for the subspace formed by their span. This is called the Graham–Schmidt Orthogonalization process. It is easiest to show it explicitly for three vectors. You can then easily generalize it to more objects. Let's assume we start with three linearly independent objects u, v and w. We will find three orthogonal objects of length 1, g_1, g_2 and g_3 to be the new basis as follows:

First Basis Object   Set

$$g_1 = \frac{u}{\|u\|}.$$

Second Basis Object

- Subtract the part of v which lies along the object g_1.

$$h = v - <v, g_1> g_1.$$

- Find the length of h and set the second new basis object as follows:

$$g_2 = \frac{h}{\|h\|}.$$

Third Basis Object

- Subtract the part of w which lies along the object g_1 and the object g_2.

$$h = w - <v, g_1> g_1 - <v, g_2> g_2.$$

- Find the length of h and set the third new basis object as follows:

$$g_3 = \frac{h}{\|h\|}.$$

It is easy to see how to generalize this to four or more objects. In fact, the procedure is the same whether we are using vectors or functions as long as the objects are linearly independent.

2.5.2 General MatLab GSO

To see how we could do GSO in computationally, we will now show some simple MatLab code. We will use the numerical integration code presented in this chapter

which uses Newton–Cotes formulae and also show you how to use the built-in quadrature codes in MatLab itself. In general, doing GSO on functions is *hard*, as you will
see, due to numerical errors. The code we have shown you so far is hard-wired to
the number of linearly independent objects. Hence, we wrote 2D and 3D GSO code
separately. Now we'll try to write more general code where the argument we send
into the function is a collection of linearly independent objects. The code ideas that
follow are specialized to functions, but it wouldn't be hard to write similar code for
sets of independent vectors in $\Re^n$.

The first code is the one that uses Newton–Cotes ideas. This is written to perform
GSO on powers of t, but it could be easily generalized to more general functions.
Some word of explanation are in order. First, we use a generic powers of t function

Listing 2.22: Generic Powers of t Function

```
  function y = Powers(t,n)
2 % t = independent variable
  % n = power
  y = t.^n;
  end
```

We then use extensively the idea of pointers to functions which in MatLab are called
function handles. We also use the idea of *anonymous functions* in MatLab. The
basic syntax for these is something like **h = @(x)x.^2;** which sets up h as
an *anonymous* function so that a session line like **h(3)** would be evaluated as 3^2.
The letter **h** here is the function handle which allows us to refer to this function in
other code. We do the GSO using nested loops of anonymous functions using syntax
like

Listing 2.23: Constructing an anonymous function as a loop

```
  for k=M+2:N+1
    %compute next orthogonal piece
    phi = @(x) 0;
    for j = M+1:k-1
5     c = ip(f{k},g{j},a,b,r);
      phi = @(x) (phi(x)+c*g{j}(x));
    end
    psi = @(x) (f{k}(x) - phi(x));
    nf = sqrt(ip(psi,psi,a,b,r));
10  g{k} = @(x) (psi(x)/nf);
```

to calculate each of the new basis objects g_k. We start with the zero anonymous
function **phi** and progressively add up the pieces we must subtract that lie along the
previous g_j objects. Once this is done, we subtract this sum from the original object
and then divide by the length. This then creates the new object g_k.

We also use an inner product implementation nested inside this code as follows:

Listing 2.24: Nested inner product code

```
      function c = ip(f,g,a,b,r)
          w = WNC(r);
          s = linspace(a,b,r)';
          u = zeros(r,1);
5         u = f(s).*g(s);
          c = (b-a)*(w'*u);
      end
```

which uses the function handles f and g we pass with the desired Newton–Cotes order to approximate $\int_a^b fg\,ds$.

Listing 2.25: A Newton–Cotes GSO Code

```
      function [t,z] = GrahamSchmidtNC(a,b,r,M,N,NumPoints)
      %
 3    % Perform Graham - Schmidt Orthogonalization
      % on a set of functions 1, t, t^2, ..., t^N
      %
      % a = start of interval
      % b = end of interval
 8    % r = Newton Cotes order -- can be as high as 9
      % M = power of t to start at
      % N = power of t to end with
      % NumPoints = the number of points to use in the t
      %             vector we will use for plotting the g_k's
13    %
      % setup t as 1xNumPoints
      t = linspace(a,b,NumPoints);
      % setup x as NumPoints rows x N+1 columns
      z = zeros(NumPoints,N+1);
18    %Setup function handles
      f = cell(N+1,1);
      g = cell(N+1,1);
      for i=1:N+1
          f{i} = @(x) Powers(x,i-1);
23    end
      nf = sqrt(ip(f{M+1},f{M+1},a,b,r));
      g{M+1} = @(x) f{M+1}(x)/nf;
      y = zeros(1,NumPoints);
      d = zeros(N+1,N+1);
28    y = g{M+1}(t);
      z(:,M+1) = y';
      for k=M+2:N+1
          %compute next orthogonal piece
          phi = @(x) 0;
33        for j = M+1:k-1
              c = ip(f{k},g{j},a,b,r);
              phi = @(x) (phi(x)+c*g{j}(x));
          end
          psi = @(x) (f{k}(x) - phi(x));
38        nf = sqrt(ip(psi,psi,a,b,r));
          g{k} = @(x) (psi(x)/nf);
          y = g{k}(t);
          z(:,k) = y';
      end
43    % find the matrix of inner products
      for i=M+1:N+1
          for j=M+1:N+1
              d(i,j) = ip(g{i},g{j},a,b,r);
          end
48    end
```

```
% Print out the matrix of inner products
d

     function c = ip(f,g,a,b,r)
53       w = WNC(r);
         s = linspace(a,b,r)';
         u = zeros(r,1);
         u = f(s).*g(s);
         c = (b-a)*(w'*u);
58   end

end
```

Here are some Newton–Cotes results. First, we try a low order 3 Newton–Cotes method. There is too much error here. In the snippet below, d is the matrix of $< g_i, g_j >$ inner products which should be an identity matrix. Note the many off diagonal terms which are close to 1 is absolute value. Much error here.

Listing 2.26: GSO with Newton–Cotes Order 3

```
    [t,z] = GrahamSchmidtNC(0,1,3,0,5,101);

    d =

5      1.0000        0      0.0000     -0.4003     -0.4003     -0.4003
            0   1.0000      0.0000     -0.1387     -0.1387     -0.1387
       0.0000   0.0000      1.0000     -0.9058     -0.9058     -0.9058
      -0.4003  -0.1387     -0.9058      1.0000      1.0000      1.0000
      -0.4003  -0.1387     -0.9058      1.0000      1.0000      1.0000
10    -0.4003  -0.1387     -0.9058      1.0000      1.0000      1.0000
```

Newton–Cotes of order 4 is somewhat better but still unacceptable.

Listing 2.27: GSO with Newton–Cotes Order 4

```
    [t,z] = GrahamSchmidtNC(0,1,4,0,5,101);
    d =
       1.0000  -0.0000      0.0000     -0.0000      0.0112     -0.0112
      -0.0000   1.0000      0.0000     -0.0000      0.0208     -0.0208
5      0.0000   0.0000      1.0000     -0.0000      0.1206     -0.1206
      -0.0000  -0.0000     -0.0000      1.0000      0.9924     -0.9924
       0.0112   0.0208      0.1206      0.9924      1.0000     -1.0000
      -0.0112  -0.0208     -0.1206     -0.9924     -1.0000      1.0000
```

We must go to Newton–Cotes of order 6 to get proper orthogonality.

Listing 2.28: GSO with Newton–Cotes Order 6

```
   [t,z] = GrahamSchmidtNC(0,1,6,0,5,101);
 2 d =
        1.0000      0.0000     -0.0000      0.0000      0.0000     -0.0000
        0.0000      1.0000     -0.0000      0.0000     -0.0000      0.0000
       -0.0000     -0.0000      1.0000      0.0000     -0.0000      0.0000
        0.0000      0.0000      0.0000      1.0000     -0.0000      0.0000
 7      0.0000     -0.0000     -0.0000     -0.0000      1.0000      0.0000
       -0.0000      0.0000      0.0000      0.0000      0.0000      1.0000
```

It is much harder also, if we want to apply GSO to more powers of t, say up to t^{20}. In general, we would need a Newton–Cotes order of about 20 which would be numerically unstable itself.

Another approach is to replace the inner product calculations with the build in quadrature codes in MatLab. Here we use the function **quadl** instead of **quad**. Do a **help quad** and **help quadl** in MatLab/Octave to figure out how these two functions differ! We will find we need to fiddle with the error tolerance to get good results. This is analogous to what we had to do before by increasing the Newton–Cotes order.

Listing 2.29: Inner products using quad

```
    function c = ip(f,g,a,b)
        v = @(x) (f(x).*g(x));
        tol = 1.0e-9;
        c = quadl(v,a,b,tol);
 5  end
```

The new GSO code is as follows:

Listing 2.30: Graham–Schmidt Orthogonalization With quadl

```
    function [t,z] = GrahamSchmidt(a,b,M,N,NumPoints)
    %
    % Perform Graham - Schmidt Orthogonalization
    % on a set of functions 1, t, t^2, ..., t^N
 5  %
    % a = start of interval
    % b = end of interval
    % M = first power of t to use
    % N = last power of t to use
10  % NumPoints = number of time points to use for time
    %               for plotting of the g_k's
    %
    % setup t as 1xNumPoints
    t = linspace(a,b,NumPoints);
15  % setup x as NumPoints rows x N+1 columns
    z = zeros(NumPoints,N+1);
    %Setup function handles
    f = cell(N+1,1);
    g = cell(N+1,1);
20  for i=1:N+1
        f{i} = @(x) Powers(x,i-1);
    end
```

```
     nf = sqrt(ip(f{M+1},f{M+1},a,b));
     g{M+1} = @(x)  f{M+1}(x)/nf;
25   y = zeros(1,NumPoints);
     d = zeros(N+1,N+1);
     y = g{M+1}(t);
     z(:,M+1) = y';
     for k=M+2:N+1
30     %compute next orthogonal piece
       phi = @(x)  0;
       for j = M+1:k-1
         c = ip(f{k},g{j},a,b);
         phi = @(x)  (phi(x)+c*g{j}(x));
35     end
       psi = @(x)  (f{k}(x) - phi(x));
       nf = sqrt(ip(psi,psi,a,b));
       g{k} = @(x)  (psi(x)/nf);
       y = g{k}(t);
40     z(:,k) = y';
     end
     for i=M+1:N+1
       for j=M+1:N+1
         d(i,j) = ip(g{i},g{j},a,b);
45     end
     end
     d

     function c = ip(f,g,a,b)
50       v = @(x)  (f(x).*g(x));
         tol = 1.0e-9;
         c = quadl(v,a,b,tol);
     end

55 end
```

We then get similar results for the GSO.

Listing 2.31: GSO with quadl

```
     [t,z] = GrahamSchmidt(0,1,0,5,101);
     d =
        1.0000   -0.0000        0    0.0000   -0.0000    0.0000
       -0.0000    1.0000    0.0000   -0.0000    0.0000   -0.0000
5            0    0.0000    1.0000   -0.0000    0.0000   -0.0000
        0.0000   -0.0000   -0.0000    1.0000    0.0000   -0.0000
       -0.0000    0.0000    0.0000    0.0000    1.0000   -0.0000
        0.0000   -0.0000   -0.0000   -0.0000   -0.0000    1.0000
```

There are other ways to do this too, but this should give the idea. The bottom line is that it is straightforward to do theoretically, but devilishly hard to do in practice!

Reference

J. Peterson, *Calculus for Cognitive Scientists: Higher Order Models and Their Analysis*, Springer Series on Cognitive Science and Technology. (Springer Science + Business Media, Singapore 2015 in press)

Chapter 3
Numerical Differential Equations

We have already discussed how to solve ordinary differential equations in Peterson (2015a) and in Peterson (2015b) so if you want to review our discussion of the error estimates for Euler's algorithm, you can review that material in those volumes. Here, we will revisit Runge–Kutta methods and add a new one: the Fehlberg variation.

3.1 Approximating Solutions Numerically

These methods extend the simple Euler method by essentially looking at Taylor expansions of $f(t_0 + \alpha h, x_0 + \beta k)$ for various choices of α, β and k and then combining them in interesting ways to estimate error. Hence, we say these methods are based on differences of multiple function evaluations. The iteration scheme generates a sequence $\{\hat{x}_n\}$ starting at $\hat{x}_0 = x_0$ using the following recursion equation:

$$\hat{x}_{n+1} = \hat{x}_n + F(t_n, \hat{x}_n, h, f)\, h$$
$$\hat{x}_0 = x_0$$

where h is the step size we use for our underlying partition of the time space giving

$$t_i = t_0 + ih$$

for appropriate indices and F is some function of the previous approximate solution, the step size and the right hand side function f. For convenience of exposition, we will switch to a different notation for the approximate values $\hat{x}_n$. We will now denote them by y_n and so we solve the recursion

$$y_{n+1} = y_n + F(t_n, y_n, h, f)\, h$$
$$y_0 = x_0$$

© Springer Science+Business Media Singapore 2016
J.K. Peterson, *Calculus for Cognitive Scientists*, Cognitive Science
and Technology, DOI 10.1007/978-981-287-880-9_3

For our purposes, we will assume F has the form

$$F(t, x, h, f) = \gamma_1 f(t, x) + \gamma_2 f(t + \alpha h, x + \beta h f(t, x)), \qquad (3.1)$$

which tells us we are using $\beta h f(t, x)$ as our generic k term and our basic difference is an interpolation between our base point $f(t, x)$ and the new point $f(t + \alpha h, x + \beta h f(t, x))$. The amount of interpolation is determined by the values of γ_1 and γ_2.

3.1.1 Expansions of F

Let's look at the *error* term

$$T_{n+1} = x_{n+1} - x_n - h\, F(t_n, x_n, h, f).$$

Rewriting, we have for $\tau_{n+1} = \frac{T_{n+1}}{h}$,

$$x_{n+1} = x_n + h\, F(t_n, x_n, h, f) T_{n+1} + h\, \tau_{n+1}.$$

To understand what to do with this expression, we need to expand various versions of f as Taylor polynomials with remainder. We have

$$f(t + \alpha h, x + \beta h\, f(t, x)) = f(t, x) + f_t\,(\alpha h) + f_x(t, x)(\beta h f(t, x))$$
$$+ \frac{1}{2} \begin{bmatrix} \alpha h, & \beta h f(t, x) \end{bmatrix} \begin{bmatrix} f_{tt}(t, x) & f_{xt}(t, x) \\ f_{tx}(t, x) & f_{xx}(t, x) \end{bmatrix} \begin{bmatrix} \alpha h \\ \beta h f(t, x) \end{bmatrix}$$
$$+ \mathcal{O}(h^3),$$

where the term $\mathcal{O}(h^3)$ consists of the third order partial derivatives of f in various combinations all evaluated at some point ξ between (t, x) and $(t + \alpha h, x + \beta h\, f(t, x))$ and multiplied by h^3. Hence, these terms go to zero like h^3 does. We can multiply out the vectors and matrices involved and assuming f has continuous second order partials, the terms f_{tx} and f_{xt} will match. Thus, we obtain

$$f(t + \alpha h, x + \beta h\, f(t, x)) = f(t, x) + f_t(t, x)\,(\alpha h) + f_x(t, x)(\beta h f(t, x))$$
$$+ h^2 \left(\frac{1}{2} \alpha^2 f_{tt}(t, x) + \alpha \beta f_{tx}(t, x) f(t, x) + \frac{1}{2} \beta^2 f_{xx}(t, x) f(t, x) \right)$$
$$+ \mathcal{O}(h^3),$$

We know also

$$x' = f$$
$$x'' = f_t + f_x\, f$$
$$x''' = f_{tt} + 2 f_{tx} f + f_{xx} f^2 + f_t f_x + f_x^2 f.$$

Now using the Taylor expansion for x, we have to fourth order

$$x_{n+1} - x_n = h\, x'(t_n) + \frac{1}{2}h^2 x''(t_n) + \frac{1}{6}h^3 x'''(t_n) + \mathcal{O}(h^4).$$

We can then use these expressions in the T_{n+1} expansion to give

$$T_{n+1} = x_{n+1} - x_n - hF(t_n, x_n, h, f)$$
$$= hx'(t_n) + \frac{1}{2}h^2 x''(t_n) + \frac{1}{6}h^3 x'''(t_n) + \mathcal{O}(h^4) - hF(t_n, x_n, h, f).$$

But, we can replace the $f(t + \alpha h, x + \beta h f(t, x))$ in F by its expansion to get

$$hF(t_n, x_n, h, f) = h\gamma_1 f^{(n)} + h\gamma_2 f(t_n + \alpha h, x_n + \beta h f^{(n)})$$
$$= h\gamma_1 f^{(n)}$$
$$+ h\gamma_2 \Big(f^{(n)} + f_t^{(n)}(\alpha h) + f_x^{(n)}(\beta h f^{(n)})$$
$$+ \frac{h^2}{2}(\alpha^2 f_{tt}^{(n)} + 2\alpha\beta f_{tx}^{(n)} f^{(n)} + \beta^2 f_{xx}^{(n)}(f^{(n)})^2 + \mathcal{O}(h^3)\Big)$$

where for ease of exposition, the terms evaluated at (t_n, x_n) are denoted with a superscript (n). Now, plug all of this into the original expression for T_{n+1}. Here, we will use the fact that $h\mathcal{O}(h^3) = \mathcal{O}(h^4)$ too. We find (you will need to do this with a pad of paper in hand!)

$$T_{n+1} = h\Big(x'(t_n) - \gamma_1 f^{(n)} - \gamma_2 f^{(n)} \Big)$$
$$+ h^2 \Big(\frac{1}{2}x''(t_n) - \alpha\gamma_2 f_t^{(n)} - \beta\gamma_2 f^{(n)} f_x^{(n)} \Big)$$
$$+ h^3 \Big(\frac{1}{6}x'''(t_n) - \frac{1}{2}\alpha^2\gamma_2 f_{tt}^{(n)} - \alpha\beta\gamma_2 f_{tx}^{(n)} f^{(n)} - \frac{1}{2}\beta^2\gamma_2 f_{xx}^{(n)}(f^{(n)})^2 \Big)$$
$$+ \mathcal{O}(h^4)$$

where we combine all the order h^4 terms into one $\mathcal{O}(h^4)$. Now, plug in all the x' etc. terms. This is indeed messy, so follow along on your paper. You will get

$$T_{n+1}$$
$$= h\Big(1 - \gamma_1 - \gamma_2 \Big) f^{(n)} + h^2 \Big(\frac{f_t^{(n)} + f_x^{(n)} f^{(n)}}{2} - \alpha\gamma_2 f_t^{(n)} - \beta\gamma_2 f^{(n)} f_x^{(n)} \Big)$$
$$+ h^3 \Big(\frac{f_{tt}^{(n)} + 2f_{tx}^{(n)} f^{(n)} + f_{xx}^{(n)}(f^{(n)})^2 + f_t^{(n)} f_x^{(n)} + (f_x^{(n)})^2 f^{(n)}}{6}$$
$$- \frac{1}{2}\alpha f_{tt}^{(n)} - \alpha\beta f_{tx}^{(n)} f^{(n)} - \frac{1}{2}\beta^2 f_{xx}^{(n)}(f^{(n)})^2 \Big)$$
$$+ \mathcal{O}(h^4)$$

Many of these terms can be combined. We obtain

$$T_{n+1} = h\left(1 - \gamma_1 - \gamma_2\right)f^{(n)} + h^2\left(\left(\frac{1}{2} - \alpha\gamma_2\right)f_t^{(n)} + \left(\frac{1}{2} - \beta\gamma_2\right)f^{(n)}f_x^{(n)}\right)$$

$$+ h^3\left(\left(\frac{1}{6} - \frac{1}{2}\alpha^2\gamma_2\right)f_{xx}^{(n)}(f^{(n)})^2 + \left(\frac{1}{3} - \alpha\beta\gamma_2\right)f_t^{(n)}f_x^{(n)} + \left(\frac{1}{6} - \frac{1}{2}\beta^2\gamma_2\right)f_{xx}^{(n)}(f^{(n)})^2\right)$$

$$+ \mathcal{O}(h^4)$$

3.1.2 Minimizing Error

From the above, we see we can make T_{n+1} have zero error in the h and h^2 terms by choosing

$$\gamma_1 + \gamma_2 = 1 \tag{3.2}$$

$$2\alpha\gamma_2 = 2\beta\gamma_2 = 1. \tag{3.3}$$

There are many choices for α, β, γ_1 and γ_2 all of which give Runge–Kutta order two methods.

- If we choose $\gamma_1 = 0$ and $\gamma_2 = 1$, we find $\alpha = \beta = \frac{1}{2}$. This gives the numerical method

$$y_{n+1} = y_n + h\, f(t_n + 0.5h, x_n + 0.5hf(t_n, x_n))$$

  which is an average slope method.
- If we choose $\gamma_1 = 1$ and $\gamma_2 = 0$, we find α and β can be anything and the h^2 error is always $0.5f_t^{(n)} + 0.5f_x(n)f(n) = \beta = \frac{1}{2}$. This gives Euler's method, of course.

$$y_{n+1} = y_n + h\, f(t_n, x_n).$$

- If we choose $\gamma_1 = \gamma_2 = 0.5$, we find $\alpha = \beta = 0.25$ and the method becomes

$$y_{n+1} = y_n + h\left(0.25f^{(n)} + f(t_n + 0.25h, x_n + 0.25hf^{(n)})\right).$$

  which is a trapezoidal method.

Note all of the Runge–Kutta order two methods have a local error proportional to h^3 and use two f evaluations at each step to generate the next approximate value.

Comment 3.1.1 *For example, using the average slope method, we build an approximate solution $\hat{x}_{n+1}$ to the true solution x_{n+1} that uses the information starting from t_n with data x_n to construct the approximation*

$$\hat{x}_{n+1} = x_n + h\, f(t_n + 0.5h, x_n + 0.5hf(t_n, x_n))$$

which has a local error proportional to h^3. Since this error is due to the fact that we truncate the Taylor series expansions for f earlier, this is known as **truncation error**.

Finally, to build a method which has local error proportional to h^4, we would need to look at an additional evaluation for f. We would have

$$F(t, x, h, f) = \gamma_1 f(t, x) + \gamma_2 f(t + \alpha_1 h, x + \beta_1 h f(t, x)) + \gamma_3 f(t + \alpha_2 h, x + \beta_2 h f(t, x))$$

and follow the same sort of discussion. We would combine the various Taylor expansions from the base point (t, x) and find how to zero the h, h^2 and h^3 error. This would then be the family of Runge–Kutta order 3 methods. All of these methods have error proportional to h^4. Note, we do 3 function evaluations using f at various places to generate this method.

Finally, if we solve a vector system, we basically apply the technique above to each component of the vector solution. So if the problem is four dimensional, we are getting a constant C and D for each component. This, of course, complicates our life as one component might have such a fast growing error that the error for it is very large compared to the others. To control this, we generally cut the step size h to bring down the error. Hence, the component whose error grows the fastest controls how big a step size we can use in this type of technique. Thus, the Runge–Kutta methods are based on multiple function evaluations.

Now in the theory of the numerical solution of ordinary differential equations (ODEs) we find that there a several sources of error in finding a numerical approximation to the solution of our model process.

1. As discussed earlier, usually we approximate the function f using a few terms of the Taylor series expansion of f around the point (t_n, y_n) at each iteration. This truncation of the Taylor series expansion is not the same as the true function f of course, so there is an error made. Depending on how many terms we use in the Taylor series expansion, the error can be large or small. If we let h denote the difference $t_{n+1} - t_n$, a fourth order method is one where this error is of the form $C h^5$ for some constant C. This means that is you use a step size which is one half h, the error decreases by a factor of 32. For a Runge–Kutta method to have a local truncation error of order 5, we would need to do 4 function evaluations. Now this error is local to this time step, so if we solve over a very long interval with say N being 100,000 or more, the global error can grow quite large due to the addition of so many small errors. So the numerical solution of an ODE can be very accurate locally but still have a lot of problems when we try to solve over a long time interval. For example, if we want to track a voltage pulse over 1000 ms, if we use a step size of 10^{-4} this amounts to 10^7 individual steps. Even a fifth order method can begin to have problems. This error is called **truncation error**.
2. We can't represent numbers with perfect precision on any computer system, so there is an error due to that which is called **round-off error** which is significant over long computations.

3. There is usually a **modeling error** also. The model we use does not perfectly represent the physics, biology and so forth of our problem and so this also introduces a mismatch between our solution and the reality we might measure in a laboratory.

3.1.3 The Matlab Implementation

The basic code to implement the Runge–Kutta methods is broken into two pieces. We have discussed this in the earlier volumes, but for completeness, we'll add it here too. The first one, **RKstep.m** implements the evaluation of the next approximation solution at point (t_n, y_n) given the old approximation at $(t_{n-1}.y_{n-1})$. Here is that code for Runge–Kutta codes of orders one to four.

Listing 3.1: Runge–Kutta Codes

```
    function  [tnew,ynew,fnew]  =  RKstep(fname,tc ,yc ,fc ,h,k)
    %
    % fname        the  name  of  the  right  hand  side  function  f(t,y)
    %              t  is  a  scalar  usually  called  time  and
 5  %              y  is  a  vector  of  size  d
    % yc           approximate  solution  to  y'(t) = f(t,y(t))  at  t=tc
    % fc           f(tc,yc)
    % h            The  time  step
    % k            The  order  of  the  Runge–Kutta  Method  1<= k <= 4
10  %
    % tnew         tc+h
    % ynew         approximate  solution  at  tnew
    % fnew         f(tnew,ynew)
    %
15  if  k==1
       k1 = h*fc ;
       ynew = yc+k1;
    elseif  k==2
       k1 = h*fc ;
20     k2 = h*feval(fname,tc+(h/2),yc+(k1/2));
       ynew = yc + k2;
    elseif  k==3
       k1 = h*fc ;
       k2 = h*feval(fname,tc+(h/2),yc+(k1/2));
25     k3 = h*feval(fname,tc+h,yc-k1+2*k2);
       ynew = yc+(k1+4*k2+k3)/6;
    elseif  k==4
       k1 = h*fc ;
       k2 = h*feval(fname,tc+(h/2),yc+(k1/2));
30     k3 = h*feval(fname,tc+(h/2),yc+(k2/2));
       k4 = h*feval(fname,tc+h,yc+k3);
       ynew = yc+(k1+2*k2+2*k3+k4)/6;
    else
       disp(sprintf('The RK method %2d order is not allowed!',k));
35  end
    tnew = tc+h;
    fnew = feval(fname,tnew,ynew);
    end
```

Once the step is implemented, we solve the system using the RK steps like this:

Listing 3.2: The Runge–Kutta Solution

```
function [tvals,yvals,fcvals] = FixedRK(fname,t0,y0,h,k,n)
%
%               Gives approximate solution to
%                 y'(t) = f(t,y(t))
%                 y(t0) = y0
%               using a kth order RK method
%
% t0            initial time
% y0            initial state
% h             stepsize
% k             RK order   1<= k <= 4
% n             Number of steps to take
%
% tvals         time values of form
%                 tvals(j) = t0 + (j-1)*h,  1 <= j <= n
% yvals         approximate solution
% %             yvals(:j) = approximate solution at
%                 tvals(j),   1 <= j <= n
%
  tc = t0;
  yc = y0;
  tvals = tc;
  yvals = yc;
  fc = feval(fname,tc,yc);
  fcvals = fc;
  for j=1:n-1
    [tc,yc,fc] = RKstep(fname,tc,yc,fc,h,k);
    yvals = [yvals yc];
    tvals = [tvals tc];
    fcvals = [fcvals fc];
  end
end
```

3.2 Runge–Kutta Fehlberg Methods

In general, the Runge–Kutta solution schemes generate a sequence $\{y_n\}$ starting at y_0 using some sort of recursion equation:

$$y_{n+1} = y_n + h_n \times F(t_n, y_n, h_n, f)$$
$$y_0 = y0$$

where h_n is the step size we use at time point t_n and the function F is some function of the previous approximate solution y_n, the step size h_n and the dynamics vector f. We usually choose a numerical method which allows us to estimate what a good step size would be at each step n and then alter the step size to that optimal choice. Our order two method discussed earlier used

$$F(t_n, y_n, h_n, f) = f\left(t_n + \frac{1}{2}h_n, x_n + \frac{1}{2}h_n f(t_n, x_n)\right)$$

A typical numerical algorithm which will do this is the Runge–Kutta–Fehlberg 45 (**RKF45**) algorithm which works like this:

- Use a Runge–Kutta order 4 method to generate a solution with local error proportional to h^5. This needs four function evaluations.
- Do one more function evaluation to obtain a Runge–Kutta order 5 method which generates a solution with local error proportional to h^6.
- We now have two ways to approximate the true solution x at $t + h$. This gives us a way to compare the two approximations and to use one more function evaluation to get an estimate of the error we are making. We can then use that error estimate to see if our step size h is too large (that is the error is too big and we can compute a new h using our information), just right (we keep h as it is) or too small (we double the step size).

Finally, just to give you the flavor of what needs to be computed, here is a outline of the standard Runge–Kutta method: in what follows K is a six dimensional vector which we use to store intermediate results. If the system we want to solve is a vector system (for example, the vector function f we use in our Hodgkin–Huxley model dynamics discussed later is at least 4 dimensional and is also computationally complex) these six function evaluations can really slow us down!!

3.2.1   The RKF5 Flowchart

First, we do the function evaluations.

Listing 3.3: RKF5 function evaluations

```
   h ⇐ h₀
   t ⇐ t₀
 3 while (t is less than final time t_f){
      Compute f(t, y)
      K[0] ⇐ hf
      z ⇐ y + 0.25K[0]

 8    Compute f(t + ¼h, z)
      K[1] ⇐ hf
      z ⇐ y + 1.0/32.0 (3.0K[0] + 9.0K[1])

      Compute f(t + 3.0/8.0 h, z)
13    K[2] ⇐ hf
      z ⇐ y + 1.0/2197.0 (1932.0K[0] − 7200.0K[1] + 7296.0K[2])

      Compute f(t + 12.0/13.0 h, z)
      K[3] ⇐ h ∗ f
18    z ⇐ y + 439.0/216.0 K[0] − 8.0K[1] + 3680.0/513.0 K[2] − 845.0/4104.0 K[3]

      Compute f(t + h, z)
      K[4] ⇐ hf
      z ⇐ y − 8.0/27.0 K[0] + 2.0K[1] − 3544.0/2565.0 K[2] + 1859.0/4104.0 K[3] − 11.0/40.0 K[4]
23
      Compute f(t + ½h, z)
      K[5] ⇐ hf
```

Then, compute the error vector e for this step size h.

Listing 3.4: Compute the error vector

```
e  ⇐  1.0/360.0 K[0] − 128.0/4275.0 K[2] − 2197.0/75240.0 K[3] + 1.0/50.0 K[4] + 2.0/55.0 K[5]
||e||∞  ⇐  max{|eᵢ|}
||y||∞  ⇐  max{|yᵢ|}
```

See if this step size is acceptable. Given the tolerances ϵ_1, the amount of error we are willing to make in our discretization of the problem, and ϵ_2, the amount of weight we wish to place on the solution we previously computed y, compute local decision parameters

Listing 3.5: See if step size is acceptable

```
    η ⇐ ε₁ + ε₂||y||∞
    if (||e||∞ < η){
        //this step size is acceptable; use it to compute the next
        //value of y}
4       δy ⇐ 16.0/135.0 K[0] + 6656.0/12825.0 K[2] + 28561.0/56430.0 K[3] − 9.0/50.0 K[4] + 2.0/55.0 K[5]
        y ⇐ y + δy
        t ⇐ t + h
```

Now although this step size is acceptable, it might be smaller than necessary; determine if $||e||_\infty$ is smaller than some fraction of η—heuristically, 0.3 is a reasonable fraction to use.

Listing 3.6: Is step size too small?

```
    if (||e||∞ < 0.3η) {
        \\Double the step size in this case
        h ⇐ 2.0h
        }
5   }
```

It is possible that the step size is too big which can be determined by checking the local error

Listing 3.7: Is step too big?

```
    if (||e||∞ ≥ η){
        if (||e||∞ > ε₁){
```

The maximum error is larger than the discretization error. A reasonable way to reset the step size to reset to $0.9h\frac{\eta}{||e||_\infty}$. The rational behind this choice is that if the biggest error in the new computed solution is 5 times the allowed tolerance sought—ϵ_1, then $\frac{\epsilon_1 + \epsilon_2||y||_\infty}{5\epsilon_1} \Leftarrow 0.2(1 + \frac{\epsilon_2}{\epsilon_1}||y||_\infty)$ Now if the tolerance ϵ_2 is r times ϵ_1, the computation above would reset h to be $0.18h(1 + r||y||_\infty)$. Hence the maximum component size of the solution y influences how the step size is reset. If $||y||_\infty$ was 10, then for r 0.1, the new step size is $0.36h$. Of course, the calculations are a bit more messy, as the $||y||_\infty$ term is constantly changing, but this should give you the idea.

Listing 3.8: Ending the loops

```
                    h ⟸ 0.9h (η/||e||∞)
                 }
   3        }
         }
```

Again, we stress that since this algorithm uses six function evaluations per time step
that is actually used, it is very costly. We also use 6 evaluations each time we reset
h even if we don't use this time step. So we trade off the cost of these computations
for the ability to widen our time step choice as much as possible to save computation
time.

3.2.2   Runge–Kutta Fehlberg MatLab Implementation

The basic RKF5 code is implemented using some dynamics function f of the form
$f(t, y)$.

Listing 3.9: RKF5 with No Parameters

```
 1
    function [tvals , yvals , fvals , hvals] = RKF5NoP( errortol , steptol , minstep
        , maxstep , fname , hinit , tinit , tend , yinit )
    %
    % Runge-Kutta Fehlberg Order 5
    % Automatically selects optimal step size
 6  %
    % fname       the name of the right hand side function f(t,y)
    %             t is time , y is a vector
    % yinit       initial y data
    % hinit       initial time step
11  % tinit =     initial time
    %
    % tvals       time vector
    % yvals       approximate solution vector
    % fvals       dynamics vector
16  % hvals       step history vector
    %
    h = hinit ;
    t = tinit ;
    y = yinit ;
21  f = feval(fname , t , y) ;

    hvals = h ;
    tvals = t ;
    yvals = y ;
```

```
26 fvals = f;

   while t < tend
     k1 = h*feval(fname,t,y);
     z = y+k1/4;
31   k2 = h*feval(fname,t+(h/4),z);
     z = y+(3.0/32.0)*k1+(9.0/32.0)*k2;
     k3 = h*feval(fname,t+3.0*h/8,z);
     z = y+(1932.0/2197.0)*k1-(7200.0/2197.0)*k2+(7296.0/2197.0)*k3;
     k4 = h*feval(fname,t+12.0*h/13.0,z);
36   z = y+(439.0/216.0)*k1-8.0*k2+(3680.0/513.0)*k3-(845.0/4104.0)*k4;
     k5 = h*feval(fname,t+h,z);
     z = y-(8.0/27.0)*k1+2*k2-(3544.0/2565.0)*k3+(1859.0/4104.0)*k4
         -(11.0/40.0)*k5;
     k6 = h*feval(fname,t+h/2,z);
     temp1 = abs(k1/360.0-(128.0/4275.0)*k3-(2197.0/75240.0)*k4+k5
             /50.0+(2.0/55.0)*k6);
41   temp2 = abs(y);
     error1 = max(temp1);
     error2 = max(temp2);
     % adjusting step size code
     % error is ok, so use stepsize h
46   % keep new t, new y from here
     decisiontol = steptol+errortol*error2;
     if error1 < decisiontol
         y = y+(16.0/135.0)*k1+(6656.0/12925.0)*k3+(28561.0/56430.0)*k4
             -(9.0/50.0)*k5+(2.0/55.0)*k6;
         t = t + h;
51       hvals = [hvals h];
         tvals = [tvals t];
         yvals = [yvals y];
         fvals = [fvals f];
         % we have computed the new state.  Should we keep
56       % the step size? If error is too small, double h
         % but don't exceed maxstep or go past final time
         if error1 <= .3*decisiontol
             hdouble = 2*h;
             htop    = min(hdouble,maxstep);
61           h = min(htop,tend-t);
         end
     % picking an optimal step size when error is too large
     else
         if error1 >= steptol
66           hnew = .9*h*decisiontol/error1;
             hclipped = max(hnew,minstep);
             h = min(hclipped,tend-t);
         end
     end
71 end
```

However, many dynamics functions depend on parameters, so we also implement RKF5 in the case where f depends on a parameter vector p also. For example, a typical dynamics function with parameters would be something like this:

Listing 3.10: Tunnel Diode Dynamics With Parameters

```
  function  f = tunneldiode(p,t,x)
  %
  % param  vector  is  p = [R;L;C;u]
4 %
  % dynamics:
  % x1' = (-h(x1) + x2)/C;
  % x2' = (-x1 -Rx2 +u)/L;
  % f = tunneldiode(1.5,5.0,2,1.2,x)
9 %
  h = x(1).*(17.76+x(1).*(-103.79+x(1).*(229.62+x(1).*(-226.31+83.72*x
      (1))))) ;
  f = [(-h+x(2))/p(3);(-x(1)-p(1)*x(2)+p(4))/p(2)];
```

So we will change the way we write our dynamics functions so that we can pass in a parameter vector. Consider the dynamics for the tunnel diode model we discussed previously. Instead of editing the dynamics file every time we want to change the inductance, resistance, capacitance and incoming voltage values, we will now pass them in as a parameter **p**. We alter the RKF5 algorithm to allow the function f to use a parameter vector p. We will pass back the time vector, **tvals**, the solution vector, **yvals**, the values of the dynamics, **fvals** and, just for educational purposes, the vector of stepsize values, **hvals**. Since this is a step size adjustment method, the time values are no longer uniformly spaced and it is interesting to see how the values of the step size change over time. We need to pass in two tolerances, **errortol** and **steptol**, that are used in determining how we adjust h, a minimum possible step size, **minstep**, and maximum possible step size, **maxstep**, the name of the dynamics function, **fname**, an initial step size guess, **hinit**, the initial and final times, **tinit** and **tend**, and the vector of initial values, **yinit**.

Listing 3.11: RKF5 Algorithm With Parameters

```
  function [tvals,yvals,fvals,hvals] = RKF5(errortol,steptol,minstep,
      maxstep ,...
                                          fname,params,hinit,tinit,
                                              tend,yinit)
  %
  % Runge-Kutta  Fehlberg  Order  5
5 % Automatically  selects  optimal  step  size
  %
  % errortol  tolerance  for  step  adjustment
  % steptol   tolerance  for  step  adjustment
  % minstep   minimum  step  size  allowed
10 % maxstep   maximum  step  size  allowed
  % fname     the  name  of  the  right  hand  side  function  f(t,y,p)
  %           t  is  time,  y  is  a  vector,  p  is  parameter
  % params    parameter  vector  for  dynamics
  % hinit     initial  time  step
15 % tinit     initial  time
  % tend      final  time
  % yinit     initial  y  data
  %
  % tvals     time  vector
20 % yvals     approximate  solution  vector
  % fvals     dynamics  vector
  % hvals     step  history  vector
  %
  h = hinit;
```

```
25  t = tinit;
    y = yinit;
    f = feval(fname,params,t,y);

    hvals = h;
30  tvals = t;
    yvals = y;
    fvals = f;

    while t < tend
35     k1 = h*feval(fname,params,t,y);
       z = y+k1/4;
       k2 = h*feval(fname,params,t+(h/4),z);
       z = y+(3.0/32.0)*k1+(9.0/32.0)*k2;
       k3 = h*feval(fname,params,t+3.0*h/8,z);
40     z = y+(1932.0/2197.0)*k1-(7200.0/2197.0)*k2+(7296.0/2197.0)*k3;
       k4 = h*feval(fname,params,t+12.0*h/13.0,z);
       z = y+(439.0/216.0)*k1-8.0*k2+(3680.0/513.0)*k3-(845.0/4104.0)*k4;
       k5 = h*feval(fname,params,t+h,z);
       z = y-(8.0/27.0)*k1+2*k2-(3544.0/2565.0)*k3+(1859.0/4104.0)*k4
           -(11.0/40.0)*k5;
45     k6 = h*feval(fname,params,t+h/2,z);
       temp1 = abs(k1/360.0-(128.0/4275.0)*k3-(2197.0/75240.0)*k4+k5
           /50.0+(2.0/55.0)*k6);
       temp2 = abs(y);
       error1 = max(temp1);
       error2 = max(temp2);
50     % adjusting step size code
       % error is ok, so use stepsize h
       % keep new t, new y from here
       decisiontol = steptol+errortol*error2;
       if error1 < decisiontol
55        y = y+(16.0/135.0)*k1+(6656.0/12925.0)*k3+(28561.0/56430.0)*k4
              -(9.0/50.0)*k5+(2.0/55.0)*k6;
          t = t + h;
          hvals = [hvals h];
          tvals = [tvals t];
          yvals = [yvals y];
60        fvals = [fvals f];
          % we have computed the new state.  Should we keep
          % the step size? If error is too small, double h
          % but don't exceed maxstep or go past final time
          if error1 <= .3*decisiontol
65           hdouble = 2*h;
             htop    = min(hdouble,maxstep);
             h = min(htop,tend-t);
          end
       % picking an optimal step size when error is too large
70     else
          if error1 >= steptol
             hnew = .9*h*decisiontol/error1;
             hclipped = max(hnew,minstep);
             h = min(hclipped,tend-t);
          end
75     end
    end
  end
```

Let's try using this in some tests. We solve the tunnel diode model for $L = 5$ micro
henrys, $C = 2$ pico fahrads, $R = 1.5$ kilovolts and $u = 1.2$ Volts. We solve this with
the following lines in MatLab:

Listing 3.12: Solving a tunnel diode model

```
  [tvals,yvals,fvals,hvals] = RKF5(1.0e-6,1.0e-8,1.0e-5,2.0,...
                              'tunneldiode',[1.5;5;2;1.2]],...
3                             .3,0,10,[-0.4,1.5]);
  plot(tvals,yvals(1,:));
  xlabel('Time nanoseconds');
  ylabel('V_C Volts');
  title('V_C vs Time');
8 plot(tvals,yvals(2,:));
  xlabel('Time nanoseconds');
  ylabel('I_L mA');
  title('I_L vs Time');
  plot(tvals,hvals);
13 xlabel('Time nanoseconds');
  ylabel('Step size');
  title('h vs Time');
```

This generates the plot of V_c versus time in Fig. 3.1 and I_L versus time in Fig. 3.2.

We can also see the step size history in Fig. 3.3. Note how the step size changes throughout the course of the solution. The step size plot shows us that to obtain reasonable error, we need the h to be as low as 0.01 sometimes. For 10 time units, this would require 1000 calculations requiring 6 function evaluations each time step for a total of 6000 separate calculations. The RKF5 code here generates 91 individual steps. We don't keep track, however, how many times we reject a step size and restart. Ignoring that, we have 546 function evaluations here. Even with restarts, this is a significant reduction in computation time.

Fig. 3.1 The tunnel diode capacitor voltage versus time

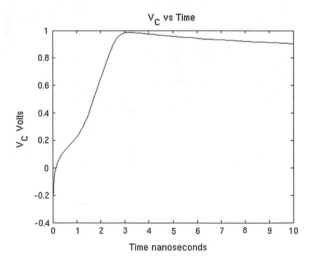

Fig. 3.2 The tunnel diode inductor current versus time

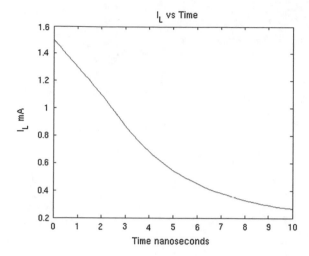

Fig. 3.3 Step size versus time

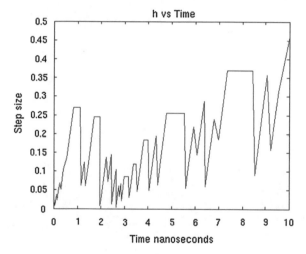

References

J. Peterson, *Calculus for Cognitive Scientists: Derivatives, Integration and Modeling*, Springer Series on Cognitive Science and Technology (Springer Science+Business Media Singapore Pte Ltd, Singapore, 2015a, in press)

J. Peterson, *Calculus for Cognitive Scientists: Higher Order Models and Their Analysis*, Springer Series on Cognitive Science and Technology (Springer Science+Business Media Singapore Pte Ltd, Singapore, 2015b, in press)

Part III
Deriving the Cable Model

Chapter 4
Biological Molecules

We will begin by going through some background material on what might be called the **chemistry of life**; we had a hard time getting all of these things straight, so for all of those **mathematician** and **computer scientist** types out there, here is the introduction we wished we had when we started out. Molecules depend on the interplay of **non covalent** and **covalent** interactions. Recall covalent bonds share electrons in a tightly coupled way. There are three fundamental covalent bonds:

- Electrostatic Bonds due to Coulomb's Law
- Hydrogen Bonds
- Vanderwaals Bonds

4.1 Molecular Bonds

The force F between two molecules is called a molecular bond. Specifically, the force between two charges is given by the formula:

$$F = \frac{1}{4\pi\epsilon_0} \frac{q_1 \, q_2}{r^2} \tag{4.1}$$

where q_1 and q_2 are the charges on the bonding units, r is the distance between the units and ϵ_0 is a basic constant which depends on the solution (air, water and so forth) that the units live within. For example, in Fig. 4.1, we see a representation of the electrostatic attraction between two common molecules; one with a carboxyl and the other with an amide group on their ends.

Hydrogen bonds occur when a hydrogen atom is shared between two other atoms as shown in Fig. 4.2: The atom to which the hydrogen is held more tightly is called the **hydrogen donor** and the other one which is less tightly linked is called the **hydrogen acceptor**. You can see this represented abstractly as follows:

© Springer Science+Business Media Singapore 2016
J.K. Peterson, *Calculus for Cognitive Scientists*, Cognitive Science
and Technology, DOI 10.1007/978-981-287-880-9_4

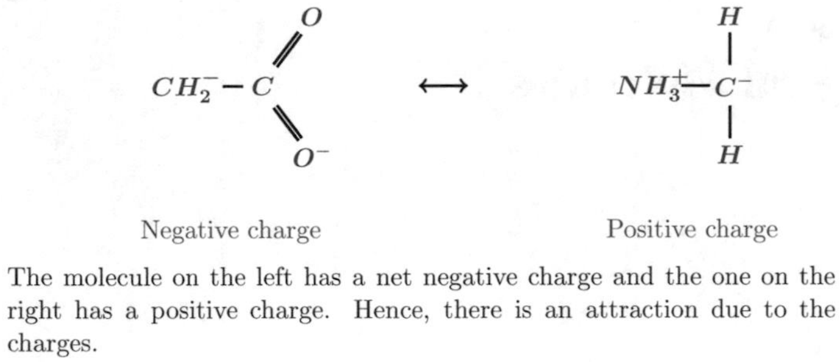

$$CH_2^- - C \underset{O^-}{\overset{O}{\lessgtr}} \qquad \longleftrightarrow \qquad NH_3^+ \overset{H}{\underset{H}{-C^-}}$$

Negative charge   Positive charge

The molecule on the left has a net negative charge and the one on the right has a positive charge. Hence, there is an attraction due to the charges.

Fig. 4.1 Electrostatic attraction between two molecules: one has a carboxyl group and the other an amide group

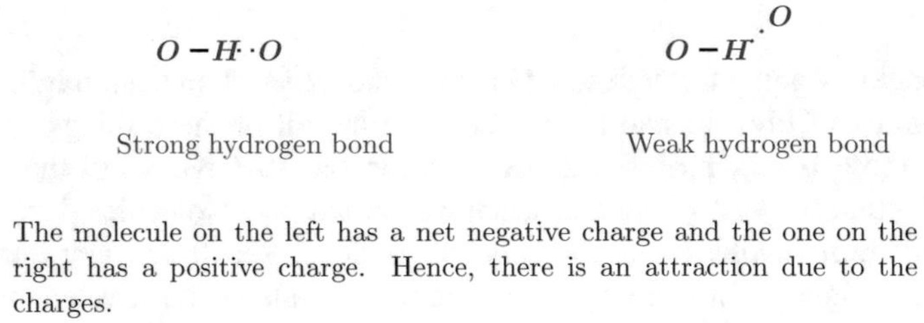

$$O - H \cdot O \qquad\qquad\qquad O - H \overset{\cdot O}{}$$

Strong hydrogen bond   Weak hydrogen bond

The molecule on the left has a net negative charge and the one on the right has a positive charge. Hence, there is an attraction due to the charges.

Fig. 4.2 Typical hydrogen bonds

Table 4.1 Hydrogen bond lengths

Example	Bond length (Å)
O–H$\cdots$O	2.70
O–H$\cdots$O$^-$	2.63
O–H$\cdots$N	2.88
N–H$\cdots$O	3.04
N$^+$–H$\cdots$O	2.93
N–H$\cdots$N	3.10

H Donor	*Length*	H Acceptor
$-O-H-$	2.88Å	$\cdots - N-$
$-N-H-$	3.04Å	$\cdots - O-$

Recall, one Angstrom, (1 Å), is 10^{-10} meters or 10^{-8} cm. Hydrogen bond lengths vary as we can see in Table 4.1.

The donor in biological systems is an oxygen or nitrogen with a covalently attached hydrogen. The acceptor is either oxygen or nitrogen. These bonds are highly directional. The strongest bonds occur when all the atoms **line up** or are **collinear** (think alignment of the planets like in the Tomb Raider movie!). In Fig. 4.3, we see an idealized helix structure with two attached carbon groups. The carbons are part of the helix backbone and the rest of the molecular group spills out from the backbone.

Fig. 4.3 Typical hydrogen bonds in a helix structure

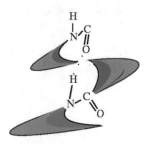

Fig. 4.4 Vanderwaals forces between molecules

Force

+

Repulsion increases

The horizontal axis measures the distance between molecules while the vertical axis gives the force between them.

Energy 0

Distance

Repulsion decreases

Maximum attraction

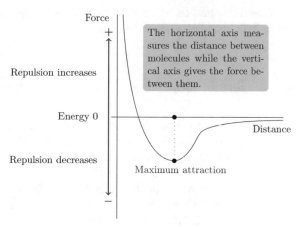

Table 4.2 Maximum attraction distances

Atom	Radius of maximum attraction (Å)
H	1.20
C	2.00
N	1.50
O	1.40
S	1.85
P	1.90

If the two groups are separated enough along the background, the oxygen of the top one is close enough physically to the nitrogen of the bottom group to allow them to share a hydrogen. As you might expect, this sort of hydrogen bonding stabilizes helical structures. This sort of upward spiral occurs in some protein configurations.

There is also a nonspecific attractive force which occurs when any two atoms are 3–4 Å apart which is called a **Vanderwaals Bond**. The basis of this bond is that the charge distribution around an atom is time dependent and at any instant is not perfectly symmetrical. This "transient" asymmetry in an atom encourages a similar asymmetry in the electron distribution around its neighboring atoms. The standard picture of this force is shown in Fig. 4.4. The distance for maximum attraction varies with the atom involved as we can see in Table 4.2.

Table 4.3 Bond distances by type

Bond	Interaction distance (Å)	Bond energy (kcal/mole)
Electrostatic	2.80	3–7
Hydrogen	2.70–3.10	3–7
Vanderwaals	2.70–3.20	1
Covalent	1.00	80

Table 4.2 shows us the $H–C$ has a 3.20 Å Vanderwaals interaction radius; $H–N$ has a 2.70 Å; and $H–P$ has a 13.10 Å.

4.1.1   Bond Comparisons

These three different types of bonds are therefore ranked according to their interaction distance as shown in Table 4.3 where although we haven't really discussed it, it should be easy to understand that to pull a bond apart you would have to exert a force or in other words do some work.

The amount of work you do can be measure in many units, but a common one in biology is the **calorie** which can be converted to the standard physics energy measure of ergs. The abbreviation **kcal** refers to 10^3 calories and the term **mole** refers to a collection of 6.02×10^{23} (Avogadro's Number) of molecules. Note that the covalent bond is far stronger than the other bonds we have discussed!

4.2   Energy Considerations

The ultimate source of energy for most life is the sun (we will neglect here those wonderful Mid-Atlantic smokers with a sulfur based life chemistry that does not require any oxygen or photosynthesis at all). Energy can be stored in various ways for later use—kind of like a battery. Some common energy sources are given in Table 4.4.

Table 4.4 Energy sources

Source	Energy stored
Green photon energy	57 kcal/mole
ATP (Adenine Triphostate) (universal currency of biochemical energy)	12 kcal/mole
Each vibrational degree of freedom in a molecule	0.6 kcal/mole at 25 °C
Covalent bond	80 kcal/mole

$$\uparrow \quad \overset{O}{\diagup \diagdown}\; 99\,\text{Å}$$

$$\overset{-}{\underset{+}{\updownarrow}} \quad H \quad 105° \quad H \qquad\qquad N \; - \; H \cdots O \; \; C$$

A Water Molecule

Polar Bonding Without Water

The water molecule shown on the left has an asymmetric charge distribution which has a profound effect on how molecules interact in liquid. There is an attraction due to this asymmetry: the positive *side* of water attracts the negative *side* of other molecules.

Fig. 4.5 Special forces act on water

Basically, energy is moved from one storage source or another via special helper molecules so that a living thing can perform its daily tasks of eating, growing, reproducing and so forth. Now for us, we are going to concentrate on what happens inside a cell. Most of us already know that the inside of a cell is a liquid solution which is mostly water but contains many ions. You probably have this picture in your head of an ion, like Na^+ sitting inside the water close to other such ions with which they can interact. The reality is more complicated. Water is made up of two hydrogens and one oxygen and is what is called a **polar** molecule. As you can see from the left side of Fig. 4.5, the geometry of the molecule means that the minus and plus charges are not equally balanced. Hence, there is a charge gradient which we show with the vertical arrow in the figure. The asymmetrical distribution of charge in water implies that water molecules have a high affinity for each other and so water will **compete** with other ions for hydrogens to share in a **hydrogen bond**.

If we had two molecules, say NH_3 and COH_2, shown on the right side of Fig. 4.5, we would see a hydrogen bond from between the oxygen of COH_2 and the central hydrogen of NH_3 due to the asymmetrical distribution of charge in these molecules. However, in an environment with water, the polarized water molecules are attracted to these asymmetrical charge distributions also and so each of these molecules will actually be surrounded by a **cage** of water molecules as shown in Fig. 4.6a. This shield of polar water molecules around the molecules **markedly** reduces the attraction between the $+$ and $-$ sites of the molecules. The same thing is true for ionized molecules. If we denote a minus ion by a triangle with a minus inside it and a plus ion by a square with a plus inside it, then this caging effect would look something like what is shown in Fig. 4.6b.

This reduction occurs because the electric field of the *shield* opposes the electric field of the ion and so weakens it. Consider common table salt, $NaCl$. This salt can ionize into Na^+ and Cl^-, the exact amount that ionizes being dependent on the solution you drop the salt into. The minus side of $NaCl$ is very attractive to the positive side of a polar water molecule which in turn is attracted to the negative side of another water molecule. The same thing can be said about the positive side of the salt. The presence of the polar water molecules actively encourages the splitting or disassociating of the salt into two charged pieces. Hence, water can dissolve many

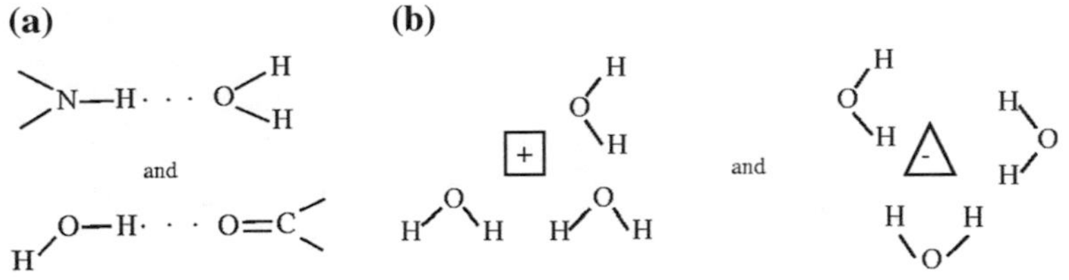

(a)   **(b)**

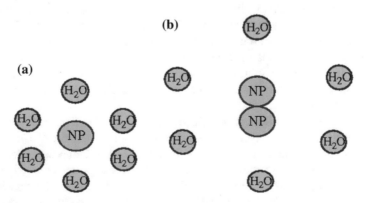

Fig. 4.6 The polarity of water induces cages to form. **a** Polar bonding with water. **b** Ion water cages

polar molecules, like the salt mentioned above, that serve as fuels, building blocks, catalysts and information carriers. This causes a problem because the caging of an ion by water molecules also inhibits ion interactions. Biological systems have solved this problem by creating **water free** micro environments where polar interactions have maximal strength. A consequence of this is that non polar groups aren't split apart by water and so it is energetically more favorable for non polar molecules to be placed into one cage rather that have a separate water cage for each one. If we denote a non polar molecule by the symbol *NP* enclosed in a circle, as a thought experiment we can add our non polar molecule to a water environment. A cavity in the water is created because the non polar molecule disrupts some hydrogen bonds of water to itself. We see a picture like the one in Fig. 4.7a. This means that the number of ways to from hydrogen bonds in the cage around the non polar molecule is smaller that the the number of ways to from hydrogen bonds without the non polar molecule present. This implies a **cost** to caging *NP* as **order** is created. Now if a second non polar molecule *NP* is added, where will it go? If a second cage is created, more hydrogen bonds are used up than if both *NP* molecules clump together inside one cage. Remember always that **order** is costly and **disorder** is energetically favored. So two non polar molecules clump together to give the picture shown in Fig. 4.7b.

Fig. 4.7 It is more efficient to group non polar molecules in cages. **a** Non polar molecule water cages. **b** Non polar molecule group water cages

4.3 Hydrocarbons

We need to learn a bit about the interesting molecules we will see in our quest to build interesting models of cell function. Any good organic chemistry book is good to have as a general reference on your shelf if you really like this stuff. We will begin with simple hydrocarbons.

The basic thing to remember is that carbon's outer shell of electrons is four shy of being filled up. The far right edge of the periodic table consists of the **noble** gases or elements because their outer most electron shells are completely populated. Hence, they are considered **special** or **noble**. For us, a quick way to understand hydrocarbons is to think of a carbon as wanting to add four more electrons so that it can be like a noble gas. Of course, there are less anthropomorphic ways to look at it and really nice physical models but looking at it in terms of a **want** really helps! We typically draw a carbon atom, denoted by C, surrounded by four lines. Each of the lines is ninety degrees apart and the lines represent a covalent bond with something. On the left side of Fig. 4.8, we see a carbon with three of its covalent bonds with hydrogen (hence, these lines go from the central C to an H) and the last covalent bond on the left goes to an unknown molecule we denote as R. The R is called a **residue** and must supply an electron for this last covalent bond. The molecule CH_4 is called **methyl** and is the simplest hydrocarbon (note the residue here is just a hydrogen); here we have the molecule by CH_3R; for example, a ionized phosphate group, PO_3^-, would supply such an electron and we would have the molecule CH_3PO_3. The next simplest hydrocarbon is one that is built from two carbons. If the two carbons bond with each other, that will leave six bonds left over to fill. If all of these bonds are filled with hydrogens, we get the molecule **ethyl** with chemical formulae C_2H_6. Usually, we think of the left most hydrogen as being replaced by a residue R which we show on the right side of Fig. 4.8.

Methyl with Residue R Ethyl with Residue R

Carbon is represented by the symbol **C** and it can form four simple bonds. In the methyl molecule, three of the bonds are taken up by hydrogen (symbol **H**) and the last one is used to attach the residue (symbol **R**). In the ethyl molecule, the right most carbon uses its last available bond to connect to the right most carbon. The right most carbon then has one last bond that can be used to add the residue. The residue group itself be quite complicated.

Fig. 4.8 Methyl and ethyl can have side chains of arbitrary complexity called residues

Clearly, these groupings can get complicated very quickly! We won't show any more specific molecules, but you can get the flavor of all this by looking at Fig. 4.8 which shows a few of the residues we will see attached to various hydrocarbons. Of course, we can also use more than two carbons and if you think about it a bit, as the number of carbons we use goes up, there is no particular reason to think our pictures will always organize the carbons in a central chain like beads on a string. Instead, the carbons may forms side chains or branches, from circular groupings where the last carbon in a group bonds to the first carbon and so forth. Also our pictures are just idealizations to help us think about things; these molecules live in solutions, so there are water cages, there are three dimensional concerns—like which carbons are in a particular plane—and so forth. We won't really discuss those things in a lot of detail. However, the next step is to look at a special class of molecules called **amino acids** and we will see some of this complication show up in our discussions there.

4.4 Amino Acids

An α amino acid consists of the following things:

- an amide group NH_2
- a carbonyl group $COOH$
- a hydrogen atom H
- a distinctive residue R

These groups are all attached to a central carbon atom which is called the α carbon. There are many common residues. First, let's look at a methyl molecule with some common residues attached.

Methyl + residue hydroxyl: The chemical formula here is OH and since O needs to add electrons so that its outer electron shell can be filled, we think of it as having a polarity or charge of -2. Oxygen shares the single electron of hydrogen to add one electron to oxygen's outer shell bringing oxygen closer to a filled outer shell. Hence, the hydrogen bond here brings the net charge of the hydroxyl group to -1 as hydroxyl still needs one more electron. In fact, it is energetically favorable for the hydroxyl group to accept an electron to fill the outermost shell. Hence, hydroxyl can act as OH^- in an ionic bond. Since carbon wants electrons to fill its outer shell, it saves energy for the hydroxyl group and the carbon to share one of carbon's outer electrons in a covalent bond. If we use the hydroxyl group as the residue for methyl, we replace one of the hydrogen's in methyl with hydroxyl giving $CH_3–OH$.

Methyl + residue amide: The chemical formula here is NH_2 and since N needs three electrons to fill its outer most electron shell, we think of it as having a polarity or charge of -3. Here N forms a single bond with the two hydrogens. This molecule can accept an electron and act as the ion NH_2^- or it can bond covalently with carbon replacing one methyl's hydrogens. The methyl plus amide residue would then be written as $CH_3–NH_2$.

Methyl + residue carbonyl: The hydroxyl OH can form a covalent bond with carbon and an oxygen can form a double covalent bond with carbon to give $COOH$. This molecule can accept an additional electron and function in an ionic bond as $COOH^-$ or it can form a covalent bond giving the group OOH which can add an electron and function in ionic bonds as OOH^-. This group can then form a double covalent bond with carbon and share two electrons. The carbon atom then can share two more electrons with another oxygen in a double covalent bond. This gives the molecule $COOH$. This molecule finds it favorable to add an electron in the outer shell so that it can form ionic bonds. In this state, we would call it $COOH^-$. The carbonyl can also form a covalent bond with another carbon. The methyl molecule plus carbonyl residue would then be written as CH_3-COOH.

Methyl + phosphate: The chemical formula here is PO_4. Phosphate, P, has three covalent bonds that can be used to fill its outer most electron shell. It carries two electrons in its $2s$ orbital and three in its $2p$ orbitals. Since oxygen needs two electrons to fill its outermost shell, the $2s$ electrons of phosphorus can be shared with one oxygen. This is still considered a single bond as only two electrons are involved (it is actually called a coordinate covalent bond). However, this bond is often drawn as a double bond in pictorial representations anyway. The remaining three electrons carbon needs to fill its $2p$ orbitals are then obtained by covalent sharing with oxygen. Each of these three oxygen's sharing a $2p$ carbon orbital, still needs an electron. Hence, phosphate can form the ionic bond using PO_4^{-3}.

The three oxygen covalently sharing $2p$ orbitals can then ionically bond with hydrogens to create the molecule PO_4H_3 which is usually written reversed as H_3PO_4. If you leave off one of the hydrogens, we have $H_2PO_4^-$. which can form a residue on methyl giving methyl phosphate $H_2PO_4CH_3$.

Phosphates in dilute water solutions exist in four forms. In a strongly basic solution, PO_4^{-3} predominates. However, in a weakly basic setting, HPO_4^{-2} is more common. In a weakly acid water solution, the dominant form is $H_2PO_4^{-1}$ and finally, in a strongly acidic solution, H_3PO_4 has the highest concentration.

Methyl + thiol: The chemical formula here is SH and since sulphur has two covalent bonds that can be used to fill its outer most electron shell, we think of it as having a polarity or charge of -2. Hence, the hydrogen bond here gives us an ion with -1. So we can denote this group as SH^-. A methyl molecule with a thiol residue would thus have the formula $SH-CH_3$.

If we replace two of the hydrogens on the methyl group with an amide group and a carboxyl group we obtain the molecule $NH_2-COOH-CH_2$. We can see how a molecule of this type will be assembled using are simple pictorial representation of covalent bonds in Fig. 4.9. As we mentioned earlier, we think of the central carbon as the α carbon so that we distinguish it from any other carbons that are attached to it. In fact, one way to look at all this is to think that since a carbon lacks four electrons in its outer shell, it is energetically favorable to seek *alliances* with other molecules so that it can *fill* these empty spots. This is, of course, a very anthropomorphic way to look at it, but it helps you see what is going on at a gut level. So draw the α carbon with four dots around it. These are the places in the outer shell that are not filled.

Fig. 4.9 Methyl with
carboxyl and amide residues
with electron shells

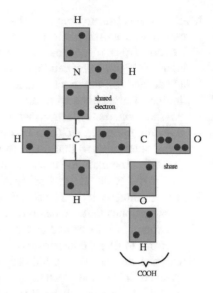

Chemical theory tells us that the outer shell of a carbon consists of four groupings
of two electrons each. So when we draw four groupings of just one electron per
group we are clearly indicating that one electron is missing in each group. Now the
amide molecule is handled in a similar way: the nitrogen is missing three electrons
in its outer shell which we indicate by three single dots placed around the nitrogen.
A hydrogen has only one electron as it has a very simple shell structure; hence it is
drawn as a single dot. So hydrogen would like one more electron to fill its outer shell
and the nitrogen would also like an additional electron to fill one of its groups. So
a good solution is for the two atoms to *share* an electron which we denote by a box
around the two single electrons. Continuing in this way, we can build an electron dot
diagram for the amide and its connection to the α carbon. The carbonyl is a little
more complicated. So far we have only looked at bonds between atoms where one
electron is shared. Another type of bond which is even stronger is one where two of
the outer shell electrons are shared between two atoms. The carbonyl group would
be written as $C = OOH$ to reflect the fact the there is such a double bond between
the carbon and the first oxygen. Oxygen is missing only two electrons in its outer
shell and so this double bond which is indicated by the double bars = completely
fills the outer shell of the oxygen and half of the outer shell of the carbon. One of the
remaining two outer shell groups is then filled by a hydroxyl group, OH. Note the
hydroxyl group is an oxygen with one of its outer shell groups filled by a hydrogen
leaving one group to fill. It does this by sharing with one of the remaining two groups
that are open on the carbon of the carbonyl group. This leaves one opening left on
the carbon of the carbonyl which is used to make a shared bond with the α carbon
of the amino acid. The entire electron dot diagram is shown in Fig. 4.9 and it is very
complex. So we generally do not use this kind of explicit notation to draw an amino
acid. All of this detail is assumed in the simple skeleton formula we see on the left

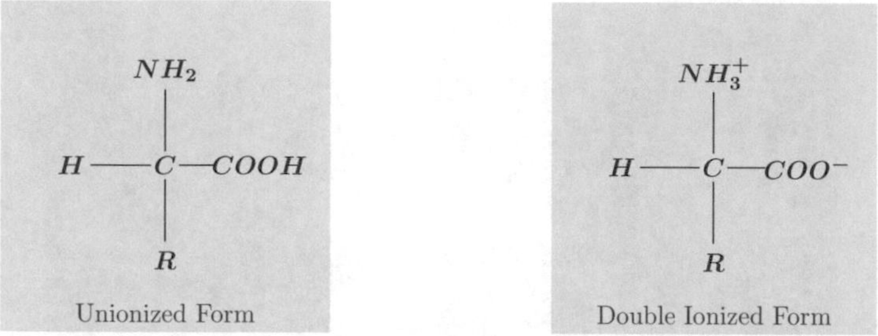

Here we see a typical amino acid. The common elements are the carboxyl and amide groups. Each amino acid then has a different residue. The carbon atom in the middle is called the central carbon and the elements attached to it can be ionized in a variety of ways. This ionization can substantially effect how the amino acid reacts with other molecules. Also, charge distribution in amino acids is not uniform and so one side of an amino acid may act more positive than the other.

Fig. 4.10 A typical amino acid in normal and ionized forms

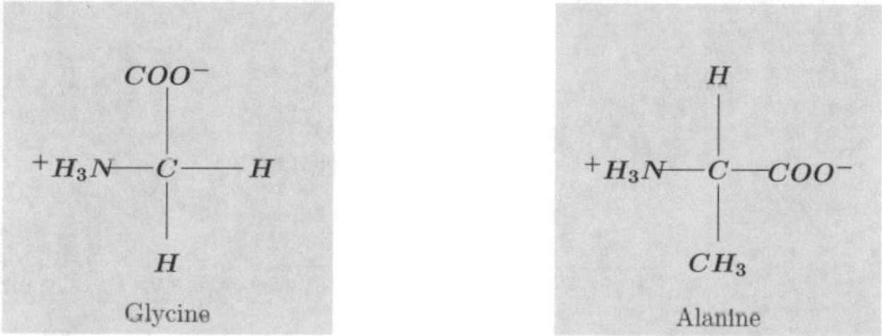

Glycine's residue is H and alanine's, is CH_3. Glycine is the simplest amino acid and is optically inactive.

Fig. 4.11 The amino acids **Glycine** and **Alanine**

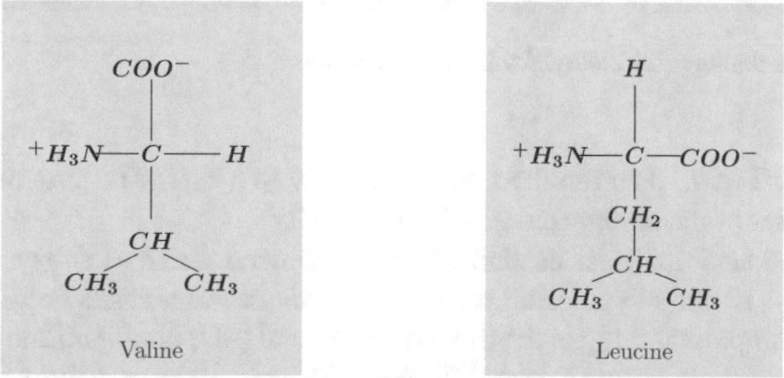

Valine's residue is $CH(CH_3)_2$ and Leucine's is $CH_2CH(CH_3)_2$. Note Valine and Leucine have a longer residue which makes them hydrophobic.

Fig. 4.12 The amino acids **Valine** and **Leucine**

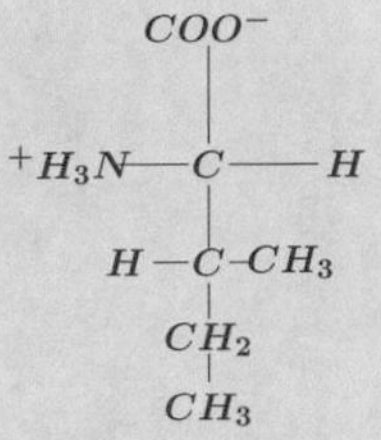

Isoleucine

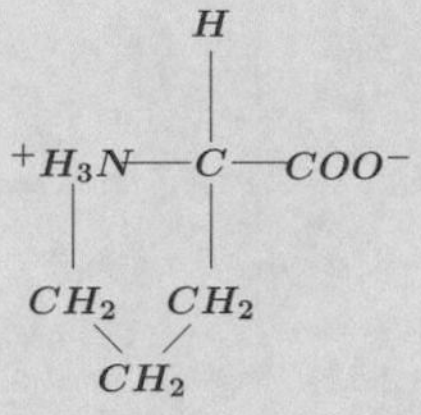

Proline

Isoleucine's residue is $HCCH_3CH_2CH_3$ and Proline's is a cyclic structure $H_2CCH_2CH_2$ attaching to both the amide and the central carbon. Isoleucine is hydrophobic but Proline's cyclic residue is indifferent to water.

Fig. 4.13 The amino acids **Isoleucine** and **Proline**

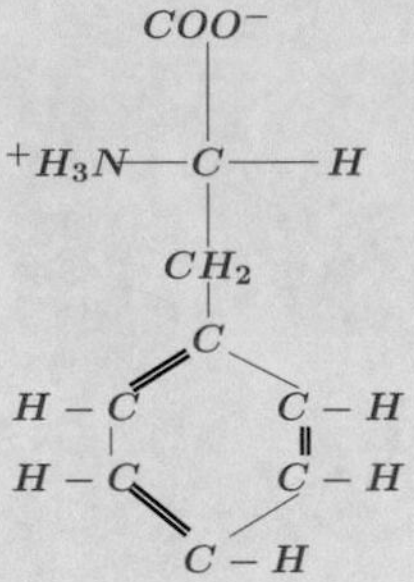

Phenylalanine

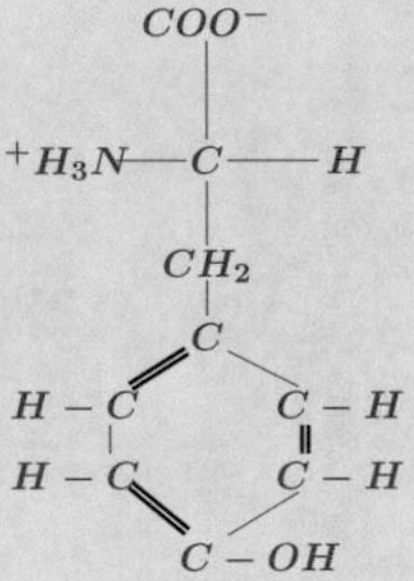

Tyrosine

Phenylalanine has a very hydrophobic phenyl group as a residue. The ring structure of the phenyl group creates a localized cloud of *pi* electrons which make it very reactive. Tyrosine replaces the bottom $C-H$ with $C-OH$. The addition of the hydroxyl group makes this amino acid hydrophilic. It is also very reactive due to the localized *pi* electron cloud.

Fig. 4.14 The amino acids **Phenylalanine** and **Tyrosine**

hand side of Fig. 4.10 and in the notation CH_2NH_2COOH. The structure shown in Fig. 4.9 is that of the *amino* acid *glycine* (**G** or **Gly**).

An amino acid can exist in ionized or non ionized forms as shown on the right side of Fig. 4.10 and of course all we have said about water cages is still relevant.

Another important thing is the three dimensional (3D) configuration of an amino acid. An amino acid occurs in two different 3D forms. To keep it simple, look at this simple representation

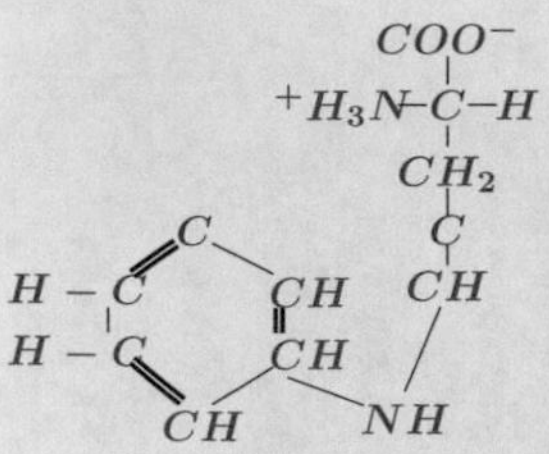

Trytophan

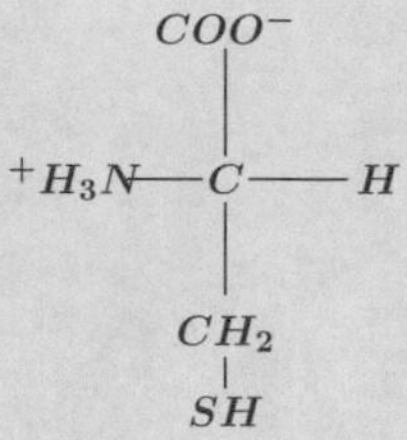

Cysteine

Tryptophan's residue is fairly complicated with a phenyl group off to the left. It is very hydrophobic. Cysteine plays a special role in biology because bonds similar to hydrogen bonds can form between sulphur atoms occurring on different cysteine molecules. These bonds are called *disulfide links.*

Fig. 4.15 The amino acids **Tryptophan** and **Cysteine**

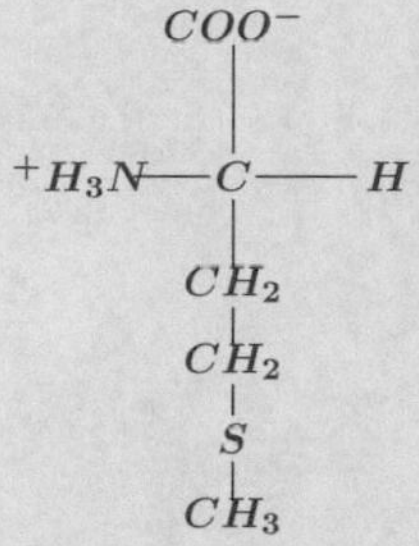

Methionine

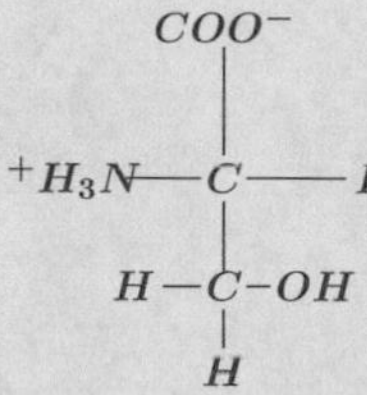

Serine

Methionine is hydrophobic. If you look at Alalanine on the right side of Figure 4.11, you'll see that Serine is formed by Adding a hydroxyl group to the methyl residue on Alalanine. This is called *hydroxylation.*

Fig. 4.16 The amino acids **Methionine** and **Serine**

$$\begin{array}{cccc} & R \ +y \text{ axis} & & \\ & \uparrow & & \\ H & \leftarrow C_\alpha \rightarrow & NH_2 \ +x \text{ axis} \\ & \downarrow & & \\ & COOH & & \end{array}$$

The R, H, NH_2 and $COOH$ are in the xy plane and the C_α carbon is along the positive z axis above the side groups. The NH_2 is on the positive x axis and the R is on the positive y axis. This is the L form as if you take your **right hand**, line up the fingers along the NH_2 line and rotate your fingers **left** towards the residue R. Note your

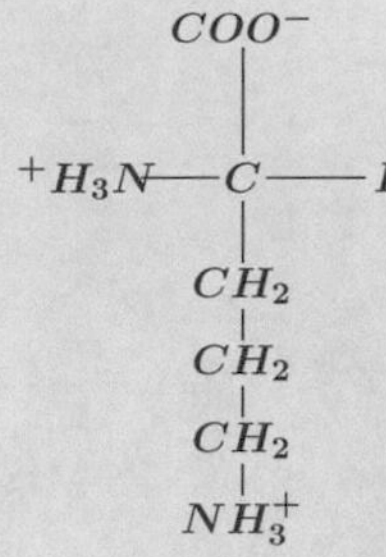

$$COO^-$$
$$^+H_3N-C-\!\!\!-H$$
$$CH_2$$
$$CH_2$$
$$CH_2$$
$$NH_3^+$$

Lysine

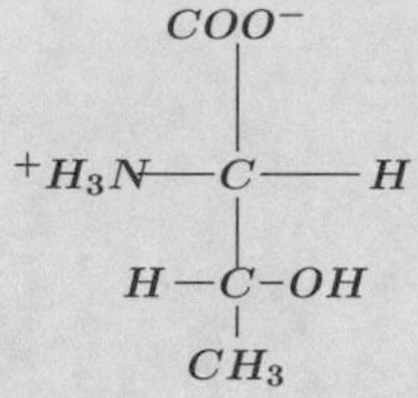

$$COO^-$$
$$^+H_3N-C-\!\!\!-H$$
$$H-C\text{-}OH$$
$$CH_3$$

Threonine

Lysine is very polar and hence, it is very hydrophobic. Threonine is a hydroxylated version of Valine (see the left side of Figure 4.12).

Fig. 4.17 The amino acids **Lysine** and **Threonine**

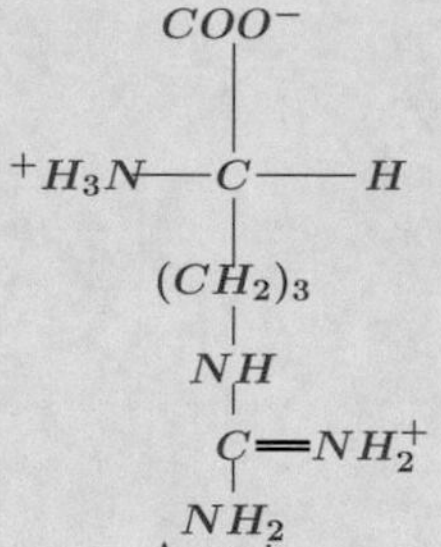

$$COO^-$$
$$^+H_3N-C-\!\!\!-H$$
$$(CH_2)_3$$
$$NH$$
$$C=NH_2^+$$
$$NH_2$$
Arganine

$$COO^-$$
$$^+H_3N\text{-}C\text{-}H$$
$$CH_2$$

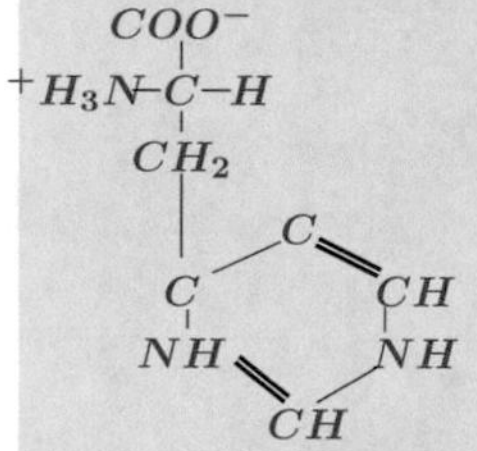

Histidine

Arganine and Histidine are very polar, hydrophobic and are positive ions at neutral pH. However, Histidine is a negative ion at physiological pH.

Fig. 4.18 The amino acids **Arganine** and **Histidine**

thumb points out of the page towards the positive z axis location of C_α. Also, it is easy to visualize by just imaging grabbing the C_α and pulling it up out of the page that the other groups lie in. The other form is called the R form and looks like this:

$$
\begin{array}{ccc}
 & H & \\
 & \uparrow & \\
COOH & \leftarrow C_\alpha \rightarrow & NH_2 \;+x\,\text{axis} \\
 & \downarrow & \\
 & R & -\,y\,\text{axis}
\end{array}
$$

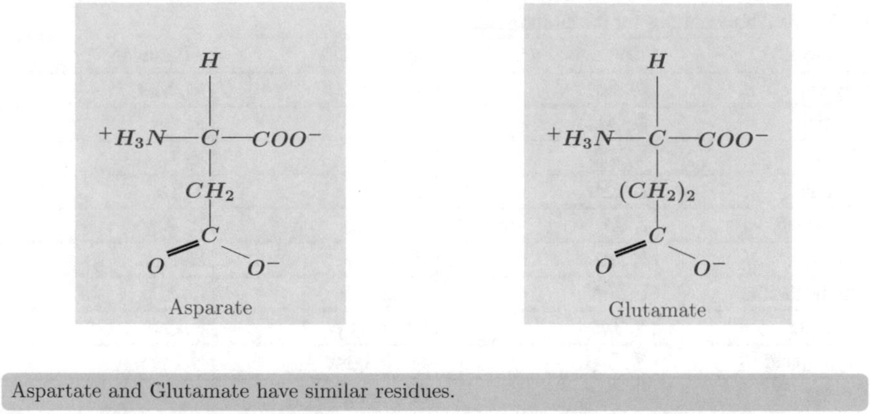

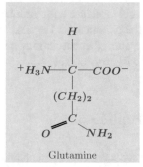

Aspartate and Glutamate have similar residues.

Fig. 4.19 The amino acids **Asparate** and **Glutamate**

The oxygen ion in Aspartate and Glutamate is replaced by the amide group in both Asparagine and Glutamine. This change makes these amino acids neutral in charge.

Fig. 4.20 The amino acids **Asparagine** and **Glutamine**

The R, NH_2, H and $COOH$ are in the xy plane and the C_α carbon is along the negative z axis below the side groups. Here the NH_2 is on the positive x axis but the R is on the negative y axis. This is the R form as if you take your **right hand**, line up the fingers along the NH_2 line and rotate your fingers **right** towards the residue R. Hence, we pull the C_α down below the page determined by the other groups here. For unknown reasons, only L-forms are used in life on earth. Now there are a total of twenty amino acids: we list them in Fig. 4.11 (glycine and alanine), Fig. 4.12 (valine and leucine), Fig. 4.13 (isoleucine and proline), Fig. 4.14 (phenylalanine and tyrosine), Fig. 4.15 (tryptophan and cysteine), Fig. 4.16 (methionine and serine), Fig. 4.17 (lysine and threonine), Fig. 4.18 (arginine and histidine), Fig. 4.19 (aspartate and glutamate) and Fig. 4.20 (asparagine and glutamine). We have organized all of these figures so that the residues are underneath the central carbon. As you can see, all have the common

Table 4.5 Abbreviations for the amino acids

Amino acid	Abbreviation	Amino acid	Abbreviation
Glycine	G, Gly	Methionine	M, Met
Alanine	A, Ala	Serine	S, Ser
Valine	V, Val	Lysine	K, Lys
Leucine	L, Leu	Threonine	T, Thr
Isoleucine	I, Ile	Arginine	R, Arg
Proline	P, Pro	Histidine	H, His
Phenylalanine	F, Phe	Aspartate	D, Asp
Tyrosine	Y, Tyr	Glutamate	E, Glu
Tryptophan	W, Trp	Asparagine	N, Asn
Cysteine	C, Cys	Glutamine	G, Gln

amino acid structure with different residues R attached. The type of residue R determines the chemical and optical reactivity of the amino acids. For convenience, we list the standard abbreviations for the names of the amino acids in table form as well as in the figures in Table 4.5.

4.5 Peptide Bonds

Amino acids can link up in chains because the $COOH$ on one can bond with the NH_2 on another as is seen in Fig. 4.21. In this figure, there is an outlined box that contains the bond between the $COOH$ and the NH_2; this is called the peptide bond and is shown in Fig. 4.21. The two amino acids that pair are connected by a rigid planar bond. There is a C_α^1 atom from amino acid one and another C_α^2 from amino acid two attached to this bond. The $COOH$ and NH_2 bond looks like this in block diagram form. The $COOH$ loses an OH and the NH_2 loses an H to form the bond. Think of bond as forming a rigid piece of cardboard and attached on the left is the amino acid built around C_α^1 and attached on the right is the amino acid build around C_α^2.

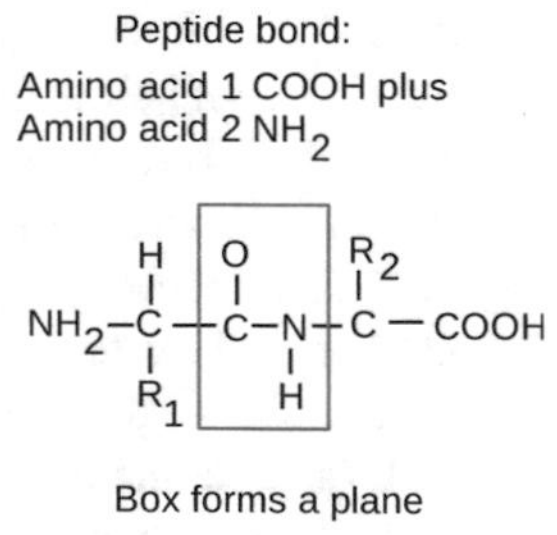

Fig. 4.21 The bond between two amino acids

Peptide bond:
Amino acid 1 COOH plus
Amino acid 2 NH_2

Box forms a plane

$$O \qquad (+y \text{ local axis})$$
$$\uparrow$$
$$C_\alpha^1 \rightarrow \leftarrow C \rightarrow N(+x \text{ local axis}) \rightarrow \leftarrow C_\alpha^2$$
$$\downarrow$$
$$H$$

Now think of C_α^1 as attached to a pencil which is plugged into the side of the peptide bond. The C_α^1 to CO bond is an axis that amino acid one is free to rotate about. Call this angle of rotation Ψ_1. We can do the same thing for the other side and talk about a rotation angle Ψ_2 for the NH to C_α^2 bond. In Fig. 4.21, R_1 is the residue or side chain for the first amino acid and R_2 is the side chain for the other. Note amino acid one starts with an N_2H group on the left and amino acid two ends with a $COOH$ group on the right

$$O \qquad (+y \text{ local axis})$$
$$\uparrow$$
$$C_\alpha^1 \rightarrow \quad \Psi_1 \quad \leftarrow C \rightarrow N(+x \text{ local axis}) \rightarrow \quad \Psi_2 \quad \leftarrow C_\alpha^2$$
$$\downarrow$$
$$H$$

The peptide bond allows amino acids to link into chains as we show in the next block diagram.

$$O \qquad (+y \text{ local axis})$$
$$\uparrow$$
$$N_2H \leftarrow C_\alpha^1 \rightarrow \quad \Psi_1 \quad \leftarrow C \rightarrow N(+x \text{ local axis}) \rightarrow \quad \Psi_2 \quad \leftarrow C_\alpha^2 \rightarrow COOH$$
$$\downarrow$$
$$H$$

We show the peptide bond with a bit more three dimensionality in Fig. 4.22. We can also draw the chain with two peptide bonds as we indicate in Fig. 4.23. You can see that the side chains, R_1, R_2 and R_3, then *hang* off of this chain of linked peptide bonds. In a longer chain, there are two rotational degrees of freedom at the central carbon of any two peptide bonds; i.e. a rotation angle with the peptide bond on the right and on

Fig. 4.22 The peptide bond between two amino acids with the rotation possibilities indicated. C_{α_1} and C_{α_2} point up in the picture

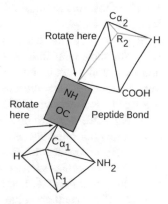

Fig. 4.23 Details of the
peptide bonds in a chain

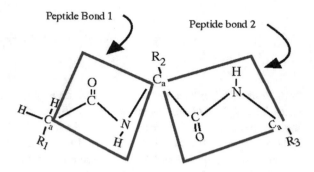

the left. This means that if we imagine the linked peptide bonds as beads on a string, there is a great deal of flexibility possible in the three dimensional configuration of these beads on the string. It isn't hard to imagine that if the beads of string were long enough, full loops could form and there could even be complicated repeated patterns or motifs. Also, these beads on a string are molecular groupings that are inside a solution that is full of various charged groups and the residues or side chains coming off of the string are also potentially charged. Hence, there are many forces that act on this string including hydrogen bonds between residues, Vanderwaals forces acting on motifs and so forth.

Now the peptide bond is always the same, so let's indicate it by a simple **PP**. Then we have

$$N_2H \leftarrow C_\alpha^1 \rightarrow \Psi_1 \leftarrow PP \rightarrow \Psi_2 \leftarrow C_\alpha^2 \rightarrow COOH$$

It is then easy to show a chain of three amino acids with three separate rotation angles. We have simplified the picture by replacing $\rightarrow \Psi_1 \leftarrow$ with just Ψ_1 and so forth giving the block diagram

$$N_2H \ C_\alpha^1 \ \Psi_1 \ PP \ \Psi_2 \ C_\alpha^2 \ PP \ \Psi_3 \ C_\alpha^3 \ COOH$$

We also show this in Fig. 4.24. In this picture, we show the chain we just laid out in block form above.

If we look at one three amino acid piece of a long chain, we would see the following molecular form as represented in Fig. 4.25. As discussed above, as more and more amino acids link up, we get a chain of peptide bonds whose geometry is very complicated in solution. To get a handle on this kind of chain at a high level, we need to abstract out of this a simpler representation.

In Fig. 4.26, we show how we can first drop most of the molecular detail and just label the peptide bond planes using the letter **P**. We see we can now represent our chain in the very compact form $-NCCNCCNCCN-$. This is called the **backbone** of the chain. The molecules such as side chains and hydrogen atoms hang off the backbone in the form that is most energetically favorable. Of course, this representation does not show the particular amino acids in the chain, so another representation for a five

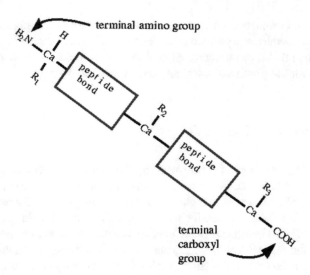

Fig. 4.24 A three amino acid chain

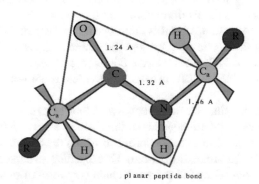

Fig. 4.25 Molecular details of a chain

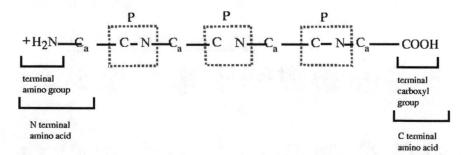

Fig. 4.26 A first chain abstraction

amino acid chain would be $A_1 A_2 A_3 A_4 A_5$ where the symbol A_i for appropriate indices i represents one of the twenty amino acids. The peptide bonds are not even mentioned as it is assumed that they are there. Also, note that in these amino acid chains, the left end is an **amino** group and the right end is a **carbonyl** group.

4.6   Chains of Amino Acids

We roughly classify chains of amino acids by their length. Hence, we say **polypeptides** are chains of amino acids less than or equal to 50 units long. Clearly, this naming is a judgment call. Further, longer chains of amino acids are called **proteins**. As we mentioned earlier, these long chains have side chains and other things that interact via weak bonds or via other sorts of special bonds. For example, the amino acid cysteine (see the right side of Fig. 4.15) has the residue CH_2SH and if the residues of two cysteine's in an amino acid chain can become physically close (this can happen even if the two cysteines are very far apart on the chain because the chain twists and bend in three dimensional space!), a $S–S$ bond can form between the sulphur in the SH groups. This is called a **disulfide** bond and it is yet another important bond for amino acid chains. An example of this bond occurs in the protein insulin as shown in Fig. 4.27. In Fig. 4.27, note we represent the amino acid chains by drawing them as beads on a string: each bead is a circle containing the abbreviation of an amino acid as we listed in Table 4.5. This is a very convenient representation even if much detail is hidden.

In general, for a protein, there are four ways to look at its structure: The **primary** structure is the sequence of amino acids in the chain as shown in Fig. 4.28a.

To know this, we need to know the order in which the amino acids occur in the chain: this is called **sequencing the protein**. Due to amino acid interactions along the chain, different regions of the full **primary** chain may form local three dimensional structures. If we can determine these, we can say we know the **secondary** structure of the protein. An example is a helix structure as shown in Fig. 4.28b. If we pack

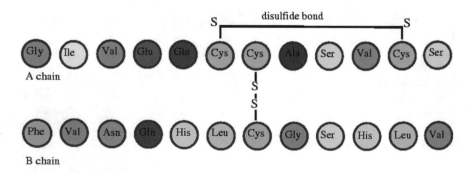

Fig. 4.27 Disulfide insulin protein

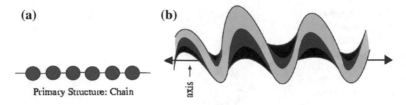

Fig. 4.28 First and second order protein foldings. **a** Primary structure of a protein. **b** Secondary structure of a protein

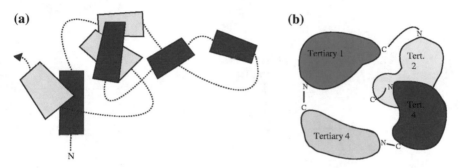

Fig. 4.29 Third and fourth order protein foldings. **a** Tertiary structure of a protein. **b** Quatenary structure of a protein

secondary structures into one or more compact globular units called domains, we obtain the **tertiary** structure an example of which is shown in Fig. 4.29a. In this figure, each rectangle represents secondary structural elements. Finally, the protein my contain several **tertiary** elements which are organized into larger structures. This way the amino acid far apart in the primary sequence structure can be brought close enough together in three dimensions to interact. This is called the **quatenary** structure of the protein. An example is shown in Fig. 4.29b

4.7 Nucleic Acids

Our genetic code is contained in linear chains of what are called nucleic acids in combination with a particular type of sugar. These nucleic acid plus sugar groups are used in a very specific way to **code** for each of the twenty amino acids we mentioned in Sect. 4.4. So our next task is to discuss sugars and nucleic acids and the way these things are used to **code** for the amino acids.

4.7.1 Sugars

Consider the cyclic hydrocarbon shown in Fig. 4.30a The ring you see in Fig. 4.30a is formed from five carbons and one oxygen. For sugars, we label the carbons with

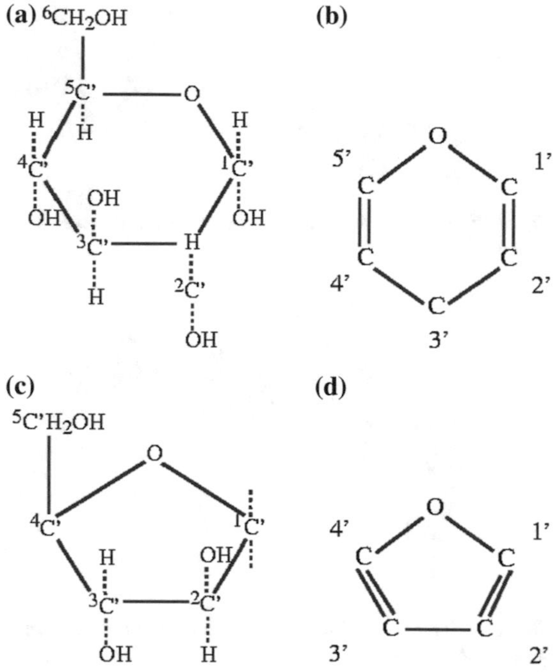

Fig. 4.30 The Pyran and
Furan structures. **a** Details of
a cyclic hydrocarbon sugar:
Pyran. **b** The schematic for
Pyran. **c** Details of a cyclic
hydrocarbon sugar: Furan. **d**
The schematic for Furan

primes as $^1C'$ to $^6C'$ because it will be important to remember which side chains
are attached to which carbons. This type of structure is called a **pyran** and can be
indicated more schematically as in Fig. 4.30b. Another common type of structure is
that shown in Fig. 4.30c which is formed from four carbons and one oxygen. We
label the carbons in a similar fashion to the way we labeled in the pyran molecule.
More symbolically, we would draw a furan as shown in Fig. 4.30d.

We will spend most of our time with the **furan** structures which have the very
particular three dimensional geometry shown in Fig. 4.31a. Note that $^3C'$ and $^5C'$ are
out of the plane formed by $O-^1C'-^2C'-^4C'$; this is called the $^3C'$ **endo** form. Another
three dimensional version of the furan molecule is the $^2C'$ **endo** form shown in
Fig. 4.31b. Here, $^2C'$ and $^5C'$ are out of the plane formed by $O-^1C'-^3C'-^4C'$. Looking
ahead some, these three dimensional forms are important because only certain ones
are used in biologically relevant structures. Later, we will define the large molecules
DNA and RNA and we will find that DNA uses the $^2C'$ **endo** and RNA, the $^3C'$
endo form. Now the particular sugar we are interested in is called **ribose** which will
come in an oxygenated and non-oxygenated (deoxy) form. Consider the formula for
a ribose sugar as shown in Fig. 4.32a. Note the $^2C'$ carbon has an hydroxyl group on
it. If we remove the oxygen from this hydroxyl, the resulting sugar is known as the
deoxy-ribose sugar (see Fig. 4.32b).

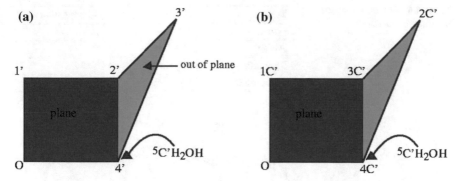

Fig. 4.31 Three dimensional Furan structures. **a** The $^3C'$ endo Furan. **b** The $^2C'$ endo Furan

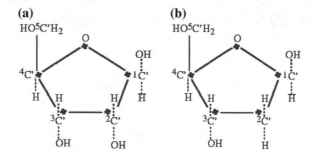

Fig. 4.32 Oxygenated and de-oxygenated ribose sugars. **a** The ribose sugar. **b** The deoxy ribose sugar

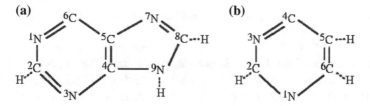

Fig. 4.33 Forms of nitrogenous bases. **a** The generic Purine. **b** The generic Pyrimidine

4.7.2 Nucleotides

There are four special nitrogenous bases which are important. They come in two flavors: **purines** and **pyrimidines**. The **purines** have the form shown in Fig. 4.33a while the **pyrimidines** have the one shown in Fig. 4.33b. There are two purines and two pyrimidines we need to know about: the purines **adenine** and **guanine** and the pyrimidines **thymine** and **cytosine**. These are commonly abbreviated as shown in Table 4.6.

Table 4.6 Abbreviations for the nitrogenous bases

Type	Name	Abbreviation
Purine	Adenine	A
Purine	Guanine	G
Pyrimidine	Thymine	T
Pyrimidine	Cytosine	C

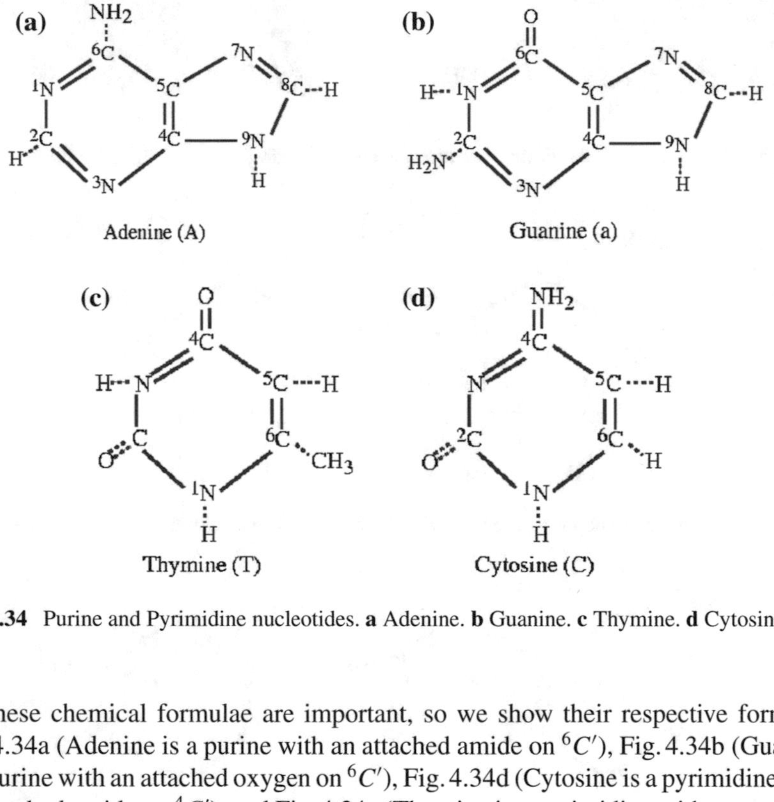

Fig. 4.34 Purine and Pyrimidine nucleotides. **a** Adenine. **b** Guanine. **c** Thymine. **d** Cytosine

These chemical formulae are important, so we show their respective forms in Fig. 4.34a (Adenine is a purine with an attached amide on $^6C'$), Fig. 4.34b (Guanine is a purine with an attached oxygen on $^6C'$), Fig. 4.34d (Cytosine is a pyrimidine with an attached amide on $^4C'$), and Fig. 4.34c (Thymine is a pyrimidine with an attached oxygen on $^4C'$). These four nitrogenous bases can bond to the ribose or deoxyribose sugars to create what are called **nucleotides**. For example, adenine plus deoxyribose would give a compound called **deoxy-adenoside** as shown in Fig. 4.35. In general, a sugar plus a purine or pyrimidine nitrogenous base give us a **nucleoside**. If we add phosphate to the $^5C'$ of the sugar, we get a new molecule called a **nucleotide** (note the change from *side* to *tide*!). In general, sugar plus phosphate plus nitrogenous base gives **nucleotide**. An example is **deoxy-adenotide** as shown in Fig. 4.36.

This level of detail is far more complicated and messy than we typically wish to show; hence, we generally draw this in the compact form shown in Fig. 4.37. There, we have replaced the base with a simple shaded box and simply labeled the primed carbons with the numerical ranking. In Fig. 4.38 we show how nucleotides can link up into chains: bond the $^5C'$ of the ribose on one nucleotide to the $^3C'$ of the ribose

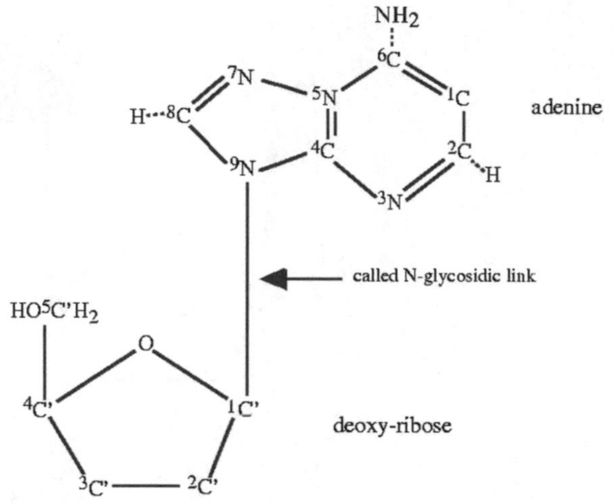

Fig. 4.35 Deoxy-adenoside

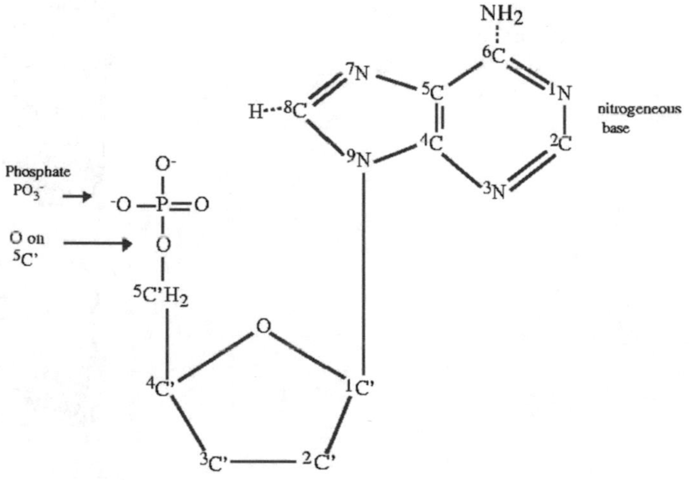

Fig. 4.36 Deoxy-adenotide

on another nucleotide with a phosphate or PO_3^- bridge. Symbolically this looks like Fig. 4.38. This chain of three nucleotides has a terminal OH on the $^5C'$ of the top sugar and a terminal OH on the $^3C'$ of the bottom sugar. We often write this even more abstractly as shown in Fig. 4.39 or just $OH-$ **Base 3** P **Base 2** P **Base 1** P $-OH$, where the P denotes a phosphate bridge. For example, for a chain with bases adenine, adenine, cytosine and guanine, we would write $OH-A-p-A-p-C-p-G-OH$ or $OHApApCpGOH$. Even this is cumbersome, so we will leave out the common phosphate bridges and terminal hydroxyl groups and simply write $AACG$. It is thus understood the left end is an OH terminated $^5C'$ and the right end an hydroxyl terminated $^3C'$.

Fig. 4.37 A general
nucleotide

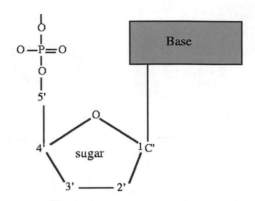

Fig. 4.38 A nucleotide
chain

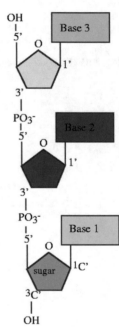

Fig. 4.39 An abstract
nucleotide chain

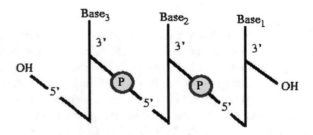

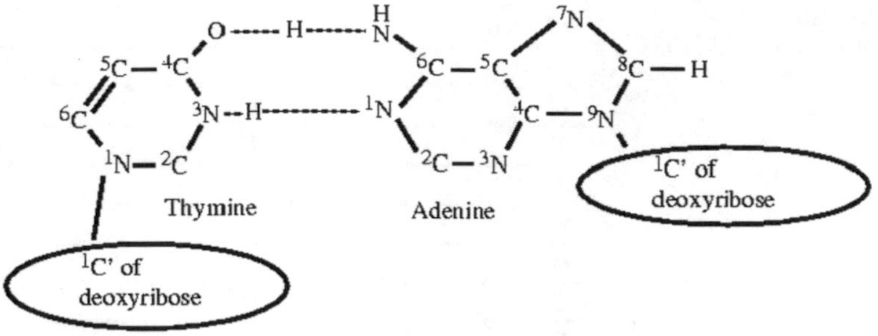

Fig. 4.40 The Tyrosine–Adenine bond

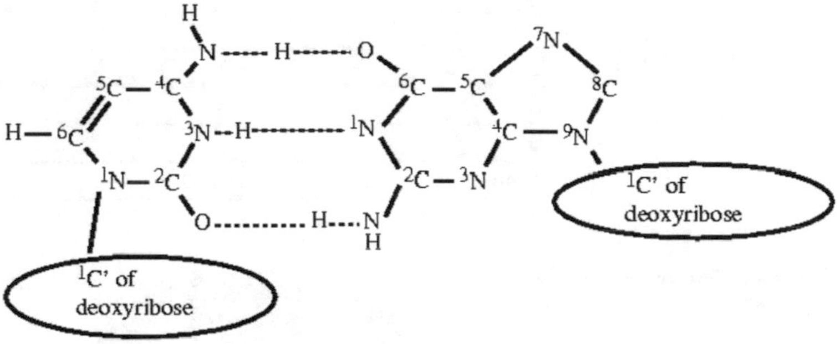

Fig. 4.41 The Cytosine–Guanine bond

4.7.3 Complementary Base Pairing

The last piece in this puzzle is the fact that the purine and pyrimidine nucleotides can bond together in the following ways: A to T or T to A and C to G or G to C. We say that adenine and thymine and cytosine and guanine are complementary nucleotides. This bonding occurs because hydrogen bonds can form between the adjacent nitrogen or between adjacent nitrogen and oxygen atoms. For example, look at the T–A bond in Fig. 4.40. Note the bases are *inside* and the sugars outside. Finally, note how the bonding is done for the cytosine and guanine components in Fig. 4.41. Now as we have said, nucleotides can link into a long chain via the phosphate bond. Each base in this chain is attracted to a complimentary base. It is energetically favorable for two chains to form: chain one and its *complement* chain 2. In the following table, Table 4.7, we see how this pairing is done for a short sequence of nucleotides. Note that the ^{5} pairs with a ^{3} and vice versa. Each pair of complimentary nucleotides is called a *complimentary base pair*. The forces that act on the residues of the nucleotides and between the nucleotides themselves coupled with the rigid nature of the peptide bond between two nucleotides induce the two chains to form a **double helix** structure under cellular conditions which in cross-section (see Fig. 4.42) has the bases inside and the sugars outside.

Table 4.7 Two
complimentary nucleotide
chains

Chain one	Chain two
^{5} end	^{3} end
C	G
T	A
A	T
C	G
G	C
G	C
C	G
T	A
A	T
T	A
T	A
C	G
G	C
^{3} end	^{5} end

Fig. 4.42 A cross-section of
a nucleotide helix

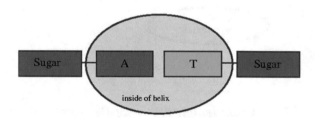

The complimentary nucleotides fit into the spiral most efficiently with 100 degrees
of rotation and 1.5 Angstroms of rise between each base pair. Thus, there are 3.6 base
pairs for every 360 degrees of rotation around the spiral with a rise of $3.6 \times 1.5 = 5.4$
Angstroms. This is, of course, hard to draw! If you were looking down at the spiral
from the top, you could imagine each base to base pair as a set of bricks. You would
see a lower set of bricks and then the next pair of bricks above that pair would be
rotated 100 degrees as shown in Fig. 4.43. To make it easier to see what is going on,
only the top pair of bases have the attached sugars drawn in. You can see that when
you look down at this spiral, all the sugars are sticking outwards. The double helix is
called **DNA** when deoxy-ribose sugars are used on the nucleotides in our alphabet.
The name **DNA** stands for *deoxy-ribose nucleic acid*. A chain structure closely related
to **DNA** is what is called **RNA**, where the **R** refers to the fact that oxy-ribose sugars or
simply ribose sugars are used on the nucleotides in the alphabet used to build **RNA**.
The **RNA** alphabet is slightly different as the nucleotide Thymine,**T**, in the **DNA**
alphabet is replaced by the similar nucleotide Uracil, **U**. The chemical structure of
uracil is shown in Fig. 4.44 right next to the formula for thymine. Note that the only
difference is that carbon $^5C'$ holds a methyl group in thymine and just a hydrogen in

Fig. 4.43 The base–base pair rotation

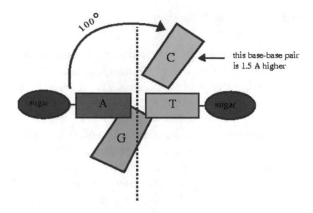

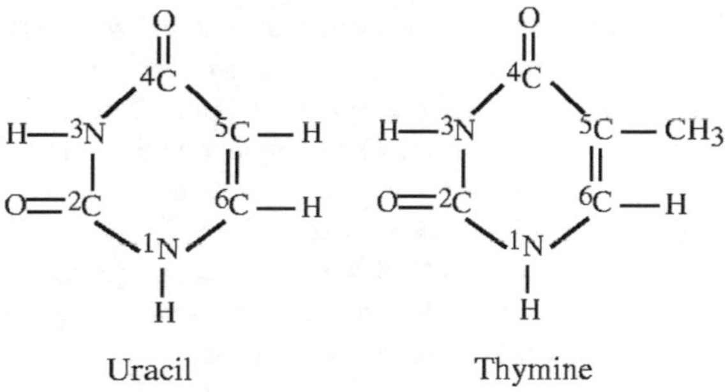

Fig. 4.44 The nucleotide Uracil and Thymine

Uracil

Thymine

uracil. Despite these differences, uracil will still bond to adenine via a complimentary bond. It is rare for the long string of oxygenated ribose nucleotides to form a double helix, although within that long chain of nucleotides there can indeed be local *hairpin* like structures and so forth. Amino acids are coded using nucleotides with what is called the triplet code. This name came about because any set of three nucleotides is used to construct one of the twenty amino acids through a complicated series of steps. We will simply say that each triplet is **mapped** to an amino acid as a shorthand for all of this detail. There are 20 amino acids and only 4 nucleotides. Hence, our alphabet here is $\{A, C, T, G\}$ The number of ways to take 3 things out of an alphabet of 4 things is 64. To see this, think of a given triplet as a set of three empty slots; there are 4 ways to fill slot 1, 4 independent ways to fill slot 2 (we know have 4×4 ways to fill the first two slots) and finally, 4 independent ways to fill slot 3. This gives a total of $4 \times 4 \times 4$ or 64 ways to fill the three slots independently. Since there are only 20 amino acids, it is clear that more than one nucleotide triplet could be mapped to a given amino acid! In a similar way, there are 64 different ways to form triplets from the RNA alphabet $\{A, C, U, G\}$. We tend to identify these two sets of triplets and the associated mapping to amino acids as it is just a matter of replacing the T in one set with an U to obtain the other set.

4.8   Making Proteins

Organisms on earth have evolved to use this nucleotide triplet to amino acid mapping (it is not clear why this is the mapping used over other possible choices!). Now **protein**s are strings of amino acids. So each amino acid in this string can be thought of as the output of a mapping from the triplet code we have been discussing. Hence, associated to each protein of length N is a long chain of nucleotides of length $3N$.

Even though the series of steps by which the triplets are mapped into a protein is very complicated, we can still get a reasonable grasp how proteins are made from the information stored in the nucleotide chain by looking at the process with the right level of abstraction. Here is an overview of the process of protein transcription. When a protein is built, certain biological machines are used to find the appropriate place in the **DNA** double helix where a long string of nucleotides which contains the information needed to build the protein is stored. This long chain of nucleotides which encodes the information to build a protein is called a **gene**. Biological machinery *unzips* the double helix at this special point into two chains as shown in Fig. 4.45. A complimentary copy of a DNA single strand fragment is made using complimentary pairing but this time adenine pairs to uracil to create a fragment of **RNA**. This fragment of RNA serves to **transfer** information encoded in the DNA fragment to other places in the cell where the actual protein can be assembled. Hence, this RNA fragment is given a special name—**Messenger RNA** or **mRNA** for short. This transfer process is called **transcription**.

For example, the DNA fragment $^5ACCGTTACCGT^3$ has the DNA complement $^3TGGCAATGGCA^5$ although in the cell, the complimentary RNA fragment

$$^3UGGCAAUGGCA^5$$

Fig. 4.45   The unzipped
double helix

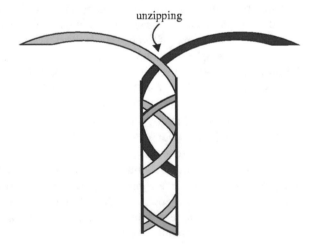

Table 4.8 The Triplet code

Amino acid	DNA triplet	RNA triplet
Alanine	GCA, GCC, GCG, GCT	GCA, GCC, GCG, GCU
Arginine	AGA, AGG, CGA, CGC CGG, CGT	AGA, AGG, CGA, CGC CGG, CGU
Asparagine	AAC, AAT	AAC, AAU
Aspartic acid	GAC, GAT	GAC, GAU
Cysteine	TAC, TAT	UAC, UAU
Glutamic acid	GAA, GAG	GAA, GAG
Glutamine	CAA, CAG	CAA, CAG
Glycine	GGA, GGC, GGG, GGT	GGA, GGC, GGG, GGU
Histidine	CAC, CAT	CAC, CAU
Isoleucine	ATA, ATC, ATT	AUA, AUC, AUU
Leucine	CTA, CTC, CTG, CTT TTA, TTG	CUA, CUC, CUG, CUU UUA, UUG
Lysine	AAA, AAG	AAA, AAG
Methionine (Start)	ATG	AUG
Phenylalanine	TTC, TTT	UUC, UUU
Proline	CCA, CCC, CCG, CCT	CCA, CCC, CCG, CCU
Serine	AGC, AGT, TCA, TCC TCG, TCT	AGC, AGU, UCA, UCC UCG, UCU
Threonine	ACA, ACC, ACG, ACT	ACA, ACC, ACG, ACU
Tryptophan	TGG	UGG
Tyrosine	TAC, TAT	UAC, UAU
Valine	GTA, GTC, GTG, GTT	GUA, GUC, GUG, GUU
Stop	TAA, TAG, TGA	UAA, UAG, UGA

is produced instead. Note again that the ^{5} pairs with a ^{3} and vice versa. There are many details of course that we are leaving out. For example, there must be a special chunk of nucleotides in the original DNA string that the specialized biological machines can locate as the place to begin the unzipping process. The mRNA is transferred to a special protein manufacturing facility called the **ribosome** where three nucleotides at a time from the mRNA string are mapped into their corresponding amino acid. From what we said earlier, there are 64 different triplets that can be made from the alphabet $\{A, C, U, G\}$ and it is this mapping that is used to assemble the protein chain a little at a time. For each chain that is unzipped, a complimentary chain is attracted to it in the fashion shown by Table 4.7. This complimentary chain will however be built from the oxygenated deoxy-ribose or simply ribose nucleotides. Hence, this complimentary chain is part of a *complimentary* RNA helix. As the amino acids encoded by mRNA are built and exit from the ribosome into the fluid inside the cell, the chain of amino acids or polypeptides begins to twist and curl into its three dimensional shape based on all the forces acting on it. We can write this whole process symbolically as **DNA → mRNA → ribosome → Protein**. This is known as the **Central Dogma of Molecular Biology**.

Hence to decode a particular gene stored in DNA which has been translated to its complimentary mRNA form all we need to know are which triplets are associated with which amino acids. These triplets are called **DNA Codons**. The DNA alphabet form of this mapping is given in Table 4.8; remember, the RNA form is the same, we just replace the thymine's (T's) by uracil's (U's).

For example, the DNA sequence,

$$\text{TAC|TAT|GTG|CTT|ACC|TCG|ATT}$$

is translated into the mRNA sequence

$$\text{AUG|AUA|CAC|GAA|UGG|AGC|UAA}$$

which corresponds to the amino acid string

Listing 4.1: Amino acid string

Start | Isoleucine | Histidine | Glutamic Acid | Tryptophan | Serine | Stop

Note that shifting the *reading* by one base to the right or left changes completely which triplets we read for coding into amino acids. This is called a **frame shift** and it can certainly lead to a very different decoded protein. Changing one base in a given triplet is a very local change and is a good example of a mutation or a kind of damage produced by the environment or by aging or disease. Since the triplet code is quite redundant, this may or may not result in a amino acid change. Even if it does, it corresponds to altering **one** amino acid in a potentially long chain.

Chapter 5
Ion Movement

We are now in a position to discuss how ions move in and out of the membranes that surround individual cells. This background will eventually enable us to understand how an excitable cell can generate an action potential.

5.1 Membranes in Cells

The functions carried out by membranes are essential to life. Membranes are highly selective permeability barriers instead of impervious containers because they contain specific **pumps** and **gates** as we have mentioned. Membranes control flow of information between cells and their environment because they contain specific receptors for external stimuli and they have mechanism by which they can generate chemical or electrical signals.

Membranes have several important common attributes. They are sheet like structures a few molecules thick (60–100 Å). They are built from specialized molecules called **lipids** together with proteins. The weight ratio of proteins to lipids is about 4 : 1. They also contain specialized molecules called **carbohydrates** (we haven't yet discussed these) that are linked to the lipids and proteins. Membrane lipids are small molecules with a **hydrophilic** (i.e. attracted to water) and a **hydrophobic** (i.e. repelled by water) part. These lipids spontaneously assemble or form into closed *bimolecular* sheets in aqueous medium. Essentially, it is energetically most favorable for the hydrophilic parts to be on the outside near the water and the hydrophobic parts to be inside away from the water. If you think about it a bit, it is not hard to see that forming a sphere is a great way to get the water hating parts away from the water by placing them inside the sphere and to get the water loving parts near the water by placing them on the outside of the sphere. This lipid sheet is of course a barrier to the flow of various kinds of molecules. Specific proteins mediate distinctive functions of these membranes. Proteins serve many functions: as pumps, pumps, gates, receptors, energy transducers and enzymes among others.

© Springer Science+Business Media Singapore 2016
J.K. Peterson, *Calculus for Cognitive Scientists*, Cognitive Science
and Technology, DOI 10.1007/978-981-287-880-9_5

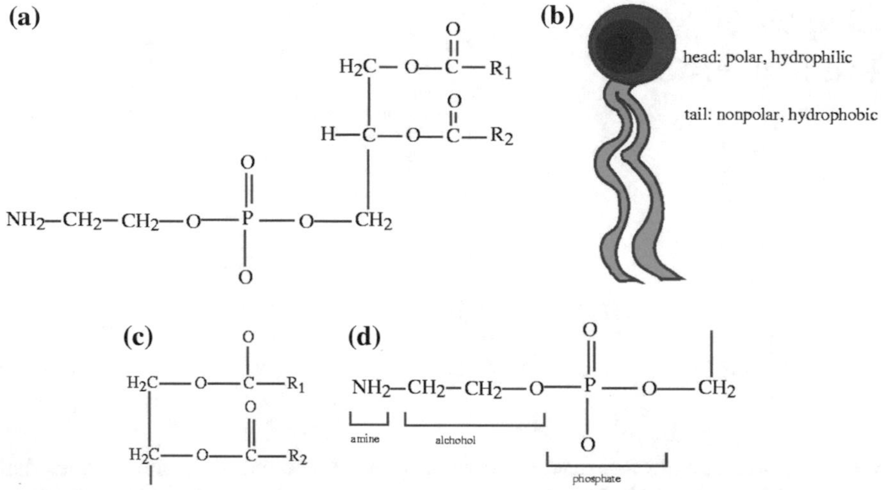

(a)

$$H_2C—O—\overset{\overset{O}{\|}}{C}—R_1$$

$$H—\overset{|}{C}—O—\overset{\overset{O}{\|}}{C}—R_2$$

$$NH_2—CH_2—CH_2—O—\overset{\overset{O}{\|}}{\underset{\underset{O}{|}}{P}}—O—CH_2$$

(b)

head: polar, hydrophilic

tail: nonpolar, hydrophobic

(c)

$$H_2C—O—\overset{\overset{O}{\|}}{C}—R_1$$

$$H_2C—O—\overset{\overset{O}{\|}}{C}—R_2$$

(d)

$$NH_2—CH_2—CH_2—O—\overset{\overset{O}{\|}}{\underset{\underset{O}{|}}{P}}—O—CH_2$$

amine   alchohol

phosphate

Fig. 5.1 The phospholipid membranes. **a** A typical phospholipid. **b** An abstract lipid. **c** The hydrophobic phospholipid end group. **d** The hydrophilic phospholipid end group

Membranes are thus structures or assemblies, whose constituent protein and lipid molecules are held together by many **non-covalent** interactions which are cooperative. Since the two faces of the membrane are different, they are called asymmetric fluid structures which can be regarded as 2-D solutions of oriented proteins and lipids. A typical membrane is built from what are called **phospholipids** which have the generic appearance shown in Fig. 5.1a. Note the group shown in Fig. 5.1d is polar and water soluble—i.e. hydrophilic and the other side, Fig. 5.1c, is water phobic. In these drawings, we are depicting the phosphate bond as a double one.

Of course, this is way too much detail to draw; hence, we use the abstraction shown in Fig. 5.1b using the term *head* for the hydrophilic part and *tail* for the hydrophobic part. Thus, these lipids will spontaneously assemble so that the heads point out and the tails point in allowing us to draw the self assembled membrane as in Fig. 5.2a.

We see the heads orient towards the water and the tails away from the water spontaneously into this sheet structure. The assembly can also form a sphere rather than a sheet as shown in Fig. 5.2b. A typical mammalian cell is 25 μm in radius where a μm is 10^{-6} meter. Since 1 Å is 10^{-10} meter or 10^{-4} μm, we see a cell's radius is around 250,000 Å. Since the membrane is only 60 Å or so in thickness, we can see the percentage of real estate of the cell concentrated in the membrane is very small. So a molecule only has to go a small distance to get through the membrane but to move through the interior of the cell (say to get to the nucleus) is a very long journey! Another way of looking at this is that the cell has room in it for a lot of things!

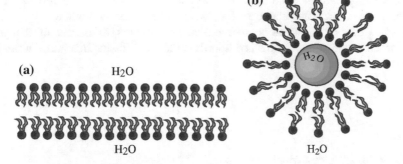

Fig. 5.2 Membrane structures. **a** The self assembled lipid sheet membrane structure. **b** The self assembled lipid sphere membrane structure

5.2 The Physical Laws of Ion Movement

We have relied on the wonderful books of Johnston and Wu (1995) and Weiss (1996) in developing this discussion. They provide even more details and you should feel free to look at these books. However, the amount of detail in them can be overwhelming, so we are trying to offer a short version with just enough detail for our mathematical/biological engineer and computer scientist audience! An ion c can move across a membrane due to several forces.

5.2.1 Ficke's Law of Diffusion

First, let's talk about what concentration of a molecule means. For an molecule b, the concentration of the ion is denoted by the symbol $[b]$ and is measured in $\frac{molecules}{liter}$. Now, we hardly ever measure concentration in molecules per unit volume; instead we use the fact that there are 6.02×10^{23} molecules in a **Mole** and usually measure concentration in the units $\frac{Moles}{cm^3} = M$ where for simplicity, the symbol M denotes the concentration in Moles per cm^3. This special number is called **Avogadro's Number** and we will denote it by N_A. In the discussions that follow, we will at first write all of our concentrations in terms of molecules, but remember what we have said about Moles as we will eventually switch to those units as they are more convenient.

The force that arises from the rate of change of the concentration of molecule b acts on the molecules in the membrane to help move them across. The amount of molecules that move across per unit area due to this force is labeled the **diffusion flux** as *flux* is defined to a rate of transfer ($\frac{something}{second}$) per unit area. Now look at a cross section of the membrane and draw a perpendicular line through it. We can then measure position of molecules in the membrane with the variable x: the membrane of thickness ℓ starts at some value x_0, and ends at $x_0 + \ell$. Hence, below x_0 we are inside the membrane and above $x_0 + \ell$, we are outside the membrane.

Ficke's Law of Diffusion is an empirical law which says the rate of change of the concentration of molecule b is proportional to the diffusion flux and is written in mathematical form as follows:

$$J_{diff} = -D\,\frac{\partial\,[b]}{\partial x} \tag{5.1}$$

where

- J_{diff} is diffusion flux which has units of $\frac{molecules}{cm^2-second}$.
- D is the diffusion coefficient which has units of $\frac{cm^2}{second}$.
- $[b]$ is the concentration of molecule b which has units of $\frac{molecules}{cm^3}$.

The *minus* sign implies that flow is from **high** to **low** concentration; hence diffusion takes place *down* the concentration gradient. Note that D is the proportionality constant in this law.

5.2.2   Ohm's Law of Drift

Ohm's Law of Drift relates the electrical field due to an charged molecule, i.e. an ion, c, across a membrane to the drift of the ion across the membrane where *drift* is the amount of ions that moves across the membrane per unit area. In mathematical form

$$J_{drift} = -\,\partial_{el}\,E \tag{5.2}$$

where it is important to define our variables and units very carefully. We have:

- J_{drift} is the drift of the ion which has units of $\frac{molecules}{cm^2-second}$.
- ∂_{el} is electrical conductivity which has units of $\frac{molecules}{volt-cm-second}$.

We know from basic physics that an electrical field is the negative gradient of the potential so if V is the potential across the membrane and x is the variable that measures our position on the membrane, we have

$$E = -\frac{\partial V}{\partial x}$$

Now the valence of ion c is the charge on the ion as an integer; i.e. the valence of Cl^- is -1 and the valence of Ca^{+2} is $+2$. We let the valence of the ion c be denoted by z. It is possible to derive the following relation between concentration $[c]$ and the electrical conductivity ∂_{el}:

$$\partial_{el} = \mu \, z \, [c]$$

where dimensional analysis shows us that the proportionality constant μ, called the **mobility** of ion c, has units $\frac{cm^2}{volt-second}$. Hence, we can rewrite Ohm's Law of Drift as

$$J_{drift} = -\mu z[c] \frac{\partial V}{\partial x} \tag{5.3}$$

We see that the drift of charged particles goes against the electrical gradient.

5.2.3 Einstein's Relation

There is a relation between the diffusion coefficient D and the mobility μ of an ion which is called Einstein's Relation. It says

$$D = \frac{\kappa T}{q} \mu \tag{5.4}$$

where

- κ is Boltzmann's constant which is $1.38 \times 10^{-23} \frac{joule}{°K}$.
- T is the temperature in degrees Kelvin.
- q is the charge of the ion c which has units of coulombs.

To see that the units work out, we have to recall some basic physics. Electrical work is $\int \boldsymbol{F} \cdot d\boldsymbol{\ell}$ where $\boldsymbol{F}$ is the Coulomb force. Hence, coulomb force times distance it acts which has units of newton − meters or joules. However, we also know the electrical field $\boldsymbol{E}$ is the coulomb force over charge, $\boldsymbol{F}/q_0$ and hence, we can rewrite $\boldsymbol{F} = q_0 \boldsymbol{E}$. We can thus use the units of coulombs − volt/meter for the Coulomb force. Thus, electrical work has units of (coulombs − volt/meter) * meter or coulombs − volts. Thus, we know that electrical work is measured in volt − coulombs or joules. Hence, we see $\frac{\kappa T}{q} \mu$ has units $\frac{volt-coulomb}{°K} \frac{°K}{coulombs} \frac{cm^2}{volt-second} = \frac{cm^2}{sec}$ which reduces to the units of D.

Further, we see that Einstein's Law says that diffusion and drift processes are additive because Ohm's Law of Drift says J_{drift} is proportional to μ which by Einstein's Law is proportional to D and hence $J_{diff}c$.

5.2.4 Space Charge Neutrality

When we look at a given volume element enclosed by a non permeable membrane, we also know that the total charge due to positively charged ions, **cations**, and negatively charged ions, **anions** is the same. If we have N cations c_i with valences z_i^+ and M

anions a_j with valences z_j^-, we have the charge due to an ion is its valence times the charge due to an electron e giving

$$\sum_{i=1}^{N} z_i^+ \, e \, [c_i] = \sum_{j=1}^{M} |z_j^-| \, e \, [c_j] \qquad (5.5)$$

Of course, in a living cell, the membrane is permeable, so Eq. 5.5 is not valid!

5.2.5 Ions, Volts and a Simple Cell

The membrane capacitance of a typical cell is one micro fahrad per unit area. Typically, we use F to denote the unit fahrads and the unit of area is cm^2. Also, recall that $1F = \frac{1\,coulomb}{volt}$. Thus, the typical capacitance is $1.0 \, \frac{\mu F}{cm^2}$. Now our simple cell will be a sphere of radius $25 \, \mu M$ with the inside and outside filled with a fluid. Let's assume the ion c is in the extracellular fluid with concentration $[c] = 0.5\,M$ and is inside the cell with concentration $[c] = 0.5\,M$. The inside and outside of the cell are separated by a biological membrane of the type we have discussed. We show our simple cell model in Fig. 5.3.

Right now in our picture, the number of ions on both sides of the membrane are the same. What if one side had more or less ions than the other? These *uncompensated* ions would produce a voltage difference across the membrane because charge is capacitance times voltage ($q = cV$). Hence, if we wanted to produce a one volt potential difference across the membrane, we can compute how many *uncompensated* ions, $\delta[c]$ would be needed:

$$\delta[c] = \frac{10^{-6}F}{cm^2} \times 1.0\,V = 10^{-6} \frac{coulombs}{cm^2}$$

Fig. 5.3 A simple cell

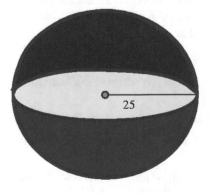

25

Now the typical voltage difference across a biological membrane is on the order of 100 millivolts or less (1 millivolt is 10^{-3} V and is abbreviated mV). The capacitance of the membrane per cm^2 multiplied by the desired voltage difference of 100 mV will give the *uncompensated* charge, n, per cm^2 we need. Thus, we find

$$n = \frac{10^{-6} \text{ coulombs}}{cm^2} \times 10^{-1} V = 10^{-7} \frac{\text{coulombs}}{cm^2}$$

For convenience, let's assume our ion c has a valence of -1. Now one electron has a charge of e of 1.6×10^{-19} coulombs, so the ratio $\frac{n}{e}$ tells us that $6.3 \times 10^{11} \frac{\text{ions}}{cm^2}$ are needed. We know the surface area, SA, and volume, Vol, of our simple cell of radius r are $SA = 4\pi r^2 = 7.854 \times 10^{-5}$ cm^2 and $Vol = \frac{4}{3}\pi r^3 = 6.545 \times 10^{-8}$ cm^3. Thus, the number of uncompensated ions per cell to get this voltage difference is $m = n \, SA$ giving

$$m = \frac{n}{e} \, SA = 4.95 \times 10^7 \text{ ions}$$

This is a very tiny fraction of the total number of ions inside or outside the cell as for a 0.5 M solution, inside the cell, there are

$$0.5 \times N_A \frac{\text{ions}}{\text{Mole}} \frac{\text{Mole}}{cm^3} \times Vol = 1.97 \times 10^{16} \text{ ions}$$

implying the percentage of uncompensated ions to give rise to a voltage difference of 100 mV is only 2.51×10^{-7} %.

5.3 The Nernst–Planck Equation

Under physiological conditions, ion movement across the membrane is influenced by both electric fields and concentration gradients. Let J denote the total flux, then we will assume that we can add linearly the diffusion due to the molecule c and the drift due to the ion c giving

$$J = J_{drift} + J_{diff}$$

Thus, applying Ohm's Law 5.3 and Ficke's Law 5.1, we have

$$J = -\mu \, z \, [c] \frac{\partial V}{\partial x} - D \frac{\partial [c]}{\partial x}$$

Next, we use Einstein's Relation 5.4 to replace the diffusion constant D to obtain what is called the Nernst Planck equation.

$$J = -\mu\, z\, [c] \frac{\partial V}{\partial x} - \frac{\kappa T}{q}\, \mu\, \frac{\partial [c]}{\partial x}$$

$$= -\mu \left(z\, [c] \frac{\partial V}{\partial x} + \frac{\kappa T}{q} \frac{\partial [c]}{\partial x} \right) \tag{5.6}$$

We can rewrite this result by moving to units that are $\frac{moles}{cm^2 - second}$. To do this, note that $\frac{J}{N_A}$ has the proper units and using the Nernst–Planck equation 5.6 we obtain

$$\frac{J}{N_A} = -\frac{\mu}{N_A} \left(z\, [c] \frac{\partial V}{\partial x} + \frac{\kappa T}{q} \frac{\partial [c]}{\partial x} \right) \tag{5.7}$$

The relationship between charge and moles is given by **Faraday's Constant**, F, which has the value $F = 96{,}480\ \frac{coulombs}{mole}$. Hence, the total charge in a mole of ions is the valence of the ion times Faraday's constant F, zF. Multiply Eq. 5.7 by zF on both sides to obtain

$$\frac{J}{N_A} zF = -\frac{\mu\, zF}{N_A} \left(z\, [c] \frac{\partial V}{\partial x} + \frac{\kappa T}{q} \frac{\partial [c]}{\partial x} \right) \tag{5.8}$$

This equation has the units of **current** per unit area because

$$\frac{J}{N_A} zF = \frac{moles}{cm^2 - second}\, \frac{coulombs}{mole} = \frac{coulombs}{cm^2 - sec}$$
$$= \frac{amps}{cm^2}$$

We can measure energy in two different units: **joules** (we use these in Boltzmann's constant κ) or **calories**. One calorie is the amount of energy needed to raise one gram of water from 25 degrees Centigrade to 26 degrees Centigrade. So it is certainly not obvious how to convert from **joules** to **calories**. An argument that is based on low level principles from physics gives us the following conversion **One Calorie** is 4.184 **joules**. From introductory level physics, there is another fundamental physical constant called the **Gas Constant** which is traditionally denoted by R. The constant can be expressed in terms of calories or joules as follows:

$$R = \frac{1.98\ calories}{^\circ K\ Mole} = \frac{8.31\ joules}{^\circ K\ Mole}$$

Hence, if we let q be the charge on one electron, e, we have for T is one degree Kelvin

$$\frac{\kappa\,(T=1)}{q=e} = (1°\mathrm{K})\,\frac{1.38\times10^{-23}\,\frac{\text{joules}}{°\mathrm{K}}}{1.6\times10^{-19}\,\text{coulombs}} = 8.614\times10^{-5}\,\frac{\text{joules}}{\text{coulomb}}$$

$$\frac{R\,(T=1)}{F} = \frac{8.31\,\frac{\text{joules}}{°\mathrm{K\,Mole}}}{96{,}480\,\frac{\text{coulomb}}{\text{Mole}}} = 8.614\times10^{-5}\,\frac{\text{joules}}{\text{coulomb}}$$

For later purposes, we will need to remember that

$$\frac{R(T=1)}{F} = 8.614\times10^{-5}\,\frac{\text{joules}}{\text{coulomb}} \tag{5.9}$$

Thus, since $\frac{\kappa T}{q}$ is the same as $\frac{RT}{F}$, they are interchangeable in Eq. 5.8 giving

$$I = \frac{J}{N_A}\,zF = -\frac{\mu}{N_A}\left(z^2 F\,[c]\,\frac{\partial V}{\partial x} + z\,RT\,\frac{\partial[c]}{\partial x}\right) \tag{5.10}$$

where the symbol I denotes this current density $\frac{\text{amps}}{\text{cm}^2}$ that we obtain with this equation. The current I is the ion current that flows across the membrane per unit area due to the forces acting on the ion c. Clearly, the next question to ask is what happens when this system is at equilibrium and the net current is zero?

5.4 Equilibrium Conditions: The Nernst Equation

The current form of the Nernst–Planck equation given in Eq. 5.10 describes ionic current flow driven by electro-chemical potentials (concentration gradients and electric fields). We know that the current I is **opposite** to $\frac{\partial V}{\partial x}$, **with** $\frac{\partial[c]}{\partial x}$ if the valence z is negative and **against** $\frac{\partial[c]}{\partial x}$ if the valence z is positive. When the net current due to all of these contributions is zero, we have $I = 0$ and by the Nernst Planck Current Equation 5.10, we have

$$0 = -\frac{\mu}{N_A}\left(z^2 F\,[c]\,\frac{\partial V}{\partial x} + z\,RT\,\frac{\partial[c]}{\partial x}\right)$$

implying

$$z^2 F\,[c]\,\frac{\partial V}{\partial x} = -z\,RT\,\frac{\partial[c]}{\partial x}$$

or since there is only one independent variable x

$$\frac{dV}{dx} = -\frac{RT}{zF}\,\frac{1}{[c]}\,\frac{d[c]}{dx}$$

Now let x_1 and x_2 be the start and end position of the membrane, respectively. Now, integrate between positions x_1 and x_2:

$$\int_{x_1}^{x_2} \frac{dV}{dx}\, dx = -\frac{RT}{zF} \int_{x_1}^{x_2} \frac{d[c]}{[c]} dx$$

Now assume that the membrane voltage and the concentration $[c]$ are functions of the position x in the membrane and hence can be written as $V(x)$ and $[c](x)$; we will then let $V(x_1) = V_1, V(x_2) = V_2, [c](x_1) = [c]_1$ and $[c](x_2) = [c]_2$. Then, upon integrating, we find

$$\int_{V_1}^{V_2} dV = -\frac{RT}{zF} \int_{[c]_1}^{[c]_2} \frac{d[c]}{[c]}$$

$$V_2 - V_1 = -\frac{RT}{zF} \ln \frac{[c]_2}{[c]_1}$$

It is traditional to define the **membrane potential** V_m of a cell to be the difference between the inside (V_{in}) and outside potential (V_{out}); hence we say

$$V_m = V_{in} - V_{out}$$

For a given ion c, the equilibrium potential of the ion is denoted by E_c and is defined as the potential across the membrane which gives a zero Nernst–Planck current. We will let the position x_1 be the place where the membrane starts and x_2, the place where the membrane ends. Here, the thickness of the membrane is not really important. So the potential at x_1 will be considered the inner potential V_{in} and the potential at x_2 will be considered the inner potential V_{out}. From our discussions above, we see that the assumption that I is zero implies that the difference $V_1 - V_2$ is $-E_c$ and so labeling $[c]_2$ and $[c]_1$ as $[c]_{out}$ and $[c]_{in}$ respectively, we arrive at the following equation:

$$E_c = \frac{RT}{zF} \ln \frac{[c]_{out}}{[c]_{in}} \tag{5.11}$$

This important equation is called the Nernst equation and is an explicit expression for the equilibrium potential of an ion species in terms of its concentrations inside and outside of the cell membrane.

5.4.1 An Example

Let's compute some equilibrium potentials. In Table 5.1, we see some typical inner and outer ion concentrations and the corresponding equilibrium voltages. Unless otherwise noted, we will assume a temperature of 70 degrees Fahrenheit—about normal room temperature—which is 21.11 Celsius and 294.11 Kelvin as Kelvin is 273 plus Celsius. Since the factor $\frac{R}{F}$ is always a constant here, note

$$\frac{R}{F} = \frac{8.31}{96,480} \frac{\text{joules}}{\text{coulomb} - \text{degrees Kelvin}} = 0.0861 \frac{\text{mV}}{\text{degrees Kelvin}} \qquad (5.12)$$

Let's look at some examples of this sort of calculation. While it is not hard to do this calculation, we have found that all the different units are confusing to students coming from the mixed background we see. Now at a temperature of 294.11 Kelvin, $\frac{RT}{F}$ becomes 25.32 mV. Hence, all of our equilibrium voltage calculations take the form

$$E_c = \frac{1}{z} 25.32 \ln \frac{[c]_{out}}{[c]_{in}} \text{ mV}$$

where all we have to do is to use the correct valence of our ion c. Also, remember that the symbol ln means we should use a natural logarithm! Here are some explicit examples for this temperature:

1. For frog muscle, typical inner and outer concentrations for potassium are $[K^+]_{out}$ is 2.25 mM (the unit mM means milliMoles) with $[K^+]_{in}$ at 124.0 mM. Then, since z is $+1$, we have

$$E_{K^+} = 25.32 \ln \frac{2.25}{124.0} \text{ mV} = 25.32 \times (-4.0094) \text{ mV} = -101.5168 \text{ mV}$$

2. For frog muscle, typical inner and outer concentrations for chlorine are $[Cl^-]_{out}$ is 77.5 mM with $[Cl^-]_{in}$ at 1.5 mM. Then, since z is -1, we have

$$E_{Cl^-} = -25.32 \ln \frac{77.5}{1.5} \text{ mV} = -25.32 \times (3.944) \text{ mV} = -99.88 \text{ mV}$$

3. For frog muscle, typical inner and outer concentrations for Calcium are $[Ca^{+2}]_{out}$ is 2.1 mM (the unit mM means milliMoles) with $[Ca^{+2}]_{in}$ at 10^{-4} mM. Then, since z is $+2$, we have

$$E_{Ca^+} = 0.5 \times 25.32 \ln \frac{2.1}{10^{-4}} \text{ mV} = 12.66 \times (9.9523) \text{ mV} = 126.00 \text{ mV}$$

We summarize the results above as well as two other sets of calculations in Table 5.1. In the first two parts of the table we use a temperature of 294.11 Kelvin and the conversion $\frac{RT}{F}$ is 25.32 mV. In the last part, the temperature is higher (310 Kelvin) and so $\frac{RT}{F}$ becomes 26.69 mV. All concentrations are given in mM.

Table 5.1 Typical inner and outer ion concentrations

	$[c]_{in}$	$[c]_{out}$	E_c
Frog muscle (Conway 1957)			
K^+	124.0	2.25	−101.52
Na^+	10.4	109.0	59.30
Cl^-	1.5	77.5	−99.88
Ca^{+2}	10^{-4}	2.1	126.00
Squid axon (Hodgkin 1964)			
K^+	400.0	20.0	−75.85
Na^+	50.0	440.0	55.06
Cl^-	40.0–150.0	560.0	−66.82−33.35
Ca^{+2}	10^{-4}	10.0	145.75
Mammalian cell			
K^+	140.0	5.0	−88.94
Na^+	5.0–15.0	145.0	89.87−60.55
Cl^-	4.0	110.0	−88.46
Ca^{+2}	10^{-4}	2.5–5.0	135.13−144.39

5.5 One Ion Nernst Computations in MatLab

Now let's do some calculations using MatLab for various ions. First, we will write a
MatLab function to compute the Nernst voltage. Here is a simple MatLab function
to do this.

Listing 5.1: Computing The Nernst Voltage simply

```
    function voltage = Nernst(valence , Temperature , InConc , OutConc)
    %
    %   compute Nernst voltage for a given ion
4   %
    R = 8.31;
    T = Temperature+273.0;
    F = 96480.0;
    Prefix = (R*T)/(valence*F);
9   %
    % output voltage in millivolts
    %
    voltage = 1000.0*( Prefix * log (OutConc/InConc) );
    end
```

It is then straightforward to compute a Nernst potential. We are computing the Nernst
potential for a potassium ion. The valence is thus 1. We will use a temperature in

Centigrade of 37 degrees C (quite hot!) and the inside concentration is 124 milliMoles with the outside concentration 2.5 milliMoles.

Listing 5.2: Finding Potassium voltage

```
  % Our function expects the argments to be
2 % entered as follows:
  %
  %   variable  valence  temperature(C)  InsideConc  OutsideConc
  %      |         |           |             |           |
  %      v         v           v             v           v
7 %    E_K   = Nernst(1,       37.0,         124,        2.5)
  %
  % so here is our line
  E_K = Nernst(1,37.0,124,2.5)
```

This function call produces the following output (edited to remove extra blank lines).

Listing 5.3: Nernst Output for Potassium

```
    valence =
        1
    Temperature =
        37
5  OutConc =
        2.5000
    InConc =
        124.5000
    T =
10     310
    F =
            96480
    Prefix =
        0.0267
15 E_K =
        -104.2400
```

5.5.1 Homework

Exercise 5.5.1 *Use the MatLab functions we have written above to generate a plot of the Nernst potential versus inside concentration for the Sodium ion at* $T = 20$ *degrees* C. *Assume the outside concentration is always* 440 *milliMoles and let the inner concentration vary from* 2 *to* 120 *in* 200 *uniformly spaced steps.*

Exercise 5.5.2 *Rewrite our Nernst and NernstMemVolt functions to accept temperature arguments in degrees Fahrenheit.*

Exercise 5.5.3 *Rewrite our NernstMemVolt function for just Sodium and Potassium ions.*

Exercise 5.5.4 *For the following outside and inside ion concentrations, calculate the Nernst voltages at equilibrium for the temperatures* 45, 55, 65 *and* 72 °F.

	$[c]_{in}$	$[c]_{out}$
K^+	130.0	5.25
Na^+	15.4	129.0
Cl^-	1.8	77.5
Ca^{+2}	10^{-5}	3.1

5.6   Electrical Signaling

The electrical potential across the membrane is determined by how well molecules get through the membrane (its *permeability*) and the concentration gradients for the ions of interest. To get a handle on this let's look at an imaginary cell which we will visualize as an array. The two vertical sides you see on each side of the array represent the cell membrane. There is cell membrane on the top and bottom of this array also, but we don't show it. We will label the part outside the box as the Outside and the part inside as Inside. If we wish to add a way for a specific type of ion to enter the cell, we will label this entry port as gates on the bottom of the array. We will assume our temperature is 70 °F which is 21.11 °C and 294.11 °K.

5.6.1   The Cell Prior to K Gates

To get started, let's assume no potential difference across the membrane and add 100 mM of KCl to both the inside and outside of the cell as shown.

Inside

Outside
100 mM   100 mM
KCl
                KCl

The KCl promptly disassociates into an equal amount of K^+ and Cl^- in both the inside and outside of the cell as shown below:

$$
\boxed{
\begin{array}{l}
\textbf{Inside} \\
\hline\hline
100 \text{ mM } K^+ \\
\\
100 \text{ mM } Cl^-
\end{array}
}
$$

Outside
100 mM K^+
100 mM Cl^-

5.6.2 The Cell with K^+ Gates

Now, we add to the cell channels that are selectively permeable to the ion K^+. These gates allow K^+ to flow back and forth across the membrane until there is a balance between the chemical diffusion force and the electric force. We don't expect a nonzero equilibrium potential because there is charge and concentration balance already. Using Nernst's Equation 5.11 we see since z is 1 that for our temperature, $\frac{RT}{zF}$ is 25.32 mV and

$$
E_K = \frac{RT}{F} \ln \frac{[K^+]_{out}}{[K^-]_{in}} = 25.32 \ln \frac{100}{100} = 0
$$

Outside
100 mM K^+
100 mM Cl^-

$$
\boxed{
\begin{array}{l}
\textbf{Inside} \\
(K^+ \text{ Gates}) \\
\hline\hline
100 \text{ mM } K^+ \\
\\
100 \text{ mM } Cl^-
\end{array}
}
$$

5.6.2.1 The Cell with Outer KCl Reduced

Now reduce the outside KCl to 10 mM giving the following cell:

Outside
10 mM K^+
10 mM Cl^-

$$
\boxed{
\begin{array}{l}
\textbf{Inside} \\
(K^+ \text{ Gates}) \\
\hline\hline
100 \text{ mM } K^+ \\
\\
100 \text{ mM } Cl^-
\end{array}
}
$$

This set up a concentration gradient with an implied chemical and electrical force. We see

$$E_K = 25.32 \ \ln \frac{10}{100}$$
$$= -58.30$$

Hence, $V_{in} - V_{out}$ is -58.30 mV and so the outside potential is 58.30 more than the inside; more commonly, we say the inside is 58.30 mV more negative than the outside.

Homework

Exercise 5.6.1 *Consider a cell permeable to K^+ with the following concentrations on ions:*

	Inside (K^+ *Gates*)
Outside 90 mM K^+ 90 mM Cl^-	120 mM K^+ 120 mM Cl^-

Find the equilibrium Potassium voltage at temperature 69 °F.

5.6.3   The Cell with NaCl Inside and Outside Changes

Now add 100 mM NaCl to the outside and 10 mM to the inside of the cell. We then have

	Inside (K^+ Gates)
Outside 10 mM K^+ 100 mM Na^+ 10 mM Cl^- 100 mM Cl^-	100 mM K^+ 10 mM Na^+ 100 mM Cl^- 10 mM Cl^-

Note there is charge balance inside and outside but since the membrane is permeable to K^+, there is still a concentration gradient for K^+. There is **no** concentration gradient for Cl^-. Since the membrane wall is not permeable to Na^+, the Na^+

concentration gradient has no effect. The K^+ concentration gradient is still the same as our previous example, the equilibrium voltage for potassium remains the same.

5.6.4 The Cell with Na^+ Gates

Next replace the potassium gates with sodium gates as shown below. We can then use the Nernst equation to calculate the equilibrium voltage for Na^+:

$$E_{Na} = 25.32 \ln \frac{100}{10}$$
$$= 58.30$$

which is the exact opposite of the equilibrium voltage for potassium.

	Inside (Na^+ Gates)
Outside 10 mM K^+ 100 mM Na^+ 10 mM Cl^- 100 mM Cl^-	100 mM K^+ 10 mM Na^+ 100 mM Cl^- 10 mM Cl^-

5.6.5 The Nernst Equation for Two Ions

Let's look at what happens if the cell has two types of gates. The cell now looks like this:

	Inside (K^+ Gates) (Na^+ Gates)
Outside 10 mM K^+ 100 mM Na^+ 10 mM Cl^- 100 mM Cl^-	100 mM K^+ 10 mM Na^+ 100 mM Cl^- 10 mM Cl^-

There are now two currents: K^+ is leaving the cell because there is more K^+ inside than outside and Na^+ is going into the cell because there is more Na^+ outside. What determines the resting potential now?

Recall Ohm's Law for a simple circuit: the current across a resistor is the voltage across the resistor divided by the resistance; in familiar terms $I = \frac{V}{R}$ using time honored symbols for current, voltage and resistance. It is easy to see how this idea fits here: there is *resistance* to the movement of an ion through the membrane. Since $I = \frac{1}{R} V$, we see that the ion current through the membrane is proportional to the resistance to that flow. The term $\frac{1}{R}$ seems to be a nice measure of how well ions flow or conduct through the membrane. We will call $\frac{1}{R}$ the *conductance* of the ion through the membrane. Conductance is generally denoted by the symbol g. Clearly, the resistance and conductance associated to a given ion are things that will require very complex modeling even if they are pretty straightforward concepts. We will develop some very sophisticated models in future chapters, but for now, for an ion c, we will use

$$I_c = g_c \, (V_m - E_c)$$

where E_c is the equilibrium voltage for the ion c that comes from the Nernst Equation, I_c is the ionic current for ion c and g_c is the conductance. Finally, we denote the voltage across the membrane to be V_m. Note the difference between the membrane voltage and the equilibrium ion voltage provides the electromotive force or emf that drives the ion. In our cell, we have both potassium and sodium ions, so we have

$$i_K = g_K \, (V_m - E_K), \quad i_{Na} = g_{Na} \, (V_m - E_{Na}).$$

At steady state, the current flows due to the two ions should sum to zero; hence

$$i_K + i_{Na} = 0$$

implying

$$0 = g_K \, (V_m - E_K) + g_{Na} \, (V_m - E_{Na})$$

which we can solve for the membrane voltage at equilibrium:

$$V_m = \frac{g_K}{g_K + g_{Na}} \, E_K + \frac{g_{Na}}{g_K + g_{Na}} \, E_{Na}$$

You will note that although we have already calculated E_K and E_{Na} here, we are not able to compute V_m because we do not know the particular ionic conductances.

5.6.5.1   Homework

Let's look at what happens if the cell has K^+ and Na^+ gates. The cell now looks like this:

	Inside
	$(K^+$ Gates)
Outside	$(Na^+$ Gates)
20 mM K^+ 130 mM Na^+	
20 mM Cl^- 130 mM Cl^-	80 mM K^+ 10 mM Na^+
	15 mM Cl^- 15 mM Cl^-

Exercise 5.6.2 *Compute E_K and E_{Na} at temperature 71 °F.*

Exercise 5.6.3 *Compute the equilibrium membrane voltage for a g_K to g_{Na} ratio of 5.0.*

Exercise 5.6.4 *Compute the equilibrium membrane voltage for a g_K to g_{Na} ratio of 3.0.*

Exercise 5.6.5 *Compute the equilibrium membrane voltage for a g_K to g_{Na} ratio of 2.0.*

5.6.6   The Nernst Equation for More Than Two Ions

Consider Fig. 5.4. We see we are thinking of a patch of membrane as a parallel circuit with one branch for each of the three ions K^+, Na^+ and Cl^- and a branch for the capacitance of the membrane. We think of this patch of membrane as having a voltage difference of V_m across it. In general, there will current that flows through each branch. We label these currents as I_K, I_{Na}, I_{Cl} and I_c where I_c denotes the capacitive current. The conductances for each ion are labeled with resistance symbols with a line through them to indicate that these conductances might be variable in a real model. For right now, we will assume all of these conductances are constant. Each of our ionic currents have the form

$$i_{ion} = g_c \left(V_m - E_c\right)$$

where V_m, as mentioned, is the actual membrane voltage, c denotes our ion, g_c is the conductance associated with ion c and E_c is the Nernst equilibrium voltage. Hence for three ions, potassium (K^+), sodium (Na^+) and chlorine (Cl^-), we have

$$i_K = g_K \left(V_m - E_K\right), \quad i_{Na} = g_{Na} \left(V_m - E_{Na}\right), \quad i_{Cl} = g_{Cl} \left(V_m - E_{Cl}\right)$$

There is also a capacitative current. We know the voltage drop across the capacitor C_m is given by $\frac{q_m}{V_m}$; hence, the charge across the capacitor is $C_m V_m$ implying the capacitative current is

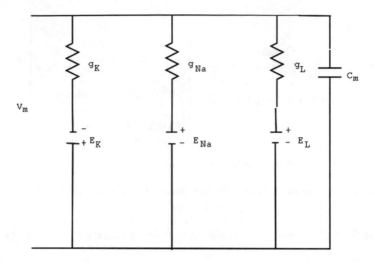

Fig. 5.4 A simple membrane model

$$i_m = C_m \frac{dV_m}{dt}$$

At steady state, i_m is zero and the ionic currents must sum to zero giving

$$i_K + i_{Na} + i_{Cl} = 0$$

Hence,

$$0 = g_K (V_m - E_K) + g_{Na} (V_m - E_{Na}) + g_{Cl} (V_m - E_{Cl})$$

leading to a Nernst Voltage equation for the equilibrium membrane voltage V_m of a membrane permeable to several ions:

$$V_m = \frac{g_K}{g_K + g_{Na} + g_{Cl}} E_K + \frac{g_{Na}}{g_K + g_{Na} + g_{Cl}} E_{Na} + \frac{g_{Cl}}{g_K + g_{Na} + g_{Cl}} E_{Cl}$$

We usually rewrite this in terms of conductance ratios: r_{Na} is the g_{Na} to g_K ratio and r_{Cl} is the g_{Cl} to g_K ratio:

$$V_m = \frac{1}{1 + r_{Na} + r_{Cl}} E_K + \frac{r_{Na}}{1 + r_{Na} + r_{Cl}} E_{Na} + \frac{r_{Cl}}{1 + r_{Na} + r_{Cl}} E_{Cl}$$

Hence, if we are given the needed conductance ratios, we can compute the membrane voltage at equilibrium for multiple ions. Some comments are in order:

- By convention, we set V_{out} to be 0 so that $V_m = V_{in} - V_{out}$ is $-58.3\,\text{mV}$ in our first K^+ gate example.

- These equations are only approximately true, but still of great importance in guiding our understanding.
- Ion currents flowing through a channel try to move the membrane potential toward the equilibrium potential value for that ion.
- If several different ion channels are open, the summed currents drive the membrane potential to a value determined by the relative conductances of the ions.
- Since an ion channel is open briefly, the membrane potentials can't stabilize and so there will always be transient ion currents. For example, if Na^+ channels pop open briefly, there will be transient Na^+ currents.

Finally, let's do a simple example: assume there are just sodium and potassium gates and the relative conductances of Na^+ and K^+ are $4 : 1$. Then r_{Na} is 4 and for the concentrations in our examples above:

$$V_m = \frac{1}{5} \, (-58.3) \text{ mV} + \frac{4}{5} \, (58.3) \text{ mV} = 34.98 \text{ mV}$$

5.6.6.1 Homework

Exercise 5.6.6 *Assume we know*

- *The temperature is* 67 *degrees* Fahrenheit.
- *The ratio* g_{Na} *to* g_K *is* 0.08.
- *The ratio of* g_{Cl} *to* g_K *is* 0.12.
- *The inside and outside concentrations for potassium are* 410 *and* 35 milliMoles *respectively.*
- *The inside and outside concentrations for sodium are* 89 *and* 407 milliMoles *respectively.*
- *The inside and outside concentrations for chlorine are* 124 *and* 450 milliMoles *respectively.*

Calculate the equilibrium membrane voltage.

5.6.7 Multiple Ion Nernst Computations in MatLab

Now let's do some calculations using MatLab for this situation. First, we will write a MatLab function to compute the Nernst voltage across the membrane using our conductance model. Here is a simple MatLab function to do this which uses our previous Nernst function. We will build a function which assumes the membrane is permeable to using potassium, sodium and chlorine ions.

Listing 5.4: Calculating the Nernst voltage: NernstMemVolt

```
function voltage = NernstMemVolt (T,gNa2gK,gCl2gK,InK,OutK,InNa,OutNa,
    InCl,OutCl)
%
% We assume three ions and compute the membrane voltage
4 %
% T is the temperature in degrees Celsius
% gNa2gK is the gNa to gK conductance ratio
% gCl2GK is the gCl to gK conductance ratio
% InK and OutK are the inside and outside potassium ion
    concentrations
9 % InNa and OutNa are the inside and outside sodium ion concentrations
% InCl and OutCl are the inside and outside chlorine ion
    concentrations
%
% Find equilibrium membrane voltages for each ion
%
14 E_K = Nernst(1,T,InK,OutK);
E_Na = Nernst(1,T,InNa,OutNa);
E_Cl = Nernst(-1,T,InCl,OutCl);
%
% Find membrane voltage for cell
19 %
denom = 1.0 + gNa2gK + gCl2gK;
voltage = ( E_K + gNa2gK * E_Na + gCl2gK * E_Cl )/denom;
end
```

Let's try this out for the following example: we assume we know

- The temperature is 20 degrees Celsius.
- The ratio g_{Na} to g_K is 0.03.
- The ratio of g_C to g_K is 0.1.
- The inside and outside concentrations for potassium are 400 and 20 milliMoles respectively.
- The inside and outside concentrations for sodium are 50 and 440 milliMoles respectively.
- The inside and outside concentrations for chlorine are 40 and 560 milliMoles respectively.

We then enter the function call into MatLab like this:

Listing 5.5: Finding the membrane voltage using conductances

```
NernstMemVolt(20,0.03,0.1,400,20,50,440,40,560)
ans =
3    -71.3414
```

If there was a explosive change in the g_{Na} to g_K ratio (this happens in a typical axonal pulse which we will discuss in later chapters), we see a large swing of the equilibrium membrane voltage from the previous -71.34 to 46.02 mV. The code below resets this ratio to 15 from its previous value of 0.03:

Listing 5.6: Increasing the Sodium to Potassium Conductance ratio

```
  NernstMemVolt (20 ,15.0 ,0.1 ,400 ,20 ,50 ,440 ,40 ,560)
2 ans =
      46.0241
```

5.7   Ion Flow

Our abstract cell is a spherical ball which encloses a fluid called **cytoplasm**. The surface of the ball is actually a membrane with an inner and outer part. Outside the cell there is a solution called the **extracellular** fluid. Both the cytoplasm and extracellular fluid contain many molecules, polypeptides and proteins disassociated into ions as well as sequestered into storage units. In this chapter, we will be interested in what this biological membrane is and how we can model the flow of ionized species through it. This modeling is difficult because some of the ions we are interested in can diffuse or drift across the membrane and others must be allowed entry through specialized holes in the membrane called **gates** or even escorted, i.e. transported or pumped, through the membrane by specialized helper molecules.

5.7.1   Transport Mechanisms

There are five general mechanisms by which molecules are moved across biological membranes. All of these methods operate simultaneously so we should know a little about all of them.

- Diffusion: here dissolved substances are transported because of concentration gradient. These substances can go right through the membrane. Some examples are water, certain molecules known as anesthetics and other large proteins which are soluble in the lipids which comprise the membrane such as the hormones known as steroids.
- Transport Through Water Channels: water will flow from a region in which a dissolved substance is at high concentration in order to equalize a high–low concentration gradient. This process is called **osmosis** and the force associated with it is called **osmotic** pressure. The biological membranes are thus semi-permeable to water and there are channels or pores in the membrane which selectively allow the passage of water molecules. In Fig. 5.5b, we see an abstract picture of the process of diffusion (part a) and the water channels (part b). The movement of water from a high to low concentration environment of the ion c is shown in Fig. 5.5a.
- Transport Through Gated Channels: some ion species are able to move through a membrane because there are special proteins embedded in the membrane which can be visualized as a cylinder whose throat can be open or blocked depending on some external signal. This external signal can be the voltage across the membrane

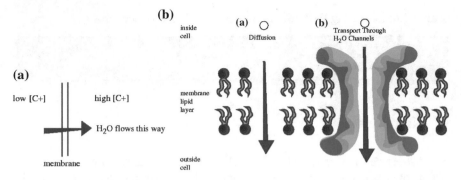

Fig. 5.5 Transport through water channels. **a** Osmosis. **b** Diffusion and water channel transport

or a molecule called a trigger. If the external signal is voltage, we call these channels **voltage gates**. When the gate is open, the ion is able to physically move through the opening.

- Carrier Mediated Transport: in this case, the dissolved substance combines with a **carrier** molecule on the one side of the membrane. The resulting **complex** that is formed moves through the membrane. At the opposite end of the membrane, the solute is released from the complex. A probable abstraction of this process is shown in Fig. 5.6.
- Ion Pumps: ions can also be transported by a mechanism which is linked to the addition of an OH group to the molecule **adenine triphosphate**. This process is called **hydrolysis** and this common molecule is abbreviated **ATP**. The addition of the hydroxyl group to ATP liberates energy. This energy is used to move ions **against** their concentration gradient.

In Fig. 5.7, parts c through e, we see abstractions of the remaining types of transport embedded in a membrane.

5.7.2   Ion Channels

In Fig. 5.8, we see a schematic of a typical voltage gate. Note that the inside of the gate shows a structure which can be in an open or closed position. The outside of the gate has a variety of molecules with sugar residues which physically extend into the extracellular fluid and carry negative charges on their tips. At the outer edge of the gate, you see a narrowing of the channel opening which is called the **selectivity filter**. As we have discussed, proteins can take on very complex three dimensional shapes. Often, their actual physical shape can *switch* from one form to another due to some external signal such as voltage. This is called a **conformational** change. In a voltage gate, the molecule which can block the inner throat of the gate moves from

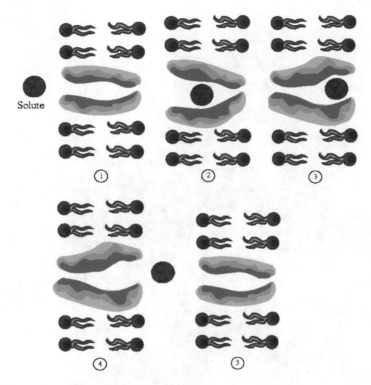

Fig. 5.6 A carrier moves a solute through the membrane

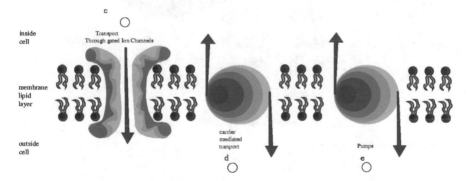

Fig. 5.7 Molecular transport mechanisms

its blocking position to its open position due to such a conformational change. In fact, this molecule can also be in between open and closed as well. The voltage gated channel is actually a protein macromolecule which is inserted into an opening in the membrane called a pore. We note that this macromolecule is quite big (1800–4000 amino acids) with one or more polypeptide chains and 100's of sugar residues hang off the extracellular face. When open, the channel is a water filled pore with a fairly

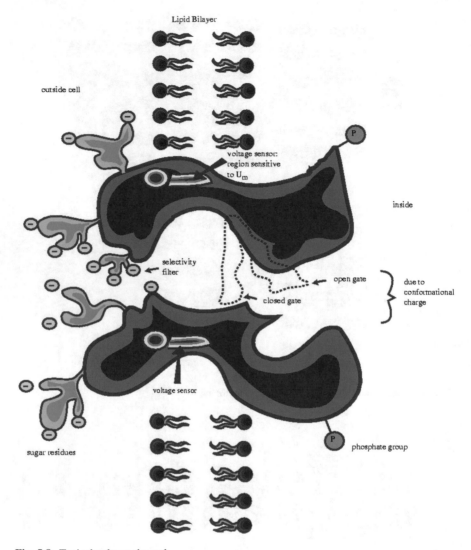

Fig. 5.8 Typical voltage channel

large inner diameter which would allow the passage of many things except that there is one narrow stretch of the channel called a selectivity filter which inhibits access. The inside of the pore is lined with hydrophilic amino acids which therefore like being near the water in the pore and the outside of the pore is lined with hydrophobic amino acids which therefore dislike water contact. These therefore lie next to the lipid bilayer. Ion concentration gradients can be maintained by selective permeabilities of the membrane to various ions. Most membranes are permeable to K^+, maybe Cl^- and much less permeable to Na^+ and Ca^{+2}. This type of passage of ions through the membrane requires no energy and so it is called the **passive distribution** of the ions. If there is no other way to transport ions, a cell membrane permeable to several ion species will reach an equilibrium potential determined by the Nernst equation. Let c^{+n} denote a cation of valence n and a^{-m} be an anion of valence $-n$. Then the Nernst equilibrium potentials for these ions are given by

$$E_c = \frac{RT}{nF} \ln \frac{[c^{+n}]_{out}}{[c^{+n}]_{in}} \quad E_a = \frac{RT}{-mF} \ln \frac{[a^{-m}]_{out}}{[a^{-m}]_{in}}$$

At equilibrium, the ionic currents must sum to zero forcing these two potentials to be the same. Hence, E_c is the same as E_a and we find

$$\frac{RT}{nF} \ln \frac{[c^{+n}]_{out}}{[c^{+n}]_{in}} = \frac{RT}{-mF} \ln \frac{[a^{-m}]_{out}}{[a^{-m}]_{in}}$$

implying

$$\frac{m}{n} \ln \frac{[c^{+n}]_{out}}{[c^{+n}]_{in}} = \ln \frac{[a^{-m}]_{in}}{[a^{-m}]_{out}}$$

Exponentiating these expressions, we find that the ion concentrations must satisfy the following expression which is known as **Donnan's Law of Equilibrium**.

$$\frac{[c^{+n}]_{out}}{[c^{+n}]_{in}}^{\frac{1}{m}} = \frac{[a^{-m}]_{in}}{[a^{-m}]_{out}}^{\frac{1}{n}}$$

5.7.2.1 K^+ and Cl^- Donnan Equilibrium

For example, for the ions K^+ and Cl^-,

Outside	Inside
$[K^+]_{out}$	$[K^+]_{in}$
$[Cl^-]_{out}$	$[Cl^-]_{in}$

and we would have that at Donnan equilibrium, the inner and outer ion concentrations would satisfy

$$\frac{[K^+]_{out}}{[K^+]_{in}} = \frac{[Cl^-]_{in}}{[Cl^-]_{out}}$$

or

$$[K^+]_{out} \, [Cl^-]_{out} = [K^+]_{in} \, [Cl^-]_{in}$$

5.7.2.2 K^+, Cl^- and A^{-m} Donnan Equilibrium

Now assume that there are other negative ions in the cell, say A^{-m}. Then, $[A^{-m}]_{out}$ is zero and we have

Outside $[K^+]_{out}$ $[Cl^-]_{out}$	**Inside** $[K^+]_{in}$ $[Cl^-]_{in}$ $[A^{-m}]_{in}$

Then because there must be charge neutrality, $[K^+]_{in} = [Cl^-]_{in} + [A^{-m}]_{in}$ and $[K^+]_{out} = [Cl^-]_{out}$. Also, we must have the usual Donnan Equilibrium state (remember the ion A^{-m} does not play a role in this):

$$[K^+]_{out} \, [Cl^-]_{out} = [K^+]_{in} \, [Cl^-]_{in}$$

Thus,

$$
\begin{aligned}
[K^+]_{in}^2 &= [K^+]_{in}([Cl^-]_{in} + [A^{-m}]_{in}) \\
&= [K^+]_{in} \, [Cl^-]_{in} + [K^+]_{in} \, [A^{-m}]_{in} \\
&= [K^+]_{out} \, [Cl^-]_{out} + [K^+]_{in} \, [A^{-m}]_{in} \\
&= [K^+]_{out}^2 + [K^+]_{in} \, [A^{-m}]_{in} \\
&> [K^+]_{out}^2
\end{aligned}
$$

Hence, if we can only use passive transport mechanisms, we must have

$$[K^+]_{in} > [K^+]_{out}$$

5.7.3 Active Transport Using Pumps

There are many pumps within a cell that move substances in or out of a cell with or against a concentration gradient.

- $Na^+ K^+$ Pump: this is a transport mechanism driven by the energy derived from the hydrolysis of ATP. Here 3 Na^+ ions are pumped out and 2 K^+ ions are pumped in. So this pump gives us sodium and potassium currents which raise $[Na^+]_{out}$ and $[K^+]_{in}$ and decrease $[Na^+]_{in}$ and $[K^+]_{out}$. From Table 5.1, we see that $[Na^+]_{out}$ is typically bigger, so this pump is pushing Na^+ out against its concentration gradient and so it costs energy.

- Na^+ Ca^{+2} Pump: this pump drives out 3 Na^+ ions inside the cell for every Ca^{+2} ion that is moved out. Hence, this pump gives us sodium and calcium currents which raise $[Na^+]_{in}$ and $[Ca^{+2}]_{out}$ and decrease $[Na^+]_{out}$ and $[Ca^{+2}]_{in}$. Here, the Na^+ movement is with its concentration gradient and so there is no energy cost.

- Plasma membrane Ca^{+2} Pump: this pump is also driven by the hydrolysis of ATP. Here, Ca^{+2} is pumped into a storage facility inside the cell called the **endoplasmic reticulum** which therefore takes Ca^{+2} out of the cytoplasm and so brings down $[Ca^{12}]_{in}$.

- HCO_3^- Cl^- Exchange: this is driven by the Na^+ concentration gradient and pumps HCO_3^- into the cell and Cl^- out of the cell.

- $Cl - Na^+$ K^+ Co transport: This is driven by the influx of Na^+ into the cell. For every one Na^+ and K^+ that are pumped into the cell via this mechanism, 2 Cl^- are driven out.

5.7.4 A Simple Compartment Model

Consider a system which has two compartments filled with ions and a membrane with K^+ and Cl^- gates between the two compartments as shown below:

Compartment One (Cl⁻ Gates) (K⁺ Gates)	Compartment Two (Cl⁻ Gates) (K⁺ Gates)
100 mM A^-	0 mM A^-
150 mM K^+	150 mM K^+
50 mM Cl^-	150 mM Cl^-

We will assume the system is held at the temperature of 70 degree Fahrenheit—about normal room temperature—which is 21.11 Celsius and 294.11 Kelvin. This implies our Nernst conversion factor is 25.32 mV for K^+ and -25.32 for Cl^-. Is this system in electrochemical equilibrium or **ECE**? In each compartment, we do have space charge neutrality as in Compartment One, 150 mM of K^+ balances 100 mM of A^- and 50 mM of Cl^-; and in Compartment Two, 150 mM of K^+ balances 150 mM of Cl^-. However, Cl^- is not concentration balanced and so Cl^- is not at ECE implying that Cl^- will diffuse into Compartment Two from Compartment One. This diffusion shifts ions in the compartments to the new values

Compartment One	Compartment Two
(Cl^- Gates)	(Cl^- Gates)
(K^+ Gates)	(K^+ Gates)
100 mM A^-	0 mM A^-
150 mM K^+	150 mM K^+
100 mM Cl^-	100 mM Cl^-

So to get concentration balance, 50 Cl^- moves Compartment Two to Compartment One. Now counting ion charges in each compartment, we see that we have lost space charge neutrality. We can regain this by shifting 50 mM of K^+ from Compartment Two to Compartment One to give:

Compartment One	Compartment Two
(Cl^- Gates)	(Cl^- Gates)
(K^+ Gates)	(K^+ Gates)
100 mM A^-	0 mM A^-
200 mM K^+	100 mM K^+
100 mM Cl^-	100 mM Cl^-

Now we see that there is a concentration imbalance in K^+! The point here is that this kind of analysis, while interesting, is so qualitative that it doesn't give us the final answer quickly! To get to the final punch line, we just need to find the Donnan Equilibrium point. This is where

$$[K^+]_{One} [Cl^-]_{One} = [K^+]_{Two} [Cl^-]_{Two}$$

From our discussion above, we know that the number of mM's of K^+ and Cl^- that move from Compartment Two to Compartment One will always be the same in order to satisfy space charge neutrality. Let x denote this amount. Then we must have

$$[K^+]_{One} = 150 + x$$
$$[K^+]_{Two} = 150 - x$$
$$[Cl^-]_{Two} = 150 - x$$
$$[Cl^-]_{One} = 50 + x$$

yielding after a little algebra:

$$(150 + x)(50 + x) = (150 - x)(150 - x)$$
$$7500 + 200x + x^2 = 22{,}500 - 300x + x^2$$
$$500x = 15{,}000$$
$$x = 30$$

So at ECE, we have

Compartment One (Cl^- Gates) (K^+ Gates)	Compartment Two (Cl^- Gates) (K^+ Gates)
100 mM A^-	0 mM A^-
180 mM K^+	120 mM K^+
80 mM Cl^-	120 mM Cl^-

What are the membrane voltages? Of course, they should match! Recall, in the derivation of the Nernst equation, the voltage difference for ion c between side 2 and 1 was

$$E_c = V_1 - V_2 = \frac{RT}{zF} \ln \frac{[c]_2}{[c]_1}$$

Here, Compartment One plays the role of side 1 and Compartment Two plays the role of side 2. So for K^+, we have

$$E_K = V_{One} - V_{Two} = 25.32 \ln \frac{[K^+]_{Two}}{[K^+]_{One}} = 25.32 \ln \frac{120}{180} = -10.27 \, \text{mV}$$

and for Cl^-,

$$E_{Cl} = V_{One} - V_{Two} = -25.32 \ln \frac{[Cl^-]_{Two}}{[Cl^-]_{One}} = -25.32 \ln \frac{120}{80} = -10.27 \, \text{mV}$$

Are we at osmotic equilibrium? To see this, we need to add up how many ions are in each compartment. In Compartment One, there are 360 ions and in Compartment Two there are 240 ions. Since there are more ions in Compartment One than Compartment Two, water will flow from Compartment One to Compartment Two to try to dilute the ionic strength in Compartment One. We are ignoring this effect here.

5.7.4.1   Homework

Exercise 5.7.1 *Consider a system which has two compartments filled with ions and a membrane with K^+ and Cl^- gates between the two compartments as shown below:*

Compartment One	Compartment Two
(Cl^- Gates)	(Cl^- Gates)
(K^+ Gates)	(K^+ Gates)
80 mM A^-	30 mM A^-
160 mM K^+	50 mM K^+
80 mM Cl^-	20 mM Cl^-

Note the ion A can't pass between compartments. Assume the system is held at the temperature of 70 degree Fahrenheit.

1. *Find the concentrations of ions that this system has at ECE.*
2. *For the ECE ion concentrations, calculate E_{Cl} and E_K.*

5.8   Movement of Ions Across Biological Membranes

We now will work our way through another way to view the movement of ions across biological membranes that was developed by Goldman in 1943 and Hodgkin and Katz in 1949. This is the **Goldman–Hodgkin–Katz** or **GHK** model. Going through all of this will give you an increased appreciation for both the kind of modeling we need to do: what to keep from the science, what to throw away and what tools to use for the model itself . We start with defining carefully what we mean by the word **permeability**. We have used this word before but always in a qualitative sense. Now we will be quantitative.

5.8.1   Membrane Permeability

Let's consider a substance c moving across a biological membrane. We will assume that the relationship between the number of moles of c moving across the membrane in unit time and unit area is proportional to the change in concentration of c. This gives

$$J_{molar} = -\mathscr{P}\,\Delta[c] \tag{5.13}$$

where

- J_{molar} is the **molar flux** which is in the units $\frac{moles}{cm^2-second}$.
- $\mathscr{P}$ is the membrane **permeability** of the ion c which has units of $\frac{cm}{second}$.
- $\Delta[]$ is the change in the concentration of c measured in the units $\frac{moles}{cm^3}$.

Ficke's Law of diffusion gives $J_{diff} = -D \frac{\partial[c]}{\partial x}$ where J_{diff} is measured in $\frac{molecules}{cm^2-second}$. Hence, we can convert to $\frac{moles}{cm^2-second}$ by dividing by Avogadro's number N_A. This gives (remember, partial differentiation becomes regular differentiation when there is just one variable)

$$\frac{J_{diff}}{N_A} = -\frac{D_m}{N_A} \frac{d[c]}{dx}$$

where we now explicitly label the diffusion constant as belonging to the membrane. These two fluxes should be the same; hence

$$J_{molar} = -\mathscr{P} \Delta[c] = -\frac{D_m}{N_A} \frac{d[c]}{dx} \tag{5.14}$$

We will want to use the above equation for the concentration of c that is in the membrane itself. So we need to distinguish from the concentration of c inside and outside the cell and the concentration in the membrane. We will let $[c]$ denote the concentration inside and outside the cell and $[c^m]$, the concentration inside the membrane. Hence, we have

$$J_{molar} = -\mathscr{P} \Delta[c] = -\frac{D_m}{N_A} \frac{d[c^m]}{dx} \tag{5.15}$$

Now a substance will probably dissolve differently in the cellular solution and in the biological membrane, let's model this quantitatively as follows: Let $[c]_{in}$ denote the concentration of c inside the cell, and $[c]_{out}$ denote the concentration of c outside the cell. The membrane has two sides; the side facing into the cell and the side facing to the outside of the cell. If we took a slice through the cell, we would see a straight line along which we can measure our position in the cell by the variable x. We have x is 0 when we are at the center of our spherical cell and x is x_0 when we reach the inside wall of the membrane (i.e. x_0 is the radius of the cell for this slice). We will assume that the thickness of the membrane is a uniform ℓ. Hence, we are at position $x_0 + \ell$ when we come to the outer boundary of the membrane. Inside the membrane, we will let $[c^m]$ denote the concentration of c. We know the membrane concentration will be some fraction of the cell concentration. So

$$[c^m](x_0 + \ell) = \beta_o [c]_{out}$$
$$[c^m](x_0) = \beta_i [c]_{in}$$

for some constants β_i and β_o. The ratio of the boundary concentrations of c and the membrane concentrations is an important parameter. From the above discussions, we see these critical ratios are given by:

$$\beta_o = \frac{[c^m]_{x_0+\ell}}{[c]_{out}}$$

$$\beta_i = \frac{[c^m]_{x_0}}{[c]_{in}}$$

For this model, we will assume that these ratios are the same: i.e. $\beta_o = \beta_i$; this common value will be denoted by β. The parameter β is called the **partition coefficient** between the solution and the membrane. We will also assume that the membrane concentration $[c^m]$ varies linearly across the membrane from $[c]_{in}$ at x_0 to $[c]_{out}$ at $x_0 + \ell$. This model is shown in Fig. 5.9: Thus, since we assume the concentration inside the membrane is linear, we have for some constants a and b:

$$[c^m](x) = a\,x + b$$
$$[c^m](x_0) = [c^m]_{in} = \beta[c]_{in}$$
$$[c^m](x_0 + \ell) = [c^m]_{out} = \beta[c]_{out}$$

This implies that

$$[c^m]_{in} = a\,x_0 + b$$
$$[c^m]_{out} = a\,(x_0 + \ell) + b$$
$$= \left(a\,x_0 + b\right) + a\ell$$
$$= [c^m]_{in} + a\ell$$

Thus, letting ΔC denote $[c]_{out} - [c]_{in}$, we have

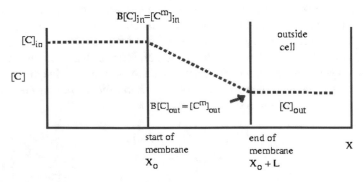

Fig. 5.9   The linear membrane concentration model

$$a = \frac{[c^m]_{out} - [c^m]_{in}}{\ell}$$

$$= \frac{\beta}{\ell}([c]_{out} - [c]_{in})$$

$$= \frac{\beta}{\ell} \Delta C$$

We can also solve for the value of b:

$$b = [c^m]_{in} - ax_0$$

$$= [c^m]_{in} - x_0 \left(\frac{\beta}{\ell}([c]_{out} - [c]_{in}) \right)$$

$$= \beta \left(1 + \frac{x_0}{\ell} \right) [c]_{in} - \beta \frac{x_0}{\ell}[c]_{out}$$

Hence,

$$\frac{d[c^m]}{dx} = \frac{\beta}{\ell}([c]_{out} - [c]_{in}) = \frac{\beta}{\ell} \Delta C$$

Now recall that

$$J_{molar} = -\mathscr{P} \, \Delta[c]$$

$$= -\frac{D_m}{N_A} \frac{d[c^m]}{dx}$$

$$= -\frac{D_m}{N_A} \frac{\beta}{\ell}([c]_{out} - [c]_{in})$$

Thus, we have a way to express the permeability $\mathscr{P}$ in terms of low level fundamental constants:

$$\mathscr{P} = \frac{D_m}{N_A} \frac{\beta}{\ell}$$

Finally, letting μ_m be the mobility within the membrane, we note that Einstein's relation tells us that $D_m = \frac{\kappa T}{q} \mu_m = \frac{RT}{F} \mu_m$, so we conclude

$$\mathscr{P} = \frac{\beta RT}{\ell F} \frac{\mu_m}{N_A} \tag{5.16}$$

5.8.2   The Goldman–Hodgkin–Katz (GHK) Model

The Goldman–Hodgkin–Huxley model is based on several assumptions about how the substance c behaves in the membrane:

1. $[c^m]$ varies linearly across the membrane as discussed in the previous subsection.
2. The electric field in the membrane is constant.
3. The Nernst–Planck equation holds inside the membrane.

We let $V^m(x)$ denote the voltage or potential and $E^m(x)$, the electrical field, in the membrane at position x. By assumption (2), we know that $E^m(x)$ is a constant we will call E^m in the membrane for all x. Now we also know from standard physics that an electric field is the negative of the derivative of the potential so

$$E^m = -\frac{dV^m}{dx}$$

implying

$$-E^m dx = \frac{dV^m}{dx} \implies -E^m \int_{x_0}^{x_0+\ell} dx = \int_{x_0}^{x_0+\ell} dV^m \implies -E^m \ell = V_{out}^m - V_{in}^m$$

The Nernst equation will hold across the membrane, so we have that the voltage due to c across the membrane is V_c^m with

$$E_c = V_c^m = V_{in}^m - V_{out}^m = \frac{RT}{zF} \ln \frac{[c]_{out}}{[c]_{in}}$$

Further, by assumption (3), the Nernst–Planck equation holds across the membrane, so the ion current in the membrane, I_m is given by

$$I^m = -\frac{\mu_m}{N_A} \left(z^2 F\,[c^m] \frac{dV^m}{dx} + z\,RT\,\frac{d[c^m]}{dx} \right)$$

Also, since

$$\frac{dV^m}{dx} = -E^m = \frac{V_{out}^m - V_{in}^m}{\ell} = \frac{-V_c^m}{\ell}$$

we have

$$
\begin{aligned}
I^m &= -\frac{\mu_m}{N_A} \left(z^2\,F\,[c^m] \frac{-V_c^m}{\ell} + z\,RT\,\frac{d[c^m]}{dx} \right) \\
&= \frac{\mu_m}{N_A} z^2\,F\,[c^m] \frac{V_c^m}{\ell} - \frac{\mu_m}{N_A} z\,RT\,\frac{d[c^m]}{dx}
\end{aligned}
$$

We then see that

$$I^m - \frac{\mu_m \, z^2 \, F \, [c^m]}{N_A \, \ell} \, V_c^m = -\frac{\mu_m}{N_A} \, z \, RT \, \frac{d[c^m]}{dx} \tag{5.17}$$

To make sense out of this complicated looking equation, we will make a change of variable: define y^m as follows:

$$y^m = I^m - \frac{\mu_m \, z^2 \, F \, [c^m]}{N_A \, \ell} \, V_c^m \tag{5.18}$$

Thus,

$$\frac{dy^m}{dx} = \frac{dI^m}{dx} - \frac{\mu_m \, z^2 \, F}{N_A \, \ell} \, V_c^m \, \frac{d[c^m]}{dx}$$

We know that at equilibrium, the current flow will be zero and so $\frac{dI^m}{dx} = 0$. Hence, at steady state, we have

$$\frac{dy^m}{dx} = -\frac{\mu_m \, z^2 \, F}{N_A \, \ell} \, V_c^m \, \frac{d[c^m]}{dx}$$

or

$$-\frac{\ell}{zF \, V_c^m} \frac{dy^m}{dx} = \frac{\mu_m \, z}{N_A} \frac{d[c^m]}{dx}$$

This gives us an expression for the term $\frac{d[c^m]}{dx}$ in Eq. 5.17. Using our change of variable y^m and substituting for the concentration derivative, we find

$$y^m = -RT \left(\frac{\mu_m z}{N_A} \frac{d[c^m]}{dx} \right) = -RT \left(-\frac{\ell}{zF \, V_c^m} \frac{dy^m}{dx} \right)$$

We have thus obtained the simpler differential equation

$$y^m = \frac{RT\ell}{zF \, V_c^m} \frac{dy^m}{dx} \tag{5.19}$$

This equation can be integrated easily:

$$dx = \frac{RT\ell}{zF \, V_c^m} \frac{dy^m}{y^m}$$

$$\int_{x_0}^{x_0+\ell} dx = \frac{RT\ell}{zF \, V_c^m} \int_{x_0}^{x_0+\ell} \frac{dy^m}{y^m}$$

which gives

$$\ell = \frac{RT\ell}{zF\,V_c^m}\,\ln\frac{y^m(x_0+\ell)}{y^m(x_0)} \tag{5.20}$$

To simplify this further, we must go back to the definition of the change of variable y^m. From Eq. 5.18, we see

$$y^m(x_0+\ell) = I^m(x_0+\ell) - \frac{\mu_m\,z^2\,F\,[c^m](x_0+\ell)}{N_A\,\ell}\,V_c^m$$

$$y^m(x_0) = I^m(x_0) - \frac{\mu_m\,z^2\,F\,[c^m](x_0)}{N_A\,\ell}\,V_c^m$$

At steady state the currents $I^m(x_0+\ell)$ and $I^m(x_0)$ are the same value. Call this constant steady state value I_0^m. Then we have

$$y^m(x_0+\ell) = I_0^m - \frac{\mu_m\,z^2\,F\,[c^m](x_0+\ell)}{N_A\,\ell}\,V_c^m$$

$$y^m(x_0) = I_0^m - \frac{\mu_m\,z^2\,F\,[c^m](x_0)}{N_A\,\ell}\,V_c^m$$

Finally, remember how we set up the concentration terms. We used the symbols $[c^m]_{out}$ to represent $[c^m](x_0+\ell)$ and $[c^m]_{in}$ to represent $[c^m](x_0)$. Thus, we have

$$y^m(x_0+\ell) = I_0^m - \frac{\mu_m\,z^2\,F\,[c^m]_{out}}{N_A\,\ell}\,V_c^m$$

$$y^m(x_0) = I_0^m - \frac{\mu_m\,z^2\,F\,[c^m]_{in}}{N_A\,\ell}\,V_c^m$$

Now, to finish our derivation, let's introduce an additional parameter ξ defined by

$$\xi = \frac{z\,F\,V_c^m}{RT} \tag{5.21}$$

Recall that we defined the permeability by

$$\mathscr{P} = \frac{D_m}{N_A}\frac{\beta}{\ell}$$

and this also equals

$$\mathscr{P} = \frac{\beta RT}{\ell F}\frac{\mu_m}{N_A}$$

Now from the definition of ξ and $\mathscr{P}$, we find that Eq. 5.20 gives

$$\ell = \frac{\ell}{\xi} \ln\left(\frac{y^m(x_0 + \ell)}{y^m(x_0)}\right)$$

Thus, after canceling ℓ from both sides, we obtain

$$\xi = \ln\left(\frac{y^m(x_0 + \ell)}{y^m(x_0)}\right)$$

Now, plug in the expressions for the y^m terms evaluated at $x_0 + \ell$ and x_0 to find

$$\xi = \ln\left(\frac{I_0^m - \frac{\mu_m z^2 F [c^m]_{out}}{N_A \ell} V_c^m}{I_0^m - \frac{\mu_m z^2 F [c^m](x_0)}{N_A \ell} V_c^m}\right) = \ln\left(\frac{I_0^m - \frac{z\mu_m}{N_A \ell} z F V_c^m [c^m]_{out}}{I_0^m - \frac{z\mu_m}{N_A \ell} z F V_c^m [c^m]_{in}}\right)$$

$$= \ln\left(\frac{I_0^m - \frac{z\mu_m}{N_A \ell} \xi RT [c^m]_{out}}{I_0^m - \frac{z\mu_m}{N_A \ell} \xi RT [c^m]_{in}}\right) = \ln\left(\frac{I_0^m - \frac{\mu_m \xi RT}{N_A} \frac{z}{\ell} [c^m]_{out}}{I_0^m - \frac{\mu_m \xi RT}{N_A} \frac{z}{\ell} [c^m]_{in}}\right)$$

But we also know that by assumption

$$[c^m]_{out} = \beta[c]_{out}, \quad [c^m]_{in} = \beta[c]_{in}$$

and so

$$\frac{\mu_m \xi RT}{N_A} \frac{z}{\ell} [c^m]_{out} = \frac{\beta RT \mu_m}{N_A \ell} \xi z [c]_{out}$$

But we know from the definition of $\mathscr{P}$ that the fractional expression is just $\mathscr{P} F$ and so we have

$$\frac{\mu_m \xi RT}{N_A} \frac{z}{\ell} [c^m]_{out} = z \mathscr{P} F \xi [c]_{out}$$

In a similar fashion, we can derive that

$$\frac{\mu_m \xi RT}{N_A} \frac{z}{\ell} [c^m]_{in} = z \mathscr{P} F \xi [c]_{in}$$

Substituting these identities into expression for ξ we derived earlier, we find

$$\xi = \ln\left(\frac{I_0^m - z \mathscr{P} F \xi [c]_{out}}{I_0^m - z \mathscr{P} F \xi [c]_{in}}\right)$$

Now exponentiate these expressions to get

$$e^{\xi} = \frac{I_0^m - z \mathscr{P} F \xi [c]_{out}}{I_0^m - z \mathscr{P} F \xi [c]_{in}}$$

$$(I_0^m - z \mathscr{P} F \xi [c]_{in}) e^{\xi} = I_0^m - z \mathscr{P} F \xi [c]_{out}$$

leading to

$$I_0^m (e^{\xi} - 1) = z \mathscr{P} F \xi ([c]_{in} e^{\xi} - [c]_{out})$$

We can solve the above equation for the steady state current term I_0^m. This gives

$$I_0^m = z \mathscr{P} F \xi \left(\frac{[c]_{in} e^{\xi} - [c]_{out}}{e^{\xi} - 1} \right)$$

When we replace ξ by its definition and reorganize the resulting expression, we obtain the GHK current equation for the ion c.

$$I_0^m = \frac{z^2 \mathscr{P} F^2 V_c^m}{RT} \left(\frac{[c]_{in} - [c]_{out} e^{-\frac{z F V_c^m}{RT}}}{1 - e^{-\frac{z F V_c^m}{RT}}} \right) \tag{5.22}$$

5.8.3 The GHK Voltage Equation

Consider a cell that is permeable to K^+, Cl^- and Na^+ ions with no active pumping. Let's calculate the resting potential of the cell. For each ion, we can apply our GHK theory to obtain an ionic current, I_K, I_{Na} and I_{Cl}. At equilibrium, the current through the membrane should be zero and so all of the ionic currents should sum to zero. Thus

$$I_K + I_{Na} + I_{Cl} = 0$$

Further, the associated Nernst potentials for the ions should all match because otherwise there would be current flow:

$$V_0^m = V_{K^+}^m = V_{Na^+} = V_{Cl^-}^m$$

where we denote this common potential by V_0^m. From the GHK current equation, we have (note the valences for potassium and sodium are $+1$ and for chlorine is -1)

$$I_K = \frac{\mathscr{P}_K F^2 V_0^m}{RT} \left(\frac{[K^+]_{in} - [K^+]_{out} e^{-\frac{FV_0^m}{RT}}}{1 - e^{-\frac{FV_0^m}{RT}}} \right)$$

$$I_{Na} = \frac{\mathscr{P}_{Na} F^2 V_0^m}{RT} \left(\frac{[Na^+]_{in} - [Na^+]_{out} e^{-\frac{FV_0^m}{RT}}}{1 - e^{-\frac{FV_0^m}{RT}}} \right)$$

$$I_{Cl} = \frac{\mathscr{P}_{Cl} F^2 V_0^m}{RT} \left(\frac{[Cl^-]_{in} - [Cl^-]_{out} e^{-\frac{-FV_0^m}{RT}}}{1 - e^{-\frac{-FV_0^m}{RT}}} \right)$$

For convenience, let ξ_0 denote the common term $\frac{FV_0^m}{RT}$. Then we have

$$I_K = \mathscr{P}_K F \xi_0 \left(\frac{[K^+]_{in} - [K^+]_{out} e^{-\xi_0}}{1 - e^{-\xi_0}} \right), \quad I_{Na} = \mathscr{P}_{Na} F \xi_0 \left(\frac{[Na^+]_{in} - [Na^+]_{out} e^{-\xi_0}}{1 - e^{-\xi_0}} \right)$$

$$I_{Cl} = \mathscr{P}_{Cl} F \xi_0 \left(\frac{[Cl^-]_{in} - [Cl^-]_{out} e^{\xi_0}}{1 - e^{\xi_0}} \right)$$

Now sum these ionic currents to obtain

$$0 = \mathscr{P}_K F \xi_0 \left(\frac{[K^+]_{in} - [K^+]_{out} e^{-\xi_0}}{1 - e^{-\xi_0}} \right) + \mathscr{P}_{Na} F \xi_0 \left(\frac{[Na^+]_{in} - [Na^+]_{out} e^{-\xi_0}}{1 - e^{-\xi_0}} \right)$$

$$- \mathscr{P}_{Cl} F \xi_0 \left(\frac{[Cl^-]_{in} e^{-\xi_0} - [Cl^-]_{out}}{1 - e^{-\xi_0}} \right)$$

or

$$0 = \frac{F \xi_0}{1 - e^{-\xi_0}} \left(\mathscr{P}_K [K^+]_{in} + \mathscr{P}_{Na} [Na^+]_{in} + \mathscr{P}_{Cl} [Cl^-]_{out} \right)$$

$$- \left(\mathscr{P}_K [K^+]_{out} + \mathscr{P}_{Na} [Na^+]_{out} + \mathscr{P}_{Cl} [Cl^-]_{in} \right) \frac{F \xi_0 e^{-\xi_0}}{1 - e^{-\xi_0}}.$$

Rewriting, we obtain

$$e^{\xi_0} \left(\mathscr{P}_K [K^+]_{in} + \mathscr{P}_{Na} [Na^+]_{in} + \mathscr{P}_{Cl} [Cl^-]_{out} \right)$$
$$= \left(\mathscr{P}_K [K^+]_{out} + \mathscr{P}_{Na} [Na^+]_{out} + \mathscr{P}_{Cl} [Cl^-]_{in} \right)$$

Hence, solving for e^{ξ_0}, we find

$$e^{\xi_0} = \frac{\mathscr{P}_K [K^+]_{out} + \mathscr{P}_{Na} [Na^+]_{out} + \mathscr{P}_{Cl} [Cl^-]_{in}}{\mathscr{P}_K [K^+]_{in} + \mathscr{P}_{Na} [Na^+]_{in} + \mathscr{P}_{Cl} [Cl^-]_{out}}$$

Finally, taking logarithms, we obtain the GHK Voltage equation for these ions:

$$V_0^m = \frac{RT}{F} \ln \left(\frac{\mathscr{P}_K[K^+]_{out} + \mathscr{P}_{Na}[Na^+]_{out} + \mathscr{P}_{Cl}[Cl]_{in}}{\mathscr{P}_K[K^+]_{in} + \mathscr{P}_{Na}[Na^+]_{in} + \mathscr{P}_{Cl}[Cl^-]_{out}} \right) \qquad (5.23)$$

Now to actually compute the GHK voltage, we will need the values of the three permeabilities. It is common to rewrite this equation in terms of permeability ratios. We choose one of the permeabilities as the reference, say $\mathscr{P}_{Na}$, and compute the ratios

$$r_K = \frac{\mathscr{P}_K}{\mathscr{P}_{Na}}, \quad r_{Cl} = \frac{\mathscr{P}_{Cl}}{\mathscr{P}_{Na}}$$

We can then rewrite the GHK voltage equation as

$$V_0^m = \frac{RT}{F} \ln \left(\frac{[Na^+]_{out} + r_K[K^+]_{out} + r_{Cl}[Cl]_{in}}{[Na^+]_{in} + r_K[K^+]_{in} + r_{Cl}[Cl^-]_{out}} \right) \qquad (5.24)$$

5.8.4 Examples

Now let's use the GHK voltage equation for a few sample calculations.

5.8.4.1 A First Example

Consider a squid giant axon. Typically, the axon at rest has the following permeability ratios:

$$\mathscr{P}_K : \mathscr{P}_{Na} : \mathscr{P}_{Cl} = 1.0 : 0.03 : 0.1$$

where we interpret this string of ratios as

$$\frac{\mathscr{P}_K}{\mathscr{P}_{Na}} = \frac{1.0}{0.03} = 33.33$$

$$\frac{\mathscr{P}_{Na}}{\mathscr{P}_{Cl}} = \frac{0.03}{0.1} = 0.3$$

Thus, we have

$$r_K = \frac{\mathscr{P}_K}{\mathscr{P}_{Na}} = 33.33$$

$$r_{Cl} = \frac{\mathscr{P}_{Cl}}{\mathscr{P}_{Na}} : 3.33$$

Now look at the part of Table 5.1 which shows the ion concentrations for the squid axon. we will use a temperature of 294.11° Kelvin and the conversion $\frac{RT}{F}$ is 25.32 mV. Also, recall that all concentrations are given in mM. Thus, the GHK voltage equation for the squid axon gives

$$
\begin{aligned}
V_0^m &= \frac{RT}{F} \ln \left(\frac{[Na^+]_{out} + r_K[K^+]_{out} + r_{Cl}[Cl]_{in}}{[Na^+]_{in} + r_K[K^+]_{in} + r_{Cl}[Cl^-]_{out}} \right) \\
&= 25.32 \ (\text{mV}) \ \ln \left(\frac{440.0 + 33.33 \times 20.0 + 3.333 \times 40.0}{50.0 + 33.33 \times 400.0 + 3.333 \times 560.0} \right) \\
&= 25.32 \ (\text{mV}) \ \ln \left(\frac{1239.9}{15248.48} \right) \\
&= -63.54 \ \text{mV}
\end{aligned}
$$

5.8.4.2   A Second Example

Now assume there is a drastic change in these permeability ratios: we now have:

$$
\mathscr{P}_K \ : \ \mathscr{P}_{Na} \ : \ \mathscr{P}_{Cl} = 1.0 \ : \ 15.0 \ : \ 0.1
$$

where we interpret this string of ratios as

$$
\begin{aligned}
r_K &= \frac{\mathscr{P}_K}{\mathscr{P}_{Na}} = \frac{1.0}{15.0} = 0.0667 \\
r_{Cl} &= \frac{\mathscr{P}_{Cl}}{\mathscr{P}_{Na}} = \frac{0.1}{15.0} = 0.00667
\end{aligned}
$$

The GHK voltage equation now gives

$$
\begin{aligned}
V_0^m &= 25.32 \ (\text{mV}) \ \ln \left(\frac{440.0 + 0.0667 \times 20.0 + 0.00667 \times 40.0}{50.0 + 0.0667 \times 400.0 + 0.00667 \times 560.0} \right) \\
&= 25.32 \ (\text{mV}) \ \ln \left(\frac{441.60}{80.4152} \right) \\
&= 43.12 \ \text{mV}
\end{aligned}
$$

We see that this large change in the permeability ratio between potassium and sodium triggers a huge change in the equilibrium voltage across the membrane. It is actually far easier to see this effect when we use permeabilities than when we use conductances. So as usual, insight is often improved by looking at a situation using different parameters! Later, we will begin to understand how this explosive switch from a negative membrane voltage to a positive membrane voltage is achieved. That's for a later chapter!

5.8.5   *The Effects of an Electrogenic Pump*

We haven't talked much about how the pumps work in our model yet, so just to give
the flavor of such calculations, let's look at a hypothetical pump for an ion c. Let I_c
denote the passive current for the ion which means no energy is required for the ion
flow (see Sect. 5.7.1 for further discussion). The ion current due to a pump requires
energy and we will let I_c^p denote the ion current due to the pump. At steady state,
we would have

$$I_{Na^+} + I_{Na^+}^p = 0$$
$$I_{K^+} + I_{K^+}^p = 0$$

Let the parameter r be the number of Na^+ ions pumped out for each K^+ ion pumped
in. This then implies that

$$r I_{K^+}^p + I_{Na^+}^p = 0$$

From the GHK current equation, we know

$$I_{K^+} = \mathscr{P}_K F \, \xi_0 \left(\frac{[K^+]_{in} - [K^+]_{out} e^{-\xi_0}}{1 - e^{-\xi_0}} \right)$$
$$I_{Na^+} = \mathscr{P}_{Na} F \, \xi_0 \left(\frac{[Na^+]_{in} - [Na^+]_{out} e^{-\xi_0}}{1 - e^{-\xi_0}} \right)$$

and thus from our steady state assumptions, we find

$$I_{Na^+} = -I_{Na^+}^p, \quad I_{K^+} = -I_{K^+}^p$$

and so

$$r I_{K^+} + I_{Na^+} = 0 \tag{5.25}$$

Substituting the GHK current expressions into Eq. 5.25, we have

$$\frac{F \xi_0}{1 - e^{-\xi_0}} \left(r \mathscr{P}_K \left([K^+]_{in} - [K^+]_{out} e^{-\xi_0} \right) + \mathscr{P}_{Na} \left([Na^+]_{in} - [Na^+]_{out} e^{-\xi_0} \right) \right) = 0$$

giving

$$r \mathscr{P}_K \left([K^+]_{in} - [K^+]_{out} e^{-\xi_0} \right) - \mathscr{P}_{Na} \left([Na^+]_{in} - [Na^+]_{out} e^{-\xi_0} \right) = 0$$
$$\left(r \mathscr{P}_K [K^+]_{in} + \mathscr{P}_{Na} [Na^+]_{in} \right) - \left(r \mathscr{P}_K [K^+]_{out} + \mathscr{P}_{Na} [Na^+]_{out} \right) e^{-\xi_0} = 0$$

We can manipulate this equation a bit to get

$$r\mathscr{P}_K[K^+]_{in} + \mathscr{P}_{Na}[Na^+]_{in} = \left(r\mathscr{P}_K[K^+]_{out} + \mathscr{P}_{Na}[Na^+]_{out}\right)e^{-\xi_0}$$

or

$$e^{\xi_0} = \frac{r\mathscr{P}_K[K^+]_{out} + \mathscr{P}_{Na}[Na^+]_{out}}{r\mathscr{P}_K[K^+]_{in} + \mathscr{P}_{Na}[Na^+]_{in}}$$

Since we know that

$$\xi_0 = \frac{FV_0^m}{RT}$$

we can solve for V_0^m and obtain

$$V_0^m = \frac{RT}{F}\left(\frac{r\mathscr{P}_K[K^+]_{out} + \mathscr{P}_{Na}[Na^+]_{out}}{r\mathscr{P}_K[K^+]_{in} + \mathscr{P}_{Na}[Na^+]_{in}}\right)$$

The $Na^+ - K^+$ pump moves out $3Na^+$ ions for every $2K+$ ions that go in. Hence the pumping ratio is r is 1.5. Thus for our usual squid axon using only potassium and sodium ions and neglecting chlorine, we find for a permeability ratio of

$$\mathscr{P}_K : \mathscr{P}_{Na} = 1.0 : 0.03$$

that

$$V_0^m = 25.32 \text{ (mV) } \ln\left(\frac{1.5 \cdot 33.33 \cdot 20.0 + 440.0}{1.5 \cdot 33.33 \cdot 400.0 + 50.0}\right)$$

$$= 25.32 \text{ (mV) } \ln\frac{1439.90}{200480.0}$$

$$= -66.68 \text{ mV}$$

Without a pump, r will be 1 and the calculation becomes

$$V_0^m = 25.32 \text{ (mV) } \ln\left(\frac{33.33 \cdot 20.0 + 440.0}{33.33 \cdot 400.0 + 50.0}\right)$$

$$= 25.32 \text{ (mV) } \ln\frac{1106.60}{13382.0}$$

$$= -63.11 \text{ mV}$$

Our sample calculation thus shows that the pump contributes -66.68 to -63.11 or -3.57 mV to the rest voltage. This is only $\frac{3.57}{66.68}$ or 5.35 percent!

5.9   Excitable Cells

There are specialized cells in most living creatures called neurons which are adapted for generating signals which are used for the transmission of sensory data, control of movement and cognition through mechanisms we don't fully understand. A neuron is an example of what is called an excitable cell and its membrane is studded with many voltage gated sodium and potassium channels. In terms of ionic permeabilities, the GHK voltage equation for the usual sodium, potassium and chlorine ions gives

$$V_0^m = 25.32 \text{ (mV) } \ln \left(\frac{\mathscr{P}_K[K^+]_{out} + \mathscr{P}_{Na}[Na^+]_{out} + \mathscr{P}_{Cl}[Cl]_{in}}{\mathscr{P}_K[K^+]_{in} + \mathscr{P}_{Na}[Na^+]_{in} + \mathscr{P}_{Cl}[Cl^-]_{out}} \right)$$

which is about -60 mV at rest but which we have already seen can rapidly increase to $+40$ mV upon a large shift in the sodium and potassium permeability ratio. Recalling our discussion in Sect. 5.6, we can write the rest voltage in terms of conductances as

$$V_0^m = \frac{g_K}{g_K + g_{Na} + g_{Cl}} E_K + \frac{g_{Na}}{g_K + g_{Na} + g_{Cl}} E_{Na} + \frac{g_{Cl}}{g_K + g_{Na} + g_{Cl}} E_{Cl}$$

Either the conductance or the permeability model allows us to understand this sudden increase in voltage across the membrane in terms of either sodium to potassium permeability or conductance ratio shifts. We will study all of this in a lot of detail later, but for right now, let's just say that under certain circumstances, the rest potential across the membrane can be stimulated just right to cause a rapid rise in the equilibrium potential of the cell, followed by a sudden drop below the equilibrium voltage and then ended by a slow increase back up to the rest potential. The shape of this wave form is very characteristic and is shown in Fig. 5.10. This type of wave form is called an action potential and is a fundamental characteristic of excitable cells. In the figure, we draw the voltage across the membrane and simultaneously we draw the conductance curves for the sodium and potassium ions. Recall that conductance is reciprocal resistance, so a spike in sodium conductance, for example, is proportional to a spike in sodium ion current. So in the figure we see that sodium current spikes first and potassium second.

Now that we have discussed so many aspects of cellular membranes, we are at a point where we can develop a qualitative understanding of how this behavior is generated. We can't really understand the dynamic nature of this pulse yet (that is it's time and spatial dependence) but we can explain in a descriptive fashion how the potassium and sodium gates physical characteristics cause the behavior we see in Fig. 5.10. The changes in the sodium and potassium conductances that we see in the action potential can be explained by molecular properties of the sodium and potassium channels. As in the example we worked through for the resting potential for the squid nerve cells, for this explanation, we will assume the chlorine conductance does not change. So all the qualitative features of the action potential will be explained solely in terms of relative sodium to potassium conductance ratios. We have already seen that if we allow a huge increase in the sodium to potassium conductance ratio,

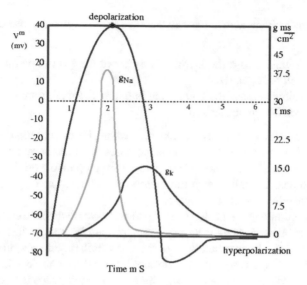

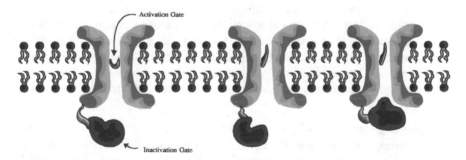

Fig. 5.10 A typical action potential

Fig. 5.11 A typical sodium channel

we generate a massive depolarization of the membrane. So now, we will try to explain how this happens. Let's look at a typical sodium channel as shown in Fig. 5.11 which is an example of a typical voltage dependent gate. The drawing of a potassium channel will be virtually identical. When you look at the drawing of the sodium channel, you'll see it is drawn in three parts. Our idealized channel has a hinged cap which can cover the part of the gate that opens into the cell. We call this the inactivation gate. It also has a smaller flap inside the gate which can close off the throat of the channel. This is called the activation gate. As you see in the drawing, these two pieces can be in one of three positions: resting (activation gate is closed and the inactivation gate is open); open (activation gate is open and the inactivation gate is open); and closed (activation gate is closed or closed and the inactivation gate is closed). Since this is a voltage activated gate, the transition from resting to open depends on the voltage across the cell membrane. We typically use the following terminology:

- When the voltage across the membrane is above the resting membrane voltage, we say the cell is depolarized.
- When the voltage across the membrane is below the resting membrane voltage, we say the cell is hyperpolarized.

These gates transition from resting to open when the membrane depolarizes. In detail, the probability that the gate opens increases upon membrane depolarization. However, the probability that the gate transitions from open to closed is NOT voltage dependent. Hence, no matter what the membrane voltage, once a gate opens, there is a fixed probability it will close again.

Hence, an action potential can be described as follows: when the cell membrane is sufficiently depolarized, there is an explosive increase in the opening of the sodium gates which causes a huge influx on sodium ions which produces a short lived rapid increase in the voltage across the membrane followed by a rapid return to the rest voltage with a typical overshoot phase which temporarily keeps the cell membrane hyperpolarized. We can explain much of this qualitatively as follows: The voltage gated channels move randomly back and forth between their open and closed states. At rest potential, the probability that a channel will move to the open position is small and hence, most channels are in the closed position. When the membrane is depolarized from some mechanism, the probability that a channel will transition to the open position increases. If the depolarization of the membrane lasts long enough or is large enough, then there is an explosive increase in the cell membrane potential fueled by

Explosive Depolarization of the Membrane The depolarization causes a few Na^+ channels to open. This opening allows Na^+ to move into the cell through the channel which increase the inside Na^+ concentration. Now our previous attempts to explain the voltage across the membrane were all based on equilibrium or steady state arguments; here, we want to think about what happens in that brief moment when the Na^+ ions flood into the cell. This is not a steady state analysis. Here, the best way to think of it is that the number of plus ions in the cell has gone up. So since the electrical field is based on minus to plus gradients, we would anticipate that the electrical field temporarily increases. We also know from basic physics that

$$E = -\frac{dV}{dx}$$

implying since the membrane starts at x_0 and ends at $x_0 + \ell$ that

$$E\ell = -V(x_0 + \ell) + V(x_0) = V_{in} - V_{out}.$$

where, as usual, we have labeled the voltage at the inside, $V(x_0)$ as V_{in} and the other as V_{out}. Hence, since we define the membrane voltage V_0^m as

$$V_0^m = V_{in} - V_{out},$$

we see that

$$E\ell = V_0^m$$

Thus if Na^+ ions come in, the electric field goes up and so does the potential across the membrane. Hence, the initial depolarizing event further depolarizes the membrane. This additional depolarization of the membrane increases the probability that the sodium gates open. Hence, more sodium gates transition from resting to closed and more Na^+ ions move into the cell causing further depolarization as outlined above. *We see depolarization induces further depolarization which is known as* **positive feedback**. Roughly speaking, the membrane voltage begins to drive towards E_{Na} as determined by the Nernst voltage equation. For biological cells, this is in the range of 70 mV. Qualitatively, this is the portion of the voltage curve that is rising from rest towards the maximum in Fig. 5.10.

Sodium Inactivation The probability that a sodium gate will move from open to closed is independent of the membrane voltage. Hence, as soon as sodium gates begin to open, there is a fixed probability they will then begin to move towards closed. Thus, even as the membrane begins its rapid depolarization phase, the sodium gates whose opening provided the trigger begin to close. This is most properly described in terms of competing rates in which one rate is highly voltage dependent, but we can easily see that essentially , the membrane continues to depolarize until the rate of gate closing exceeds the rate of opening and the net flow of sodium ions is out of the cell rather than into the cell. At this point, the membrane voltage begins to fall. The sodium gate therefore does not remain open for long. It's action provides a brief spike in sodium conductance as shown in Fig. 5.12.

Potassium Channel Effects The potassium gate is also a voltage dependent channel whose graphical abstraction is similar to that shown in Fig. 5.11. These channels respond much slower to the membrane depolarization induced by the sodium channel activations. As the K^+ channels open, K^+ inside the cell flows outward and thereby begins to restore the imbalance caused by increased Na^+ in the cytoplasm. The explanation for this uses the same electrical field and potential connection we invoked earlier to show why the membrane depolarizes. This outward flow, however, moves

Fig. 5.12 Spiking due to sodium ions

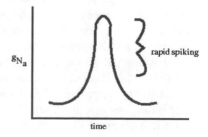

g_{Na}

rapid spiking

time

the membrane voltage the other way and counteracts the depolarization and tries to move the membrane voltage back to rest potential. The K^+ channels stay open a long time compared to the Na^+ channels. Since the sodium channels essentially pop open and then shut again quickly, this long open time of the potassium channel means the potassium conductance, g_K falls very slowly to zero. As the potassium outward current overpowers the diminishing sodium inward current, the potential accelerates its drop towards rest which can be interpreted as a drive towards the Nernst potential E_K which for biological cells is around -80 mV.

Chlorine Ion Movement The rest potential is somewhere in between the sodium and potassium Nernst potentials and the drive to reach it is also fueled by several other effects. There is also a chlorine current which is caused by the movement of chlorine through the membrane without gating. The movement of negative ions into the cell is qualitatively the same as the movement of positive potassium ions out of the cell—the membrane voltage goes down. Thus, chlorine current in the cell is always acting to bring the cell back to rest. We see the cell would return to its rest voltage even without the potassium channels, but the potassium channels accelerate the movement. One good way to think about it is this: the potassium channels shorten the length of time the membrane stays depolarizes by reducing the maximum height and the width of the pulse we see in Fig. 5.10.

The $Na^+ - K^+$ Pump There is also an active sodium–potassium pump that moves 3 sodium ions out of the cell for each 2 potassium ions that are moved in. This means that the cell's interior plus charge goes down by one. This means the membrane voltage goes down. Hence, this pump acts to bring the membrane potential back to rest also.

The Hyperpolarization Phase In Fig. 5.10, you can see that after the downward voltage pulse crosses the rest potential line, the voltage continues to fall before it bottoms out and slowly rises in an asymptotic fashion back up to the rest potential. This is called the hyperpolarization phase and the length of time a cell spends in this phase and the shape of the this phase of the curve are important to how this potential change from this cell can be used by other cells for communication. As the potassium channels begin to close, the K^+ outward current drops and the voltage goes up towards rest.

The Refractory Period During the hyperpolarization phase, many sodium channels are inactivated. The probability a channel will reopen simultaneously is small and most sodium gates are closed. So for a short time, even if there is an initial depolarizing impulse event, only a small number of sodium channels are in a position to move to open. However, if the magnitude of the depolarization event is increased, more channels will open in response to it. So it is possible that with a large enough event, a new positive feedback loop of depolarization could begin and generate a new action potential. However, without such an event, the membrane is hyperpolarized, the probability of opening is a function of membrane voltage and so is very low, and hence most gates remain closed. The membrane is continuing to move towards rest though and the closer this potential is to the rest value, the higher the probability of opening for the sodium gates. Typically, this period where the cell has a hard time initiating a new action potential is on the order of 1–2 mS. This period is called the refractory period of the cell.

With all this said, let's go back to the initial depolarizing event. If this event is a small pulse of Na^+ current that is injected into a resting cell, there will be a small depolarization which does not activate the sodium channels and so the ECE force acting on the K^+ and Cl^- ions increase. K^+ goes out of the cell and Cl^- comes in. This balances the inward Na^+ ion jet and the cell membrane voltage returns to rest. If the initial depolarization is larger, some Na^+ channels open but if the K^+ and Cl^- net current due to the resulting ECE forces is larger than the inward Na^+ current due to the jet and the sodium channels that are open, the membrane potential still goes back to rest. However, if the depolarization is sufficiently large, the ECE derived currents can not balance the inward Na^+ current due to the open sodium gates and the explosive depolarization begins.

References

D. Johnston, S. Miao-Sin Wu, *Foundations of Cellular Neurophysiology* (MIT Press, Cambridge, 1995)

T. Weiss, *Cellular Biophysics: Volume 2, Electrical Properties*, chapter The Hodgkin–Huxley model (MIT Press, Cambridge, 1996), pp. 163–292

Chapter 6
Lumped and Distributed Cell Models

We can now begin to model a simple biological cell. We can think of a cell as having an input line (this models the dendritic tree), a cell body (this models the soma) and an output line (this models the axon). We could model all these elements with cables—thin ones for the dendrite and axon and a fat one for the soma. To make our model useful, we need to understand how current injected into the dendritic cable propagates a change in the membrane voltage to the soma and then out across the axon. Our simple model is very abstract and looks like that of Fig. 6.1.

Note that the dendrite in Fig. 6.1 is of length L and its position variable is w and the soma also has length L and its position variable is w. Note that each of these cables has two endcaps—the right and left caps on the individual cylinders—and we eventually will have to understand the boundary conditions we need to impose at these endcaps when current is injected into one of these cylinders. We could also model the soma as a spherical cell, but for the moment let's think of everything as the cylinders you see here. So to make progress on the full model, we first need to understand a general cable model.

6.1 Modeling Radial Current

Now a fiber or cable can be modeled by some sort of cylinder framework. In Fig. 6.2a, we show the cylinder model. In our cylinder, we will first look at currents that run out of the cylinder walls and ignore currents that run parallel or longitudinally down the cylinder axis. We label the interesting variables as follows:

- The inner cylinder has radius a.
- L is the length of the wire.
- V is the voltage across the wall or membrane.
- I_r is the radial current flow through the cylinder. This means the current is flowing through the walls. This is measured in amps.

© Springer Science+Business Media Singapore 2016
J.K. Peterson, *Calculus for Cognitive Scientists*, Cognitive Science
and Technology, DOI 10.1007/978-981-287-880-9_6

Fig. 6.1 A simple cell
model

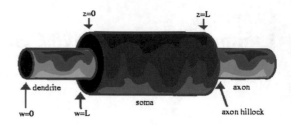

Fig. 6.2 Cable cylinders.
a The cable cylinder model.
b 2D cross-section

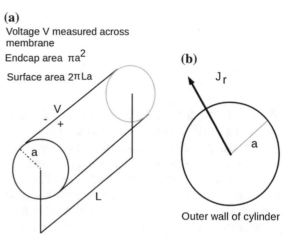

(a)
Voltage V measured across
membrane
Endcap area πa^2
Surface area $2\pi La$

(b)

Outer wall of cylinder

- J_r is the radial current flow density which has units of amps per unit area—for
  example, $\frac{amps}{cm^2}$.
- K_r is the radial current per unit length—typical units would be $\frac{amps}{cm}$.

The 2D cross-section, shown in Fig. 6.2b, further illustrates that the radial current
J_r moves out of the cylinder. Clearly, since the total surface area of our cylinder is
$2\pi a L$, we have the following conversions between the various current variables.

$$I_r = 2\pi a L \, J_r$$
$$K_r = \frac{I_r}{L} = 2\pi a \, J_r$$

Now, imagine our right circular cylinder as being hollow with a very thin skin filled
with some conducting solution, as shown in Fig. 6.3a. Further, imagine the wall is
actually a bilipid layer membrane as shown in Fig. 6.3b. We now have a simple model
of a biological cable we can use for our models. This is called the annular model of
a cable. In other words, if we take a piece of dendritic or axonal fiber, we can model
it as two long concentric cylinders. The outer cylinder is filled with a fluid which
is usually considered to be the extracellular fluid outside the nerve fiber itself. The
outer cylinder walls are idealized to be extremely thin and indeed we don't usually
think about them much. In fact, the outer cylinder is actually just a way for us to

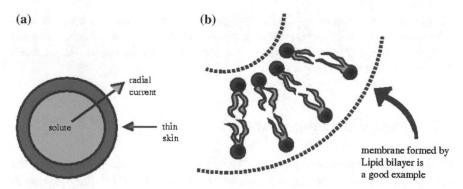

Fig. 6.3 Cable details. **a** The cable cylinder model cross section. **b** Part of the cable bilipid membrane structure

handle the fact that the dendrite or axon lives in an extracellular fluid bath. The inner cylinder has a thin membrane wall wrapped around an inner core of fluid. This inner fluid is the intracellular fluid and the membrane wall around it is the usual membrane we have discussed already. The real difference here is that we are now looking at a specific geometry for our membrane structure.

6.2 Modeling Resistance

We also will think of these cables as much as possible in terms of simple circuits. Hence, our finite piece of cable should have some sort of resistance associated with it. We assume from standard physics that the current density per unit surface area, J_r, is proportional to the electrical field density per length $\mathscr{E}$. The proportionality constant is called conductivity and is denoted by σ_e. The resulting equation is thus

$$J_r = \sigma_e \mathscr{E}$$

Another traditional parameter is called the resistivity which is reciprocal conductivity and is labeled ρ.

$$\rho = \frac{1}{\sigma_e}$$

The units here are

- $\mathscr{E}$ is measured in $\frac{\text{Volts}}{\text{cm}}$.
- σ_e is measured in $\frac{1}{\text{ohm}-\text{cm}}$.
- ρ is measured in ohm − cm.

This gives us the units of $\frac{\text{Volts}}{\text{ohm-cm}^2}$ or $\frac{\text{amps}}{\text{cm}^2}$ for the current density J_r which is what we expect them to be! The parameters ρ and σ_e measure properties that tell us what to do for one cm of our cable but do not require us to know the radius. Hence, these parameters are material properties.

6.3 Longitudinal Properties

Now let's look at currents that run parallel to the axis of the cable. These are called longitudinal currents which we will denote by the symbol I_a. Now that we want to look at currents down the fiber, we need to locate at what position we are on the fiber. We will let the variable z denote the cable position and we will assume to cable has length L. We show this situation in Fig. 6.4. The surface area of each endcap of the cable cylinder is πa^2. This implies that if we cut the cable perpendicular to its longitudinal axis, the resulting slice will have cross-sectional area πa^2. Our currents are thus:

- I_a is the longitudinal current flow which has units of amps.
- J_a is the longitudinal current flow density which has units of amps per unit area— $\frac{\text{amps}}{\text{cm}^2}$.

with the usual conversion

$$I_a = \pi a^2 \, J_a$$

Since the inside of the cable is filled with the same conducting fluid we used in our radial current discussions, we know

$$J_a = \sigma_e \, \mathscr{E}$$

and so

$$I_a = \pi a^2 \, \sigma_e \, \mathscr{E} = \frac{\pi a^2 \, \mathscr{E}}{\rho}$$

using the definition of resistivity. From standard physics, it follows that

$$\mathscr{E} = -\frac{d\Psi}{dz}$$

where Ψ is the potential in the conductor. Hence, we have

$$\frac{\rho I_a}{\pi a^2} = -\frac{d\Psi}{dz}$$

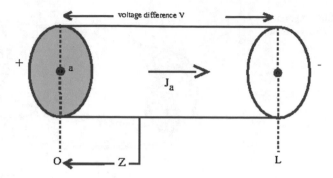

Fig. 6.4 Longitudinal cable currents

implying

$$V = \Psi(0) - \Psi(L) = \frac{\rho I_a}{\pi a^2} L$$

Now if we use Ohm's law to relate potential, current and resistance as usual, we would expect a relationship of the form

$$V = \text{Resistance } I_a = \frac{\rho L}{\pi a^2} I_a$$

This suggests that the term in front of I_a plays the role of a resistance. Hence, we will define $\mathscr{R}$ to be the resistance of the cable by

$$\mathscr{R} = \frac{\rho L}{\pi a^2}$$

The resistance per unit length would then be $r = \frac{\mathscr{R}}{L}$ which is $\frac{\rho}{\pi a^2}$. This notation is a bit unfortunate as we usually use the symbol r in other contexts to be a resistance measured in ohms; here it is a resistance per unit length measured in $\frac{\text{ohms}}{\text{cm}}$. Note that the resistance per unit length here is of the form $\frac{\rho}{A}$.

6.4 Current in a Cable with a Thin Wall

So far, we haven't actually thought of the cable wall as having a thickness. Now let's let the wall thickness be ℓ and consider the picture shown in Fig. 6.5. We let the variable s indicate where we are at inside the membrane wall. Hence, s is a variable which ranges from a value of a (the inner side of the membrane) to the value $a + \ell$ (the outer side of the membrane). Let $I_r(s)$ denote the radial current at s; then the radial current density is defined by

Fig. 6.5 Cable membrane
with wall thickness shown

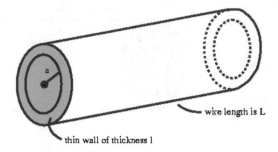

wire length is L

thin wall of thickness l

$$J_r(s) = \frac{I_r(s)}{2\pi s L}$$

We will assume the radial current is constant; i.e. I_r is independent of s. Thus, we have

$$J_r(s) = \frac{I_r}{2\pi s L}.$$

Now following the arguments we used before, we assume the current density is still proportional to the electric field density and that the electrical field density $\mathscr{E}$ is constant. Thus, we have $J_r(s) = \sigma_e \mathscr{E}$. Hence

$$\mathscr{E} = \frac{J_r(s)}{\sigma_e} = -\frac{d\Psi}{ds}$$

This implies

$$\frac{I_r}{2\pi s L} = -\sigma_e \frac{d\Psi}{ds}$$

Integrating, we find

$$\int_a^{a+\ell} \frac{I_r}{2\pi s L} = -\sigma_e \int_{\Psi(a)}^{\Psi(a+\ell)} \frac{d\Psi}{ds}$$

$$\frac{I_r}{2\pi L} \ln\left(\frac{a+\ell}{a}\right) = -\sigma_e \left(\Psi(a+\ell) - \Psi(a)\right)$$

Now $\Psi(a + \ell)$ is the voltage outside the cable at the end of the cell wall (V_{out}) and $\Psi(a)$ is the voltage inside the cable at the start of the cell wall (V_{in}). As usual, we represent the potential difference $V_{in} - V_{out}$ by V_r. Thus, using the definition of ρ, we have

$$\ln\left(\frac{a+\ell}{a}\right) = \frac{2\pi L \sigma_e V_r}{I_r} = \frac{2\pi L V_r}{\rho I_r}$$

We conclude

$$V_r = \frac{\rho I_r}{2\pi L} \ln\left(1 + \frac{\ell}{a}\right)$$

Recall that ℓ is the thickness of the cell wall which in general is very small compared to the radius a. Hence the ratio of ℓ to a is usually quite small. Recall from basic calculus that for a twice differentiable function f defined near the base point x_0 that we can replace f near x_0 as follows:

$$f(x) = f(x_0) + f'(x_0)(x - x_0) + \frac{1}{2} f''(c)(x - x_0)^2$$

where the number c lies in the interval (x_0, x). Unfortunately, this number c depends on the choice of x, so this equality is not very helpful in general. The straight line $L(x, x_0)$ given by

$$L(x, x_0) = f(x_0) + f'(x_0)(x - x_0)$$

is called the tangent line approximation or first order approximation to the function f near x_0. Hence, we can say

$$f(x) = L(x, x_0) + \frac{1}{2} f''(c)(x - x_0)^2$$

This tells us the error we make in replacing f by its first order approximation, $e(x, x_0)$, can be defined by

$$e(x, x_0) = |f(x) - L(x, x_0)| = \frac{1}{2} |f''(c)|(x - x_0)^2$$

Now $\ln(1 + x)$ is two times differentiable on $(0, \infty)$ so we have the first order approximation to $\ln(1 + x)$ near the base point 0 is given by

$$L(x, x_0) = \ln(1) + \left.\frac{1}{1 + x}\right|_{x=0} (x - 0) = x$$

with error

$$e(x, 0) = |\ln(1 + x) - x| = \frac{1}{2} \frac{1}{(1 + c)^2} x^2$$

where c is some number between 0 and x. Even though we can't say for certain where c is in this interval, we can say that

$$\frac{1}{(1 + c)^2} \leq 1$$

no matter what x is! Hence, the error in making the first order approximation on the interval $[0, x]$ is always bounded above. We have shown

$$e(x, 0) \leq \frac{1}{2} x^2$$

Now for our purposes, let's replace $\ln(1 + \frac{\ell}{a})$ with a first order approximation on the interval $(0, \frac{\ell}{a})$. Then the x above is just $\frac{\ell}{a}$ and we see that the error we make is

$$e(x, 0) \leq \frac{1}{2} \left(\frac{\ell}{a} \right)^2$$

If the ratio of the cell membrane to radius is small, the error we make is negligible. For example, a biological membrane is about 70 nM thick and the cell is about 20,000 nM in radius. Hence the membrane to radius ratio is $3.5e - 3$ implying the error is on the order of 10^{-5} which is relatively small. Since in biological situations, this ratio is small enough to permit a reasonable first order approximation with negligible error, we replace $\ln(1 + \frac{\ell}{a})$ by $\frac{\ell}{a}$. Thus, we have to first order

$$V_r = \frac{\rho I_r}{2\pi L} \frac{\ell}{a} \implies \mathscr{R} = \frac{V_r}{I_r} = \frac{\rho \ell}{2\pi L a}$$

We know $\mathscr{R}$ is measured in ohms and so the resistance R of the entire cable at the inner wall is $\mathscr{R}$ times the inner surface area of the cable which is $2\pi L a$. Hence, $R = \rho \ell$ which has units of ohms$-$cm^2. We know the wall has an inner and an outer surface area. The outer surface area is $2\pi a(1 + \frac{\ell}{a})L$ which is a bit bigger. In order to define the resistance of a unit surface area, we need to know which surfaced area we are talking about. So here, since we have already approximated the logarithm term around the base point 0 (which means the inside wall as $1 + 0$ corresponds to the inside wall!) we choose to use the surface area of the inside wall. Note how there are many approximating ideas going on here behind the scenes if you will. We must always remember these assumptions in the models we build! The upshot of all of this discussion is that for a cable model with a thin membrane, the resistance per unit length, r, with units $\frac{\text{ohms}}{\text{cm}}$, should be defined to be

$$r = \frac{\mathscr{R}}{\ell} = \frac{\rho}{2\pi L a}$$

6.5 The Cable Model

In a uniform isolated cell, the potential difference across the membrane depends on where you are on the cell surface. For example, we could inject current into a cylindrical cell at position z_0 as shown in Fig. 6.7a. In fact, in the laboratory, we

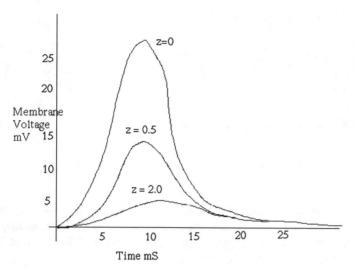

Fig. 6.6 Potential difference versus position *curves*

could measure the difference between V^m (the membrane potential) and V_0^m (the rest potential) that results from the current injection at $z = 0$ at various positions z downstream and we would see potential difference curves versus position that have the appearance of Fig. 6.6. The $z = 0$ curve is what we would measure at the site of the current injection; the $z = \frac{1}{2}$ and $z = 2$ curves are what we would measure $\frac{1}{2}$ or 2 units downstream from the injection site respectively. Note the spike in potential we see is quite localized to the point of injection and falls rapidly off as we move to the right or left away from the site. Our simple cylindrical model gave us

$$V_r = \frac{\rho I_r \ell}{2\pi L a}$$

as the voltage across the membrane or cell wall due to a radial current I_r flowing uniformly radially across the membrane along the entire length of cable. So far our model do not allow us to handle dependence on position so we can't reproduce voltage versus position curves as shown in Fig. 6.6. We also currently can't model explicit time dependence in our models!

Now we wish to find a way to model V^m as a function of the distance downstream from the current injection site and the time elapsed since the injection of the current. This model will be called the Core Conductor Model.

6.5.1 The Core Conductor Model Assumptions

Let's start by imagining our cable as a long cylinder with another cylinder inside it. The inner cylinder has a membrane wall of some thickness small compared to

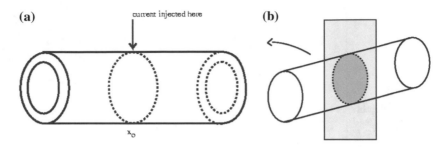

Fig. 6.7 Cylindrical cell details. **a** Current injection. **b** Radial cross section

the radius of the inner cylinder. The outer cylinder simply has a skin of negligible thickness. If we take a radial cross section as shown in Fig. 6.7b we see in this radial cross section, let's label the important currents and voltages as shown in Fig. 6.8a. As you can see, we are using the following conventions:

- t is time usually measured in milli-seconds or mS.
- z is position usually measured in cm.
- (r, θ) are the usual polar coordinates we could use to label points in any given radial cross section.
- $K_e(z, t)$ is the current per unit length across the outer cylinder due to external sources applied in a cylindrically symmetric fashion. This will allow us to represent current applied to the surface through external electrodes. This is usually measured in $\frac{amp}{cm}$
- $K_m(z, t)$ is the membrane current per unit length from the inner to outer cylinder through the membrane. This is also measure in $\frac{amp}{cm}$.
- $V_i(z, t)$ is the potential in the inner conductor measured in milli-volts or mV.
- $V_m(z, t)$ is the membrane potential measured in milli-volts or mV.
- $V_o(z, t)$ is the potential in the outer conductor measured in milli-volts or mV.

We can also look at a longitudinal slice of the cable as shown in Fig. 6.8b. The longitudinal slice allows us to see the two main currents of interest, I_i and I_o as shown in Fig. 6.9, we see
where

- $I_o(z, t)$ is the total longitudinal current flowing in the $+z$ direction in the outer conductor measured in amps.
- $I_i(z, t)$ is the total longitudinal current flowing in the $+z$ direction in the inner conductor measured in amps.

The Core Conductor Model is built on the following assumptions:

1. The cell membrane is a cylindrical boundary separating two conductors of current called the intracellular and extracellular solutions. We assume these solutions are homogeneous, isotropic and obey Ohm's Law.
2. All electrical variables have cylindrical symmetry which means the variables do not depend on the polar coordinate variable θ.

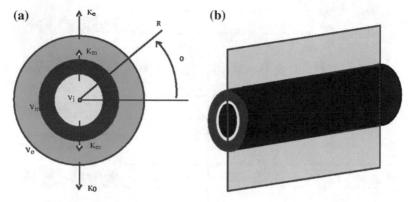

Fig. 6.8 Radial and longitudinal details. **a** Currents and voltages in the radial cross section. **b** A longitudinal slice of the cable

Fig. 6.9 Longitudinal currents

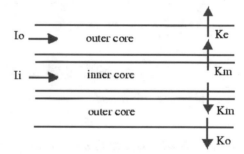

3. A circuit theory description of currents and voltages is adequate for our model.
4. Inner and outer currents are axial or longitudinal only. Membrane currents are radial only.
5. At any given position longitudinally (i.e. along the cylinder) the inner and outer conductors are equipotential. Hence, potential in the inner and outer conductors is constant radially. The only radial potential variation occurs in the membrane.

Finally, we assume the following geometric parameters:

- r_0 is the resistance per unit length in the outer conductor measured in $\frac{ohm}{cm}$.
- r_i is the resistance per unit length in the inner conductor measured in $\frac{ohm}{cm}$.
- a is the radius of the inner cylinder measured in cm.

It is also convenient to define the current per unit area variable J_m:

- $J_m(z, t)$ is the membrane current density per unit area measured in $\frac{amp}{cm^2}$.

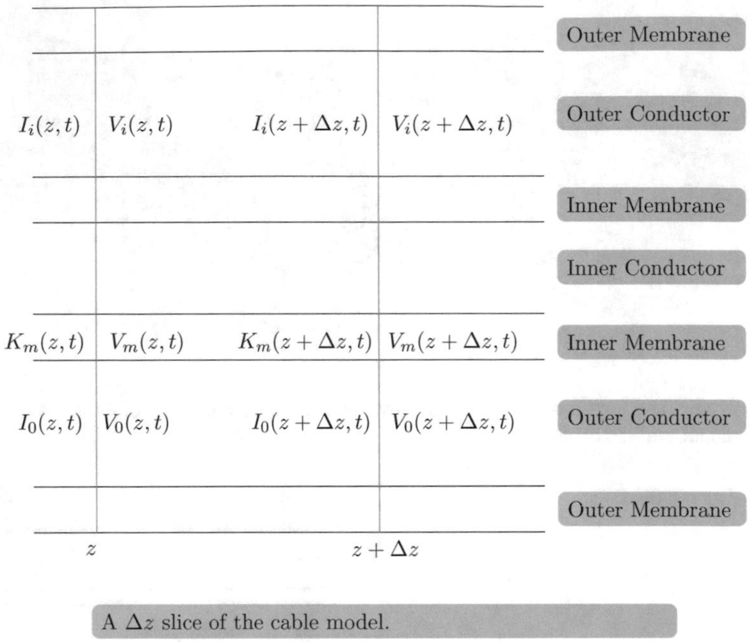

				Outer Membrane
$I_i(z,t)$	$V_i(z,t)$	$I_i(z+\Delta z,t)$	$V_i(z+\Delta z,t)$	Outer Conductor
				Inner Membrane
				Inner Conductor
$K_m(z,t)$	$V_m(z,t)$	$K_m(z+\Delta z,t)$	$V_m(z+\Delta z,t)$	Inner Membrane
$I_0(z,t)$	$V_0(z,t)$	$I_0(z+\Delta z,t)$	$V_0(z+\Delta z,t)$	Outer Conductor
				Outer Membrane
z		$z+\Delta z$		

A Δz slice of the cable model.

Fig. 6.10 The two slice model

6.5.2 Building the Core Conductor Model

Now let's look at a slice of the model between positions z and $z + \Delta z$. In Fig. 6.10, we see an abstraction of this. The vertical lines represent the circular cross sections through the cable at the positions z and $z + \Delta z$. We cut through the outer shell first giving two horizontal lines labeled *outer membrane* at the bottom of the figure. Then as we move upward, we encounter the *outer conductor*. We then move through the inner cable which has a bottom membrane, an inner conductor and a top membrane. Finally, we enter the outer conductor again and exit through the outer membrane. At each position z, there are values of I_i, I_o, V_i, V_o, K_m and V_m that are shown. Between z and $z + \Delta z$, we thus have a cylinder of inner and outer cable of length Δz.

From the Δz slice in Fig. 6.10, we can abstract the electrical network model we see in Fig. 6.11: Now in the inner conductor, we have current $I_i(z, t)$ entering the face of the inner cylinder. At that point the inner cylinder is at voltage $V_i(z, t)$. A distance Δz away, we see current $I(z + \Delta z, t)$ leaving the cylinder and we note the voltage of the cylinder is now at $V(z + \Delta z, t)$. Finally, there is a radial membrane current coming out of the cylinder uniformly all through this piece of cable. We illustrate this in Fig. 6.12: Now we know that the total current I through the membrane is given by

$$I(z, t) = 2\pi a \, J_m(z, t) \, \Delta z = K_m(z, t) \, \Delta z$$

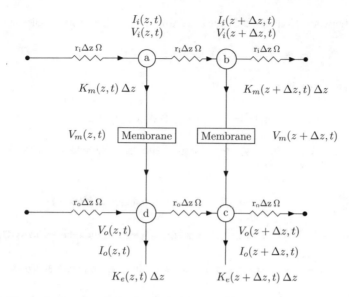

Fig. 6.11 The equivalent electrical network model

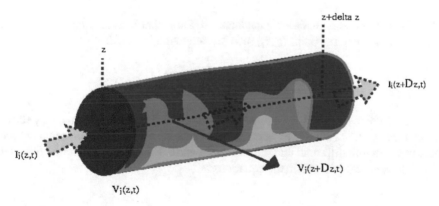

Fig. 6.12 Inner cylinder currents

and at Node d, the currents going into the node should match the currents coming out of the node:

$$I_o(z, t) + K_m(z, t)\,\Delta z = I_o(z + \Delta z, t) + K_e(z, t)\,\Delta z$$

Also, the voltage drop across the resistance $r_i\,\Delta z$ between Node a and Node b satisfies

$$r_i\,\Delta z\,I_i(z + \Delta z, t) = V_i(z, t) - V_i(z + \Delta z, t)$$

Similarly, between Node d and Node c we find the voltage drop satisfies

$$r_o \, \Delta z \, I_o(z + \Delta z, t) = V_o(z, t) - V_o(z + \Delta z, t)$$

Also at Node a, we find

$$I_i(z, t) = I_i(z + \Delta z, t) + K_m(z, t) \, \Delta z$$

Next, note that V_m is $V_i - V_o$. Rearranging the relationships we have found, we summarize as follows:

$$\frac{I_i(z + \Delta z, t) - I_i(z, t)}{\Delta z} = -K_m(z, t) \quad \text{(current balance at Node a)}$$

$$\frac{I_o(z + \Delta z, t) - I_o(z, t)}{\Delta z} = K_m(z, t) - K_e(z, t) \quad \text{(current balance at Node d)}$$

$$\frac{V_i(z + \Delta z, t) - V_i(z, t)}{\Delta z} = -r_i I_i(z + \Delta z, t) \quad \text{(voltage drop between Nodes a and b)}$$

$$\frac{V_o(z + \Delta z, t) - V_o(z, t)}{\Delta z} = -r_o I_o(z + \Delta z, t) \quad \text{(voltage drop between Nodes c and d)}$$

Now from standard multivariable calculus, we know that is a function $f(z, t)$ has a partial derivative at the point (z, t), then the following limit exists

$$\lim_{\Delta z \to 0} \frac{f(z + \Delta z, t) - f(z, t)}{\Delta z}$$

and the value of this limit is denoted by the symbol $\frac{\partial f}{\partial z}$. Now the equations above apply for any choice of Δz. Physically, we expect the voltages and currents we see here to be smooth differentiable functions of z and t. Hence, we expect that if we let Δz go to zero, we will obtain the equations:

$$\frac{\partial I_i}{\partial z} = -K_m(z, t) \tag{6.1}$$

$$\frac{\partial I_o}{\partial z} = K_m(z, t) - K_e(z, t) \tag{6.2}$$

$$\frac{\partial V_i}{\partial z} = \lim_{\Delta z \to 0} (-r_i I_i(z + \Delta z, t) = -r_i I_i(z, t) \tag{6.3}$$

$$\frac{\partial V_o}{\partial z} = \lim_{\Delta z \to 0} (-r_o I_o(z + \Delta z, t)) = -r_o I_o(z, t) \tag{6.4}$$

$$V_m = V_i - V_0 \tag{6.5}$$

We call Eqs. 6.1 and 6.2 the Core Equations. Note we can manipulate these equations as follows: Eq. 6.4 implies that

$$\frac{\partial V_m}{\partial z} = \frac{\partial V_i}{\partial z} - \frac{\partial V_0}{\partial z}$$

From Eqs. 6.3 and 6.4, it then follows that

$$\frac{\partial V_m}{\partial z} = -r_i I_i + r_o I_o$$

Thus, using Eqs. 6.1 and 6.2, we find

$$\frac{\partial^2 V_m}{\partial z^2} = -r_i \frac{\partial I_i}{\partial z} + r_o \frac{\partial I_0}{\partial z} = r_i K_m + r_o K_m - r_o K_e$$

Thus, the core equations imply that the membrane voltage satisfies the partial differential equation

$$\frac{\partial^2 V_m}{\partial z^2} = (r_i + r_o) K_m - r_o K_e \qquad (6.6)$$

Note that the units here do work out. The second partial derivative of V_m with respect to z involves ratios of first order partials of V_m with respect to z. The first order terms, by the definition of the partial derivative, have units of $\frac{\text{volt}}{\text{cm}}$; hence, the second order terms have units of $\frac{\text{volt}}{\text{cm}^2}$. Each of the r_i or r_o terms have units $\frac{\text{ohm}}{\text{cm}}$ and the K_m or K_e are current per length terms with units $\frac{\text{amps}}{\text{cm}}$. Thus the products have units $\frac{\text{amp}-\text{ohm}}{\text{cm}^2}$ or $\frac{\text{volt}}{\text{cm}}$. This partial differential equation is known as the **Core Conductor Equation**.

6.6 The Transient Cable Equations

Normally, we find it useful to model stuff that is happening in terms of how far things deviate or move away from what are called nominal values. We can use this idea to derive a new form of the Core Conductor Equation which we will call the Transient Cable Equation. Let's denote the rest values of voltage and current in our model by

- V_m^0 is the rest value of the membrane potential.
- K_m^0 is the rest value of the membrane current per length density.
- K_e^0 is the rest value of the externally applied current per length density.
- I_i^0 is the rest value of the inner current.
- I_o^0 is the rest value of the inner current.
- V_i^0 is the rest value of the inner voltage.
- V_o^0 is the rest value of the inner voltage.

It is then traditional to define the transient variables as perturbations from these rest values using the same variables but with lower case letters:

- v_m is the deviation of the membrane potential from rest.
- i_i is the deviation of the current in the inner fluid from rest.
- i_o is the deviation of the current in the outer fluid from rest.
- v_i is the deviation of the voltage in the inner fluid from rest.
- v_o is the deviation of the voltage in the outer fluid from rest.
- k_m is the deviation of the membrane current density from rest.

These variables are defined by

$$v_m(z, t) = V_m(z, t) - V_m^0$$
$$v_i(z, t) = V_i(z, t) - V_i^0$$
$$v_o(z, t) = V_o(z, t) - V_o^0$$
$$k_m(z, t) = K_m(z, t) - K_m^0$$
$$i_i(z, t) = I_i(z, t) - I_i^0$$
$$i_o(z, t) = I_o(z, t) - I_o^0$$

6.6.1   Deriving the Transient Cable Equation

Now in our core conductor model so far, we have not modeled the membrane at all. For this transient version, we need to think more carefully about the membrane boxes we showed in Fig. 6.11. We will replace our empty membrane box by a parallel circuit model. Now this box is really a chunk of membrane that is Δz wide. We will assume our membrane has a constant resistance and capacitance. We know that conductance is reciprocal resistance, so our model will consist to a two branch circuit: one branch is contains a capacitor and the other, the conductance element. We will let c_m denote the membrane capacitance density per unit length (measured in $\frac{\text{fahrad}}{\text{cm}}$). Hence, in our membrane box, since the box is Δz wide, we see the value of capacitance is $c_m \Delta z$. Similarly, we let g_m be the conductance per unit length (measured in $\frac{1}{\text{ohm}-\text{cm}}$) for the membrane. The amount of conductance for the box element is thus $gm \Delta z$. In Fig. 6.13, we illustrate our new membrane model. Since this is a resistance—capacitance parallel circuit, it is traditional to call this an RC membrane model. In Fig. 6.13, the current going into the element is $K_m(z, t)\Delta z$ and we draw the rest voltage for the membrane as a battery of value V_m^0. What happens when the membrane is at rest? At rest, all of the transients must be zero. At rest, there can be no capacitative current and so all the current flows through the conductance branch. Thus applying Ohm's law,

$$K_m^0(z, t) \, \Delta z = g_m \, \Delta z \, V_m^0 \tag{6.7}$$

On the other hand, when we are not in a rest position, we have current through both branches. Recall that for a capacitor C, the charge q held in a capacitor due to a

Fig. 6.13 The RC
membrane model

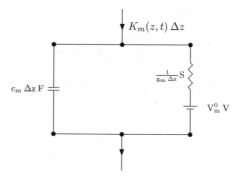

voltage V is $q = CV$ which implies that the current through the capacitor due to a time varying voltage is $i = \frac{dq}{dt}$ given by

$$i = C \frac{\partial V}{\partial t}$$

From Kirchhoff's current law, we see

$$\left(K_m^0 + k_m(z, t) \right) \, \Delta z = g_m \, \Delta z \, \left(V_m^0 + v_m(z, t) \right) + \frac{\partial}{\partial t} \left(V_m^0 + v_m(z, t) \right) c_m \, \Delta z$$

Using Eq. 6.7 and taking the indicated partial derivative, we simplify this to

$$k_m(z, t) \, \Delta z = g_m \, \Delta z \, v_m(z, t) + c_m \Delta z \, \frac{\partial v_m}{\partial t}$$

Upon canceling the common Δz term, we find the fundamental identity

$$k_m(z, t) = g_m \, v_m(z, t) + c_m \frac{\partial v_m}{\partial t} \tag{6.8}$$

Now the core conductor equation in terms of our general variables is

$$\frac{\partial^2 V_m}{\partial z^2} = (r_i + r_o) K_m - r_o K_e$$

$$\frac{\partial^2}{\partial z^2} \left(V_m^0 + v_m(z, t) \right) = (r_i + r_o) \left(K_m^0 + k_m(z, t) \right) - r_o \left(K_e^0 + k_e(z, t) \right)$$

Thus, in terms of transient variables, we have

$$\frac{\partial^2 v_m}{\partial z^2} = (r_i + r_o) K_m^0 - r_o K_e^0 + (r_i + r_o) k_m - r_o k_e$$

This leads to

$$\frac{\partial^2 v_m}{\partial z^2} - (r_i + r_o)k_m + r_o k_e = (r_i + r_o)K_m^0 - r_o K_e^0$$

Now at steady state, both sides of the above equation must be zero. This gives us the identities:

$$\frac{\partial^2 v_m}{\partial z^2} = (r_i + r_o)k_m - r_o k_e$$

$$(r_i + r_o)K_m^0 = r_o K_e^0$$

However, Eq. 6.8, allows us to replace k_m by an equivalent relationship. We obtain

$$\frac{\partial^2 v_m}{\partial z^2} = (r_i + r_o)\left(g_m v_m + c_m \frac{\partial v_m}{\partial t}\right) - r_o k_e$$

$$(r_i + r_o)K_m^0 = r_o K_e^0$$

The **Transient Cable Equation** or just **Cable Equation** is then

$$\frac{\partial^2 v_m}{\partial z^2} - (r_i + r_o)c_m \frac{\partial v_m}{\partial t} - (r_i + r_o)g_m v_m = -r_o k_e \qquad (6.9)$$

6.6.2   The Space and Time Constant of a Cable

The Cable Eq. 6.9 can be further rewritten in terms of two new constants, the space constant of the cable, λ_c, and the time constant, τ_m. Note, we can rewrite 6.9 as

$$\frac{1}{(r_i + r_o)g_m}\frac{\partial^2 v_m}{\partial z^2} - \frac{c_m}{g_m}\frac{\partial v_m}{\partial t} - v_m = -\frac{r_o}{(r_i + r_o)g_m}k_e$$

Define the new constants

$$\lambda_c = \sqrt{\frac{1}{(r_i + r_o)g_m}}$$

$$\tau_m = \frac{c_m}{g_m}$$

Then

$$\frac{r_o}{(r_i + r_o)g_m} = r_o \lambda_c^2$$

and the Cable Eq. 6.9 can be written in a new form as

$$\lambda_c^2 \frac{\partial^2 v_m}{\partial z^2} - \tau_m \frac{\partial v_m}{\partial t} - v_m = -r_o \lambda_c^2 k_e \tag{6.10}$$

The new constants τ_m and λ_c are very important to understanding how the solutions to this equation will behave. We call τ_m the time constant and λ_c the space constant of our cable.

6.6.2.1 The Time Constant

Consider the ratio $\frac{c_m}{g_m}$. Note that the units of this ratio are fahrad $-$ ohm. Recall that charge deposited on a capacitor of value C fahrads is $q = CV$ where V is the voltage across the capacitor. Hence, a dimensional analysis shows us that coulombs equal fahrad-volts. But we also know from Ohm's law that voltage is current times resistance; hence, dimensional analysis tells us that volts equal $\frac{coulomb-ohm}{sec}$. We conclude that fahrad $-$ ohm equals $\frac{coulomb}{volt}$ times ohm. But a $\frac{coulomb}{volt}$ is a $\frac{sec}{ohm}$ by Ohm's law as discussed above. Hence, simplifying, we see the ration $\frac{c_m}{g_m}$ has unit of seconds. Hence, this ratio is a time variable. This is why we define this constant to be the time constant of the cable, τ_m. Note that τ_m is a constant whose value is independent of the size of the cell; hence it is a membrane property. Also, if we let G_m be the conductance of a square centimeter of cell membrane. Then G_m has units of $\frac{1}{ohm-cm}^2$. The conductance of our Δz box of membrane in our model was $g_m \Delta z$ and the total conductance of the box is also G_m times the surface area of the box or $G_m 2\pi a \Delta z$. Equating expressions, we see that

$$g_m = 2\pi a \, G_m$$

In a similar manner, if C_m is the membrane capacitance per unit area, we have

$$c_m = 2\pi a \, C_m$$

We see the time constant can thus be expressed as $\frac{C_m}{G_m}$ also.

6.6.2.2 The Space Constant

Next, consider the dimensional analysis of the term $\frac{1}{(r_i+r_o)g_m}$. The sum of the resistances per length have units $\frac{ohm}{cm}$ and g_m has units $\frac{1}{ohm-cm}$. Hence, the denominator of this fraction has units $\frac{ohm}{cm}$ times $\frac{1}{ohm-cm}$ or cm^{-2}. Hence, this ratio has units of cm^2. This is why the square root of the ratio functions as a length parameter. We can look at this more closely. Consider again our approximate model where we divide the cable up into pieces of length Δz. We know the inner current per unit area density is J_i and it is defined by

$$J_i(z, t) = \frac{I_i(z, t)}{\pi a^2} = \sigma_i \, \mathscr{E}_i = -\sigma_i \frac{d\Psi}{dz}$$

where z is the position in the inner cylinder, σ_i is the conductivity and $\mathscr{E}_i$ is the electrical field density per length of the inner solution, respectively and Ψ is the potential at position z in the inner cylinder. It then follows that

$$\frac{I_i(z, t)}{\pi a^2 \sigma_i} = -\frac{d\Psi}{dz}$$

Now let's assume the inner longitudinal current from position z to position $z + \Delta z$ is constant with value $I(z, t)$. Then, integrating we obtain

$$\frac{I_i(z, t)}{\pi a^2 \sigma_i} \Delta z = -\int_z^{z+\Delta z} \frac{d\Psi}{dz} \, dz$$
$$= \Psi(z) - \Psi(z + \Delta z).$$

But this change in potential from z to $z+\Delta z$ can be approximated by the inner cylinder voltage at z at time t, $V_i(z, t)$. Thus, noting the resistivity of the inner solution, ρ_i, is the reciprocal of the conductivity, our approximation allows us to say

$$\frac{\rho_i \, I_i(z, t)}{\pi a^2} \Delta z = V_i(z, t)$$

This implies that resistance of this piece of cable can be modeled by

$$\mathscr{R}_i(z, t) = \frac{V_i(z, t)}{I_i(z, t)} = \frac{\rho_i \Delta z}{\pi a^2}$$

and so we conclude that since $\mathscr{R}_i$ must be the same as $r_i \Delta z$, we have

$$r_i = \frac{\rho_i}{\pi a^2}$$

Now in biological settings, we typically have that r_i is very large compared to r_o. Hence, the term $r_i + r_o$ is nicely approximated by just r_i. In this case, since g_m is $2\pi a \, G_m$, we see

$$\lambda_c = \sqrt{\frac{1}{r_i \, g_m}}$$

$$= \sqrt{\frac{1}{\frac{\rho_i}{\pi a^2} \, 2\pi a G_m}}$$

$$= \sqrt{\frac{a}{2\rho_i G_m}}$$

Now ρ_i and G_m are membrane constants independent of cell geometry. So we see that the space constant is proportional to the square root of the fiber radius. Note also that the space constant decreases as the fiber radius shrinks.

Chapter 7
Time Independent Solutions to Infinite Cables

In Fig. 7.1, we see a small piece of cable as described in Chap. 6. We are injecting current I_e into the cable via an external current source at $z = 0$. We assume the current that is injected in uniformly distributed around the cable membrane. We will begin by assuming that the cable extends to infinity both to the right and to the left of the current injection site. This is actually easier to handle mathematically, although you will probably find it plenty challenging! The picture we see in Fig. 7.1 is thus just a short piece of this infinitely long cable. Now if a dendrite or axon was really long, this would probably not be a bad model, so there is a lot of utility in examining this case as an approximation to reality. We also assume the other end of the line that delivers the current I_e is attached some place so far away it has no effect on our cable. We now begin our discussion of how to solve this problem.

7.1 The Infinite Cable

The time dependent cable equation is just the full cable Eq. 6.10 from Sect. 6.6. Recall that k_e is the external current per unit length. Hence, $\lambda_c k_e$ is a current which is measured in amps. We also know that $\lambda_c\, r_o$ is a resistance measured in ohms. Hence, the product $r_o\, \lambda_c$ times $\lambda_c\, k_e$ is a voltage as from Ohm's law, resistance times current is voltage. Thus, the right hand side of the cable equation is the voltage due to the current injection. On the left hand side, a similar analysis shows that each term represents a voltage.

- $\lambda_c^2 \frac{\partial^2 v_m}{\partial z^2}$ is measured in cm^2 times $\frac{\text{volt}}{\text{cm}^2}$ or volts.
- $\tau_m \frac{\partial v_m}{\partial t}$ is measured in seconds times $\frac{\text{volt}}{\text{second}}$ or volts.
- v_m is measured volts.

Now if we were interested only in solutions that did not depend on time, then the term $\frac{\partial v_m}{\partial t}$ would be zero. Also, we could write all of our variables as position dependent

© Springer Science+Business Media Singapore 2016
J.K. Peterson, *Calculus for Cognitive Scientists*, Cognitive Science
and Technology, DOI 10.1007/978-981-287-880-9_7

Fig. 7.1 Current injection into a cable

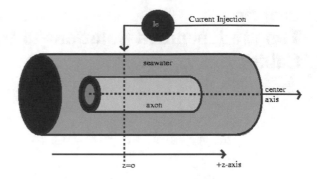

only; i.e. $v_m(z, t)$ as just $v_m(z)$ and so on. In this case, the partial derivatives are not necessary and we obtain an ordinary differential equation:

$$\lambda_c^2 \frac{d^2 v_m}{dz^2} - v_m(z) = -r_o \lambda_c^2 k_e(z) \tag{7.1}$$

7.2 Solving the Time Independent Infinite Cable Equation

Equation 7.1 as written does not impose any boundary or initial conditions on the solution. Eventually, we will have to make a decision about these conditions, but for the moment, let's solve this general differential equation as it stands. The typical solution to such an equation is written in two parts: the homogeneous part and the particular part. The homogeneous part or solution is the function ϕ_h that solves

$$\lambda_c^2 \frac{d^2 v_m}{dz^2} - v_m(z) = 0 \tag{7.2}$$

This means that if we plug ϕ_h into 7.2, we would find

$$\lambda_c^2 \frac{d^2 \phi_h}{dz^2} - \phi_h(z) = 0$$

The particular part or solution is any function ϕ_p that satisfies the full Eq. 7.1; i.e.

$$\lambda_c^2 \frac{d^2 \phi_p}{dz^2} - \phi_p(z) = -r_o \lambda_c^2 k_e(z)$$

It is implied in the above discussions that ϕ_h and ϕ_p must be functions that have two derivatives for all values of z that are interesting. Since this first case concerns a cable of infinite length, this means here that ϕ_h and ϕ_p must be twice differentiable on the entire z axis. We can also clearly see that adding the homogeneous and particular part together will always satisfy the full time dependent cable equation. Let ϕ denote the general solution. Then

$$\phi(z) = \phi_h(z) + \phi_p(z)$$

and ϕ will satisfy the time independent cable equation. If the external current k_e is continuous in z, then since the right hand side must equal the left hand side, the continuity of the right hand side will force the left hand side to be continuous also. This will force the solution ϕ we seek to be continuous in the second derivative. So usually we are looking for solutions that are very nice: they are continuous in the second derivative. This means that there are no corners in the second derivative of voltage.

7.2.1 Variation of Parameters

Since we don't know the explicit function k_e we wish to use in the non-homogeneous equation, the common technique to use to find the particular solution is the one called Variation of Parameters. We use the fact that the solutions to our model are linearly independent. Here the vector space is $C^2[a, b]$ for some interval $[a, b]$ and we use the linear independence of the two solutions of the homogeneous linear second order model to build a solution to the nonhomogeneous model. Rather than doing this in general, we will focus on a specific model:

$$\beta^2 \frac{d^2x}{dt^2} - x(t) = f(t)$$
$$x(0) = -1$$
$$x(5) = 4.$$

where β is a nonzero number and we have switched to using the independent variable t instead of z for convenience. Our cable models without time dependence will give us equations to solve with the independent variable being space, z, but other models can use time, t. The homogeneous solution x_h here solves

$$\beta^2 \frac{d^2x}{dt^2} - x(t) = 0$$

and has the form

$$x_h(t) = Ae^{-\frac{t}{\beta}} + Be^{\frac{t}{\beta}}$$

We want to find a particular solution, called x_p, to the model. Hence, we want x_p to satisfy

$$\beta^2 x_p''(t) - x_p(t) = f(t)$$

Since we don't know the explicit function f we wish to use in the nonhomogeneous equation, a common technique to find the particular solution is the one called *Variation of Parameters*. In this technique, we take the homogeneous solution and replace the constants A and B by unknown functions $u_1(t)$ and $u_2(t)$. Then we see if we can derive conditions that the unknown functions u_1 and u_2 must satisfy in order to work. So we start by assuming

$$x_p(t) = u_1(t)e^{-\frac{t}{\beta}} + u_2(t)e^{\frac{t}{\beta}}$$

Using the chain and product rule for differentiation, the first derivative of x thus gives (we use the *prime* notation for derivatives with respect to t to clean up the notation):

$$
\begin{aligned}
x_p' &= u_1'(t)e^{-\frac{t}{\beta}} - u_1(t)\frac{1}{\beta}e^{-\frac{t}{\beta}} + u_2'(t)e^{\frac{t}{\beta}} + u_2(t)\frac{1}{\beta}e^{\frac{t}{\beta}} \\
&= \left(u_1'(t)e^{-\frac{t}{\beta}} + u_2'(t)e^{\frac{t}{\beta}} \right) \\
&\quad - u_1(t)\frac{1}{\beta}e^{-\frac{t}{\beta}} + u_2(t)\frac{1}{\beta}e^{\frac{t}{\beta}}
\end{aligned}
$$

We know there are two solutions to the model, $e^{-\frac{t}{\beta}}$ and $e^{\frac{t}{\beta}}$ which are *linearly independent* functions. Hence, there associated Wronskian is never zero for all s. Letting $x_1(t) = e^{-\frac{t}{\beta}}$ and $x_2(t) = e^{t\beta}$, consider the system

$$
\begin{bmatrix} x_1(t) & x_2(t) \\ x_1'(t) & x_2'(t) \end{bmatrix}
\begin{bmatrix} \phi(t) \\ \psi(t) \end{bmatrix}
= \begin{bmatrix} 0 \\ g(t) \end{bmatrix}
$$

for any appropriately nice function g. For each fixed t, the determinant of this system is the value of the Wronskian which is not zero. Hence, there is a unique solution for the value of $\phi(t)$ and $\psi(t)$ for each t. In particular, we can use this idea with the functions $\phi(t) = u_1'(t)$ and $\psi(t) = u_2'(t)$ so the first row gives us a condition we must impose on our unknown functions $u_1'(t)$ and $u_2'(t)$. We must have

$$u_1'(t)e^{-\frac{t}{\beta}} + u_2'(t)e^{\frac{t}{\beta}} = 0$$

This simplifies the first derivative of x_p to be

$$x_p'(t) = -u_1(t)\frac{1}{\beta}e^{-\frac{t}{\beta}} + u_2(t)\frac{1}{\beta}e^{\frac{t}{\beta}}$$

Now take the second derivative to get

$$x_p''(t) = -u_1'(t)\frac{1}{\beta}e^{-\frac{t}{\beta}} + u_2'(t)\frac{1}{\beta}e^{\frac{t}{\beta}} + u_1(t)\frac{1}{\beta^2}e^{-\frac{t}{\beta}} + u_2(t)\frac{1}{\beta^2}e^{\frac{t}{\beta}}$$

Now plug these derivative expressions into the nonhomogeneous equation to find

$$f(t) = \beta^2\left(u_2'(t)\frac{1}{\beta}e^{\frac{t}{\beta}} - u_1'(t)\frac{1}{\beta}e^{-\frac{t}{\beta}} + u_2(t)\frac{1}{\beta^2}e^{\frac{t}{\beta}} + u_1(t)\frac{1}{\beta^2}e^{-\frac{t}{\beta}}\right)$$
$$-\left(u_2(t)e^{\frac{t}{\beta}} + u_1(t)e^{-\frac{t}{\beta}}\right)$$

Now factor out the common $u_1(t)$ and $u_2(t)$ terms to find after a bit of simplifying that

$$f(t) = \beta^2\left(u_2'(t)\frac{1}{\beta}e^{\frac{t}{\beta}} - u_1'(t)\frac{1}{\beta}e^{-\frac{t}{\beta}}\right) + u_2(t)(e^{\frac{t}{\beta}} - e^{\frac{t}{\beta}}) + u_1(t)(e^{-\frac{t}{\beta}} - e^{-\frac{t}{\beta}})$$

We see the functions u_1 and u_2 must satisfy

$$\frac{f(t)}{\beta^2} = u_2'(t)\frac{1}{\beta}e^{\frac{t}{\beta}} - u_1'(t)\frac{1}{\beta}e^{-\frac{t}{\beta}}$$

This gives us a second condition on the unknown functions u_1 and u_2. Combining we have

$$u_1'(t)e^{-\frac{t}{\beta}} + u_2'(t)e^{\frac{t}{\beta}} = 0$$
$$-u_1'(t)\frac{1}{\beta}e^{-\frac{t}{\beta}} + u_2'(t)\frac{1}{\beta}e^{\frac{t}{\beta}} = \frac{f(t)}{\beta^2}$$

This can be rewritten in a matrix form:

$$\begin{bmatrix} e^{-\frac{t}{\beta}} & e^{\frac{t}{\beta}} \\ -\frac{1}{\beta}e^{-\frac{t}{\beta}} & \frac{1}{\beta}e^{\frac{t}{\beta}} \end{bmatrix}\begin{bmatrix} u_1'(t) \\ u_2'(t) \end{bmatrix} = \begin{bmatrix} 0 \\ \frac{f(t)}{\beta^2} \end{bmatrix}$$

We then use Cramer's Rule to solve for the unknown functions u_1' and u_2'. Let W denote the matrix

$$W = \begin{bmatrix} e^{-\frac{t}{\beta}} & e^{\frac{t}{\beta}} \\ -\frac{1}{\beta}e^{-\frac{t}{\beta}} & \frac{1}{\beta}e^{\frac{t}{\beta}} \end{bmatrix}$$

Then the determinant of W is $det(W) = \frac{2}{\beta}$ and by Cramer's Rule

$$u_1'(t) = \frac{\begin{bmatrix} 0 & e^{\frac{t}{\beta}} \\ \frac{f(t)}{\beta^2} & \frac{1}{\beta}e^{\frac{t}{\beta}} \end{bmatrix}}{det(W)} = -\frac{1}{2\beta}f(t)e^{\frac{t}{\beta}}$$

and

$$u_2'(t) = \frac{\begin{bmatrix} e^{-\frac{t}{\beta}} & 0 \\ -\frac{1}{\beta}e^{-\frac{t}{\beta}} & \frac{f(t)}{\beta^2} \end{bmatrix}}{det(W)} = \frac{1}{2\beta}f(t)e^{-\frac{t}{\beta}}$$

Thus, integrating, we have

$$u_1(t) = -\frac{1}{2\beta}\int_0^t f(u)\, e^{\frac{u}{\beta}}\, du$$

$$u_2(t) = \frac{1}{2\beta}\int_0^t f(u)\, e^{-\frac{u}{\beta}}\, du$$

where 0 is a convenient starting point for our integration. Hence, the particular solution to the nonhomogeneous time independent equation is

$$x_p(t) = u_1(t)\, e^{-\frac{t}{\beta}} + u_2(t)\, e^{\frac{t}{\beta}}$$

$$= \left(-\frac{1}{2\beta}\int_0^t f(u)\, e^{\frac{u}{\beta}}\, du\right)e^{-\frac{t}{\beta}} + \left(\frac{1}{\beta}\int_0^t f(u)\, e^{-\frac{u}{\beta}}\, du\right)e^{\frac{t}{\beta}}.$$

The general solution is thus

$$x(t) = x_h(t) + x_p(t)$$

$$= A_1 e^{-\frac{t}{\beta}} + A_2 e^{\frac{t}{\beta}} - \frac{1}{2\beta}\, e^{-\frac{t}{\beta}}\int_0^t f(u)\, e^{\frac{u}{\beta}}\, du + \frac{1}{2\beta}\, e^{\frac{t}{\beta}}\int_0^t f(u)\, e^{-\frac{u}{\beta}}\, du$$

for any real constants A_1 and A_2. Finally, note we can rewrite these equations as

$$x(t) = A_1 e^{-\frac{t}{\beta}} + A_2 e^{\frac{t}{\beta}} - \frac{1}{2\beta}\int_0^t f(u)\, e^{\frac{u-t}{\beta}}\, du + \frac{1}{2\beta}\int_0^t f(u)\, e^{-\frac{u-t}{\beta}}\, du$$

or

$$x(t) = A_1 e^{-\frac{t}{\beta}} + A_2 e^{\frac{t}{\beta}} - \frac{1}{\beta}\int_0^t f(u)\left(\frac{e^{\frac{u-t}{\beta}} - e^{-\frac{u-t}{\beta}}}{2}\right)du$$

In applied modeling work, the function $\frac{e^w - e^{-w}}{2}$ arises frequently enough to be given a name. It is called the *hyperbolic sine function* and is denoted by the symbol $\sinh(w)$. Hence, we can write once more to see

$$x(t) = A_1 e^{-\frac{t}{\beta}} + A_2 e^{\frac{t}{\beta}} - \frac{1}{\beta} \int_0^t f(u) \sinh\left(\frac{u-t}{\beta}\right) du.$$

Finally, sinh is an odd function, so we can pull the minus side inside by reversing the argument into the sinh function. This gives

$$x(t) = A_1 e^{-\frac{t}{\beta}} + A_2 e^{\frac{t}{\beta}} + \frac{1}{\beta} \int_0^t f(u) \sinh\left(\frac{t-u}{\beta}\right) du$$

Now let's use the initial conditions $x(0) = -1$ and $x(5) = 4$. Hence,

$$-1 = x(0) = A_1 + A_2$$

$$4 = x(5) = A_1 e^{-\frac{5}{\beta}} + A_2 e^{\frac{5}{\beta}} + \frac{1}{\beta} \int_0^5 f(u) \sinh\left(\frac{5-u}{\beta}\right) du$$

which is solvable (albeit with an forbidding algebraic struggle!) for A_1 and A_2. We can use this technique on lots of models and generate similar solutions. If we have boundary conditions involving derivatives at a point, we will need a result called Leibnitz's Rule. We will only state a specific version of this; there are other more general versions.

Theorem 7.2.1 (Leibnitz's Rule)
Let $f(s, t)$ be continuous on the rectangle $[a, b] \times [c, d]$ and let u and v be continuous functions on some intervals whose range is a subset of $[a, b]$. This means $u(t)$ is in $[a, b]$ for each t in its domain and $v(t)$ is in $[a, b]$ for each t in its domain. Then, define F on the interval $[c, d]$ by

$$F(t) = \int_{u(t)}^{v(t)} f(s, t) \, ds.$$

Note that because of our assumptions, the lower and upper limit of integration always give points in the interval $[a, b]$ where $f(s, t)$ is defined and so the Riemann integral makes sense to calculate. Then, we have

$$F'(t) = \int_{u(t)}^{v(t)} \frac{\partial f}{\partial t}(s, t) \, ds + f(v(t), t) \, v'(t) - f(u(t), t) \, u'(t).$$

Proof The proof of this is essentially similar to the proof of the Cauchy Fundamental Theorem of Calculus. It is found in a good book on analysis (you can ask for our notes on these matters if you are adventuresome!). ∎

The first example below has an initial condition on the derivative of the solution at 0; hence, Leibnitz's rule will be needed.

7.2.1.1 Examples

Example 7.2.1 Solve

$$u''(t) + 9u(t) = 2t$$
$$u(0) = 1$$
$$u'(3) = -1$$

Solution *The characteristic equation is $r^2 + 9 = 0$ which has the complex roots $\pm 3i$. Hence, the two linearly independent solutions to the homogeneous equation are $\cos(3t)$ and $\sin(3t)$. We set the homogeneous solution to be*

$$u_h(t) = A\cos(3t) + B\sin(3t)$$

where A and B are arbitrary constants. The nonhomogeneous solution is of the form

$$u_p(t) = \phi(t)\cos(3t) + \psi(t)\sin(3t).$$

where this time the functions ϕ and ψ play the role of the functions u_1 and u_2 we used earlier. Since the solution we seek is $u(t)$ it is confusing to have the extra u_1 and u_2 and so we switch to the variables ϕ and ψ. We know the functions ϕ and ψ must then satisfy

$$\begin{bmatrix} \cos(3t) & \sin(3t) \\ -3\sin(3t) & 3\cos(3t) \end{bmatrix} \begin{bmatrix} \phi'(t) \\ \psi'(t) \end{bmatrix} = \begin{bmatrix} 0 \\ 2t \end{bmatrix}$$

Applying Cramer's rule, we have since the Wronskian here is 3 that

$$\phi'(t) = \frac{1}{3}\begin{bmatrix} 0 & \sin(3t) \\ 2t & 3\cos(3t) \end{bmatrix} = -\frac{2}{3}t\sin(3t)$$

and

$$\psi'(t) = \frac{1}{3}\begin{bmatrix} \cos(3t) & 0 \\ -3\sin(3t) & 2t \end{bmatrix} = \frac{2}{3}t\cos(3t)$$

Thus, integrating, we have

$$\phi(t) = -\frac{2}{3}\int_0^t u\sin(3u)\,du$$

$$\psi(t) = \frac{2}{3} \int_0^t u \cos(3u) \, du.$$

The general solution is therefore

$$u(t) = A \cos(3t) + B \sin(3t) - \left(\frac{2}{3} \int_0^t u \sin(3u) \, du \right) \cos(3t)$$
$$+ \left(\frac{2}{3} \int_0^t u \cos(3u) \, du \right) \sin(3t).$$

This can be simplified to

$$u(t) = A \cos(3t) + B \sin(3t) + \frac{2}{3} \int_0^t u\{\sin(3t)\cos(3u) - \sin(3u)\cos(3t)\} \, du$$
$$= A \cos(3t) + B \sin(3t) + \frac{2}{3} \int_0^t u \, \sin(3t - 3u) \, du.$$

Applying Leibnitz's rule, we find

$$u'(t) = -3A \sin(3t) + 3B \cos(3t) + \frac{2}{3} t \, \sin(3t - 3t) + \int_0^t 2u \cos(3t - 3u) \, du$$
$$= -3A \sin(3t) + 3B \cos(3t) + \int_0^t 2u \cos(3t - 3u) \, du$$

Applying the boundary conditions, we find

$$1 = u(0) = A$$
$$-1 = u'(3) = -3A \sin(9) + 3B \cos(9) + \int_0^3 2u \cos(9 - 3u) \, du$$

Hence, $A = 1$ and

$$B = \frac{-1 + 3\sin(9) - \int_0^3 2u \cos(9 - 3u) \, du}{3 \cos(9)}.$$

We can then assemble the solution using these constants.

Example 7.2.2 Solve

$$u''(t) + 9u(t) = 2t$$
$$u(0) = 1$$
$$u(4) = -1$$

Solution *The model is the same as the previous example except for the boundary conditions. We have*

$$1 = u(0) = A$$

$$-1 = u(4) = A\cos(12) + B\sin(12) + \frac{2}{3}\int_0^4 u\,\sin(12 - 3u)\,du.$$

Thus, since $A = 1$, we have

$$B\sin(12) = -1 - \cos(12) - \frac{2}{3}\int_0^4 u\,\sin(12 - 3u)\,du.$$

and so

$$B = -\frac{1 + \cos(12) + \frac{2}{3}\int_0^4 u\,\sin(12 - 3u)\,du}{\sin(12)}.$$

As usual, we can then assemble the solution using these constants.

7.2.1.2 Homework

Exercise 7.2.1 *Solve*

$$u''(t) + 4u(t) = 2t$$
$$u(0) = 2$$
$$u(4) = -6$$

expressing the particular solution in integral form. Note, if you are brave, you can calculate this integral, but be warned it needs integration by parts.

Exercise 7.2.2 *Solve*

$$u''(t) - u'(t) - 6u(t) = 2t$$
$$u(0) = 10$$
$$u(4) = -5$$

expressing the particular solution in integral form.

Exercise 7.2.3 *Solve*

$$u''(t) - 4u(t) = 2t$$
$$u(0) = 2$$
$$u'(0) = -1$$

expressing the particular solution in integral form.

7.3 Modeling Current Injections

The time independent Cable equation has the form

$$\lambda_c^2 \frac{d^2 v_m}{dz^2} - v_m(z) = -r_o \lambda_c^2 \, k_e(z)$$

and we solve this model using Variation of Parameters. We find the general solution is

$$\phi(z) = A_1 e^{-\frac{z}{\lambda_c}} + A_2 e^{\frac{z}{\lambda_c}} + r_o \lambda_c \int_0^z k_e(s) \sinh\left(\frac{s - z}{\lambda_c}\right) ds.$$

We are interested in understanding what the membrane voltage solution should look like in the event of a current injection at say $z = 0$ for a very short period of time. We could then use this idealized solution to understand the response of the cable model to current injections occurring over very brief time periods of various magnitudes at various spatial locations. Now let's specialize as follows: we assume

- The current injection k_e is symmetric about 0.
- The current $k_e(z)$ is zero on $(-\infty, -C) \cup (C, \infty)$ for some nonzero C.
- The current smoothly varies to zero at C and $-C$; i.e. k_e is at least differentiable on the z axis.
- The area under the curve, which is the current applied to the membrane, $\int_{-C}^{C} k_e(s)ds$ is I.

Given this type of current injection, we see we are looking for a solution to the problem

$$\lambda_c^2 \frac{d^2 v_m}{dz^2} - v_m = \begin{cases} 0 & z < -C \\ -r_o \lambda_c^2 \, k_e & -C \le z \le C \\ 0 & C < z \end{cases}$$

This amounts to solving three differential equations and then recognizing that the total solution is the sum of the three solutions with the condition that the full solution is smooth, i.e. continuous, at the points $-C$ and C where the solutions must connect.

Now we know that membrane voltages are finite. In the first and third region we seek a solution, we are simply solving the homogeneous equation. We know then the solution is of the form $Ae^{-w} + Be^{w}$ in both of these regions where $w(z) = z/\lambda_c$. However, in the $(-\infty, -C)$ region, the finiteness of the potential means that the Be^w solution is the only one possible and in the (C, ∞) region, the only solution is therefore of the form Ae^{-w}. In the middle region, the solution is given by the general solution we found from the Variation of Parameters method. Thus, for each choice of positive C, we seek a solution of the form

$$\phi_C(z) = \begin{cases} Be^{w(z)} & z < -C \\ A_1 e^{-w(z)} + A_2 e^{w(z)} & \\ +\frac{r_0\lambda_c}{2} e^{-w(z)} \int_0^z k_e(s) e^{\frac{s}{\lambda_c}} ds - \frac{r_0\lambda_c}{2} e^{w(z)} \int_0^z k_e(s) e^{-\frac{s}{\lambda_c}} ds & -C \le z \le C \\ Ae^{-w(z)} & C < z \end{cases}$$

Our solution and its derivative should be continuous at $-C$ and C. Between $-C$ and C, we compute the derivative of the solution to be

$$\phi'_C(z) = -\frac{A_1}{\lambda_c} e^{-w(z)} + \frac{A_2}{\lambda_c} e^{w(z)}$$
$$+ \frac{r_0\,\lambda_c}{2} k_e(z) e^{-w(z)} e^{w(z)} - \frac{r_0}{2} e^{-w(z)} \int_0^z k_e(s) e^{\frac{s}{\lambda_c}} ds$$
$$- \frac{r_0\,\lambda_c}{2} k_e(z) e^{-w(z)} e^{w(z)} - \frac{r_0}{2} e^{w(z)} \int_0^z k_e(s) e^{-\frac{s}{\lambda_c}} ds$$

This simplifies to

$$\phi'_C(z) = -\frac{A_1}{\lambda_c} e^{-w(z)} + \frac{A_2}{\lambda_c} e^{w(z)}$$
$$- \frac{r_0}{2} e^{-w(z)} \int_0^z k_e(s) e^{w(s)} ds$$
$$- \frac{r_0}{2} e^{w(z)} \int_0^z k_e(s) e^{-w(s)} ds \qquad (7.3)$$

7.3.1   Continuity in the Solution

Now, let $\omega = \frac{C}{\lambda_c}$. Continuity in the solution at the point $-C$ gives:

$$Be^{-\omega} = A_1 e^{\omega} + A_2 e^{-\omega} + \frac{r_0\lambda_c}{2} e^{\omega} \int_0^{-C} k_e(s) e^{w(s)} ds - \frac{r_0\lambda_c}{2} e^{-\omega} \int_0^{-C} k_e(s) e^{-w(s)} ds$$

This can be rewritten as

$$Be^{-\omega} = A_1 e^{\omega} + A_2 e^{-\omega} - \frac{r_0 \lambda_c}{2} e^{\omega} \int_{-C}^{0} k_e(s) e^{ws)} ds + \frac{r_0 \lambda_c}{2} e^{-\omega} \int_{-C}^{0} k_e(s) e^{-w(s)} ds$$

Then, the continuity condition at C gives

$$Ae^{-\omega} = A_1 e^{-\omega} + A_2 e^{\omega} + \frac{r_0 \lambda_c}{2} e^{-\omega} \int_{0}^{C} k_e(s) e^{w(s)} ds - \frac{r_0 \lambda_c}{2} e^{\omega} \int_{0}^{C} k_e(s) e^{-w(s)} ds.$$

7.3.2 Continuity in the Derivative

Continuity in the derivative at $-C$ gives:

$$\frac{B}{\lambda_c} e^{-\omega} = -\frac{A_1}{\lambda_c} e^{\omega} + \frac{A_2}{\lambda_c} e^{-\omega} - \frac{r_0}{2} e^{\omega} \int_{0}^{-C} k_e(s) e^{w(s)} ds - \frac{r_0}{2} e^{-\omega} \int_{0}^{-C} k_e(s) e^{-w(s)} ds$$

This simplifies to

$$\frac{B}{\lambda_c} e^{-\omega} = -\frac{A_1}{\lambda_c} e^{\omega} + \frac{A_2}{\lambda_c} e^{-\omega} + \frac{r_0}{2} e^{\omega} \int_{-C}^{0} k_e(s) e^{w(s)} ds + \frac{r_0}{2} e^{-\omega} \int_{-C}^{0} k_e(s) e^{-w(s)} ds$$

Finally, the continuity condition in the derivative at the point C gives

$$-\frac{A}{\lambda_c} e^{-\omega} = -\frac{A_1}{\lambda_c} e^{-\omega} + \frac{A_2}{\lambda_c} e^{\omega} - \frac{r_0}{2} e^{-\omega} \int_{0}^{C} k_e(s) e^{w(s)} ds - \frac{r_0}{2} e^{\omega} \int_{0}^{C} k_e(s) e^{-w(s)} ds$$

7.3.3 Forcing Continuity at the Boundaries

To simplify the exposition, define

$$J_1^+ = \int_{0}^{C} k_e(s) e^{w(s)} ds$$

$$J_1^- = \int_{-C}^{0} k_e(s) e^{w(s)} ds$$

$$J_2^+ = \int_{0}^{C} k_e(s) e^{-w(s)} ds$$

$$J_2^- = \int_{-C}^{0} k_e(s) e^{-w(s)} ds$$

and $M = \frac{r_0 \lambda_c}{2}$. We can then rewrite our continuity conditions as

$$Be^{-\omega} = A_1 e^{\omega} + A_2 e^{-\omega} - \frac{r_0 \lambda_c}{2} e^{\omega} \int_{-C}^{0} k_e(s) e^{w(s)} ds + \frac{r_0 \lambda_c}{2} e^{-\omega} \int_{-C}^{0} k_e(s) e^{-w(s)} ds$$

$$Ae^{-\omega} = A_1 e^{-\omega} + A_2 e^{\omega} + \frac{r_0 \lambda_c}{2} e^{-\omega} \int_{0}^{C} k_e(s) e^{w(s)} ds - \frac{r_0 \lambda_c}{2} e^{\omega} \int_{0}^{C} k_e(s) e^{-w(s)} ds$$

$$\frac{B}{\lambda_c} e^{-\omega} = -\frac{A_1}{\lambda_c} e^{\omega} + \frac{A_2}{\lambda_c} e^{-\omega} + \frac{r_0}{2} e^{\omega} \int_{-C}^{0} k_e(s) e^{w(s)} ds + \frac{r_0}{2} e^{-\omega} \int_{-C}^{0} k_e(s) e^{-w(s)} ds$$

$$-\frac{A}{\lambda_c} e^{-\omega} = -\frac{A_1}{\lambda_c} e^{-\omega} + \frac{A_2}{\lambda_c} e^{\omega} - \frac{r_0}{2} e^{-\omega} \int_{0}^{C} k_e(s) e^{w(s)} ds - \frac{r_0}{2} e^{\omega} \int_{0}^{C} k_e(s) e^{-w(s)} ds$$

With a little manipulation, we then have

$$Be^{-\omega} = A_1 e^{\omega} + A_2 e^{-\omega} - M e^{\omega} J_1^- + M e^{-\omega} J_2^-$$
$$Ae^{-\omega} = A_1 e^{-\omega} + A_2 e^{\omega} + M e^{-\omega} J_1^+ - M e^{\omega} J_2^+$$
$$\frac{B}{\lambda_c} e^{-\omega} = -\frac{A_1}{\lambda_c} e^{\omega} + \frac{A_2}{\lambda_c} e^{-\omega} + \frac{r_0}{2} e^{\omega} J_1^- + \frac{r_0}{2} e^{-\omega} J_2^-$$
$$-\frac{A}{\lambda_c} e^{-\omega} = -\frac{A_1}{\lambda_c} e^{-\omega} + \frac{A_2}{\lambda_c} e^{\omega} - \frac{r_0}{2} e^{-\omega} J_1^+ - \frac{r_0}{2} e^{\omega} J_2^+$$

Multiplying the third and fourth equation by λ_c, we obtain our final form:

$$Be^{-\omega} = A_1 e^{\omega} + A_2 e^{-\omega} - M e^{\omega} J_1^- + M e^{-\omega} J_2^-$$
$$Ae^{-\omega} = A_1 e^{-\omega} + A_2 e^{\omega} + M e^{-\omega} J_1^+ - M e^{\omega} J_2^+$$
$$Be^{-\omega} = -A_1 e^{\omega} + A_2 e^{-\omega} + M e^{\omega} J_1^- + M e^{-\omega} J_2^-$$
$$-Ae^{-\omega} = -A_1 e^{-\omega} + A_2 e^{\omega} - M e^{-\omega} J_1^+ - M e^{\omega} J_2^+$$

This gives us the equations:

$$(B - A_2 - M J_2^-)e^{-w} + (-A_1 + M J_1^-)e^{w} = 0 \tag{7.4}$$
$$(A - A_1 - M J_1^+)e^{-w} + (-A_2 + M J_2^+)e^{w} = 0 \tag{7.5}$$
$$(B - A_2 - M J_2^-)e^{-w} + (A_1 - M J_1^-)e^{w} = 0 \tag{7.6}$$
$$(-A + A_1 + M J_1^+)e^{-w} + (-A_2 + M J_2^+)e^{w} = 0 \tag{7.7}$$

Solving the four new Eqs. $(7.4) - (7.6)$, $(7.4) + (7.6)$, $(7.5) - (7.7)$ and $(7.5) + (7.7)$, we find

$$(-2A_1 + 2M J_1^-)e^{w} = 0$$
$$(2B - 2A_2 - 2M J_2^-)e^{-w} = 0$$
$$(2A - 2A_1 - 2M J_1^+)e^{-w} = 0$$
$$(-2A_2 + 2M J_2^+)e^{w} = 0$$

Since the exponentials can never be zero, we have

$$-2A_1 + 2M J_1^- = 0$$
$$2B - 2A_2 - 2M J_2^- = 0$$
$$2A - 2A_1 - 2M J_1^+ = 0$$
$$-2A_2 + 2M J_2^+ = 0$$

Hence, the solution we seek is

$$A_1 = M J_1^-$$
$$A_2 = M J_2^+$$
$$A = A_1 + M J_1^+ = M J_1^- + M J_1^+ = M(J_1^+ + J_1^-)$$
$$B = A_2 + M J_2^- = M J_2^+ + M J_2^- = M(J_2^+ + J_2^-)$$

Next note that

$$J_1^- + J_1^+ = \int_{-C}^{C} k_e(s)\, e^{w(s)}\, ds$$
$$J_2^- + J_2^+ = \int_{-C}^{C} k_e(s)\, e^{-w(s)}\, ds$$

Then the solution to this sort of current injection is thus

$$\phi_C(z) = \begin{cases} M e^{w} \int_{-C}^{C} k_e(s)\, e^{-w(s)}\, ds & z < -C \\ M(J_1^- e^{-w} + J_2^+ e^{w}) + 2M \int_{0}^{z} k_e(s) \sinh\left(\frac{s-z}{\lambda_c}\right) ds & -C \leq z \leq C \\ M e^{-w} \int_{-C}^{C} k_e(s)\, e^{w(s)}\, ds & C < z. \end{cases}$$

7.4 Modeling Instantaneous Current Injections

We can perform the analysis we did in the previous section for any pulse of that form. What happens as the parameter C approaches 0? As before, we look at pulses which have these properties for each positive C:

- The current injection k_e^C is symmetric about 0.
- The current $k_e^C(z)$ is zero on $(-\infty, -C) \cup (C, \infty)$
- The current smoothly varies to zero at $\pm C$ so that k_e^C is differentiable on the z axis.
- The area under the curve, which is the current applied to the membrane, $\int_{-C}^{C} k_e^C(s) ds$ is I for all C. This means that as the width of the symmetric pulse goes to zero, the height of the pulse goes to infinity but in a controlled way: the

area under the pulses is always the same number I. So no matter what the base of the pulse, the same amount of current is delivered to the membrane.

The solution is

$$\phi_C(z) = \begin{cases} M e^w \int_{-C}^{C} k_e^C(s) \, e^{-w(s)} \, ds & z < -C \\ M(J_{1C}^- e^{-w} + J_{2C}^+ e^w) + 2M \int_0^z k_e^C(s) \sinh\left(\frac{s-z}{\lambda_c}\right) ds & -C \le z \le C \\ M e^{-w} \int_{-C}^{C} k_e^C(s) \, e^{w(s)} \, ds & C < z. \end{cases}$$

where we now use

$$J_{1C}^+ = \int_0^C k_e^C(s) \, e^{w(s)} \, ds$$

$$J_{1C}^- = \int_{-C}^0 k_e^C(s) \, e^{w(s)} \, ds$$

$$J_{2C}^+ = \int_0^C k_e^C(s) \, e^{-w(s)} \, ds$$

$$J_{2C}^- = \int_{-C}^0 k_e^C(s) \, e^{-w(s)} \, ds$$

Let's look at the limit as C goes to 0.

7.4.1 What Happens Away from 0?

Pick a point z_0 that is not zero. Since z_0 is not zero, there is a value of positive number C^* so that z_0 is not inside the interval $(-C^*, C^*)$. This means that z_0 is outside of $(-C, C)$ for all C smaller that C^*. Hence, either the first part or the third part of the definition of ϕ_C applies. We will argue the case for z_0 is positive which implies that the third part of the definition is the one that is applicable. The case where z_0 is negative would be handled in a similar fashion. Since we assume that z_0 is positive, we have for all C smaller than C^*:

$$\phi_C(z_0) = M e^{-w_0} \int_{-C}^{C} k_e^C(s) \, e^{w(s)} \, ds$$

where $w_0 = \frac{z_0}{\lambda_c}$. Next, we can prove as C goes to 0, we obtain

Lemma 7.4.1 (Limiting Current Densities)

$$\lim_{C \to 0} J_{1C}^+ = \frac{I}{2}$$

$$\lim_{C \to 0} J_{1C}^{-} = -\frac{I}{2}$$

$$\lim_{C \to 0} J_{2C}^{+} = \frac{I}{2}$$

$$\lim_{C \to 0} J_{2C}^{-} = -\frac{I}{2}$$

Proof It suffices to show this for just one of these four cases. We will show the first one. In mathematics, if the limit as C goes to 0 of J_{1C}^{+} exists and equals $\frac{I}{2}$, this means that given any positive number ϵ, we can find a C^{**} so that if $C < C^{**}$, then

$$\left| J_{1C}^{+} - \frac{I}{2} \right| < \epsilon$$

whenever $C < C^{**}$. We will show how to do this argument below. Consider

$$\left| J_{1C}^{+} - \frac{I}{2} \right| = \left| \int_{0}^{C} k_{e}^{C}(s) \, e^{\frac{s}{\lambda_C}} \, ds - \frac{I}{2} \right|$$

However, we know that the area under the curve is always I; hence for any C,

$$I = \int_{-C}^{C} k_{e}^{C}(s) ds$$

Since our pulse is symmetric, this implies that

$$\frac{I}{2} = \int_{0}^{C} k_{e}^{C}(s) ds$$

Substituting into our original expression, we find

$$\left| J_{1C}^{+} - \frac{I}{2} \right| = \left| \int_{0}^{C} k_{e}^{C}(s) \, e^{-\frac{s}{\lambda_C}} \, ds - \int_{0}^{C} k_{e}^{C}(s) ds \right|$$

$$\leq \int_{0}^{C} k_{e}^{C}(s) \, |e^{-\frac{s}{\lambda_C}} - 1| ds$$

Now we know that this exponential function is continuous at 0; hence, given $\frac{2\epsilon}{I}$, there is a δ so that

$$|e^{-\frac{s}{\lambda_C}} - 1| < \frac{2\epsilon}{I} \quad \text{if } |s| < \delta$$

Since, C goes to zero, we see that there is a positive number C^{**}, which we can choose smaller than our original C^{*}, so that $C^{**} < \delta$. Thus, if we integrate over

[0, C] for any $C < C^{**}$, all the s values inside the integral are less than δ. So we can conclude

$$|J_{1C}^+ - \frac{I}{2}| < \left(\int_0^C k_e^C(s)\, ds \right) \frac{2\epsilon}{I} = \epsilon.$$

This shows that the limit as C goes to 0 of J_{1C}^+ is $\frac{I}{2}$.   ∎

Hence, for z_0 positive, we find

$$\lim_{C \to 0} \phi_C(z_0) = Me^{-w_0} \lim_{C \to 0} \int_{-C}^C k_e^C(s)\, e^{\omega(s)}\, ds$$

$$= Me^{-w_0} \lim_{C \to 0} \left(J_{1C}^- + J_{1C}^+ \right)$$

$$= Me^{-w_0} I = \frac{r_0 \lambda_c I}{2} e^{\frac{-z_0}{\lambda_c}}.$$

In a similar fashion, for z_0 negative, we find

$$\lim_{C \to 0} \phi_C(z_0) = \frac{r_0 \lambda_c I}{2} e^{\frac{z_0}{\lambda_c}}.$$

7.4.2   What Happens at Zero?

When we are at zero, since 0 is in $(-C, C)$ for all C, we must use the middle part of the definition of ϕ_C always. Now at 0, we see

$$\phi_C(0) = M(J_{1C}^- e^{-0} + J_{2C}^+ e^0) + 2M \int_0^0 k_e^C(s) \sinh\left(\frac{s-0}{\lambda_c} \right) ds$$

$$= M(J_{1C}^- + J_{2C}^+)$$

From Lemma 7.4.1, we then find that

$$\lim_{C \to 0} \phi_C(0) = MI = \frac{r_0 \lambda_c I}{2}.$$

Combining all of the above parts, we see we have shown that as C goes to 0, the solutions ϕ_C converge at each point z (mathematicians say this sequence of functions converges pointwise on $(-\infty, \infty)$) to the limiting solution ϕ defined by

$$\phi(z) = \frac{r_0 \lambda_c}{2} I\, e^{-|\frac{z}{\lambda_c}|}$$

Now if the symmetric pulse sequence was centered at z_0 with pulse width C, we can do a similar analysis (it is yucky though and tedious!) to show that the limiting solution would be

$$\phi(z) = \frac{r_o \lambda_c}{2} I \, e^{-\frac{|z - z_0|}{\lambda_c}}.$$

7.5 Idealized Impulse Currents

Now as C gets small, our symmetric pulses have very narrow base but very large magnitudes. Clearly, these pulses are not defined in the limit as C gocs to 0 as the limit process leads to a current $I_\delta(z)$ which is the limit of the base pulses we have described:

$$I_\delta(z) = \lim_{C \to 0} k_e^C(z).$$

Note, in the limit as C gets small, we can think of this process as delivering an instantaneous current value. This **idealized** current thus delivers a fixed amount I at $z = 0$ instantly! Effectively, this means that the value of the current density keeps getting larger while its spatial window gets smaller. A good example (though not quite right as this example is not continuous at $\pm 1/n$) is a current density of size nI which on the spatial window $[-1/n, 1/n]$. The current for any n that is applied is the product of the current density and the distance it is applied: $nI \times 1/n = I$. From a certain point of view, we can view the limiting current density as depositing an "infinite" amount of current in zero time: crazy, but easy to remember! Hence, we can use the definition below as a short hand reminder of what is going on:

$$I_\delta(z) = \begin{cases} 0 & z \neq 0 \\ \infty & z = 0 \end{cases}$$

In fact, since the total current deposited over the whole cable is I, and integration can be interpreted as giving the area under a curve, we can use the integration symbol $\int_{-\infty}^{\infty}$ to denote that the total current applied over the cable is always I and write

$$\int_{-\infty}^{\infty} I_\delta(z) dz = I.$$

Again, this is simply a mnemonic to help us remember complicated limiting current density behavior! Again, think of I_δ as an amount of current I which is delivered instantaneously at z is 0. Of course, the only way to really understand this idea is to do what we have done and consider a sequence of constant current density pulses k_e^C.

Summarizing, it is convenient to think of a unit idealized impulse called δ functions defined by

$$\delta(z - 0) = \begin{cases} 0 & z \neq 0 \\ \infty & z = 0 \end{cases}$$

$$\int_{-\infty}^{\infty} \delta(u - 0)du = 1$$

Then using this notation, we see our idealized current I_δ can be written as

$$I_\delta(z - 0) = I\,\delta(z - 0).$$

Moreover, if an idealized pulse is applied at z_0 rather than 0, we can abuse this notation to define

$$\delta(z - z_0) = \begin{cases} 0 & z \neq z_0 \\ \infty & z = z_0 \end{cases}$$

$$\int_{-\infty}^{\infty} \delta(u - z_0)du = 1$$

and hence $I\,\delta(z - z_0)$ is a idealized pulse applied at z_0. Note that the notion of an idealized pulse allows us to write the infinite cable model for an idealized pulse applied at z_0 of magnitude I in a compact form. We let the applied current density be $k_e(z) = I\,\delta(z - z_0)$ giving us the differential equation

$$\lambda_c^2 \frac{d^2 v_m}{dz^2} - v_m(z) = -r_o \lambda_c^2 I \delta(z - z_0)$$

which we know has solution

$$\phi(z) = \frac{r_0 \lambda_c}{2} I\, e^{-\left|\frac{z - z_0}{\lambda_c}\right|}$$

In all of our analysis, we use a linear ordinary differential equation as our model. Hence, the linear superposition principle applies and hence for N idealized pulses applied at differing centers and with different magnitudes, the underlying differential equation is

$$\lambda_c^2 \frac{d^2 v_m}{dz^2} - v_m(z) = -r_o \lambda_c^2 \sum_{i=0}^{N} I_i\, \delta(z - z_i)$$

with solution

$$\phi(z) = \frac{r_0 \lambda_c}{2} \sum_{i=0}^{N} I_i \, e^{-|\frac{z-z_i}{\lambda_c}|}$$

Note that the membrane voltage solution is a nonlinear summation of the applied idealized current injections. For example, let's inject current of magnitudes I at 0 and $2I$ at λ_c respectively. The solution to the current injection at 0 is v_m^1 and at λ_c, v_m^2. Thus,

$$v_m^1(z) = \frac{r_0 \lambda_c}{2} I \, e^{-\frac{|z|}{\lambda_c}}$$

$$v_m^2(z) = \frac{r_0 \lambda_c}{2} 2I \, e^{-\frac{|z-\lambda_c|}{\lambda_c}}$$

We get a lot of information by looking at these two solutions in terms of units of the space constant λ_c. If you look at Fig. 7.2, you'll see the two membrane voltages plotted together on the same access. Now the actual membrane voltage is, of course, v_m^1 added to v_m^2 which we don't show. However, you can clearly see how quickly the peak voltages fall off as you move away from the injection site.

Note that one space constant away from an injection site z_0, the voltage falls by a factor of $\frac{1}{e}$ or 37 %. This is familiar exponential decay behavior.

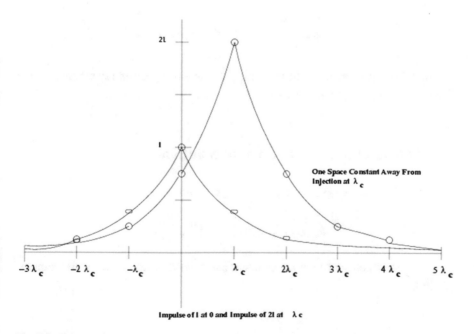

Fig. 7.2   Voltages due to two current impulses

7.5.1 *Integrals of Impulse Functions*

We often need to evaluate integrals of the form $\int_a^b \delta(x - x_0)g(x)dx$ where x_0 is in the interval $[a, b]$. Let k_e^C be our usual set of pulses which we use to approximate the impulse function except this time we center the pulses at x_0. Then, we have

$$\int_a^b k_e^C(x)g(x)dx = \int_{C-x_0}^{C+x_0} k_e^C g(x)dx.$$

Now as long as g is continuous on $[a, b]$, g has a minimum and maximum value on the interval $[C - x_0, C + x_0]$; call these g_m^C and g_M^C. Then we have the inequalities

$$g_m^C \int_{C-x_0}^{C+x_0} k_e^C(x)dx \le \int_{C-x_0}^{C+x_0} k_e^C(x)g(x)dx \le g_M^C \int_{C-x_0}^{C+x_0} k_e^C(x)dx.$$

But the $\int_{C-x_0}^{C+x_0} k_e^C(x)dx = 1$ so we have

$$g_m^C \le \int_{C-x_0}^{C+x_0} k_e^C(x)g(x)dx \le g_M^C.$$

Finally, as $C \to 0$, $g_m^C \to g(x_0)$ and $g_M^C \to g(x_0)$ because g is continuous on $[a, b]$. We conclude

$$\lim_{C \to 0} \int_a^b k_e^C(x)g(x)dx = g(x_0).$$

This limit is what we mean by the integral $\int_a^b \delta(x - x_0)g(x)dx$ and so we have shown that $\int_a^b \delta(x - x_0)g(x)dx = g(x_0)$.

7.6 The Inner and Outer Current Solutions

Recall that the membrane current density k_m satisfies

$$k_m(z, t) = g_m v_m(z, t) + c_m \frac{\partial v_m}{\partial t}$$

Here, we are interested in time independent solutions, so we have the simpler equation

$$k_m(z) = g_m v_m(z)$$

Thus, we have for an idealized current impulse injection at 0

$$k_m(z) = g_m \frac{r_o \lambda_c}{2} I\, e^{-\frac{|z|}{\lambda_c}}$$

From our core conductor equations, we know that

$$\frac{\partial I_i}{\partial z} = -K_m(z)$$

and thus, using our transient variables

$$I_o(z) = I_o^0 + i_0(z), \quad I_i(z) = I_i^0 + i_i(z), \quad K_m(z) = K_m^0 + k_m(z)$$

we see that

$$\frac{\partial i_i}{\partial z} = -K_m^0 - k_m(z)$$

Integrating, we see

$$i_i(z) = -\int_{-\infty}^{z} (K_m^0 + k_m(u))du$$

We expect that the internal current is finite and this implies that the initial current density K_m^0 must be zero as otherwise we obtain unbounded current. We thus have

$$\frac{\partial i_i}{\partial z} = -k_m(z) \tag{7.8}$$

and so

$$i_i(z) - i_i(-\infty) = -\int_{-\infty}^{z} k_m(u)du = -\int_{-\infty}^{z} g_m v_m(u)du$$

$$= -\frac{r_o \lambda_c}{2} g_m I \begin{cases} \int_{-\infty}^{z} e^{\frac{u}{\lambda_c}} & z < 0 \\ \int_{-\infty}^{0} e^{\frac{u}{\lambda_c}}\,du + \int_{0}^{z} e^{-\frac{u}{\lambda_c}}\,du & z \geq 0 \end{cases}$$

Also, for an impulse current applied at zero, we would expect that the inner current vanish at both ends of the infinite cable. Hence, $i_i(-\infty)$ must be zero. We conclude

$$i_i(z) = -\frac{r_o \lambda_c}{2} g_m I \begin{cases} \lambda_c e^{\frac{u}{\lambda_c}}\Big|_{-\infty}^{z} = \lambda_c e^{\frac{z}{\lambda_c}} & z < 0 \\ \lambda_c e^{\frac{u}{\lambda_c}}\Big|_{-\infty}^{0} - \lambda_c e^{-\frac{u}{\lambda_c}}\Big|_{0}^{z} = \lambda_c(2 - e^{-\frac{z}{\lambda_c}}) & z \geq 0 \end{cases}$$

Using the definition of the space constant λ_c, we note the identity

$$\lambda_c^2 \, g_m = \frac{r_o}{r_i + r_o}$$

allowing us to rewrite the inner current solution as

$$i_i(z) = -\frac{r_o}{r_i + r_o} \frac{I}{2} \begin{cases} e^{\frac{z}{\lambda_c}} & z < 0 \\ (2 - e^{-\frac{z}{\lambda_c}}) & z \geq 0 \end{cases}$$

Now, from Sect. 6.5, Eqs. 6.1 and 6.2, we know in our time independent case

$$\frac{\partial I_i}{\partial z} = -K_m(z), \quad \frac{\partial I_o}{\partial z} = K_m(z) - K_e(z)$$

which implies

$$\frac{\partial i_i}{\partial z} + \frac{\partial i_o}{\partial z} = -K_e(z) = -K_e^0 - k_e(z)$$

Integrating, we have

$$\int_{-\infty}^{z} \frac{\partial i_i}{\partial z} du + \int_{-\infty}^{z} \frac{\partial i_o}{\partial z} du = -\int_{-\infty}^{z} (K_e^0 + k_e(u)) du$$

In order for this integration to give us finite currents, we see the constant K_e^0 must be zero implying

$$i_i(z) - i_i(-\infty) + i_o(z) - i_o(-\infty) = -\int_{-\infty}^{z} k_e(u) du$$

We already know that i_i is zero at $z = -\infty$ for our idealized current impulse $k_e = I\delta(u)$. Further, we know from our lengthy analysis of sequences of current pulses of constant area I, k_e^n, that integrating from $-\infty$ to $z > 0$ gives I and 0 if $z \leq 0$. Hence, the inner and outer transient currents for an idealized pulse must satisfy

$$i_i(z) + i_o(z) - i_o(-\infty) = \begin{cases} -I & z > 0 \\ 0 & z \leq 0 \end{cases}$$

The argument to see this can be sketched as follows. If z is positive, for small enough C, the impulse current k_e^C is active on the interval $[-C, C]$ with C less than z. For such values of C, the integral becomes

$$\int_{-C}^{C} k_e^C(u) du = I$$

On the other hand, if z is negative, eventually the interval where k_e^C is non zero lies outside the region on integration and so we get the value of the integral must be zero. Physically, since we are using an idealized injected current, we expect that the outer current satisfies $i_o(-\infty)$ is zero giving

$$i_i(z) + i_o(z) = \begin{cases} -I & z > 0 \\ 0 & z \leq 0 \end{cases}$$

We can rewrite our current solutions more compactly by defining the signum function sgn and the unit step function u as follows:

$$sgn(z) = \begin{cases} -1 & z < 0 \\ +1 & z \geq 0 \end{cases}, \quad u(z) = \begin{cases} 0 & z < 0 \\ 1 & z \geq 0 \end{cases}$$

Then we have

$$i_i(z) = \frac{r_o I}{2(r_i + r_o)} \left(e^{-\frac{|z|}{\lambda_c}} sgn(z) - 2u(z) \right)$$

Next, we can solve for i_o to obtain

$$\begin{aligned} i_o(z) &= -Iu(z) - i_i(z) \\ &= -Iu(z) - \frac{r_o I}{2(r_i + r_o)} \left(e^{-\frac{|z|}{\lambda_c}} sgn(z) - 2u(z) \right) \\ &= -u(z) \left[1 - \frac{r_o}{2(r_i + r_o)} \right] I - \frac{r_o I}{2(r_i + r_o)} e^{-\frac{|z|}{\lambda_c}} sgn(z) \\ &= -u(z) \frac{r_i}{r_i + r_o} I - \frac{r_o I}{2(r_i + r_o)} e^{-\frac{|z|}{\lambda_c}} sgn(z) \end{aligned}$$

This further simplifies to the final forms

$$\begin{aligned} i_o(z) &= -\frac{r_o I}{2(r_i + r_o)} \left[e^{-\frac{|z|}{\lambda_c}} sgn(z) - 2\frac{r_i}{r_o} u(z) \right] \\ &= -\frac{r_o \lambda_c^2 g_m I}{2} \left[e^{-\frac{|z|}{\lambda_c}} sgn(z) - 2\frac{r_i}{r_o} u(z) \right] \end{aligned}$$

7.7 The Inner and Outer Voltage Solutions

From Sect. 6.5, Eqs. (6.3) and (6.4), we see that for our time independent case

$$\frac{\partial V_i}{\partial z} = -r_i I_i(z), \quad \frac{\partial V_o}{\partial z} = -r_o I_o(z)$$

Rewriting in terms of transient variables, we have

$$\frac{\partial v_i}{\partial z} = -r_i[I_i^0 + i_i(z)], \quad \frac{\partial v_o}{\partial z} = -r_o[I_o^0 + i_0(z)]$$

We expect our voltages to remain bounded and so we must conclude that I_i^0 and I_o^0 are zero, giving

$$v_i(z) - v_i(-\infty) = -r_i \int_{-\infty}^{z} i_i(u)du$$

$$v_o(z) - v_o(-\infty) = -r_o \int_{-\infty}^{z} i_o(u)du$$

To finish this step of our work, we must perform these messy integrations. They are not hard, but are messy! After a bit of arithmetic, we find

$$v_i(z) - v_i(-\infty) = \frac{r_i r_o I \lambda_c}{2(r_i + r_o)} \left[e^{-\frac{|z|}{\lambda_c}} + 2\frac{z}{\lambda_c} u(z) \right]$$

$$v_o(z) - v_o(-\infty) = \frac{r_o^2 I \lambda_c}{2(r_i + r_o)} \left[-e^{-\frac{|z|}{\lambda_c}} + 2\frac{r_i}{r_o}\frac{z}{\lambda_c} u(z) \right]$$

Finally, note that after a bit of algebraic magic

$$v_i(z) - v_o(z) = v_i(-\infty) - v_o(-\infty) + \frac{r_o I \lambda_c}{2} e^{-\frac{|z|}{\lambda_c}}$$

Recall that v_m is precisely the last term in the equation above; hence we have

$$v_i(z) - v_o(z) = v_i(-\infty) - v_o(-\infty) + v_m(z)$$

The usual convention is that the voltages at infinity vanish; hence $v_i(-\infty)$ and $v_o(-\infty)$ are zero and we have the membrane voltage solution we expect:

$$v_i(z) - v_o(z) = v_m(z)$$

7.8 Summarizing the Infinite Cable Solutions

We have shown that the solutions here are

$$v_m(z) = \frac{r_o \lambda_c I}{2} e^{-\frac{|z|}{\lambda_c}}$$

$$i_i(z) = \frac{r_o \lambda_c^2 g_m I}{2} (e^{\frac{-|z|}{\lambda_c}} sgn(z) - 2u(z))$$

$$i_o(z) = -\frac{r_o \lambda_c^2 g_m I}{2}(e^{\frac{-|z|}{\lambda_c}} sgn(z) + 2\frac{r_i}{r_o} u(z))$$

$$v_i(z) = -\frac{r_i r_o \lambda_c^3 g_m I}{2}(e^{\frac{-|z|}{\lambda_c}} + 2\frac{z}{\lambda_c} u(z))$$

$$v_o(z) = -\frac{r_o^2 \lambda_c^3 g_m I}{2}(- e^{\frac{-|z|}{\lambda_c}} + 2\frac{r_i}{r_o}\frac{z}{\lambda_c} u(z))$$

$$v_m(z) = v_i(z) - v_o(z)$$

where we assume v_i and v_o are zero at negative infinity. Since λ_c^2 can be expressed in terms of r_i and r_o, we have also shown that

$$v_m(z) = \frac{r_o \lambda_c I}{2} e^{-\frac{|z|}{\lambda_c}}$$

$$i_i(z) = \frac{r_o I}{2(r_i + r_o)}(e^{\frac{-|z|}{\lambda_c}} sgn(z) - 2u(z))$$

$$i_o(z) = -\frac{r_o I}{2(r_i + r_o)}(e^{\frac{-|z|}{\lambda_c}} sgn(z) + 2\frac{r_i}{r_o} u(z))$$

$$v_i(z) = -\frac{r_i r_o I}{2(r_i + r_o)}(e^{\frac{|z|}{\lambda_c}} + 2\frac{z}{\lambda_c} u(z))$$

$$v_o(z) = -\frac{r_o^2 I}{2(r_i + r_o)}(- e^{\frac{-|z|}{\lambda_c}} + 2\frac{r_i}{r_o}\frac{z}{\lambda_c} u(z))$$

7.8.1 Homework

Exercise 7.8.1 *Write Matlab functions to implement all the infinite cable transient variable solutions using as many arguments to the functions as are needed.*

1. v_m
2. i_i
3. i_o
4. v_i
5. v_o.

Exercise 7.8.2 *Generate a parametric plot of each of these variables versus the space variable z on a reasonable size range of z for the parameters λ_c and I_e^*.*

1. v_m
2. i_i
3. i_o
4. v_i
5. v_o.

Chapter 8
Time Independent Solutions to Finite and Half-Infinite Space Cables

We are actually interested in a model of information processing that includes a dendrite, a cell body and an axon. Now we know that the cables that make up dendrites and axons are not infinite in extent. So although we understand the currents and voltages change in our infinite cable model, we still need to figure out how these solutions change when the cable in only finite in extent. A first step in this direction is to consider a half-infinite cable such as shown in Fig. 8.1 and then a finite length cable like in Fig. 8.2. In both figures, we think of a real biological dendrite or axon as an inner cable surrounded by a thin cylindrical sheath of seawater. So the outer resistance r_o will be the resistance of seawater. At first, we think of the cable as extending to the right forever; i.e. the soma is infinitely far from the front end cap of the cable. This is of course not realistic, but we can use this thought experiment as a vehicle towards understanding how to handle a finite cable attached to a soma. Before we had an impulse current of magnitude I injected at some point on the outer cylinder and uniformly distributed around the outer cylinder wall. Now we inject current directly into the front face of our cables. In the finite length L cable case, there will also be a back end cap at $z = L$. This back endcap will be attached to the soma. Then, although we could have the membrane properties of the cable endcap and the soma itself be different, a reasonable assumption is to make them identical. Hence the back endcap of the cable is a portion of the cell soma. At any rate, in both situations, the front endcap of the cable is a logical place to think of current as being injected.

We will begin our modeling with a half-infinite cable. Once we know how to model the front endcap current injection in this case, we will move directly to the finite cable model.

8.1 The Half-Infinite Cable Solution

Earlier, we injected an idealized pulse of current into the inner cable at $z = 0$ but now we will inject into the current into the front face of the cable. Since the cable is of radius a, this means we are injecting current into a membrane cap of surface

© Springer Science+Business Media Singapore 2016
J.K. Peterson, *Calculus for Cognitive Scientists*, Cognitive Science and Technology, DOI 10.1007/978-981-287-880-9_8

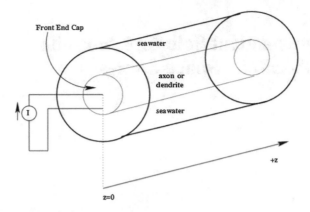

Fig. 8.1 A half-infinite dendrite or axon model

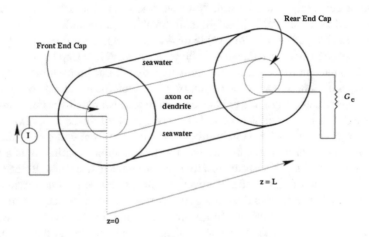

Fig. 8.2 Finite cable

area πa^2. Note there is no external current source here and hence k_e is zero. The underlying cable equation is

$$\lambda_c^2 \frac{d^2 v_m}{dz^2} - v_m = 0, \quad z \geq 0$$

with solution

$$v_m(z) = A\, e^{-\frac{z}{\lambda_c}}, \quad z \geq 0$$

for some value of A which we will determine in a moment. We know that the membrane current density is given by

$$k_m(z) = g_m v_m(z)$$

and further, from Eq. 7.8, the time independent transient inner current satisfies

$$\frac{di_i}{dz} = -k_m(z) = -g_m v_m(z)$$

Since the cable starts at $z = 0$ and moves off to infinity, let's integrate this differential equation from z to R for progressively large values of R and see what happens in the limit as $R \to \infty$. We find

$$\int_z^R \frac{di_i}{dz} dz = -g_m A \int_z^R e^{-\frac{s}{\lambda_c}} ds$$

$$i_i(R) - i_i(z) = g_m \lambda_c A \left(e^{-\frac{R}{\lambda_c}} - e^{-\frac{z}{\lambda_c}} \right)$$

Thus,

$$\lim_{R \to \infty} i_i(R) = i_i(z) - g_m \lambda_c A e^{-\frac{z}{\lambda_c}}$$

However, we would expect the inner current to become negligible as we move to the end of the cable. Hence, $\lim_{R \to \infty} i_i(R) = 0$ and we have

$$i_i(z) = g_m \lambda_c A e^{-\frac{z}{\lambda_c}}$$

We also know that current of magnitude I is being injected into the front face of the inner cable; hence $i_i(0)$ must be I. This tells us that

$$i_i(0) = g_m \lambda_c A = I$$

Combining, we have for positive z:

$$v_m(z) = \frac{I}{g_m \lambda_c} e^{-\frac{z}{\lambda_c}}, \quad i_i(z) = I e^{-\frac{z}{\lambda_c}}$$

From Ohm's Law, we know that the ratio of current to voltage is conductance. The work above suggests that we can define the conductance of the inner cable at $z = 0$ by the ratio $\frac{i_i(0)}{v_m(0)}$. Therefore the ratio of this current to voltage at $z = 0$, defines an idealized conductance for the end cap of the cable. This conductance is called the

Th'evenin Equivalent Conductance of the half-infinite cable. It is denoted by $\mathscr{G}_\infty$. The ratio is easy to compute

$$\mathscr{G}_\infty = \frac{i_i(0)}{v_m(0)} = g_m \, \lambda_c$$

We can show that $\mathscr{G}_\infty$ is dependent on the geometry of the cable; indeed, $\mathscr{G}_\infty$ is proportional to a to the three-halves power when r_i is much bigger than r_o. To see this, recall that the definition of λ_c gives us that

$$\mathscr{G}_\infty = \sqrt{\frac{g_m}{(r_i + r_o)}}$$

and so if r_i is much bigger than r_o, we find that

$$\mathscr{G}_\infty \approx \sqrt{\frac{g_m}{r_i}}$$

For a cylindrical cell of radius a, we know

$$r_i = \frac{\rho_i}{\pi a^2}, \quad g_m = 2\pi a G_m$$

and so

$$\mathscr{G}_\infty \approx \pi \sqrt{\frac{2G_m}{\rho_i}} \, a^{\frac{3}{2}}$$

which tells us that $\mathscr{G}_\infty$ is proportional to the $3/2$ power of a. Thus, larger fibers have larger characteristic conductances!

If a cable is many space constants long (remember the space constant is proportional to the square root of a, then the half-infinite model we discuss here may be appropriate. We will be able to show this is a reasonable thing to do after we handle the true finite cable model. Once that is done, we will see that the solutions there approach the half-infinite model solutions as L goes to infinity.

8.2   The Finite Cable Solution: Current Initialization

If the cable becomes a piece of length L as shown in Fig. 8.2, then there are now two faces to deal with; the input face through which a current pulse of size I is delivered into some conductance and an output face at z is L which has a output load conductance of $\mathscr{G}_e$. $\mathscr{G}_e$ represents either the conductance of the membrane that caps the cable or the conductance of another cable or cell soma attached at this point.

We again have no external source and so the cable equation is

$$\lambda_c^2 \frac{d^2 v_m}{dz^2} - v_m(z) = 0, \quad 0 \le z \le L$$

The general solution to this *homogeneous* equation has been discussed before. The solution we seek will also need two boundary conditions to be fully specified. The general form of the membrane potential solution is

$$v_m(z) = A_1 e^{\frac{z}{\lambda_c}} + A_2 e^{-\frac{z}{\lambda_c}}, \quad 0 \le z \le L$$

We will rewrite this in terms of the new functions hyperbolic sine and hyperbolic cosine defined as follows:

$$\cosh\left(\frac{z}{\lambda_c}\right) = \frac{e^{\frac{z}{\lambda_c}} + e^{-\frac{z}{\lambda_c}}}{2}, \quad \sinh\left(\frac{z}{\lambda_c}\right) = \frac{e^{\frac{z}{\lambda_c}} - e^{-\frac{z}{\lambda_c}}}{2}$$

leading to the new form of the homogeneous solution

$$v_m(z) = A_1 \cosh\left(\frac{z}{\lambda_c}\right) + A_2 \sinh\left(\frac{z}{\lambda_c}\right), \quad 0 \le z \le L$$

It is convenient to reorganize this yet again and rewrite in another equivalent form as

$$v_m(z) = A_1 \cosh\left(\frac{L - z}{\lambda_c}\right) + A_2 \sinh\left(\frac{L - z}{\lambda_c}\right), \quad 0 \le z \le L$$

Now the membrane current density satisfies

$$k_m(z) = g_m v_m(z) = g_m A_1 \cosh\left(\frac{L - z}{\lambda_c}\right) + g_m A_2 \sinh\left(\frac{L - z}{\lambda_c}\right)$$

Further, since

$$\frac{d}{dz} \cosh(z) = \sinh(z), \quad \frac{d}{dz} \sinh(z) = \cosh(z)$$

we can use the internal current equation to find

$$\frac{di_i}{dz} = -k_m(z) = -g_m A_1 \cosh\left(\frac{L - z}{\lambda_c}\right) - g_m A_2 \sinh\left(\frac{L - z}{\lambda_c}\right)$$

Using the simplest possible antiderivative (i.e. the integration constant is 0), upon integrating, we find an inner current solution to be

$$i_i(z) = g_m \lambda_c A_1 \sinh\left(\frac{L-z}{\lambda_c}\right) + g_m \lambda_c A_2 \cosh\left(\frac{L-z}{\lambda_c}\right)$$

Since the conductance at $z = L$ is $\mathscr{G}_e$, the current at L must be $i_i(L) = \mathscr{G}_e v_m(L)$ or

$$i_i(L) = \mathscr{G}_e \, v_m(L) = \mathscr{G}_e \left(A_1 \cosh\left(\frac{0}{\lambda_c}\right) + A_2 \sinh\left(\frac{0}{\lambda_c}\right)\right) = \mathscr{G}_e \, A_1.$$

Next, we have

$$i_i(0) = g_m \lambda_c A_1 \sinh\left(\frac{L}{\lambda_c}\right) + g_m \lambda_c A_2 \cosh\left(\frac{L}{\lambda_c}\right)$$

$$i_i(L) = g_m \lambda_c A_1 \sinh\left(\frac{0}{\lambda_c}\right) + g_m \lambda_c A_2 \cosh\left(\frac{0}{\lambda_c}\right) = A_2 \, g_m \lambda_c.$$

Combining, since $i_i(0) = I$, we have

$$I = g_m \lambda_c \left(A_1 \sinh\left(\frac{L}{\lambda_c}\right) + A_2 \cosh\left(\frac{L}{\lambda_c}\right)\right), \quad A_1 \mathscr{G}_e = A_2 \, g_m \lambda_c$$

This implies using the definition of $\mathscr{G}_\infty$

$$I = \mathscr{G}_\infty \left(A_1 \sinh\left(\frac{L}{\lambda_c}\right) + A_2 \cosh\left(\frac{L}{\lambda_c}\right)\right), \quad A_1 \mathscr{G}_e = A_2 \mathscr{G}_\infty$$

Thus,

$$I = A_1 \mathscr{G}_\infty \left(\sinh\left(\frac{L}{\lambda_c}\right) + \frac{\mathscr{G}_e}{\mathscr{G}_\infty} \cosh\left(\frac{L}{\lambda_c}\right)\right)$$

giving us

$$A_1 = \frac{I}{\mathscr{G}_\infty} \left(\frac{1}{\sinh\left(\frac{L}{\lambda_c}\right) + \frac{\mathscr{G}_e}{\mathscr{G}_\infty} \cosh\left(\frac{L}{\lambda_c}\right)}\right)$$

$$A_2 = \frac{I \mathscr{G}_e}{\mathscr{G}_\infty^2} \left(\frac{1}{\sinh\left(\frac{L}{\lambda_c}\right) + \frac{\mathscr{G}_e}{\mathscr{G}_\infty} \cosh\left(\frac{L}{\lambda_c}\right)}\right)$$

This leads to the solution we are looking for

$$v_m(z) = \frac{I}{\mathscr{G}_\infty} \left(\frac{\cosh\left(\frac{L-z}{\lambda_c}\right) + \frac{\mathscr{G}_e}{\mathscr{G}_\infty} \sinh\left(\frac{L-z}{\lambda_c}\right)}{\sinh\left(\frac{L}{\lambda_c}\right) + \frac{\mathscr{G}_e}{\mathscr{G}_\infty} \cosh\left(\frac{L}{\lambda_c}\right)} \right)$$

$$i_i(z) = \frac{I}{\mathscr{G}_\infty} \left(\frac{\mathscr{G}_\infty \sinh\left(\frac{L-z}{\lambda_c}\right) + \mathscr{G}_e \cosh\left(\frac{L-z}{\lambda_c}\right)}{\sinh\left(\frac{L}{\lambda_c}\right) + \frac{\mathscr{G}_e}{\mathscr{G}_\infty} \cosh\left(\frac{L}{\lambda_c}\right)} \right)$$

Note that at 0, we find

$$v_m(0) = \frac{I}{\mathscr{G}_\infty} \left(\frac{\cosh\left(\frac{L}{\lambda_c}\right) + \frac{\mathscr{G}_e}{\mathscr{G}_\infty} \sinh\left(\frac{L}{\lambda_c}\right)}{\sinh\left(\frac{L}{\lambda_c}\right) + \frac{\mathscr{G}_e}{\mathscr{G}_\infty} \cosh\left(\frac{L}{\lambda_c}\right)} \right)$$

From Ohm's Law, the conductance we see at 0 is given by $\frac{i_i(0)}{v_m(0)}$. We will call this the Th'evenin Equivalent Conductance looking into the cable of length L at 0 or simply the Th'evenin Input Conductance of the Finite Cable and denote it by the symbol $\mathscr{G}_T(L)$ since its value clearly depends on L. It is given by

$$\mathscr{G}_T(L) = \mathscr{G}_\infty \left(\frac{\sinh\left(\frac{L}{\lambda_c}\right) + \frac{\mathscr{G}_e}{\mathscr{G}_\infty} \cosh\left(\frac{L}{\lambda_c}\right)}{\cosh\left(\frac{L}{\lambda_c}\right) + \frac{\mathscr{G}_e}{\mathscr{G}_\infty} \sinh\left(\frac{L}{\lambda_c}\right)} \right)$$

The hyperbolic function tanh is defined by

$$\tanh(u) = \frac{\sinh(u)}{\cosh(u)}$$

and it is easy to show that as u goes to infinity, $\tanh(u)$ goes to 1. We can rewrite the formula for $\mathscr{G}_T(L)$ to be

$$\mathscr{G}_T(L) = \mathscr{G}_\infty \left(\frac{\frac{\mathscr{G}_e}{\mathscr{G}_\infty} + \tanh\left(\frac{L}{\lambda_c}\right)}{\frac{\mathscr{G}_e}{\mathscr{G}_\infty} \tanh\left(\frac{L}{\lambda_c}\right) + 1} \right)$$

and so as L goes to infinity, we find

$$\lim_{L \to \infty} \mathcal{G}_T(L) = \mathcal{G}_\infty \frac{\frac{\mathcal{G}_e}{\mathcal{G}_\infty} + 1}{\frac{\mathcal{G}_e}{\mathcal{G}_\infty} + 1} = \mathcal{G}_\infty$$

Hence, the Th'evenin input conductance of the cable approaches the idealized Th'evenin input conductance of the half-infinite cable. There are several interesting cases: for convenience of exposition, let's define

$$Z = \frac{L - z}{\lambda_c}, \quad H = \frac{\mathcal{G}_e}{\mathcal{G}_\infty},$$

$$\mathcal{L} = \frac{L}{\lambda_c}, \quad D = \sinh(\mathcal{L}) + H \cosh(\mathcal{L})$$

These symbols allow us to rewrite our solutions more compactly as

$$v_m(z) = \frac{I}{\mathcal{G}_\infty} \frac{\cosh(Z) + H \sinh(Z)}{\sinh(\mathcal{L}) + H \cosh(\mathcal{L})}$$

$$= \frac{I}{\mathcal{G}_\infty} \frac{\cosh(Z) + H \sinh(Z)}{D}$$

$$i_i(z) = \frac{I}{\mathcal{G}_\infty} \frac{\mathcal{G}_\infty \sinh(Z) + \mathcal{G}_e \cosh(Z)}{\sinh(\mathcal{L}) + H \cosh(\mathcal{L})}$$

$$= \frac{I}{\mathcal{G}_\infty} \frac{\mathcal{G}_\infty \sinh(Z) + \mathcal{G}_e \cosh(Z)}{D}$$

$$\mathcal{G}_T(L) = \mathcal{G}_\infty \frac{\sinh(\mathcal{L}) + H \cosh(\mathcal{L})}{\cosh(\mathcal{L}) + H \sinh(\mathcal{L})}$$

$\mathcal{G}_e = 0$ If the conductance of the end cap is zero, then H is zero and no current flows through the endcap. We see

$$\mathcal{G}_T(L) = \mathcal{G}_\infty \frac{\sinh(\mathcal{L})}{\cosh(\mathcal{L})} = \mathcal{G}_\infty \tanh(\mathcal{L})$$

$\mathcal{G}_e = \mathcal{G}_\infty$ If the conductance of the end cap is $\mathcal{G}_\infty$, then H is one and we see the finite cable acts like the half-infinite cable:

$$\mathcal{G}_T(L) = \mathcal{G}_\infty \frac{\sinh(\mathcal{L}) + \cosh(\mathcal{L})}{\cosh(\mathcal{L}) + \sinh(\mathcal{L})} = \mathcal{G}_\infty$$

$\mathcal{G}_e = \infty$ If the conductance of the end cap is ∞, the end of the cable acts like it is short-circuited, H is infinity. Divide our original conductance solution by H top and bottom and letting K denote the reciprocal of H, we see:

$$\mathscr{G}_T(L) = \mathscr{G}_\infty \frac{K \sinh(\mathscr{L}) + \cosh(\mathscr{L})}{K \cosh(\mathscr{L}) + \sinh(\mathscr{L})}$$

Now K is zero here so we get

$$\mathscr{G}_T(L) = \mathscr{G}_\infty \frac{\cosh(\mathscr{L})}{\sinh(\mathscr{L})} = \mathscr{G}_\infty \coth(\mathscr{L})$$

8.2.1 Parametric Studies

We can calculate that

$$\frac{v_m(L)}{v_m(0)} = \frac{1}{\cosh\left(\frac{L}{\lambda_c}\right) + \frac{\mathscr{G}_e}{\mathscr{G}_\infty} \sinh\left(\frac{L}{\lambda_c}\right)}$$

For convenience, let's think of $\mathscr{L}$ as $\frac{L}{\lambda_c}$ and ρ as $\frac{\mathscr{G}_e}{\mathscr{G}_\infty}$. Then the ratio of the voltage at the end of the cable to the voltage at the beginning can be expressed as

$$\frac{v_m(L)}{v_m(0)} = \frac{1}{\cosh(\mathscr{L}) + \rho \sinh(\mathscr{L})}$$

This ratio measures the attenuation of the initial voltage as we move down the cable toward the end. We can plot a series of these attenuations for a variety of values of ρ. In Fig. 8.3, the highest ρ value is associated with the bottom most plot and the lowest value is associated with the top plot. Note as ρ increases, there is more conductivity through the end cap and the voltage drops faster. The top plot is for ρ is zero which is the case of no current flow through the end cap. We can also look how the Th'evenin equivalent conductance varies with the value of ρ. We can easily show that

$$\mathscr{G}_T^*(L) = \frac{\mathscr{G}_T(L)}{\mathscr{G}_\infty} = \frac{\tanh(\mathscr{L}) + \rho}{1 + \rho \tanh(\mathscr{L})}$$

In Fig. 8.4, we see that the ρ is one is the plot to which the other choices approach. When ρ is above one, the input conductance ratio curve starts above the ρ is one curve; when ρ is below one, the ratio curve approaches from below.

8.2.2 Some MatLab Implementations

We want to compare the membrane voltages we see in the infinite cable case to the ones we see in the finite cable case. Now in the infinite cable case, I current is deposited at $z = 0$ and the current injection spreads out uniformly to both sides of 0

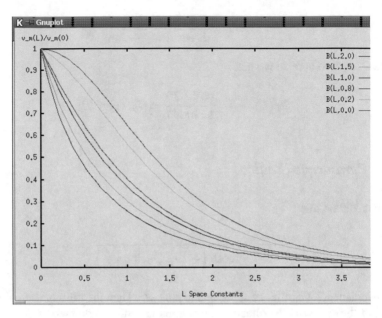

Fig. 8.3   Attenuation increases with ρ

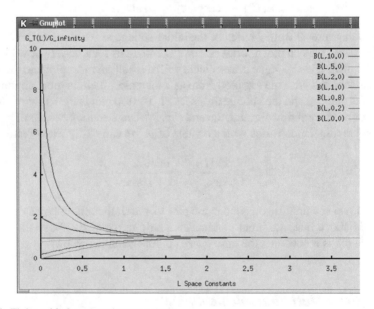

Fig. 8.4   Finite cable input conductance study

leading to the solution

$$v_m(z) = \frac{r_0 \, I \, \lambda_c}{2} \, e^{\frac{-|z|}{\lambda_c}}$$

where the finite cable case is missing the division by 2 because in a sense, the current is not allowed to spread backwards. Hence to compare solutions, we will inject 2 I into the infinite cable at 0 and I into the finite cable. Now at $z = 0$, the finite cable solution gives

$$v_m(0) = \frac{I}{\mathscr{G}_\infty} \, \frac{\cosh(\mathscr{L}) + H \sinh(\mathscr{L})}{\sinh(\mathscr{L}) + H \cosh(\mathscr{L})} = \frac{I}{\mathscr{G}_\infty} \, \frac{1 + H \tanh(\mathscr{L})}{H + \tanh(\mathscr{L})}$$

In the examples below, we set $\mathscr{G}_\infty$, λ_c and r_0 are one and set the cable length to 3. Hence, since for $L = 3$, $\tanh(\mathscr{L})$ is very close to one, we have

$$v_m(0) \approx I \, \frac{1 + H}{H + 1} \approx I$$

So for these parameters, we should be able to compare the infinite and finite cable solutions nicely.

MatLab code to implement the infinite cable voltage solution is given below:

Listing 8.1: Computing The Infinite Cable Voltage

```
  function  t = IniniteMembraneVoltage(Iehat,r0,lambda_c,z)
2 %
  %  compute  membrane  voltage  inside  a  finite  fiber
  %
  Prefix = (Iehat*r0*lambda_c)/2.0;
  t   = Prefix*exp(-1.0*abs(z)/lambda_c);
```

It is straightforward to modify the code above to handle the finite cable case:

Listing 8.2: Computing The Finite Cable Voltage

```
  function  t = FiniteMembraneVoltage(Iehat,Ginfinity,...
                                      Ge,L,lambda_c,z)
  %
4 %  compute  membrane  voltage  inside  a  finite  fiber
  %
  Prefix = Iehat/Ginfinity;
  FixedArg = L/lambda_c;
  Ratio = Ge/Ginfinity;
9 Denominator = sinh(FixedArg)+Ratio*cosh(FixedArg);
  Arg = (L-z)/lambda_c;
  Numerator = cosh(Arg)+ Ratio*sinh(Arg);
  t   = (Iehat/Ginfinity)*(Numerator/Denominator);
  end
```

8.2.3 Run-Time Results

We try out these new functions in the following MatLab sessions:

Listing 8.3: Sample Matlab session

```
    Ge = 0.1;
 2  Iehat = 1.0;
    Ginfinity = 1.0;
    L = 3.0;
    lambda_c = 1.0;
    r0 = 1.0;
 7  Z = linspace(0,L,200);
    % Find Voltage for infinite cable for 2*Iehat
    V_minfinity = InfiniteMembraneVoltage(2*Iehat,r0,lambda_c,Z);
    % Find Voltage for finite cable for Iehat, Ge = 0.1
    V_m0 = FiniteMembraneVoltage(Iehat,Ginfinity,Ge,L,lambda_c,Z);
12  % Find Voltage for finite cable for Iehat, Ge = 1.0
    V_m8 = FiniteMembraneVoltage(Iehat,Ginfinity,1.0,L,lambda_c,Z);
    % Find Voltage for finite cable for Iehat, Ge = 5.0
    V_m6 = FiniteMembraneVoltage(Iehat,Ginfinity,5.0,L,lambda_c,Z);
    % Find Voltage for finite cable for Iehat, Ge = 50.0
17  V_m7 = FiniteMembraneVoltage(Iehat,Ginfinity,50.0,L,lambda_c,Z);
    % Plot Infinite cable and finite cable Voltages for Iehat, Ge =
        0.1
    plot(Z,V_minfinity,'g-',Z,V_m0,'r-');
    % Plot Infinite cable and finite cable Voltages for Iehat, Ge =
        1.0
    plot(Z,V_minfinity,'g-',Z,V_m8,'r-.');
22  % Plot Infinite cable and finite cable Voltages for Iehat, Ge =
        5.0
    plot(Z,V_minfinity,'g-',Z,V_m6,'r-');
    % Plot Infinite cable and finite cable Voltages for Iehat, Ge =
        50.0
    plot(Z,V_minfinity,'g-',Z,V_m7,'r-.');
```

We are choosing to look at these solutions for a cable length of 3 with all the other parameters set to 1 for convenience except for the cable length which will be 3 and the endcap load conductance $\mathcal{G}_e$ which we will vary. We are also injecting $2\,I_e^*$ into the infinite cable as we mentioned we would do. Note the finite cable response attenuated more quickly than the infinite cable unless $\mathcal{G}_e$ is $\mathcal{G}_\infty$! You can see a variety of results in Figs. 8.5a, b and 8.6a, b.

8.2.4 Homework

Exercise 8.2.1 *Write Matlab functions to implement the finite cable transient variable solutions using as many arguments to the functions as are needed.*

1. v_m
2. i_i

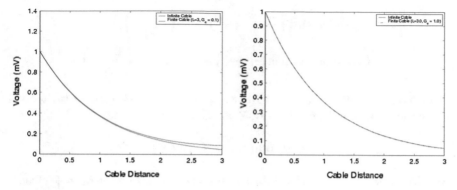

Fig. 8.5 Low end cap loads. **a** End cap load is 0.1, **b** End cap load is 1.0

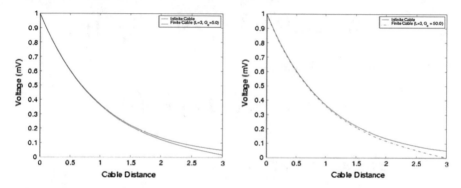

Fig. 8.6 High end cap loads. **a** End cap load is 5.0, **b** End cap load is 50.0

Exercise 8.2.2 *Generate a parametric plot of each of these variables versus the space variable z on a reasonable size range of z for the parameters λ_c, I_e^*, cable length L and $\mathscr{G}_e$. The ratio of $\mathscr{G}_e$ to $\mathscr{G}_\infty$ is a very reasonable parameter to use.*

1. v_m
2. i_i

Exercise 8.2.3 *Plot $\mathscr{G}_T(L)$ versus L and the horizontal line $\mathscr{G}_\infty$ on the same plot and discuss what is happening.*

8.3 The Finite Cable: Voltage Initialization

We can redo what we have discussed in the above using a different initial condition. Instead of specifying initial current, we will specify initial voltage. As you might expect, this will generate slightly different solutions. We use the same start:

$$v_m(z) = A_1 \cosh\left(\frac{L-z}{\lambda_c}\right) + A_2 \sinh\left(\frac{L-z}{\lambda_c}\right)$$

$$i_i(z) = A_1 \mathscr{G}_\infty \sinh\left(\frac{L-z}{\lambda_c}\right) + A_2 \mathscr{G}_\infty \cosh\left(\frac{L-z}{\lambda_c}\right)$$

Our boundary conditions are now $v_m(0) = V_0$ and $i_i(L) = \mathscr{G}_e\, v_m(L)$. Also, we will use the abbreviations from before: H, Z and $\mathscr{L}$ but change D to

$$E = \cosh(\mathscr{L}) + H \sinh(\mathscr{L})$$

Now let V_L denote $v_m(L)$. Then it follows that $A_1 = V_L$. If we set $B_L = \frac{A_2}{V_L}$, then

$$v_m(z) = V_L \left(\cosh\left(\frac{L-z}{\lambda_c}\right) + B_L \sinh\left(\frac{L-z}{\lambda_c}\right)\right)$$

$$i_i(z) = V_L \mathscr{G}_\infty \left(\sinh\left(\frac{L-z}{\lambda_c}\right) + B_L \cosh\left(\frac{L-z}{\lambda_c}\right)\right)$$

We know that $i_i(L) = \mathscr{G}_e\, V_L$ and so

$$i_i(L) = V_L \mathscr{G}_\infty \left(\sinh\left(\frac{0}{\lambda_c}\right) + B_L \cosh\left(\frac{0}{\lambda_c}\right)\right)$$

$$\mathscr{G}_e\, V_L = V_L\, B_L\, \mathscr{G}_\infty$$

The above implies $B_L = H$. Finally, note that

$$v_m(0) = V_L \left(\cosh\left(\frac{L}{\lambda_c}\right) + H \sinh\left(\frac{L}{\lambda_c}\right)\right)$$

$$= V_L \left(\sinh\left(\mathscr{L}\right) + H \cosh\left(\mathscr{L}\right)\right) = V_L\, E$$

Hence, $\frac{V_0}{E} = V_L$ and from this we obtain our final expression for the solution:

$$v_m(z) = \frac{V_0}{E} \left(\cosh(Z) + H \sinh(Z)\right)$$

$$i_i(z) = \frac{V_0}{E} \mathscr{G}_\infty \left(\sinh(Z) + H \cosh(Z)\right)$$

We will find that the Th'evenin equivalent conductance at the end cap is still the same. We have the same calculation as before

$$\mathscr{G}_T(L) = \frac{i_i(0)}{v_m(0)} = \mathscr{G}_\infty \frac{\sinh(\mathscr{L}) + H \cosh(\mathscr{L})}{\cosh(\mathscr{L}) + H \sinh(\mathscr{L})}$$

Finally, it is easy to see that the relationship between the current and voltage initialization condition is given by

$$V_0 \, \mathcal{G}_T(L) = i_i(0)$$

We can then find voltage and current equations for various interesting end cap conductance loads:

$\mathcal{G}_e = 0$: If the conductance of the end cap is zero, then H is zero and no current flows through the endcap. We see

$$v_m(z) = V_0 \, \frac{\cosh(Z)}{\cosh(\mathcal{L})}, \quad i_i(z) = V_0 \, \mathcal{G}_\infty \frac{\sinh(Z)}{\cosh(\mathcal{L})}, \quad \mathcal{G}_T(L) = \mathcal{G}_\infty \tanh(\mathcal{L}).$$

$\mathcal{G}_e = \mathcal{G}_\infty$: If the conductance of the end cap is $\mathcal{G}_\infty$, then H is one and we see the finite cable acts like the half-infinite cable:

$$v_m(z) = V_0 \, \frac{\cosh(Z) + \sinh(Z)}{\cosh(\mathcal{L}) + \sinh(\mathcal{L})} = V_0 \, e^{\frac{-z}{\lambda_c}},$$

$$i_i(z) = V_0 \, \mathcal{G}_\infty \frac{\sinh(Z) + \cosh(Z)}{\cosh(\mathcal{L}) + \sinh(\mathcal{L})} = V_0 \, \mathcal{G}_\infty \, e^{\frac{-z}{\lambda_c}}$$

$$\mathcal{G}_T(L) = \mathcal{G}_\infty.$$

$\mathcal{G}_e = \infty$: If the conductance of the end cap is ∞, the end of the cable acts like it is short-circuited, H is infinity. We see

$$v_m(z) = V_0 \, \frac{\sinh(Z)}{\sinh(\mathcal{L})}, \quad i_i(z) = V_0 \, \mathcal{G}_\infty \frac{\cosh(Z)}{\sinh(\mathcal{L})}, \quad \mathcal{G}_T(L) = \mathcal{G}_\infty \coth(\mathcal{L})$$

8.3.1 Homework

Exercise 8.3.1 *Write Matlab functions to implement v_m, i_i and $\mathcal{G}_T(L)$ using as many arguments to the functions as are needed. The arguments will be L, λ_c, z, V_0, $\mathcal{G}_\infty$ and $\mathcal{G}_e$.*

Exercise 8.3.2 *Using $V_0 = 1$, $\lambda_c = 5$, $L = 10$, and $\mathcal{G}_\infty = 2.0$ and generate a parametric plot of each of these variables versus the space variable z on $[0, 10]$ for a reasonable size range of $\mathcal{G}_e$. Discuss how the special conductance cases above fit into these plots.*

Exercise 8.3.3 *Assume $L = 1$ and $\lambda_c = 1$ also. Suppose sufficient current is injected to give $V_0 = 10$ mV. Let $\mathcal{G}_\infty = 2.0$.*

1. Let $\mathcal{G}_e = 0.5$

(a) *Compute v_m at $z = 0.6$ and 1.0. Compute v_m for the infinite cable model too at these points.*

2. *Let $\mathcal{G}_e = 2.0$*

   (a) *Compute v_m at $z = 0.6$ and 1.0. Compute v_m for the infinite cable model too at these points.*

3. *Let $\mathcal{G}_e = 50.0$*

   (a) *Compute v_m at $z = 0.6$ and 1.0. Compute v_m for the infinite cable model too at these points.*

4. *Discuss the results.*

8.4 Synaptic Currents

We can have two types of current injection. The first is injected through the outer membrane and is modeled by the pulses k_e. We know the outer cylinder is a theoretical abstraction and so there is really no such physical membrane. The second is injected into the front face of the cable in either a finite or half-infinite case. This current is modeled as an initialization of the inner current i_i. So consider the model we see in Fig. 8.7. The differential equation we would solve here would be

$$\lambda_c^2 \frac{d^2 v_m}{dz^2} - v_m = -r_o \lambda_c^2 I_0 \delta(z - z_0) - r_o \lambda_c^2 I_1 \delta(z - z_1), \quad 0 \le z \le L$$

$$i_i(0) = I$$

$$i_i(L) = \mathcal{G}_e v_m(L) \tag{8.1}$$

Here current I is injected into the front face and two impulse currents are injected into the outer cylinder. If we are thinking of this as a model of an axon, we could throw away the external sources and think of the current I injected into the front face as the current that arises from membrane voltage changes that propagate forward from the dendrite and soma system. Hence, the front face current I is a lumped sum approximation of the entire dendrite and soma subsystem response. On the other hand, if we are modeling a dendrite, we could think of the front face current I as a lumped sum model of the synaptic voltages that have propagated forward up to that point from the the rest of the dendrite system we are not modeling. The external current sources are then currents induced by synaptic interactions or currents flowing through pores in the membrane. Of course, voltage modulated gates would be handled differently!

Since our differential equation system is linear, to solve a problem like (8.1), we can simply add together the solutions to individual problems. This is called

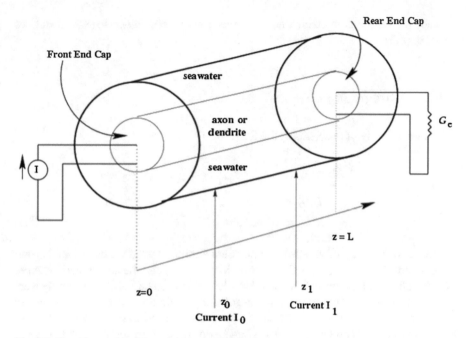

Fig. 8.7 Dendrite model with synaptic current injections

superposition of solutions and it is a very important tool in our work. Hence, to solve (8.1) we solve

$$\lambda_c^2 \frac{d^2 v_m}{dz^2} - v_m = 0, \quad 0 \le z \le L$$
$$i_i(0) = I$$
$$i_i(L) = \mathscr{G}_e\, v_m(L) \tag{8.2}$$

and

$$\lambda_c^2 \frac{d^2 v_m}{dz^2} - v_m = -r_o \lambda_c^2\, I_0 \delta(z - z_0), \quad 0 \le z \le L$$
$$i_i(0) = 0$$
$$i_i(L) = \mathscr{G}_e\, v_m(L) \tag{8.3}$$

and

$$\lambda_c^2 \frac{d^2 v_m}{dz^2} - v_m = -r_o \lambda_c^2\, I_1 \delta(z - z_1), \quad 0 \le z \le L$$
$$i_i(0) = 0$$
$$i_i(L) = \mathscr{G}_e\, v_m(L) \tag{8.4}$$

and add the solutions together. Since we already know the solution to (8.2), it suffices to solve (8.3).

8.4.1 A Single Impulse

Consider a family of problems of the form (8.5)

$$\lambda_c^2 \frac{d^2 v_m}{dz^2} - v_m = -r_o \lambda_c^2 I_0 k_e^C (z - z_0), \quad 0 \le z \le L$$

$$i_i(0) = 0$$

$$i_i(L) = \mathscr{G}_e v_m(L) \tag{8.5}$$

where the family $\{k_e^C\}$ of impulses are modeled similar to what we have done before: each is zero off $[z_0 - C, z_0 + C]$, symmetric around z_0 and the area under the curve is 1 for all C. So the only difference here is that our pulse family always delivers a constant 1 amp of current and we control the magnitude of the delivered current by the multiplier I_0. This way we can do the argument just once and know that our results hold for different magnitudes I_1 and so on. We assume for now that the site of current injection is z_0 which is in $(0, L)$. The case of z_0 being 0 or L will then require only a slight modification of our arguments which we will only sketch leaving the details to the reader. As usual, each current pulse delivers I_0 amps of current even though the base of the pulse is growing smaller and smaller. Physically, we expect, like in the infinite cable problem, that the voltage acts like a decaying exponential on either side of z_0. To see this is indeed the idealized solution we obtain when we let C go to 0, we resort to the following model:

$$v_m^C(z) = \begin{cases} A_1^C e^{\frac{z-z_0}{\lambda_c}}, & 0 \le z \le z_0 - C \\ \alpha^C e^{\frac{z-z_0}{\lambda_c}} + \beta^C e^{-\frac{z-z_0}{\lambda_c}} + \phi_p^C(z), & z_0 - C \le z \le z_0 + C \\ A_2^C e^{-\frac{z-z_0}{\lambda_c}}, & z_0 + C \le z \le L \end{cases}$$

The parts of the model before $z_0 - C$ and after $z_0 + C$ are modeled with exponential decay, while the part where the pulse is active is modeled with the full general solution to the problem having the form $\phi_h(z) + \phi_p^C(z)$, where ϕ_h^C is the homogeneous solution to the problem and ϕ_p^C is the particular solution obtained from the variations of parameters technique for a pulse k_e^C. Since the pulse k_e^C is smooth, we expect the voltage solution to be smooth also; hence our solution and its derivative must be continuous at the points $z_0 - C$ and $z_0 + C$. This will give us four equations in four unknowns we can solve for the constants A_1^C, A_2^C, α^C and β^C. Recall the auxiliary equation for this differential equation is

$$\lambda_c^2 r^2 - 1 = 0$$

with roots $\frac{1}{\lambda_c}$ and $-\frac{1}{\lambda_c}$. The homogeneous solution is then

$$\phi_h(z) = B_1 e^{\frac{z-z_0}{\lambda_c}} + B_2 e^{-\frac{z-z_0}{\lambda_c}}$$

Using the method of Variation of Parameters, we search for a particular solution of the form

$$\phi_p^n(z) = U_1(z) e^{\frac{z-z_0}{\lambda_c}} + U_2(z) e^{-\frac{z-z_0}{\lambda_c}}$$

where the coefficient functions U_1 and U_2 satisfy

$$\begin{bmatrix} e^{\frac{z-z_0}{\lambda_c}} & e^{-\frac{z-z_0}{\lambda_c}} \\ \frac{1}{\lambda_c} e^{\frac{z-z_0}{\lambda_c}} & -\frac{1}{\lambda_c} e^{-\frac{z-z_0}{\lambda_c}} \end{bmatrix} \begin{bmatrix} \frac{dU_1}{dz} \\ \frac{dU_2}{dz} \end{bmatrix} = \begin{bmatrix} 0 \\ -r_o I_0 k_e^C (z - z_0) \end{bmatrix}$$

This is easily solved using Cramer's rule to give

$$\frac{dU_1}{dz} = -\frac{r_o \lambda_c I_0}{2} k_e^C (z - z_0) e^{-\frac{z-z_0}{\lambda_c}}$$

$$\frac{dU_2}{dz} = \frac{r_o \lambda_c I_0}{2} k_e^C (z - z_0) e^{\frac{z-z_0}{\lambda_c}}$$

We can integrate then to obtain

$$U_1(z) = -\frac{r_o \lambda_c I_0}{2} \int_{z_0}^{z} k_e^C (s - z_0) e^{-\frac{s-z_0}{\lambda_c}} ds$$

$$U_2(z) = \frac{r_o \lambda_c I_0}{2} \int_{z_0}^{z} k_e^C (s - z_0) e^{\frac{s-z_0}{\lambda_c}} ds$$

giving

$$\phi_p^C(z) = U_1(z) e^{\frac{z-z_0}{\lambda_c}} + U_2(z) e^{-\frac{z-z_0}{\lambda_c}}$$

$$+ \left(-\frac{r_o \lambda_c I_0}{2} \int_{z_0}^{z} k_e^C (s - z_0) e^{-\frac{s-z_0}{\lambda_c}} ds \right) e^{\frac{z-z_0}{\lambda_c}}$$

$$+ \left(\frac{r_o \lambda_c I_0}{2} \int_{z_0}^{z} k_e^C (s - z_0) e^{\frac{s-z_0}{\lambda_c}} ds \right) e^{-\frac{z-z_0}{\lambda_c}}$$

Now we rewrite this a bit more compactly as follows:

$$\phi_p^C(z) = -\frac{r_o \lambda_c I_0}{2} \int_{z_0}^{z} k_e^C (s - z_0) e^{\frac{s-z}{\lambda_c}} ds + \frac{r_o \lambda_c I_0}{2} \int_{z_0}^{z} k_e^C (s - z_0) e^{-\frac{s-z}{\lambda_c}} ds$$

$$= -r_o \lambda_c I_0 \int_{z_0}^{z} k_e^C (s - z_0) \sinh\left(\frac{z - x}{\lambda_c} \right) ds$$

Hence, the particular solution for pulse k_e^C is

$$\phi_p^C(z) = -r_0 \lambda_c I_0 \int_{z_0}^z k_e(s - z_0) \, \sinh\left(\frac{z - s}{\lambda_c}\right) dx$$

8.4.2 Forcing Continuity

When we modeled current injections before in Sect. 7.3, we forced continuity in the solution and the derivative of the solution to find the unknown constants. To do this, we need the derivative terms which are given below:

$$\frac{dv_m^C}{dz} = \begin{cases} \dfrac{A_1^C}{\lambda_c} e^{\frac{z - z_0}{\lambda_c}}, & 0 \le z \le z_0 - C \\[3mm] \dfrac{\alpha^C}{\lambda_c} e^{\frac{z - z_0}{\lambda_c}} - \dfrac{\beta^C}{\lambda_c} e^{-\frac{z - z_0}{\lambda_c}} + \dfrac{d\phi_p^C}{dz}, & z_0 - C \le z \le z_0 + C \\[3mm] -\dfrac{A_2^C}{\lambda_c} e^{-\frac{z - z_0}{\lambda_c}}, & z_0 + C \le z \le L \end{cases}$$

Then, continuity at $z_0 - C$ and $z_0 + C$ gives

$$A_1^C e^{-C} = \alpha^C e^{-C} + \beta^C e^C + \phi_p^C(z_0 - C)$$

$$A_2^C e^{-C} = C_n e^C + D_n e^{-C} + \phi_p^C(z_0 + C)$$

$$\frac{A_1^C}{\lambda_c} e^{-C} = \frac{\alpha^C}{\lambda_c} e^{-C} - \frac{\beta^C}{\lambda_c} e^C + \frac{d\phi_p^n}{dz}(z_0 - C)$$

$$-\frac{A_2^C}{\lambda_c} e^{-C} = \frac{\alpha^C}{\lambda_c} e^C - \frac{\beta^C}{\lambda_c} e^{-C} + \frac{d\phi_p^n}{dz}(z_0 + C)$$

which can be rewritten in the form

$$A_1^C e^{-C} - \alpha^C e^{-C} - \beta^C e^C = \phi_p^C(z_0 - C) \tag{8.6}$$

$$A_2^C e^{-C} - \alpha^C e^C - \beta^C e^{-C} = \phi_p^C(z_0 + C) \tag{8.7}$$

$$A_1^C e^{-C} - \alpha^C e^{-C} + \beta^C e^C = \lambda_c \frac{d\phi_p^C}{dz}(z_0 - C) \tag{8.8}$$

$$-A_2^C e^{-C} - \alpha^C e^C + \beta^C e^{-C} = \lambda_c \frac{d\phi_p^C}{dz}(z_0 + C) \tag{8.9}$$

Computing (Eq. 8.7 + Eq. 8.8), (Eq. 8.7 − Eq. 8.8), (Eq. 8.9 and Eq. 8.9) and (Eq. 8.9 − Eq. 8.9), we find

$$2A_1^C e^{-C} - 2\alpha^C e^{-C} = \phi_p^C(z_0 - C) + \lambda_c \frac{d\phi_p^C}{dz}(z_0 - C)$$

$$-2\beta^C e^C = \phi_p^C(z_0 - C) - \lambda_c \frac{d\phi_p^C}{dz}(z_0 - C)$$

$$-2\alpha^C e^C = \phi_p^C(z_0 + C) + \lambda_c \frac{d\phi_p^C}{dz}(z_0 + C)$$

$$2A_2^C e^{-C} - 2\beta^C e^{-C} = \phi_p^C(z_0 + C) - \lambda_c \frac{d\phi_p^C}{dz}(z_0 + C)$$

Although very messy, this can be solved (not so easy!) to give

$$2A_1^C e^{-C} = 2\alpha^C e^{-C} + \phi_p^C(z_0 - C) + \lambda_c \frac{d\phi_p^C}{dz}(z_0 - C)$$

$$= -\left(\phi_p^C(z_0 + C) + \lambda_c \frac{d\phi_p^C}{dz}(z_0 + C)\right)e^{-2C} + \phi_p^C(z_0 - C)$$

$$+ \lambda_c \frac{d\phi_p^C}{dz}(z_0 - C)$$

$$2A_2^C e^{-C} = 2D_n e^{-C} + \phi_p^C(z_0 + C) - \lambda_c \frac{d\phi_p^C}{dz}(z_0 + C)$$

$$= -\left(\phi_p^n(z_0 - C) - \lambda_c \frac{d\phi_p^C}{dz}(z_0 + C)\right)e^{-2C} + \phi_p^C(z_0 + C)$$

$$- \lambda_c \frac{d\phi_p^C}{dz}(z_0 + C)$$

or

$$A_1^C = -\frac{1}{2}\left(\phi_p^C(z_0 + C) + \lambda_c \frac{d\phi_p^C}{dz}(z_0 + C)\right)e^{-C}$$

$$+ \frac{1}{2}\left(\phi_p^C(z_0 - C) + \lambda_c \frac{d\phi_p^C}{dz}(z_0 - C)\right)e^C$$

$$A_2^C = -\frac{1}{2}\left(\phi_p^C(z_0 - C) - \lambda_c \frac{d\phi_p^C}{dz}(z_0 + C)\right)e^{-C}$$

$$+ \frac{1}{2}\left(\phi_p^C(z_0 + C) - \lambda_c \frac{d\phi_p^C}{dz}(z_0 + C)\right)e^C$$

8.4.3   The Limiting Solution

The above solutions work for all positive C. We note

$$\phi_p^C(z_0 + C) = -r_0\lambda_c I_0 \int_{z_0}^{z_0+C} k_e^n(s - z_0)\sinh\left(\frac{z_0 + C - s}{\lambda_c}\right)$$

and so, using arguments very similar to those presented in Lemma 7.4.1, we can show

$$\lim_{C\to 0}\phi_p^C(z_0 + C) = -r_0\lambda_c I_0\,\frac{1}{2}\,\sinh(0)$$
$$= 0$$

In a similar fashion, we can show

$$\lim_{C\to 0}\phi_p^C(z_0 - C) = r_0\lambda_c I_0\,\frac{1}{2}\,\sinh(0)$$
$$= 0$$

Finally, since

$$\frac{d\phi_p^C}{dz}(z) = -r_0\lambda_c I_0\,k_e^C(z - z_0)\sinh(0) - r_0 I_0\int_{z_0}^{z} k_e^C(s - z_0)\cosh\left(\frac{z - s}{\lambda_c}\right)ds$$

we see

$$\frac{d\phi_p^C}{dz}(z_0 + C) = -r_0 I_0\int_{z_0}^{z_0+C} k_e^C(s - z_0)\cosh\left(\frac{z - s}{\lambda_c}\right)ds$$

which gives

$$\lim_{C\to 0}\frac{d\phi_p^C}{dz}(z_0 + C) = -r_0 I_0\lambda_c\,\frac{1}{2}\,\cosh(0) = -\frac{r_0 I_0\lambda_c}{2}.$$

and similarly

$$\lim_{C\to 0}\frac{d\phi_p^C}{dz}(z_0 - C) = r_0 I_0\lambda_c\,\frac{I_0}{2}\,\cosh(0) = \frac{r_0 I_0\lambda_c}{2}.$$

Thus in the limit, we obtain the limiting constants

$$A_1 = \lim_{C\to 0} A_1^C = \frac{r_0\lambda_c I_0}{2}$$
$$A_2 = \lim_{C\to 0} A_2^C = \frac{r_0\lambda_c I_0}{2}$$

This gives the limiting solution

$$v_m(z) = \begin{cases} \frac{r_0 \lambda_c I_0}{2} e^{\frac{z - z_0}{\lambda_c}}, & 0 \leq z \leq z_0 \\ \frac{r_0 \lambda_c I_0}{2} e^{-\frac{z - z_0}{\lambda_c}}, & z_0 \leq z \leq L \end{cases}$$

which is essentially our usual idealized impulse solution from the infinite cable model

$$v_m(z) = \frac{r_0 \lambda_c I_0}{2} e^{-\frac{|z - z_0|}{\lambda_c}}, \quad 0 \leq z \leq L$$

Note the values of α^C and β^C are not important for the limiting solution.

8.4.4 Satisfying Boundary Conditions

Since

$$\frac{di_i}{dz} = -g_m v_m$$

we see that for z below $z_0 - C$, we have, for integration constant γ^C,

$$i_i(z) = -g_m \lambda_c A_1^C e^{\frac{z - z_0}{\lambda_c}} + \gamma^C$$
$$= -\mathcal{G}_\infty A_1^C e^{\frac{z - z_0}{\lambda_c}} + \gamma^C$$

and the boundary condition $i_i(0) = 0$ then implies

$$\gamma^C = \mathcal{G}_\infty A_1^C e^{-\frac{z_0}{\lambda_c}}$$

and for the second boundary condition, $i_i(L) = \mathcal{G}_e v_m(L)$, we use the last part of the definition of v_m to give us, for a new integration constant ζ^C,

$$i_i(z) = \mathcal{G}_\infty A_2^C e^{-\frac{z - z_0}{\lambda_c}} + \zeta^C$$

and so

$$\mathcal{G}_\infty A_2^n e^{-\frac{L - z_0}{\lambda_c}} + \zeta^C = \mathcal{G}_e A_2^C e^{-\frac{L - z_0}{\lambda_c}}$$

or

$$\zeta^C = A_2^C e^{-\frac{L - z_0}{\lambda_c}} (\mathcal{G}_e - \mathcal{G}_\infty)$$

Hence, from the above, we see that the limiting inner current solutions satisfy

$$\gamma = \lim_{C \to 0} \gamma^C = \mathscr{G}_\infty \frac{r_0 \lambda_c I_0}{2} e^{-\frac{z_0}{\lambda_c}}$$

$$\zeta = \lim_{C \to 0} \zeta^C = (\mathscr{G}_e - \mathscr{G}_\infty) \frac{r_0 \lambda_c I_0}{2} e^{-\frac{L - z_0}{\lambda_c}}$$

giving for z below z_0

$$i_i(z) = -\mathscr{G}_\infty \frac{r_0 \lambda_c I_0}{2} e^{\frac{z - z_0}{\lambda_c}} + \mathscr{G}_\infty \frac{r_0 \lambda_c I_0}{2} e^{-\frac{z_0}{\lambda_c}}$$

and for z above z_0,

$$i_i(z) = \mathscr{G}_\infty \frac{r_0 \lambda_c I_0}{2} e^{-\frac{z - z_0}{\lambda_c}} + (\mathscr{G}_e - \mathscr{G}_\infty) \frac{r_0 \lambda_c I_0}{2} e^{-\frac{L - z_0}{\lambda_c}}$$

8.4.5   Some Results

We can now see the membrane voltage solution for the stated problem

$$\lambda_c^2 \frac{d^2 v_m}{dz^2} - v_m = 0, \quad 0 \le z \le L$$

$$i_i(0) = I$$

$$i_i(L) = \mathscr{G}_e \, v_m(L)$$

in Fig. 8.8. Note here there are no impulses applied. If we add the two impulses at z_0 and z_1 as described by the differential equations (8.3) and (8.1), the solution to the full problem for somewhat weak impulses at two points on the cable can be seen in Fig. 8.9. For another choice of impulse strengths, we see the summed solution in Fig. 8.10. For ease of comparison, we can plot the solutions for no impulses, two impulse of low strength and two impulses of higher strength simultaneously in Fig. 8.11.

8.5   Implications

Even though we have still not looked into the case of time dependent solutions, we can still say a lot about the nature of the solutions we see in the time independent case.

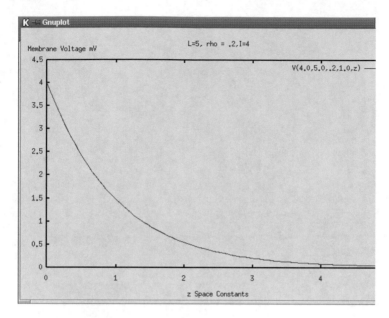

Fig. 8.8 Finite cable initial endcap current

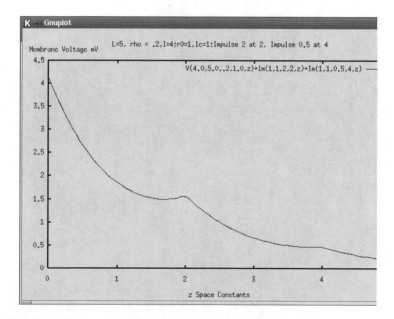

Fig. 8.9 Finite cable current initializations: strong impulses

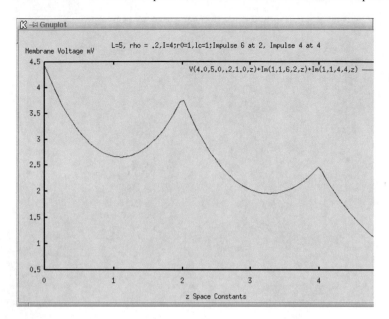

Fig. 8.10 Finite cable current initializations: strong impulses

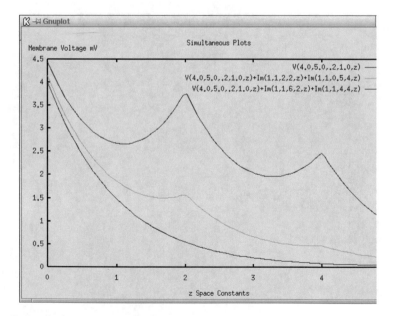

Fig. 8.11 Three finite cable current initializations

Table 8.1 Typical cable constants

Species	a	C_m	G_m	r_i	λ_c	τ_m
Squid (Loligo Peali)	250	1	1	0.015	6.5	1
Lobster (Homanus Vulgaris)	37.5	1	0.5	1.4	2.5	2
Crab (Carcinus Maenas)	15	1	0.14	13.0	2.3	7
Lobster (Homanus Americanus)	50	–	0.13	1.0	5.1	–
Earthworm (Lumbricus Terrestris)	52.5	0.3	0.083	2.3	4.0	3.6
Marine worm (Myxicula Infundibulum)	280	0.75	0.012	0.023	5.4	0.9
Cat (Motoneuron)	–	2	0.4	–	–	5.0
Frog muscle fiber (Sartorius)	37.5	2.5–6.0	0.25	4.5	2.0	10.0–24.0

- The space constant λ_c tells us how far we must be from the site of input current to see significant attenuation of the resulting potential. Thus, if $\lambda_c \gg 1$, the cable's membrane potential is close to position independent.
- An electrode used to input current or measure voltage can be thought of as infinitesimal in tip size if the tip size is very small compared to λ_c.
- Our underlying cable equation is a linear PDE. Hence, the superposition principle applies and we can use it to handle arbitrary arrangements of current sources.
- We already know that

$$\lambda_c \approx \sqrt{\frac{a}{2\rho_i G_m}} \approx \sqrt{\frac{1}{2\rho_i G_m}} \sqrt{a}$$

implying λ_c decreases as the cable fiber inner radius decreases.

Now let a be the cable radius in μm, C_m be membrane capacitance in $\frac{\mu F}{cm^2}$, G_m be membrane conductance in $\frac{mS}{cm^2}$, r_i be internal resistance of the protoplasm per unit length in $\frac{M\Omega}{cm}$, λ_c be the cable space constant in mm and τ_m be the cable membrane time constant in m sec. Then consider a table of these typical cable constants as shown in Table 8.1 which can be found in many reference texts.

From Table 8.1, we see that the earthworm gives $\frac{1000\,\lambda_c\,\mu m}{a\,\mu m}$ is $\frac{4000}{52.5}$ or 76.2. If we assume this ratio holds, then for a cable with a equal to 0.5 μm, we see that the space constant for this cable would be $76.2 \times 0.5\,\mu$m or $38.1\,\mu$m which is 0.038 mm. The earthworm cable fiber is unmylenated and so signals are not insulated from transmission loss. This extrapolation shows that in this unmylenated case, we would expect the fiber to transmit signal poorly as the fiber radius drops. Of course, many species protect themselves against this transmission loss by shielding the fibers using mylenin, but that is another story.

Part IV
Excitable Neuron Models

Chapter 9
A Primer on Series Solutions

We now discuss what we mean by a series of functions and what we mean by convergence of this series. This is a complicated topic and so we'll ease into it slowly. We'll start with a long motivating example. The cable equation can also be solved for solutions that are functions of space and time. To do this we need to introduce a new technique called the Separation of Variables Method. This will lead to a type of solution which is written in infinite series form. Hence, new ideas and new mathematics! We have discussed Separation of Variables earlier in Peterson (2015).

9.1 The Separation of Variables Method

We have been discussing the equation whose solutions describe the voltage on a dendritic cable. Recall the general model is

$$\beta^2 \frac{\partial^2 \Phi}{\partial x^2} - \Phi - \alpha \frac{\partial \Phi}{\partial t} = 0, \ \text{for } 0 \leq x \leq L, \ t \geq 0,$$

$$\frac{\partial \Phi}{\partial x}(0, t) = 0,$$

$$\frac{\partial \Phi}{\partial x}(L, t) = 0,$$

$$\Phi(x, 0) = f(x).$$

for positive constants α and β. The domain is the usual half infinite $[0, L] \times [0, \infty)$ where the spatial part of the domain corresponds to the length of the dendritic cable. The boundary conditions $u_x(0, t) = 0$ and $u_x(L, t) = 0$ are called *Neumann Boundary conditions*. The conditions $u(0, t) = 0$ and $u(L, t) = 0$ are known as *Dirichlet Boundary conditions*. The ballstick neuron model we solve in Chap. 11 uses a different set of boundary conditions from these because the cable end cap membrane must

© Springer Science+Business Media Singapore 2016
J.K. Peterson, *Calculus for Cognitive Scientists*, Cognitive Science
and Technology, DOI 10.1007/978-981-287-880-9_9

interact with the cell body membrane. We discuss this solution later. The solution to
a model such as this is a function $\Phi(x, t)$ which is sufficiently smooth to have partial
derivatives with respect to the needed variables continuous for all the orders required.
For these problems, the highest order we need is the second order partials. One way
to find the solution is to assume we can separate the variables so that we can write
$\Phi(x, t) = u(x)w(t)$. If we make this separation assumption, we will find solutions
that must be written as what are called infinite series and to solve the boundary conditions, we will have to be able to express boundary functions as series expansions.
Hence, we will have to introduce some new ideas in order to understand these things.
Let's motivate what we need to do by applying the separation of variables technique
to the cable equation. This will shows the ideas we need to use in a specific example.
Then we will step back and go over the new mathematical ideas of series and then
return to the cable model and finish finding the solution. In Chap. 10 we apply the
method to other linear PDE models.

We assume a solution of the form $\Phi(x, t) = u(x)\, w(t)$ and compute the needed
partials. This leads to a the new equation

$$\beta^2 \frac{d^2u}{dx^2}\, w(t) - u(x)w(t) - \alpha u(x)\frac{dw}{dt} = 0.$$

Rewriting, we find for all x and t, we must have

$$w(t)\left(\beta^2 \frac{d^2u}{dx^2} - u(x)\right) = \alpha u(x)\frac{dw}{dt}.$$

Rewriting, we have

$$\frac{\beta^2 \frac{d^2u}{dx^2} - u(x)}{u(x)} = \frac{\alpha \frac{dw}{dt}}{w(t)}, \quad 0 \le x << L, \ t > 0.$$

The only way this can be true is if both the left and right hand side are equal to the
separation constant Θ. This leads to the decoupled Eqs. 9.1 and 9.2.

$$\alpha \frac{dw}{dt} = \Theta\, w(t), \quad t > 0, \tag{9.1}$$

$$\beta^2 \frac{d^2u}{dx^2} = (1 + \Theta)\, u(x), \quad 0 \le x \le L, \tag{9.2}$$

We also have boundary conditions. Our assumption leads to the following boundary
conditions in x:

$$\frac{du}{dx}(0)\, w(t) = 0, \quad t > 0,$$

$$\frac{du}{dx}(L)\, w(t) = 0, \quad t > 0.$$

Since these equations must hold for all t, this forces

$$\frac{du}{dx}(0) = 0, \tag{9.3}$$

$$\frac{du}{dx}(L) = 0. \tag{9.4}$$

Equations 9.1–9.4 give us the boundary value problem in $u(x)$ we need to solve. Then, we can find w.

9.1.1 Determining the Separation Constant

The model is then

$$u'' - \frac{1 + \Theta}{\beta^2} u = 0$$

$$\frac{du}{dx}(0) = 0,$$

$$\frac{du}{dx}(L) = 0.$$

We have already solved a system like this when we looked at the solution to the wave equation; hence, the work we do here is quite familiar. We are looking for nonzero solutions, so any choice of separation constant Θ that leads to a zero solution will be rejected.

9.1.1.1   Case I: $1 + \Theta = \omega^2, \omega \neq 0$

The model to solve is

$$u'' - \frac{\omega^2}{\beta^2} u = 0$$

$$u'(0) = 0,$$

$$u'(L) = 0.$$

with characteristic equation $r^2 - \frac{\omega^2}{\beta^2} = 0$ with the real roots $\pm \frac{\omega}{\sqrt{D}}$. The general solution of this second order model is given by

$$u(x) = A \cosh\left(\frac{\omega}{\beta}x\right) + B \sinh\left(\frac{\omega}{\beta}x\right)$$

which tells us

$$u'(x) = A\,\frac{\omega}{\beta}\,\sinh\left(\frac{\omega}{\beta}x\right) + B\,\frac{\omega}{\beta}\,\cosh\left(\frac{\omega}{\beta}x\right)$$

Next, apply the boundary conditions, $u'(0) = 0$ and $u'(L) = 0$. Hence,

$$u'(0) = 0 = B$$
$$u'(L) = 0 = A\,\sinh\left(L\frac{\omega}{\beta}\right)$$

Hence, $B = 0$ and $A\,\sinh\left(L\frac{\omega}{\beta}\right) = 0$. Since sinh is never zero when ω is not zero, we see $A = 0$ also. Hence, the only u solution is the trivial one and we can reject this case.

9.1.1.2   Case II: $1 + \Theta = 0$

The model to solve is now

$$u'' = 0$$
$$u'(0) = 0,$$
$$u'(L) = 0.$$

with characteristic equation $r^2 = 0$ with the double root $r = 0$. Hence, the general solution is now

$$u(x) = A + B\,x$$

Applying the boundary conditions, $u(0) = 0$ and $u(L) = 0$. Hence, since $u'(x) = B$, we have

$$u'(0) = 0 = B$$
$$u'(L) = 0 = BL$$

Hence, $B = 0$ but the value of A can't be determined. Hence, any arbitrary constant which is not zero is a valid non zero solution. Choosing $A = 1$, let $u_0(x) = 1$ be our chosen nonzero solution for this case. We now need to solve for w in this case. Since $\Theta = -1$, the model to solve is

$$\frac{dw}{dt} = -\frac{1}{\alpha}w(t), \ \ 0 < t,$$
$$w(L) = 0.$$

The general solution is $w(t) = Ce^{-\frac{1}{\alpha}t}$ for any value of C. Choose $C = 1$ and we set

$$w_0(y) = e^{-\frac{1}{\alpha}t}.$$

Hence, the product $\phi_0(x, t) = u_0(x)\, w_0(t)$ solves the boundary conditions. That is

$$\phi_0(x, t) = e^{-\frac{1}{\alpha}t}.$$

is a solution.

9.1.1.3   Case III: $1 + \Theta = -\omega^2$, $\omega \neq 0$

$$u'' + \frac{\omega^2}{\beta^2}u = 0$$
$$u'(0) = 0,$$
$$u'(L) = 0.$$

The general solution is given by

$$u(x) = A\,\cos\left(\frac{\omega}{\beta}x\right) + B\,\sin\left(\frac{\omega}{\beta}x\right)$$

and hence

$$u'(x) = -A\frac{\omega}{\beta}\sin\left(\frac{\omega}{\beta}x\right) + B\frac{\omega}{\beta}\cos\left(\frac{\omega}{\beta}x\right)$$

Next, apply the boundary conditions to find

$$u'(0) = 0 = B$$
$$u'(L) = 0 = A\sin\left(L\frac{\omega}{\beta}\right)$$

Hence, $B = 0$ and $A\sin\left(L\frac{\omega}{\beta}\right) = 0$. Thus, we can determine a unique value of A only if $\sin\left(L\frac{\omega}{\beta}\right) \neq 0$. If $\omega \neq \frac{n\pi\beta}{L}$, we can solve for A and find $A = 0$, but otherwise, A can't be determined. So the only solutions are the trivial or zero solutions unless $\omega L = n\pi\beta$. Letting $\omega_n = \frac{n\pi\beta}{L}$, we find a non zero solution for each nonzero value of A of the form

$$u_n(x) = A\cos\left(\frac{\omega_n}{\beta}x\right) = A\cos\left(\frac{n\pi}{L}x\right).$$

For convenience, let's choose all the constants $A = 1$. Then we have an infinite family of nonzero solutions $u_n(x) = \cos\left(\frac{n\pi}{L}x\right)$ and an infinite family of separation constants $\Theta_n = -1 - \omega_n^2 = -1 - \frac{n^2\pi^2 D}{L^2}$. We can then solve the w equation. We must solve

$$\frac{dw}{dt} = -\frac{(1 + \omega_n^2)}{\alpha} w(t), \quad t \geq 0.$$

The general solution is

$$w(t) = B_n \, e^{-\frac{1+\omega_n^2}{\alpha}t} = B_n \, e^{-\frac{1+n^2\pi^2\beta^2}{\alpha L^2}t}$$

Choosing the constants $B_n = 1$, we obtain the w_n functions

$$w_n(t) = e^{-\frac{n^2\pi^2\beta^2}{\alpha L^2}t}$$

Hence, any product

$$\phi_n(x, t) = u_n(x) \, w_n(t)$$

will solve the model with the x boundary conditions and any finite sum of the form, for arbitrary constants A_n

$$\Psi_N(x, t) = \sum_{n=1}^{N} A_n\phi_n(x, t) = \sum_{n=1}^{N} A_n u_n(x) \, w_n(t)$$

$$= \sum_{n=1}^{N} A_n \cos\left(\frac{n\pi}{L}x\right) e^{-\frac{1+n^2\pi^2\beta^2}{\alpha L^2}t}$$

Adding in the $1 + \Theta = 0$ case, we find the most general finite term solution has the form

$$\Phi_N(x, t) = A_0\phi_0(x, t) + \sum_{n=1}^{N} A_n\phi_n(x, t) = A_0 u_0(x)w_0(t) + \sum_{n=1}^{N} A_n u_n(x) \, w_n(t)$$

$$= A_0 e^{-\frac{1}{\alpha}t} + \sum_{n=1}^{N} A_n \cos\left(\frac{n\pi}{L}x\right) e^{-\frac{1+n^2\pi^2\beta^2}{\alpha L^2}t}.$$

Now these finite term solutions do solve the boundary conditions $\frac{\partial\Phi}{\partial x}(0, t) = 0$ and $\frac{\partial\Phi}{\partial x}(L, t) = 0$, but how do we solve the remaining condition $\Phi(x, 0) = f(x)$? To do this, we note since we can assemble the finite term solutions for any value of N, no matter how large, it is clear we should let $N \to \infty$ and express the solution as

$$\Phi(x, t) = A_0 \phi_0(x, t) + \sum_{n=1}^{\infty} A_n \phi_n(x, t) = A_0 u_0(x) w_0(t) + \sum_{n=1}^{\infty} A_n u_n(x) \, w_n(t)$$

$$= A_0 e^{-\frac{1}{\alpha}t} + \sum_{n=1}^{\infty} A_n \cos\left(\frac{n\pi}{L}x\right) e^{-\frac{1+n^2\pi^2\beta^2}{\alpha L^2}t}.$$

This is the form that will let us solve the remaining boundary condition. These solutions are written in what is called infinite series form we need to back up and discuss carefully what we mean by letting $N \to \infty$.

9.1.1.4   Homework

Exercise 9.1.1 *Use separation of variables to find the general infinite series solution to the cable equation with Dirichlet boundary data given below.*

$$\beta^2 \frac{\partial^2 \Phi}{\partial x^2} - \Phi - \alpha \frac{\partial \Phi}{\partial t} = 0, \; for \; 0 \le x \le L, \; t \ge 0,$$

$$\Phi(0, t) = 0,$$

$$\Phi(L, t) = 0,$$

$$\Phi(x, 0) = f(x).$$

9.2   Infinite Series

To discuss infinite series, we need to understand limiting processes better. We can then add on the new ideas we need.

9.2.1   Simple Limits from Calculus

In your earlier calculus courses, you learned about *limits* and *convergence* to a limit in certain situations. For example, we know if a function f is continuous at the point x_0 in its domain, then we can say

$$\lim_{x \to x_0} f(x) = f(x_0)$$

which is short hand for saying the limit of $f(x)$ as x approaches x_0 exists and that the value of the limit is $f(x_0)$. More precisely, we can use a different language to express this: the $\epsilon - \delta$ form. We would say the limit of $f(x)$ as x approaches x_0 exists if there is a number A so that given any positive tolerance ϵ there is a positive radius δ so that

$$|x - x_0| < \delta \implies |f(x) - A| < \epsilon.$$

This seems like a hard statement to understand, but it simply says in very precise language that as x gets close to x_0, the corresponding function values $f(x)$ get close to A also. There are all kinds of ways to make this statement seem even more forbidding. For example, saying it the way we did above, assumes the function f is defined on all the x values within δ of x_0. So we could adjust our statement to say

$$|x - x_0| < \delta \text{ and } x \text{ is in the domain of } f \implies |f(x) - A| < \epsilon.$$

but usually, we know that this statement is dependent on x being in the domain of f without saying and so we write the statement in the more relaxed form. Now, to finish with the idea of continuity, we then have to assume the value of A is the same as $f(x_0)$. If we restricted our attention to x values moving towards x_0 only from the left, we could call this the *left hand* limit as x goes to x_0 and denote it by $\lim_{x \to x_0^-} f(x)$. Similarly, If we restricted our attention to x values moving towards x_0 only from the right, we could call this the *right hand* limit as x goes to x_0 and denote it by $\lim_{x \to x_0^+} f(x)$. If you think about it a bit, the limit only exists if and only if the two one sided limits both exist and match the value A.

A couple of standard examples will help you understand this. Let the function f be defined like this:

$$f(t) = \begin{cases} t^2, & 0 \le t < 2, \\ -t^2, & 2 \le t \le 4 \end{cases}$$

Then, it is easy to see $\lim_{x \to 2^-} f(x) = 4$ and $\lim_{x \to 2^+} f(x) = -4$ and since the right and left hand limits don't match, f is not continuous at $t = 2$. If you graph this function, you can see what this means visually: the graph of f has a jump at the point $t = 2$. We can redefine f so that the right and left hand limits match but f is still not continuous at $t = 2$ as follows. Let f be defined as

$$f(t) = \begin{cases} t^2, & 0 \le t < 2, \\ 10, & t = 2, \\ t^2, & 2 < t \le 4 \end{cases}$$

This function f has both the right and left hand limits equal to 4, so the limit of f as t approaches 2 does exist. However, the limit value 4 does not match the function value $f(2) = 10$ and so this function is not continuous at $t = 2$. Of course, this lack of continuity is there simply because we defined the function badly at $t = 2$. Since this lack of continuity can easily be removed by just redefining f as

$$f(t) = \begin{cases} t^2, & 0 \le t < 2, \\ 4, & t = 2, \\ t^2, & 2 < t \le 4 \end{cases}$$

i.e., $f(t) = t^2$ on $[0, 4]$, we call this a *removeable discontinuity*. A more sophisticated example is the function $g(t) = \frac{\sin(t)}{t}$ which is undefined at $t = 0$. It is well-known that the limit as $t \to 0$ of $g(t)$ is 1. Hence, we can define g so that is it continuous for all t as follows:

$$g(t) = \begin{cases} \frac{\sin(t)}{t}, & t \neq 0, \\ 1, & t = 0. \end{cases}$$

In a similar way, if f has a derivative at x_0 in its domain, it means that the limit of the slopes of the lines through the pairs $(x_0, f(x_0))$ and $(x, f(x))$ exists and equals some number B. More formally, we say f is differentiable at x_0 if there is a number B so that as x approaches x_0 the slopes $\frac{f(x)-f(x_0)}{x-x_0}$ approach the value B. In our more technical $\epsilon - \delta$ language, if a positive tolerance ϵ is chosen, then there is a positive δ so that

$$|x - x_0| < \delta \implies \left| \frac{f(x)-f(x_0)}{x-x_0} - B \right| < \epsilon.$$

The value B is usually denoted by the symbol $f'(x_0)$ and the right and left hand limits are then denoted by $(f')^+(x_0)$ and $(f')^-(x_0)$, respectively. It is possible for the derivative f' to exist on an interval, but fail to be continuous at some point in the interval. Consider this function f

$$f(t) = \begin{cases} t^2 \sin\left(\frac{1}{t}\right), & t \neq 0, \\ 1, & t = 0 \end{cases}$$

The function $\lim_{t \to 0} f(t) = 0$ and so we have removed the discontinuity f would normally have at $t = 0$ by defining it to have that value at 0. Then f is continuous at 0 and we have for $t \neq 0$ that

$$f'(t) = 2t \sin\left(\frac{1}{t}\right) - 2\cos\left(\frac{1}{t}\right).$$

Then, we find

$$f'(0) = \lim_{t \to 0} \frac{t^2 \sin\left(\frac{1}{t}\right) - 0}{t}$$

$$= \lim_{t \to 0} t \sin\left(\frac{1}{t}\right)$$

Now make a change of variable and let $y = 1/t$. Then the above limit changes to

$$f'(0) = \lim_{y \to \infty} \frac{\sin(y)}{y}$$

which we clearly see is 0 as the numerator is bounded between -1 and 1 yet the denominator grows without bound. Hence, $f'(0)$ exists and equals 0. However, the limit as t approaches 0 from either direction gives

$$\lim_{t \to 0} 2t \sin\left(\frac{1}{t}\right) - 2\cos\left(\frac{1}{t}\right)$$

which does not exist as the second term $-2\cos\left(\frac{1}{t}\right)$ oscillates between -2 and 2 in any circle $(-r, r)$ around 0 no matter how small r is chosen. Hence, the limit can not exist. Hence, since the limit does not exist, the definition of continuity for f' fails. We would need the limit to exist and for the value of the limit to match the value of the function. So here, we have f' exists for all t but f' fails to be continuous at $t = 0$. We see strange things can happen with functions!

9.2.2 Sequences

A sequence of real numbers is just a list such as this

$$\{c_1, c_2, c_3, \ldots\}$$

The numbers in this list do not have to have any pattern at all such as the sequence of sine values using radians for the argument:

$$\{c_1 = \sin(1), \; c_2 = \sin(2), \; c_3 = \sin(3), \ldots, \; c_n = \sin(n), \; \ldots\}$$

The sequence could be increasing such as

$$\{c_1 = 1, \; c_2 = 2, \; c_3 = 3, \ldots, \; c_n = n, \ldots\}$$

which simply increases without bound or it could be a sequence which alternates between to fixed values such as

$$\{c_1 = (-1)^1, \; c_2 = (-1)^2, \; c_3 = (-1)^3, \ldots, \; c_n = (-1)^n, \; \ldots\}.$$

A more interesting simple example is the collection

$$\left\{ c_1 = 1, \; c_2 = \frac{1}{2}, \; c_3 = \frac{1}{3}, \ldots \right\}$$

which can be written using the formula $c_n = \frac{1}{n}$. As n grows without bound, i.e. as $n \to \infty$, we can see that $c_n \to 0$. Next, note the sequence

$$\left\{ c_1 = 2, \; c_2 = 1 + \frac{1}{2}, \; c_3 = 1 + \frac{1}{3}, \ldots \right\}$$

can be written using the formula $c_n = 1 + \frac{1}{n}$ and clearly $c_n \to 1$ as $n \to \infty$. We need a more precise way of saying the values of a sequence approach a fixed number as for larger indices n. So let's state this idea more generally. If (c_n) is a sequence of real numbers, we say the sequence (c_n) converges to the number A and write $\lim_{n \to \infty} c_n = A$. We can state this using the more formal language of tolerances as follows. We say the sequence (c_n) converges to the number A if for any positive tolerance ϵ there is a positive integer N so that

$$n > N \implies |c_n - A| < \epsilon.$$

Less precisely, this means that the difference between the numbers c_n and the limiting value A is arbitrarily small once n is large enough. Now, for a simple sequence of real numbers, there are special types of *lack of convergence* which we call *divergence*. The sequence (c_n) might never stop growing, the example is $c_n = n$ which moves to ∞ as $n \to \infty$; the sequence (c_n) has limit $-\infty$, the example is $c_n = -n$; and the sequence oscillates between a finite number of fixed choices. One example of this is $c_n = (-1)^n$ which flips between -1 and 1 and so there is no fixed number A it can have as a limiting value. Another example is $c_n = \sin\left(\frac{n\pi}{3}\right)$ which moves between the values $\{\sin\left(\frac{\pi}{3}\right), 0, \sin\left(\frac{2\pi}{3}\right)\}$ none of which can be the limiting value A.

9.2.3 Series of Positive Terms

If you add up positive numbers, it is common sense to realize the partial sums you get never go down in value. Hence, if (a_n) is some sequence of positive numbers, then the partial sums $s_n = \sum_{i=1}^{n} a_i$ are always going up in value. If we allow some of the terms to be zero, then it is possible for these sums to plateau until a new positive term is added in, so if we look at the partial sums of non negative terms, we would say they are non decreasing to reflect the fact that the partial sums might plateau as we said. The behavior of these partial sums is fairly easy to predict. The partial sums either never stop growing or they grow towards some upper bound. Let's look at the growth towards an upper bound idea first. Take the open interval $(0, 2)$. Notice that if x is in $(0, 2)$, x is always bounded above by 2. Further, it is easy to see that although other numbers such as 3, 2.5, 2.1 and so forth are also upper bounds for this interval, 2 is the smallest or least such upper bound. We call such a least upper bound, the *supremum* of this set. We can be more formal with these ideas but this is enough for us here. For another example, consider the function $\tanh(x)$ which has the two horizontal asymptotes -1 and 1. We know for any real number x, $\tanh(x)$ is always above -1 and below 1. So the supremum of the set $\{\tanh(x) : -\infty < x < \infty\}$ is 1. A little thought shows the number -1 is the greatest possible lower bound. We call such a number the *infimum* of the set. Now, the big idea here is that a *supremum* is not the same as a *maximum*. Look at the example of the interval $(0, 2)$. The value 2 is not in the interval, so we can't say that there is a maximum value of x in this interval. We *can* say that 2 is the supremum of the values of x from this interval though, even though there is no maximum value. In the second example, 1 is the supremum of the set of $\tanh(x)$ values but there is no value of x where $\tanh(x) = 1$. Hence, the value of 1 can not be achieved using any value of x. So again, it is true 1 is the supremum of the set of possible values $\tanh(x)$ even though a maximum value is not achieved. Let's look at a third example using the function values $-x^2$. The values $\{-x^2 : -\infty < x < \infty\}$ have a supremum value of 0 and this supremum value is achieved at the point 0; hence, we know this set of values does have a maximum value.

Let's be a bit more precise and prove a small result. Suppose (a_n) is a sequence of non negative terms and let (S_n) be its corresponding sequence of partial sums. Since the partial sums are always non decreasing in this case, we can see that the sequence of partial sums either increases to ∞ or it has a finite supremum, S. If it increases to ∞, we say the sequence of partial sums *diverges* to ∞. We can prove that if the sequence of partial sums has a finite supremum S, the $S_n \to S$ is the formal way we defined earlier.

To do that, pick a positive tolerance ϵ. Is it possible that every partial sum S_n lies below or equals the supremum minus this tolerance, $S - \epsilon$? Suppose that was true. Then, we would have the inequality $S_n \le S - \epsilon$ for n. But that would say that the value $S - \epsilon$ was an upper bound also. However, when we defined the supremum of a set of numbers, we noted that the supremum is the least such upper bound. So, since S is the least upper bound value, we also have the inequality $S \le S - \epsilon$! But this makes no sense as S can not be less than itself. So we must conclude that our assumption is wrong. All the values S_n can't be below or equal $S - \epsilon$. There has to be at least one value, S_n, so that we have the inequalities $S - \epsilon < S_n \le S$.

Now we can show the partial sums of a sequence of non negative terms must converge to its supremum value as long as the supremum is finite. Pick any positive tolerance ϵ. By the argument above, there is at least one S_N satisfying $S - \epsilon < S_N \le S$. However, the partial sums are always non decreasing, so we can say more. Since $S_N \le S_n$ for all $n > N$, we have $S - \epsilon < S_N \le S_n \le S$ for all $n > N$. Now just rearrange this a bit and we see

$$n > N \implies 0 \le S - S_n < \epsilon.$$

This is precisely how we defined the convergence of a sequence. We can say now $S_n = \sum_{i=1}^{n} a_n \to S$ when the supremum S is finite. We use a special notation for this limiting value S. We write $S = \sum_{i=n_0}^{\infty} a_i$ to denote the supremum or limiting value of the sequence of partial sums when the underlying sequence is $\{a_{n_0}, a_{n_0} + 1, \ldots, \}$. This is called an *infinite series* of real numbers and this compact notation is very useful. Another way of writing the convergence statement then is as

$$n > N \implies 0 \leq \sum_{i=1}^{\infty} a_i - \sum_{i=1}^{n} a_i < \epsilon.$$

where for expositional convenience we let the starting value $n_0 = 1$, although of course, it could be 2 or 20 etc. The term $\sum_{i=1}^{\infty} a_i - \sum_{i=1}^{n} a_i$ occurs so frequently, that it is often written as the symbol $\sum_{i=n+1}^{\infty} a_i$. This is often called the *remainder of the series* from the index $n + 1$ on. Of course, this is just a symbol, like many of the symbols we have in calculus, for a fairly complicated limit idea.

Hence, in this section, we have found that if $\sum_{i=1}^{\infty} a_i < \infty$, the sequence of partial sums converges.

9.2.4 Series of Functions

Now let's look at sequences of functions made up of building blocks of the form $u_n(x) = \cos\left(\frac{n\pi}{L}\right)$ or $v_n(x) = \sin\left(\frac{n\pi}{L}\right)$ for various values of the integer n. The number L is a fixed value here. We can combine these functions into finite sums: let $U_N(x)$ and $V_N(x)$ be defined as follows:

$$U_N(x) = \sum_{n=1}^{N} a_n \sin\left(\frac{n\pi}{L}x\right).$$

and

$$V_N(x) = b_0 + \sum_{n=1}^{N} b_n \cos\left(\frac{n\pi}{L}x\right).$$

If we fixed the value of x to be say, x_0, the collection of numbers

$$\{U_1(x_0), \ U_2(x_0), \ U_3(x_0), \ldots, \ U_n(x_0), \ \ldots\}$$

and

$$\{V_0(x_0), \ V_1(x_0), \ V_2(x_0), \ V_3(x_0), \ldots, \ V_n(x_0), \ \ldots\}$$

are the partial sums formed from the sequences of cosine and sine numbers. However, the underlying sequences can be negative, so these are not sequences of non negative terms like we previously discussed. These sequences of partial sums may or may not have a finite supremum value. Nevertheless, we still represent the supremum using the same notation: i.e. the supremum of $\left(U_i(x_0)\right)_{i=1}^{\infty}$ and the supremum of $\left(V_i(x_0)\right)_{i=0}^{\infty}$ can be written as $\sum_{n=1}^{\infty} a_n \sin\left(\frac{n\pi}{L}x\right)$ and $b_0 + \sum_{n=1}^{\infty} b_n \cos\left(\frac{n\pi}{L}x\right)$.

Let's consider the finite sequence

$$\{U_1(x_0),\; U_2(x_0),\; U_3(x_0), \ldots, \; U_n(x_0)\}.$$

This sequence of real numbers converges to a possibly different number for each x_0; hence, let's call this possible limit $S(x_0)$. Now the limit may not exist, of course. We will write $\lim_{n\to} U_n(x_0) = S(x_0)$ when the limit exists. If the limit does not exist for some value of x_0, we will understand that the value $S(x_0)$ is not defined in some way. Note, from our discussion above, this could mean the limiting value flips between a finite set of possibilities, the limit approaches ∞ or the limit approaches $-\infty$. In any case, the value $S(x_0)$ is not defined as a finite value. We would say this precisely as follows: given any positive tolerance ϵ, there is a positive integer N so that

$$n > N \implies \left| \sum_{i=1}^{n} a_i \sin\left(\frac{i\pi}{L}\right) - S(x_0) \right| < \epsilon.$$

We use the notation of the previous section and write this as

$$\lim_{n\to\infty} \sum_{i=1}^{n} a_i \sin\left(\frac{i\pi}{L}\right) = S(x_0).$$

with limiting value, $S(x_0)$ written as

$$S(x_0) = \sum_{i=1}^{\infty} a_i \sin\left(\frac{i\pi}{L}\right),$$

As before, this symbol is called an *infinite series* and we see we get a potentially different series at each point x_0. The error term $S(x_0) - U_n(x_0)$ is then written as

$$S(x_0) - \sum_{i=1}^{n} a_i \sin\left(\frac{i\pi}{L}x_0\right) = \sum_{i=n+1}^{\infty} a_i \sin\left(\frac{i\pi}{L}x_0\right),$$

which you must remember is just a short hand for this error. Now that we have an infinite series notation defined, we note the term $U_n(x_0)$, which is the sum of n terms, is also called the *n*th partial sum of the series $\sum_{i=1}^{\infty} a_i \sin\left(\frac{i\pi}{L} x_0\right)$. Note we can define the convergence at a point x_0 for the partial sums of the cos functions in a similar manner.

9.2.5 Fourier Series

A general trigonometric series $S(x)$ has the following form

$$S(x) = b_0 + \sum_{i=1}^{\infty} \left(a_n \sin\left(\frac{i\pi}{L} x\right) + b_n \cos\left(\frac{i\pi}{L} x\right) \right)$$

for any numbers a_n and b_n. Of course, there is no guarantee that this series will converge at any x! If we start with a function f which is continuous on the interval $[0, L]$, we can define the trigonometric series associated with f as follows

$$S(x) = \frac{1}{L} <f, 1>$$
$$+ \sum_{i=1}^{\infty} \left(\frac{2}{L} \left\langle f(x), \sin\left(\frac{i\pi}{L} x\right) \right\rangle \sin\left(\frac{i\pi}{L} x\right) + \frac{2}{L} \left\langle f(x), \cos\left(\frac{i\pi}{L} x\right) \right\rangle \cos\left(\frac{i\pi}{L} x\right) \right).$$

This series is called the Fourier Series for f and the coefficients in the Fourier series for f are called the *Fourier coefficients* of f. Since these coefficients are based on inner products with the normalized sin and cos functions, we could call these the *normalized* Fourier coefficients. Let's be clear about this and a bit more specific. The *n*th *Fourier* sin *coefficient*, $n \geq 1$, of f is as follows:

$$a_n(f) = \frac{2}{L} \int_0^L f(x) \sin\left(\frac{i\pi}{L} x\right) dx$$

The *n*th *Fourier* cos *coefficient*, $n \geq 0$, of f are defined similarly:

$$b_0(f) = \frac{1}{L} \int_0^L f(x)\, dx$$
$$b_n(f) = \frac{2}{L} \int_0^L f(x) \cos\left(\frac{i\pi}{L} x\right) dx, \quad n \geq 1.$$

9.3   The Cable Model Infinite Series Solution

We now know series of the form

$$A_0 e^{-\frac{1}{\alpha}t} + \sum_{n=1}^{N} A_n \cos\left(\frac{n\pi}{L}x\right) e^{-\frac{1+n^2\pi^2\beta^2}{\alpha L^2}t}.$$

are like Fourier series although in terms of two variables. We will show these series converge pointwise for x in $[0, L]$ and all t. We can show that we can take the partial derivative of this series solutions term by term to obtain

$$\sum_{n=1}^{N} -A_n \frac{n\pi}{L} \sin\left(\frac{n\pi}{L}x\right) e^{-\frac{1+n^2\pi^2\beta^2}{\alpha L^2}t}.$$

This series evaluated at $x = 0$ and $x = L$ gives 0 and hence the Neumann conditions are satisfied. Hence, the solution $\Phi(x, t)$ given by

$$\Phi(x, t) = A_0 e^{-\frac{1}{\alpha}t} + \sum_{n=1}^{N} A_n \cos\left(\frac{n\pi}{L}x\right) e^{-\frac{1+n^2\pi^2\beta^2}{\alpha L^2}t}.$$

for the arbitrary sequence of constants (A_n) is a well-behaved solution on our domain. The remaining boundary condition is

$$\Phi(x, 0) = f(x), \ \ \text{for } 0 \leq x \leq L$$

and

$$\Phi(x, 0) = A_0 + \sum_{n=1}^{\infty} A_n \cos\left(\frac{n\pi}{L}x\right).$$

Rewriting in terms of the series solution, for $0 \leq x \leq L$, we find

$$A_0 + \sum_{n=1}^{\infty} A_n \cos\left(\frac{n\pi}{L}x\right) = f(x)$$

The Fourier series for f is given by

$$f(x) = B_0 + \sum_{n=1}^{\infty} B_n \cos\left(\frac{n\pi}{L}x\right)$$

with

$$B_0 = \frac{1}{L} \int_0^L f(x),$$

$$B_n = \frac{2}{L} \int_0^L f(x) \cos\left(\frac{n\pi}{L}x\right) dx.$$

Then, setting these series equal, we find that the solution is given by $A_n = B_n$ for all $n \geq 0$.

9.3.1 Homework

Solve the following cable models using separation of variables. Recall from Sect. 7.5.1 that we can calculate the Fourier series coefficients for a delta function of the form $H\delta(x - x_0)$ easily to obtain

$$B_0 = \frac{1}{L} \int_0^L H\delta(x - x_0)\, dx = \frac{1}{L}$$

$$B_n = \frac{2}{L} \int_0^L H\delta(x - x_0) \sin\left(\frac{i\pi}{L}x\right) dx = \frac{2}{L} \sin\left(\frac{i\pi}{L}x_0\right), \quad n \geq 1.$$

You can use these results to work the exercises below.

Exercise 9.3.1

$$0.0009\frac{\partial^2 \Phi}{\partial x^2} - \Phi - 0.006\frac{\partial \Phi}{\partial t} = 0, \text{ for } 0 \leq x \leq 10, \ t > 0,$$

$$\frac{\partial \Phi}{\partial x}(0, t) = 0,$$

$$\frac{\partial \Phi}{\partial x}(10, t) = 0,$$

$$\Phi(x, 0) = f(x).$$

where $f(x) = 100\delta(x - 3)$, *the pulse of magnitude* 100 *applied instantaneously at position* $x = 3$.

Exercise 9.3.2

$$0.25\frac{\partial^2 \Phi}{\partial x^2} - \Phi - 0.0008\frac{\partial \Phi}{\partial t} = 0, \text{ for } 0 \leq x \leq 100, \ t > 0,$$

$$\frac{\partial \Phi}{\partial x}(0, t) = 0,$$

$$\frac{\partial \Phi}{\partial x}(100, t) = 0,$$

$$\Phi(x, 0) = f(x).$$

where $f(x) = 200\delta(x - 4)$, the pulse of magnitude 100 applied instantaneously at position $x = 4$.

Exercise 9.3.3

$$0.25\frac{\partial^2 \Phi}{\partial x^2} - \Phi - 0.0008 \frac{\partial \Phi}{\partial t} = 0, \ for \ 0 \le x \le 100, \ t > 0,$$

$$\Phi(0, t) = 0,$$

$$\Phi(10, t) = 0,$$

$$\Phi(x, 0) = f(x).$$

where $f(x) = 300\delta(x - 3)$, the pulse of magnitude 200 applied instantaneously at position $x = 3.5$.

Exercise 9.3.4

$$0.125\frac{\partial^2 \Phi}{\partial x^2} - \Phi - 0.004 \frac{\partial \Phi}{\partial t} = 0, \ for \ 0 \le x \le 15, \ t > 0,$$

$$\Phi(0, t) = 0,$$

$$\Phi(15, t) = 0,$$

$$\Phi(x, 0) = f(x).$$

where $f(x) = 500\delta(x - 8)$, the pulse of magnitude 300 applied instantaneously at position $x = 9.5$.

9.4   The Membrane Voltage Cable Equation

Now our interest is the cable equation which we have derived using ideas from biology and physics. For the cable equation, we think of the dendritic cylinder as having end caps made of the same membrane as its walls. For boundary conditions, we will impose what are called **zero-rate end cap** potentials–this means that each end cap is at equipotential: hence the rate of change of the potential v_m with respect to the space variable λ is zero at both ends. Thus, we look the solution to the homogeneous full transient cable equation

$$\lambda_C^2 \frac{\partial^2 v_m(z, t)}{\partial z^2} = v_m(z, t) + \tau_M \frac{\partial v_m}{\partial t}, \ 0 \le z \le \ell, \ t \ge 0, \tag{9.5}$$

$$\frac{\partial v_m(0, t)}{\partial z} = 0, \tag{9.6}$$

$$\frac{\partial v_m(\ell, t)}{\partial z} = 0, \tag{9.7}$$

$$v_m(z, 0) = f(z), \ \ 0 \le z \le \ell. \tag{9.8}$$

This is the general cable equation with $\beta = \lambda_C$ and $\alpha = \tau$. For convenience, we can replace Eqs. 9.5–9.8 with a normalized form using the transformations

$$\tau = \frac{t}{\tau_M}, \ \ \lambda = \frac{z}{\lambda_C}, \ \ L = \frac{\ell}{\lambda_C},$$
$$\hat{v}_m = v_m(\lambda_C \lambda, \tau_M \tau).$$

This gives us the system

$$\frac{\partial^2 \hat{v}_m}{\partial \lambda^2} = \hat{v}_m + \frac{\partial \hat{v}_m}{\partial \tau}, \ \ 0 \le \lambda \le L, \ \tau \ge 0 \tag{9.9}$$

$$\frac{\partial \hat{v}_m(0, \tau)}{\partial \lambda} = 0 \tag{9.10}$$

$$\frac{\partial \hat{v}_m(L, \tau)}{\partial \lambda} = 0, \tag{9.11}$$

$$\hat{v}_m(\lambda, 0) = \hat{f}(z), \ \ 0 \le \lambda \le L. \tag{9.12}$$

where $\hat{f} = f(\lambda_C \lambda)$. This is the general cable equation with $\beta = 1$ and $\alpha = 1$. As usual, we assume a solution of the form $\hat{v}_m(\lambda, \tau) = u(\lambda) w(\tau)$. We have already solved this in Sect. 7.2.1 so we can just state the results now. The solution is

$$\hat{v}_m(\lambda, \tau) = A_0 e^{-\tau} + \sum_{n=1}^{\infty} A_n \cos\left(\frac{n\pi}{L}\lambda\right) e^{-\frac{1+n^2\pi^2}{L^2}\tau}.$$

for the arbitrary sequence of constants (A_n). The constants A_n are then chosen to be the Fourier coefficients of $\hat{f}$. The Fourier series for f is given by

$$f(x) = B_0 + \sum_{n=1}^{\infty} B_n \cos\left(\frac{n\pi}{L}x\right)$$

with

$$B_0 = \frac{1}{L}\int_0^L f(x)dx,$$

$$B_n = \frac{2}{L}\int_0^L f(x)\cos\left(\frac{n\pi}{L}x\right)dx.$$

The solution is given by choosing $A_n = B_n$ for all $n \ge 0$. We now need to look at the convergence of Fourier series as our solutions require it to make sense!

9.5 Fourier Series Components

To prove that Fourier series converge pointwise for the kind of data functions f we want to use, requires a lot more work. So let's get started. Let's go back and look at the sin and cos functions we have been using. They have many special properties.

9.5.1 Orthogonal Functions

We will now look at some common sequences of functions on the domain $[0, L]$ that are very useful in solving models: the sequences of functions are $\left(\sin\left(\frac{i\pi}{L}x\right)\right)$ for integers $i \geq 1$ and $\left(\cos\left(\frac{i\pi}{L}x\right)\right)$ for integers $i \geq 0$. The second sequence is often written $\{1, \cos\left(\frac{\pi}{L}x\right), \cos\left(\frac{2\pi}{L}x\right), \ldots\}$. Here L is a positive number which often is the length of a cable or spatial boundary.

9.5.1.1 The Sine Sequence

Let's look carefully at the interval $[0, L]$. Using the definition of inner product on $C[0, L]$, by direct integration, we find for $i \neq j$

$$\left\langle \sin\left(\frac{i\pi}{L}x\right), \sin\left(\frac{j\pi}{L}x\right)\right\rangle = \int_0^L \sin\left(\frac{i\pi}{L}x\right)\sin\left(\frac{j\pi}{L}x\right)dx$$

Now use the substitution, $y = \frac{\pi x}{L}$ to rewrite this as

$$\left\langle \sin\left(\frac{i\pi}{L}x\right), \sin\left(\frac{j\pi}{L}x\right)\right\rangle = \frac{L}{\pi}\int_0^\pi \sin(iy)\sin(jy)dy$$

Now the trigonometric substitutions

$$\cos(u + v) = \cos(u)\,\cos(v) - \sin(u)\,\sin(v),$$
$$\cos(u - v) = \cos(u)\,\cos(v) + \sin(u)\,\sin(v)$$

imply

$$\sin(iy)\sin(jy) = \frac{1}{2}\Big(\cos((i-j)y) - \cos((i+j)y)\Big).$$

Using this identity in the integration, we see

$$\frac{L}{\pi} \int_0^\pi \sin(iy)\sin(jy)dy = \frac{L}{\pi} \int_0^\pi \left(\cos\Big((i-j)y\Big) - \cos\Big((i+j)y\Big) \right) dy$$

$$= \frac{L}{\pi} \left(\frac{\sin\Big((i-j)y\Big)}{i-j} \Big|_0^\pi + \frac{\sin\Big((i+j)y\Big)}{i+j} \Big|_0^\pi \right)$$

$$= 0.$$

Hence, the functions $\sin\left(\frac{i\pi}{L}x\right)$ and $\sin\left(\frac{j\pi}{L}x\right)$ are orthogonal on $[0, L]$ if $i \neq j$. On the other hand, if $i = j$, we have, using the same substitution $y = \frac{\pi x}{L}$, that

$$\left\langle \sin\left(\frac{i\pi}{L}x\right), \sin\left(\frac{i\pi}{L}x\right) \right\rangle \frac{L}{\pi} \int_0^\pi \sin(iy)\sin(iy)dy = \frac{L}{\pi} \int_0^\pi \left(\frac{1-\cos(2iy)}{2}\right)dy$$

using the identify $\cos(2u) = 1 - 2\sin^2(u)$. It then follows that

$$\frac{L}{\pi} \int_0^\pi \sin(iy)\sin(iy)dy = \frac{L}{\pi} \left(\frac{y}{2} - \frac{\sin(2y)}{4}\right)\Big|_0^\pi$$

$$= \frac{L}{2}.$$

Hence, letting $u_n(x) = \sin\left(\frac{n\pi}{L}x\right)$, we have shown that

$$< u_i, u_j > = \begin{cases} \frac{L}{2}, & i = j \\ 0, & i \neq j. \end{cases}$$

Now define the new functions $\hat{u}_n$ by $\sqrt{\frac{2}{L}}u_n$. Then, $< \hat{u}_i, \hat{u}_j >= \delta_i^j$ and the sequence of functions $(\hat{u}_n)$ are all mutually orthogonal. Next, define the length of a function f by the symbol $\|f\|$ where

$$\|f\| = \sqrt{\int_0^L f^2(t)dt}.$$

Using this definition of length, it is clear $\|\hat{u}_n\| = 1$ always. So the sequence of functions $(\hat{u}_n)$ are all mutually orthogonal and length one.

9.5.1.2   The Cosine Sequence

Let's look carefully at on the interval $[0, L]$. Using the definition of inner product on $C[0, L]$, by direct integration, we find for $i \neq j$ with both i and j at least 1, that

$$\left\langle \cos\left(\frac{i\pi}{L}x\right), \cos\left(\frac{j\pi}{L}x\right) \right\rangle = \int_0^L \cos\left(\frac{i\pi}{L}x\right) \cos\left(\frac{j\pi}{L}x\right) dx$$

Again use the substitution, $y = \frac{\pi x}{L}$ to rewrite this as

$$\left\langle \cos\left(\frac{i\pi}{L}x\right), \cos\left(\frac{j\pi}{L}x\right) \right\rangle = \frac{L}{\pi} \int_0^\pi \cos(iy) \cos(jy) dy$$

Again, the trigonometric substitutions

$$\cos(u + v) = \cos(u)\,\cos(v) - \sin(u)\,\sin(v),$$
$$\cos(u - v) = \cos(u)\,\cos(v) + \sin(u)\,\sin(v)$$

imply

$$\cos(iy) \cos(jy) = \frac{1}{2}\left(\cos((i + j)y) + \cos((i - j)y) \right).$$

Using this identity in the integration, we see

$$\frac{L}{\pi} \int_0^\pi \cos(iy) \cos(jy) dy = \frac{L}{\pi} \int_0^\pi \left(\cos\left((i + j)y\right) + \cos\left((i - j)y\right) \right) dy$$

$$= \frac{L}{\pi} \left(\frac{\sin\left((i + j)y\right)}{i + j} \Big|_0^\pi + \frac{\sin\left((i - j)y\right)}{i - j} \Big|_0^\pi \right)$$

$$= 0.$$

Hence, the functions $\cos\left(\frac{i\pi}{L}x\right)$ and $\cos\left(\frac{j\pi}{L}x\right)$ are orthogonal on $[0, L]$ if $i \neq j$, $i, j \geq 1$. Next, we consider the case of the inner product of the function 1 with a $\cos\left(\frac{i\pi}{L}x\right)$ for $j \geq 1$. This gives

$$\left\langle 1, \cos\left(\frac{j\pi}{L}x\right) \right\rangle = \int_0^L \cos\left(\frac{j\pi}{L}x\right) dx$$

$$= \frac{L}{\pi} \int_0^\pi \cos(jy) dy$$

$$= \frac{L}{\pi} \frac{\sin(jy)}{j} \Big|_0^\pi$$

$$= 0.$$

Thus, the functions 1 and $\cos\left(\frac{i\pi}{L}x\right)$ for any $j \geq 1$ are also orthogonal. On the other hand, if $i = j$, we have for $i \geq 1$,

$$\left\langle \cos\left(\frac{i\pi}{L}x\right), \cos\left(\frac{i\pi}{L}x\right) \right\rangle = \frac{L}{\pi} \int_0^\pi \cos(iy)\cos(iy) dy$$

$$= \frac{L}{\pi} \int_0^\pi \left(\frac{1 + \cos(2iy)}{2}\right) dy$$

using the identify $\cos(2u) = 2\cos^2(u) - 1$. It then follows that

$$\frac{L}{\pi} \int_0^\pi \cos(iy)\cos(iy) dy = \frac{L}{\pi} \left(\frac{y}{2} + \frac{\sin(2y)}{4}\right) \Big|_0^\pi$$

$$= \frac{L}{2}.$$

We also easily find that $< 1, \; 1 >= L$. Hence, on $[0, L]$, letting

$$v_0(x) = 1$$

$$v_n(x) = \cos\left(\frac{n\pi}{L}x\right), \; n \geq 1,$$

we have shown that

$$< v_i, v_j > = \begin{cases} \frac{1}{L}, & i = j = 1, \\ \frac{L}{2}, & i = j, \; i \geq 1 \\ 0, & i \neq j. \end{cases}$$

Now define the new functions

$$\hat{v}_0(x) = \sqrt{\frac{1}{L}}$$

$$\hat{v}_n(x) = \sqrt{\frac{2}{L}} \cos\left(\frac{n\pi}{L}x\right), \; n \geq 1,$$

$\hat{v}_n$ by $\sqrt{\frac{2}{L}} v_n$. Then, $< \hat{v}_i, \hat{v}_j >= \delta_i^j$ and the sequence of functions $(\hat{v}_n)$ are all mutually orthogonal with length $||v_n|| = 1$.

9.5.1.3 Fourier Coefficients Revisited

Let f be any continuous function on the interval $[0, L]$. Then we know f is Riemann integrable and so $||f|| = \sqrt{\int_0^L f(t)^2 dt}$ is a finite number. From Sect. 9.5.1.1, recall the sequence of functions $\{\sin\left(\frac{\pi}{L}x\right), \sin\left(\frac{2\pi}{L}x\right), \ldots\}$ and $\{1, \cos\left(\frac{\pi}{L}x\right), \cos\left(\frac{2\pi}{L}x\right), \ldots\}$ are mutually orthogonal sequences on the interval $[0, L]$. Further, we know that we can divide each of the functions by their length to create the orthogonal sequences of length one we called $(\hat{u}_n)$ for $n \geq 1$ and $(\hat{v}_n)$ for $n \geq 0$, respectively. Consider for any n

$$0 \leq \; <f - \sum_{i=1}^{n} <f, \hat{u}_i> \hat{u}_i, f - \sum_{j=1}^{n} <f, \hat{u}_j> \hat{u}_j >$$

$$= <f,f> -2\sum_{i=1}^{n}(<f, \hat{u}_i>)^2 + \sum_{i=1}^{n}\sum_{j=1}^{n} <f, \hat{u}_i><f, \hat{u}_j> <\hat{u}_i, \hat{u}_j>$$

$$= <f,f> -2\sum_{i=1}^{n}(<f, \hat{u}_i>)^2 + \sum_{i=1}^{n}(<f, \hat{u}_i>)^2$$

$$= <f,f> - \sum_{i=1}^{n}(<f, \hat{u}_i>)^2,$$

where we use the fact that the $<\hat{u}_i, \hat{u}_j> \; = \; \delta_i^j$. Hence, we conclude that for all n, $\sum_{i=1}^{n}(<f, \hat{u}_i>)^2 \leq ||f||^2$. A similar argument shows that for all n, $\sum_{i=0}^{n}(<f, \hat{v}_i>)^2 \leq ||f||^2$. The numbers $<f, \hat{u}_i>$ and $<f, \hat{v}_i>$ are called the ith **Fourier sine** and ith **Fourier cosine** coefficients of f, respectively. Note this is similar to the Fourier coefficients defined earlier but we are writing them in terms of the normalized sin and cos functions. In fact, in the context of the sums, they are the *same*; i.e.

$$<f, \hat{u}_i> \hat{u}_i = \left\langle f, \sqrt{\frac{2}{L}}u_i \right\rangle \sqrt{\frac{2}{L}}u_i = \frac{2}{L} <f, u_i> u_i$$

is the same term in both summations. We can do a similar expansion for the cos functions. So depending on where you read about these things, you could see them defined either way. Thus, we can rewrite the finite sums here in terms of the original sin and cos functions. We have

$$\sum_{i=1}^{n} <f, \hat{u}_i> \hat{u}_i = \sum_{i=1}^{n} \frac{2}{L}\left\langle f(x), \sin\left(\frac{i\pi}{L}x\right) \right\rangle \sin\left(\frac{i\pi}{L}x\right)$$

and

$$\sum_{i=0}^{n} <f, \hat{v}_i > \hat{v}_i = \frac{1}{L} <f(x), 1 > 1 \sum_{i=1}^{n} \frac{2}{L}\left\langle f(x), \cos\left(\frac{i\pi}{L}x\right)\right\rangle \cos\left(\frac{i\pi}{L}x\right).$$

We often write sums like this in terms of the original sine and cosine functions: the coefficients $\frac{2}{L}\left\langle f(x), \sin\left(\frac{i\pi}{L}x\right)\right\rangle$ are thus also called the ith **Fourier sine coefficients** of f and the coefficients $\frac{1}{L}\left\langle f(x), 1\right\rangle$ and $\frac{2}{L}\left\langle f(x), \cos\left(\frac{i\pi}{L}x\right)\right\rangle$, the ith **Fourier cosine coefficients** of f. You can think of these as the *un normalized* Fourier coefficients for f, if you want as they are based on the *un normalized* sin and cos functions. This ambiguity in the definition is easy to remember and it comes about because it is easy to write these sorts of finite expansions of f in terms of its Fourier coefficients in terms of the original sine and cosine functions. We are now ready to discuss the convergence of what are called Fourier series. We start with the sin and cos sequences discussed above. We can say more about them.

9.5.1.4 Fourier Coefficients Go to Zero

We will now show the nth Fourier sine and cosine coefficients must go to zero. From the previous section, we know that $\sum_{i=1}^{n} | <f, \hat{u}_i > |^2 \le \|f\|^2$ and $\sum_{i=0}^{n} | <f, \hat{v}_i > |^2 \le \|f\|^2$ which tells us that the series $\sum_{i=1}^{\infty} | <f, \hat{u}_i > |^2 \le \|f\|^2$ and $\sum_{i=0}^{\infty} | <f, \hat{v}_i > |^2 \le \|f\|^2$. Hence, since these series of non negative terms are bounded above, they must converge. Pick any positive tolerance ϵ. Let's focus on the sin series first.

Since this series converges to say SS, there is a positive integer N so that $n > N \implies |S_n - SS| < \frac{\epsilon}{2}$, where S_n is the usual partial sum. Thus, we can say for any $n > N + 1$, we have

$$\begin{aligned}
|S_n - S_{n-1}| &= |S_n - SS + SS - S_{n-1}| \\
&\le |S_n - SS| + |SS - S_{n-1}| \\
&< \frac{\epsilon}{2} + \frac{\epsilon}{2} \\
&= \epsilon.
\end{aligned}$$

From the definition of partial sums, we know $S_n - S_{n-1} = (<f, \hat{u}_n >)^2$. Hence, we know $n > N \implies (<f, \hat{u}_n >)^2 < \epsilon$. But this implies $\lim_{n\to\infty} <f, \hat{u}_n >= 0$ which is the result we wanted to show. It is clear this is the same as

$$\lim_{n\to\infty} \frac{2}{L} \int_0^L f(x) \sin\left(\frac{n\pi}{L}x\right) = 0.$$

Now, a similar argument will show that $\lim_{n\to\infty} <f, \hat{v}_n> = 0$ or equivalently

$$\lim_{n\to\infty} \frac{2}{L} \int_0^L f(x) \cos\left(\frac{n\pi}{L}x\right) = 0.$$

9.6   The Convergence of Fourier Series

The Fourier series associated to the continuous function f on $[0, L]$ is the series

$$S(x) = \frac{1}{L} <f, \mathbf{1}>$$
$$+ \sum_{i=1}^{\infty} \left(\frac{2}{L}\left\langle f(x), \sin\left(\frac{i\pi}{L}x\right)\right\rangle \sin\left(\frac{i\pi}{L}x\right) + \frac{2}{L}\left\langle f(x), \cos\left(\frac{i\pi}{L}x\right)\right\rangle \cos\left(\frac{i\pi}{L}x\right)\right).$$

We will sneak up on the analysis of this series by looking instead at the Fourier series of f on the interval $[0, 2L]$. The sin and cos functions we have discussed in Sects. 9.5.1.1 and 9.5.1.2 were labeled u_n and v_n respectively. On the interval $[0, L]$, their lengths were $\sqrt{\frac{2}{L}}$ for each u_n and v_n for $n \geq 1$ and $\sqrt{\frac{1}{L}}$ for v_0. We used these lengths to define the normalized functions $\hat{u}_n$ and $\hat{v}_n$ which formed mutually orthogonal sequences. On the interval $[0, 2L]$ the situation is virtually the same. The only difference is that on $[0, 2L]$, we still have

$$u_n(x) = \sin\left(\frac{n\pi}{L}x\right), \ n \geq 1$$
$$v_0(x) = 1$$
$$v_n(x) = \cos\left(\frac{n\pi}{L}x\right), \ n \geq 1$$

but now, although we still have orthogonality, the lengths change. We find

$$<u_i, u_j> = \begin{cases} L, & i=j \\ 0, & i\neq j. \end{cases}$$

$$<v_i, v_j> = \begin{cases} \frac{1}{2L}, & i=j=1, \\ L, & i=j, \ i\geq 1 \\ 0, & i\neq j. \end{cases}$$

The normalized functions are now

$$\hat{u}_n(x) = \sqrt{\frac{1}{L}} \sin\left(\frac{n\pi}{L}x\right)$$

$$\hat{v}_0(x) = \sqrt{\frac{1}{2L}}$$

$$\hat{v}_n(x) = \sqrt{\frac{1}{L}} \cos\left(\frac{n\pi}{L}x\right), \ n \geq 1.$$

The argument given in Sect. 9.5.1.4 still holds with just obvious changes. So these Fourier coefficients go to zero as $n \to \infty$ also. This series on $[0, 2L]$ does not necessarily converge to the value $f(x)$ at each point in $[0, 2L]$; in fact, it is known that there are continuous functions whose Fourier series does not converge at all. Still, the functions of greatest interest to us are typically functions that have continuous derivatives except at a finite number of points and for those sorts of functions, $S(x)$ and $f(x)$ usually match. We are going to prove this in the work below. Consider the difference between a typical partial sum $S_N(x)$ and our possible target $f(x)$.

$$|S_N(x) - f(x)| = \left| \frac{1}{2L} <f, 1> \right.$$

$$+ \sum_{i=1}^{N} \left(\frac{1}{L}\left\langle f(x), \sin\left(\frac{i\pi}{L}x\right)\right\rangle \sin\left(\frac{i\pi}{L}x\right) \right.$$

$$\left. + \frac{1}{L}\left\langle f(x), \cos\left(\frac{i\pi}{L}x\right)\right\rangle \cos\left(\frac{i\pi}{L}x\right) \right) - f(x) \bigg|$$

$$= \left| \frac{1}{2L} \int_0^L f(t)\, dt \right.$$

$$+ \sum_{i=1}^{N} \frac{1}{L} \int_0^{2L} f(t) \left(\sin\left(\frac{i\pi}{L}t\right) \sin\left(\frac{i\pi}{L}x\right) \right.$$

$$\left. \left. + \cos\left(\frac{i\pi}{L}t\right) \cos\left(\frac{i\pi}{L}x\right) \right) dt - f(x) \right|$$

Now, $\sin(u)\sin(v) + \cos(u)\cos(v) = \cos(u-v)$ and hence we can rewrite the above as follows:

$$|S_N(x) - f(x)| = \left| \frac{1}{2L} \int_0^{2L} f(t)\, dt + \sum_{i=1}^{N} \frac{1}{L} \int_0^{2L} f(t) \left(\cos\left(\frac{i\pi}{L}(t-x)\right) \right) dt - f(x) \right|$$

$$= \left| \frac{1}{L} \int_0^{2L} f(t) \left(\frac{1}{2} + \sum_{i=1}^{N} \cos\left(\frac{i\pi}{L}(t-x)\right) \right) dt - f(x) \right|$$

9.6.1 Rewriting $S_N(x)$

Next, we use another identity. We know from trigonometry that

$$\cos(iy)\sin\left(\frac{y}{2}\right) = \sin\left(\left(i+\frac{1}{2}\right)y\right) - \sin\left(\left(i-\frac{1}{2}\right)y\right)$$

and so

$$\left(\frac{1}{2} + \sum_{i=1}^{N}\cos\left(\frac{i\pi}{L}(t-x)\right)\right)\sin\left(\frac{\pi}{2L}(t-x)\right)$$

$$= \frac{1}{2}\sin\left(\frac{\pi}{2L}(t-x)\right) + \sum_{i=1}^{N}\left(\sin\left(\left(i+\frac{1}{2}\right)\frac{\pi}{L}(t-x)\right) - \sin\left(\left(i-\frac{1}{2}\right)\frac{\pi}{L}(t-x)\right)\right)$$

$$= \frac{1}{2}\sin\left(\frac{\pi}{2L}(t-x)\right) + \frac{1}{2}\sin\left(\frac{3\pi}{2L}(t-x)\right) - \frac{1}{2}\sin\left(\frac{\pi}{2L}(t-x)\right)$$

$$+ \frac{1}{2}\sin\left(\frac{5\pi}{2L}(t-x)\right) - \frac{1}{2}\sin\left(\frac{3\pi}{2L}(t-x)\right)$$

$$+ \frac{1}{2}\sin\left(\frac{7\pi}{2L}(t-x)\right) - \frac{1}{2}\sin\left(\frac{5\pi}{2L}(t-x)\right)$$

$$\cdots$$

$$+ \frac{1}{2}\sin\left(\frac{(2N+1)\pi}{2L}(t-x)\right) - \frac{1}{2}\sin\left(\frac{(2N-1)\pi}{2L}(t-x)\right)$$

$$= \frac{1}{2}\sin\left(\left(N+\frac{1}{2}\right)\frac{\pi}{L}(t-x)\right)$$

We have found

$$\left(\frac{1}{2} + \sum_{i=1}^{N}\cos\left(\frac{i\pi}{L}(t-x)\right)\right) = \frac{\frac{1}{2}\sin\left(\left(N+\frac{1}{2}\right)\frac{\pi}{L}(t-x)\right)}{\sin\left(\frac{\pi}{2L}(t-x)\right)} \qquad (9.13)$$

Note the argument $t-x$ is immaterial and for any y, the identity can also be written as

$$\left(\frac{1}{2} + \sum_{i=1}^{N}\cos\left(\frac{i\pi}{L}y\right)\right) = \frac{\frac{1}{2}\sin\left(\left(N+\frac{1}{2}\right)\frac{\pi}{L}y\right)}{\sin\left(\frac{\pi}{2L}y\right)} \qquad (9.14)$$

Now, we use this to rewrite $S_N(x)$. We can then use this identity to find

$$S_N(x) = \frac{1}{L} \int_0^{2L} f(t) \left(\frac{\frac{1}{2} \sin\left(\left(N + \frac{1}{2}\right) \frac{\pi}{L}(t - x) \right)}{\sin\left(\frac{\pi}{2L}(t - x) \right)} \right) dt$$

Making the change of variable $y = t - x$, we have

$$\frac{1}{L} \int_0^{2L} f(t) \left(\frac{\frac{1}{2} \sin\left(\left(N + \frac{1}{2}\right) \frac{\pi}{L}(t - x) \right)}{\sin\left(\frac{\pi}{2L}(t - x) \right)} \right) dt = \frac{1}{L} \int_{-x}^{2L-x} f(y + x) \left(\frac{\frac{1}{2} \sin\left(\left(N + \frac{1}{2}\right) \frac{\pi}{L} y \right)}{\sin\left(\frac{\pi}{2L} y \right)} \right) dy.$$

Now, switch the integration variable y back to t to obtain the form we want:

$$S_N(x) = \frac{1}{L} \int_{-x}^{2L-x} f(t + x) \left(\frac{\frac{1}{2} \sin\left(\left(N + \frac{1}{2}\right) \frac{\pi}{L} t \right)}{\sin\left(\frac{\pi}{2L} t \right)} \right) dt.$$

We can rewrite this even more. All of the individual sin terms are periodic over the interval $[0, 2L]$. We extend the function f to be periodic also by defining

$$\hat{f}(x + 2nL) = f(x), \quad 0 < x < 2L.$$

and defining what happens at the endpoints $0, \pm 2L, \pm 4L$ and so on using one sided limits. Since the original f is continuous on $[0, 2L]$, we know $f(0^+)$ and $f(2L^-)$ both exists and match $f(0)$ and $f(2L)$, respectively. Because these two values need not be the same, the periodic extension will always have a potential discontinuity at the point $2nL$ for all integers n. We will define the periodic extension at these points as

$$\hat{f}(2nL^-) = f(2L^-), \quad \text{and} \quad \hat{f}(2nL^+) = f(0^+)$$
$$\hat{f}(2nL) = \frac{1}{2}(f(0^+) + f(2L^-)).$$

For example, if f is the square wave

$$f(x) = \begin{cases} H, & 0 \le x \le L, \\ 0, & L < 2L \end{cases}$$

which has a discontinuity at L, then the periodic extension will have discontinuities at each multiple $2nL$ as $\hat{f}(2L^-) = 0, f(0^+) = H$ and $f(2nL)$ is the average value. However, the value of the integral $\int_{-x}^{2L-x}$ and $\int_0^{2L}$ are still the same. Hence, $S_N(x)$ can be written as

$$S_N(x) = \frac{1}{L} \int_0^{2L} \hat{f}(t + x) \left(\frac{\frac{1}{2} \sin\left(\left(N + \frac{1}{2}\right) \frac{\pi}{L} t \right)}{\sin\left(\frac{\pi}{2L} t \right)} \right) dt.$$

9.6.2   Rewriting $S_N - F$

It would be nice if we could pull the $f(x)$ at the end of this expression into the integral sign. This is how we can do that. Consider

$$\frac{1}{L}\int_0^{2L}\left(\frac{1}{2}+\sum_{i=1}^{N}\cos\left(\frac{i\pi}{L}(t)\right)\right)dt = \frac{1}{L}\left(\frac{1}{2}t+\sum_{i=1}^{N}\frac{L}{i\pi}\sin\left(\frac{i\pi}{L}(t)\right)\right)\Bigg|_0^{2L}$$

$$= \frac{1}{L}\left(L+\sum_{i=1}^{N}\frac{L}{i\pi}\left\{\sin\left(\frac{i\pi}{L}(2L)\right)-\sin\left(\frac{i\pi}{L}(0)\right)\right\}\right).$$

$$= 1.$$

Hence, we can say

$$f(x) = f(x) \times \mathbf{1}$$

$$= \frac{1}{L}\int_0^{2L}\left(\frac{1}{2}+\sum_{i=1}^{N}\cos\left(\frac{i\pi}{L}(t)\right)\right)f(x)\,dt$$

However, using Eq. 9.14, we can rewrite again as

$$f(x) = f(x) \times \mathbf{1}$$

$$= \frac{1}{L}\int_0^{2L}\left(\frac{\frac{1}{2}\sin\left((N+\frac{1}{2})\frac{\pi}{L}t\right)}{\sin\left(\frac{\pi}{2L}t\right)}\right)f(x)\,dt$$

Using the periodic extension $\hat{f}$, the equation above is still valid. Hence, $S_N(x)-f(x) = S_N(x) - \hat{f}(x)$ and we find

$$|S_N(x) - f(x)|$$

$$= \left|\frac{1}{L}\int_0^{2L}\hat{f}(t+x)\left(\frac{\frac{1}{2}\sin((N+\frac{1}{2})\frac{\pi}{L}t)}{\sin(\frac{\pi}{2L}t)}\right)dt - \frac{1}{L}\int_0^{2L}\left(\frac{\frac{1}{2}\sin((N+\frac{1}{2})\frac{\pi}{L}t)}{\sin(\frac{\pi}{2L}t)}\right)\hat{f}(x)\,dt\right|$$

$$= \left|\frac{1}{2L}\int_0^{2L}\left(\hat{f}(t+x)-\hat{f}(x)\right)\left(\frac{\sin((N+\frac{1}{2})\frac{\pi}{L}t)}{\sin(\frac{\pi}{2L}t)}\right)\right|$$

$$= \left|\frac{1}{2L}\int_0^{2L}\frac{(\hat{f}(t+x)-\hat{f}(x))}{\sin\left(\frac{\pi}{2L}t\right)}\sin\left(\left(N+\frac{1}{2}\right)\frac{\pi}{L}t\right)\right|$$

We can package this in a convenient form by defining the function h on $[0, 2L]$ by $h(t) = \left(\hat{f}(t+x)-\hat{f}(x)\right)/\sin\left(\frac{\pi}{2L}t\right)$ Hence, the convergence of the Fourier series associated with f on $[0, 2L]$ is shown by establishing that

$$\lim_{N \to \infty} |S_N(x) - f(x)| = \lim_{N \to \infty} \left| \frac{1}{2L} \int_0^{2L} h(t) \, \sin\left(\left(N + \frac{1}{2}\right) \frac{\pi}{L} t\right) \right| = 0$$

We can simplify our arguments a bit more by noticing the function $t / \sin\left(\frac{\pi}{2L} t\right)$ has a removeable discontinuity at $t = 0$. This follows from a simple L'Hopital's rule argument. Hence, this function is continuous on $[0, 2L]$ and so there is a constant C so that $|t / \sin\left(\frac{\pi}{2L} t\right)| \le C$. This implies $|1 / \sin\left(\frac{\pi}{2L} t\right)| \le C \, 1/t$ on the interval and so we can establish the convergence we want by showing

$$\lim_{N \to \infty} \left| \frac{1}{2L} \int_0^{2L} \frac{\left(\hat{f}(t + x) - \hat{f}(x)\right)}{t} \, \sin\left(\left(N + \frac{1}{2}\right) \frac{\pi}{L} t\right) \right| = 0.$$

Hence, we need to look at the function $H(t) = \left(\hat{f}(t+x) - \hat{f}(x)\right)/t$ in our convergence discussions.

9.6.3 Convergence

It is straightforward to shown that the sequence of functions $w_N(x) = \sin\left(\left(N + \frac{1}{2}\right) \frac{\pi}{L} t\right)$ are mutually orthogonal with length L. Hence, the functions $\hat{w}_N(x) = \sqrt{\frac{1}{L}} w_N(x)$ have length one and are orthogonal. Hence, we already know that if H is continuous on $[0, 2L]$, then the Fourier coefficients $< H, \hat{w}_N > \to 0$ using arguments just like we did in Sect. 9.5.1.4. For our purposes, we will only look at two kinds of functions f: the first is differentiable at x and the second has f has right and left sided derivatives at x.

9.6.3.1 f Is Differentiable

On the interval $[0, 2L]$, we can see that $H(t)$ is continuous at every point except possible $t = 0$. At $t = 0$, continuity of H comes down to whether or not H has a removeable discontinuity at 0. First, note

$$\lim_{t \to 0} H(t) = \lim_{t \to} \frac{\hat{f}(t + x) - \hat{f}(x)}{t} = f'(x)$$

since we assume f is differentiable on $[0, 2L]$. So H has a removeable discontinuity at 0 as long as we define $H(0) = f'(x)$. Hence at each point where f has a derivative, we have shown that $S_N(x) \to f(x)$; i.e. the Fourier series of f on the interval $[0, 2L]$ converges to $f(x)$.

9.6.3.2 f Has One Sided Derivatives

Now we assume f is differentiable to the left of x and to the right of f. This would include the case where f has a jump discontinuity at x as well as the case where f has a corner there. Hence $(f'(x))^+$ and $(f'(x))^-$ exist and are finite numbers. We can handle this case by going back to our estimates and replacing $f(x)$ by the new function $\frac{1}{2}(f(x^+) + f(x^-))$ where $f(x^+)$ is the right hand value and $f(x^-)$ is the left hand value of the jump, respectively. Then, note

$$
\frac{1}{L}\int_0^L \left(\frac{1}{2} + \sum_{i=1}^N \cos\left(\frac{i\pi}{L}(t)\right)\right) dt = \frac{1}{L}\left(\frac{1}{2}t + \sum_{i=1}^N \frac{L}{i\pi}\sin\left(\frac{i\pi}{L}(t)\right)\right)\Big|_0^L
$$
$$
= \frac{1}{L}\left(\frac{L}{2} + \sum_{i=1}^N \frac{L}{i\pi}\left\{\sin\left(\frac{i\pi}{L}(L)\right) - \sin\left(\frac{i\pi}{L}(0)\right)\right\}\right).
$$
$$
= \frac{1}{2}.
$$

and

$$
\frac{1}{L}\int_L^{2L} \left(\frac{1}{2} + \sum_{i=1}^N \cos\left(\frac{i\pi}{L}(t)\right)\right) dt = \frac{1}{L}\left(\frac{1}{2}t + \sum_{i=1}^N \frac{L}{i\pi}\sin\left(\frac{i\pi}{L}(t)\right)\right)\Big|_L^{2L}
$$
$$
= \frac{1}{L}\left(\frac{L}{2} + \sum_{i=1}^N \frac{L}{i\pi}\left\{\sin\left(\frac{i\pi}{L}(2L)\right) - \sin\left(\frac{i\pi}{L}(L)\right)\right\}\right).
$$
$$
= \frac{1}{2}.
$$

Hence, we can say

$$
f(x^+) = 2f(x^+) \times \frac{1}{2} = \frac{2}{L}\int_0^L \left(\frac{1}{2} + \sum_{i=1}^N \cos\left(\frac{i\pi}{L}(t)\right)\right) f(x^+)\, dt
$$
$$
= \frac{2}{L}\int_0^L \left(\frac{\frac{1}{2}\sin((N+\frac{1}{2})\frac{\pi}{L}t)}{\sin(\frac{\pi}{2L}t)}\right) f(x^+) dt
$$

and

$$f(x^-) = 2f(x^-) \times \frac{1}{2}$$

$$= \frac{2}{L} \int_L^{2L} \left(\frac{1}{2} + \sum_{i=1}^N \cos\left(\frac{i\pi}{L}(t)\right) \right) f(x^+) \, dt$$

$$= \frac{2}{L} \int_L^{2L} \left(\frac{\frac{1}{2}\sin\left((N+\frac{1}{2})\frac{\pi}{L}t\right)}{\sin\left(\frac{\pi}{2L}t\right)} \right) f(x^-) \, dt$$

Thus, $\frac{1}{2}(f(x^+) + f(x^-))$ can be rewritten as

$$\frac{1}{2}(f(x^+) + f(x^-)) = \frac{1}{L} \int_0^L \left(\frac{\frac{1}{2}\sin\left((N+\frac{1}{2})\frac{\pi}{L}t\right)}{\sin\left(\frac{\pi}{2L}t\right)} \right) f(x^+) \, dt$$

$$+ \frac{1}{L} \int_L^{2L} \left(\frac{\frac{1}{2}\sin\left((N+\frac{1}{2})\frac{\pi}{L}t\right)}{\sin\left(\frac{\pi}{2L}t\right)} \right) f(x^-) \, dt$$

We can also rewrite the $S_N(x)$ terms as two integrals, $\int_0^L$ and $\int_L^{2L}$ giving

$$S_N(x) = \frac{1}{L} \int_0^L \hat{f}(t+x) \left(\frac{\frac{1}{2}\sin\left((N+\frac{1}{2})\frac{\pi}{L}t\right)}{\sin\left(\frac{\pi}{2L}t\right)} \right) dt + \frac{1}{L} \int_L^{2L} \hat{f}(t+x) \left(\frac{\frac{1}{2}\sin\left((N+\frac{1}{2})\frac{\pi}{L}t\right)}{\sin\left(\frac{\pi}{2L}t\right)} \right) dt.$$

Combining, we obtain

$$\left| S_N(x) - \frac{1}{2}(f(x^+) + f(x^-)) \right| \le \left| \frac{1}{L} \int_0^L \left(\hat{f}(t+x) - \hat{f}(x^+) \right) \left(\frac{\frac{1}{2}\sin\left((N+\frac{1}{2})\frac{\pi}{L}t\right)}{\sin\left(\frac{\pi}{2L}t\right)} \right) dt \right|$$

$$+ \left| \frac{1}{L} \int_L^{2L} \left(\hat{f}(t+x) - \hat{f}(x^-) \right) \left(\frac{\frac{1}{2}\sin\left((N+\frac{1}{2})\frac{\pi}{L}t\right)}{\sin\left(\frac{\pi}{2L}t\right)} \right) dt \right|.$$

Using the triangle inequality and using the same overestimates as before, we find the series will converge as long as

$$\lim_{t \to 0^+} \left| \frac{1}{2L} \int_0^L \left(\frac{\hat{f}(t+x) - \hat{f}(x^+)}{t} \right) \sin\left((N+\frac{1}{2})\frac{\pi}{L}t\right) dt \right| = 0,$$

and

$$\lim_{t \to (2L)^-} \left| \frac{1}{2L} \int_0^L \left(\frac{\hat{f}(t+x) - \hat{f}(x^+)}{t} \right) \sin\left((N+\frac{1}{2})\frac{\pi}{L}t\right) dt \right| = 0.$$

The integrand in the first estimate is continuous on $[0, L]$ if the integrand has a removeable discontinuity at $t = 0$. This is true as when $t \to 0^+$, the limiting value is $(f(x))^+$. Hence, using the arguments of Sect. 9.5.1.4, these terms go to 0 in the limit as $N \to \infty$. To understand the second limit, remember $\hat{f}$ is the periodic extension of f on $[0, 2L]$. Hence, the limit as $t \to 2L$ of the integrand is the same as the limit as $t \to 0^-$ of the integrand. By assumption, this limit is $(f'(x))^-$ and thus, the second integrand has a removeable discontinuity as well. We conclude that at any point where f has a jump discontinuity with f differentiable from the right and the left at this point, the series converges to the average of the jump values: $S_N(x) \to \frac{1}{2}(f(x^+) + f(x^-))$.

9.7 Fourier Sine Series

Let f be defined only on the interval $[0, L]$. Extend f to be an odd function f_o on $[0, 2L]$ as follows:

$$f_o(x) = \begin{cases} f(x), & 0 \le x \le L, \\ -f(2L - x), & L < x \le 2L. \end{cases}$$

Then extend f_o periodically as usual to $\hat{f}_o$. The Fourier coefficient for the sin terms are then

$$\frac{1}{L} \int_0^{2L} f_o(t) \sin\left(\frac{i\pi}{L} t\right) dt = \frac{1}{L} \int_0^L f(t) \sin\left(\frac{i\pi}{L} t\right) dt + \frac{1}{L} \int_L^{2L} (-f(2L - t)) \sin\left(\frac{i\pi}{L} t\right) dt.$$

Consider the second integration. Making the change of variable $y = 2L - t$, we find

$$\frac{1}{L} \int_L^{2L} (-f(2L - t)) \sin\left(\frac{i\pi}{L} t\right) dt = \frac{1}{L} \int_L^0 (-f(y)) \sin\left(\frac{i\pi}{L}(2L - y)\right) (-dy).$$

But $\frac{i\pi}{L}(2L - y) = 2i\pi - \frac{i\pi}{L} y$ and since sin term 2π periodic, we have

$$\frac{1}{L} \int_0^L (-f(y)) \sin\left(\frac{i\pi}{L}(2L - y)\right) dy = \frac{1}{L} \int_L^0 (-f(y)) \sin\left(\frac{i\pi}{L}(-y)\right) (-dy)$$

$$= \frac{1}{L} \int_0^L f(y) \sin\left(\frac{i\pi}{L}(y)\right) dy.$$

This is the same as the first integral. Hence, we have shown

$$\frac{1}{L} \int_0^{2L} f_o(t) \sin\left(\frac{i\pi}{L} t\right) dt = \frac{2}{L} \int_0^L f(t) \sin\left(\frac{i\pi}{L} t\right) dt.$$

The terms corresponding to the cos parts will all then be zero. The argument is straightforward.

$$\frac{1}{L} \int_0^{2L} f_o(t) \cos\left(\frac{i\pi}{L} t\right) dt = \frac{1}{L} \int_0^L f(t) \cos\left(\frac{i\pi}{L} t\right) dt + \frac{1}{L} \int_L^{2L} (-f(2L - t)) \cos\left(\frac{i\pi}{L} t\right) dt.$$

Consider the second integration. Making the change of variable $y = 2L - t$, we find

$$\frac{1}{L} \int_L^{2L} (-f(2L-t)) \cos\left(\frac{i\pi}{L}t\right) dt = \frac{1}{L} \int_L^0 (-f(y)) \cos\left(\frac{i\pi}{L}(2L-y)\right)(-dy).$$

Again, $\frac{i\pi}{L}(2L - y) = 2i\pi - \frac{i\pi}{L}y$ and since cos term 2π periodic and cos is an even function, we have

$$\frac{1}{L} \int_L^{2L} (-f(2L-t)) \cos\left(\frac{i\pi}{L}t\right) dt = \frac{1}{L} \int_L^0 f(y) \cos\left(\frac{i\pi}{L}y\right) dy$$
$$\frac{1}{L} \int_0^L -f(y) \cos\left(\frac{i\pi}{L}y\right) dy$$

which is the negative of the first integral. So all of these coefficients are zero. The last one is $\int_0^{2L} f_o(t)dt = 0$ because f_0 is odd on this interval. Thus, all the cos based terms in the Fourier series vanish. The Fourier series on the interval $[0, 2L]$ of the odd extension f_o becomes the standard Fourier sine series on the interval $[0, L]$ of the function f.

$$\sum_{i=1}^{\infty} <f, \hat{u}_i> \hat{u}_i = \sum_{i=1}^{\infty} \frac{2}{L}\left\langle f(x), \sin\left(\frac{i\pi}{L}x\right)\right\rangle \sin\left(\frac{i\pi}{L}x\right).$$

We know now from Sect. 9.6.3 this converges to $f(x)$ at each point x where f is differentiable and converges to the average $\frac{1}{2}(f(x^+) + f(x^-))$ at each point x where f has one sided finite derivatives. Note because the sin functions are always 0 at the endpoints 0 and L, this series must converge to 0 at those points.

9.8 Fourier Cosine Series

Let f be defined only on the interval $[0, L]$. Extend f to be an even function f_e on $[0, 2L]$ as follows:

$$f_e(x) = \begin{cases} f(x), & 0 \le x \le L, \\ f(2L-x), & L < x \le 2L. \end{cases}$$

Then extend f_e periodically as usual to $\hat{f}_e$. The Fourier coefficient for the sin terms are now

$$\frac{1}{L} \int_0^{2L} f_e(t) \sin\left(\frac{i\pi}{L}t\right) dt = \frac{1}{L} \int_0^L f(t) \sin\left(\frac{i\pi}{L}t\right) dt + \frac{1}{L} \int_L^{2L} f(2L-t) \sin\left(\frac{i\pi}{L}t\right) dt.$$

Consider the second integration. Making the change of variable $y = 2L - t$, we find

$$\frac{1}{L}\int_L^{2L} f(2L - t)) \, \sin\left(\frac{i\pi}{L}t\right) dt = \frac{1}{L}\int_L^0 f(y) \, \sin\left(\frac{i\pi}{L}(2L - y)\right) (-dy)$$
$$= \frac{1}{L}\int_0^L f(y) \, \sin\left(\frac{i\pi}{L}(-y)\right) dy.$$

However, sin is an odd function and thus the second integral is the negative of the first and these coefficients vanish. Next, consider the first Fourier cos coefficient. This is

$$\frac{1}{2L}\int_0^{2L} f_e(t)dt = \frac{1}{2L}\int_0^L f(t)dt + \frac{1}{L}\int_L^{2L} f(2L - t)dt$$
$$= \frac{1}{2L}\int_0^L f(t)dt + \frac{1}{L}\int_L^0 f(y)(-dy) = \frac{1}{L}\int_0^L f(t)dt.$$

Now let's look at the other cos based coefficients. We have

$$\frac{1}{L}\int_0^{2L} f_e(t) \cos\left(\frac{i\pi}{L}t\right) dt = \frac{1}{L}\int_0^L f(t) \cos\left(\frac{i\pi}{L}t\right) dt + \frac{1}{L}\int_L^{2L} f(2L - t) \, \cos\left(\frac{i\pi}{L}t\right) dt.$$

Consider the second integration. Making the change of variable $y = 2L - t$, we find

$$\frac{1}{L}\int_L^{2L} f(2L - t) \, \cos\left(\frac{i\pi}{L}t\right) dt = \frac{1}{L}\int_L^0 f(y) \, \cos\left(\frac{i\pi}{L}(2L - y)\right) (-dy)$$
$$= \frac{1}{L}\int_0^L f(y) \, \cos\left(\frac{i\pi}{L}(2L - y)\right) dy$$

Again, $\frac{i\pi}{L}(2L - y) = 2i\pi - \frac{i\pi}{L}y$ and since cos term 2π periodic and cos is an even function, we have

$$\frac{1}{L}\int_L^{2L} (-f(2L - t)) \, \cos\left(\frac{i\pi}{L}t\right) dt = \frac{1}{L}\int_0^L f(y) \, \cos\left(\frac{i\pi}{L}y\right) dy$$

This is the same as the first integral. Thus, the Fourier series on the interval $[0, 2L]$ of the even extension f_e becomes the standard Fourier cosine series on the interval $[0, L]$ of the function f

$$\sum_{i=0}^{\infty} <f, \hat{v}_i> \hat{v}_i = \frac{1}{L} <f(x), 1> + \sum_{i=1}^{\infty} \frac{2}{L}\left\langle f(x), \cos\left(\frac{i\pi}{L}x\right)\right\rangle \cos\left(\frac{i\pi}{L}x\right).$$

We know from Sect. 9.6.3 this series converges to $f(x)$ at each point x where f is differentiable and converges to the average $\frac{1}{2}(f(x^+) + f(x^-))$ at each point x where f has one sided finite derivatives.

9.9 MatLab Implementation

We will need to approximate Fourier series expansions for arbitrary functions f when we solve the cable equation. All of these approximations require that we find inner products of the form $< f, u_n >$ for some function u_n which is a sin or cos term. In Chap. 2, we discussed Newton–Cotes methods for integration and in Sect. 2.5, we went over the Graham-Schmidt Orthogonalization method that takes linearly independent vectors/functions $\{f_1, \ldots, f_N\}$ and outputs orthogonal linearly independent vectors of length one, $\{g_1, \ldots, g_N\}$. A standard test to see if our numerical inner product calculations are sufficiently accurate is to compute the $N \times N$ matrix $d = (< g_i, g_j >)$ which should be the essentially the $N \times N$ identify matrix as the off diagonal entries should all be zero. However, if the inner product computations are inaccurate, off diagonal values need not be zero and the GSO method will fail. So we begin with a search for a better inner product.

9.9.1 Inner Products Revisited

The errors we get when we use the Newton-Cotes formulae can be unacceptable. Hence, we will handle our inner product calculations using a Riemann sum approximation to the needed integral. This is not very sophisticated, but it works well and is easy to program. The function **innerproduct** to do this is shown below. This code approximates $\int_a^b f(t)g(t)dt$ using a Riemann sum with N terms formed from a uniform partition based on the subinterval length $\frac{b-a}{N}$. This performs well in practice for us.

Listing 9.1: Riemann sum innerproduct

```
function c = innerproduct(f,g,a,b,N)
%
%
h = @(t) f(t)*g(t);

delx = (b-a)/N;
x = linspace(a,b,N+1);
c = 0;
for i=1:N
   c = c+ h(x(i))*delx;
end

end
```

We can check how accurate we are with these inner product calculation by using them to do a Graham-Schmidt orthogonalization on the functions $1, t, t^2, \ldots, t^N$ on the interval $[a, b]$. The code to do this is in the function **GrahamSchmidtTwo**. This function returns the new orthogonal of length one functions g_i and also prints out the matrix with terms $< g_i, g_j >$ which should be an identity. The input **NIP** is the number of terms to use in the Riemann sum approximations to the inner product and **N** is size of the number of functions we perform GSO on.

Listing 9.2: Using Riemann sums for GSOs: Graham–SchmidtTwo

```
    function  g = GrahamSchmidtTwo(a,b,N,NIP)
    %
    % Perform  Graham − Schmidt  Orthogonalization
    % on  a  set  of  functions  t^0,...,  t^N
 5  %
    %Setup  function  handles
    f = SetUpFunctions(N);
    g = cell(N+1,1);

10  nf = sqrt(innerproduct(f{1},f{1},a,b,NIP));
    g{1} = @(x)  f{1}(x)/nf;
    d = zeros(N+1,N+1);
    for k=2:N+1
        %compute  next  orthogonal  piece
15      phi = @(x)  0;
        for j = 1:k−1
            c = innerproduct(f{k},g{j},a,b,NIP);
            phi = @(x)  (phi(x)+c*g{j}(x));
        end
20      psi = @(x)  (f{k}(x) − phi(x));
        nf = sqrt(innerproduct(psi,psi,a,b,NIP));
        g{k} = @(x)  (psi(x)/nf);
    end

25  for i=1:N+1
        for j=1:N+1
            d(i,j) = innerproduct(g{i},g{j},a,b,NIP);
        end
    end
30  d
    end
```

Here is a typical run.

Listing 9.3: Sample GrahamSchmidtTwo run

```
    g = GrahamSchmidtTwo(0,2,2,5000);
```

This computes the GSO of the functions $1, t, t^2$ on the interval $[0, 2]$ using Riemann sum approximations with 5000 points. The matrix $< g_i, g_j >$ which is calculated is

Listing 9.4: GSO Orthogonality results: first 3 powers of t

```
    d =

        1.0000e+00   −1.3459e−13   1.7656e−13
 4     −1.3459e−13    1.0000e+00   6.1171e−13
        1.7656e−13    6.1171e−13   1.0000e+00
```

This is the 3×3 identify that we expect. Next, we do GSO on the functions $1, t, \ldots, t^6$.

Listing 9.5: Running GSO on the first 5 powers of t

```
g = GrahamSchmidtTwo(0,2,6,5000);
```

This also generates the identity matrix we expect for $< g_i, g_j >$.

Listing 9.6: GSO Orthogonality results: first 5 powers of t

```
d =

     1.0000e+00   -1.3459e-13    1.7656e-13   -2.0684e-13    1.8885e
     -13
4   -1.3459e-13    1.0000e+00    6.1171e-13   -1.8011e-12    3.6344e
     -12
     1.7656e-13    6.1171e-13    1.0000e+00   -1.0743e-11    4.8201e
     -11
    -2.0684e-13   -1.8011e-12   -1.0743e-11    1.0000e+00    3.8416e
     -10
     1.8885e-13    3.6344e-12    4.8201e-11    3.8416e-10    1.0000e
     +00
```

To compute the first N terms of the Fourier Cosine series of a function f on the interval $[0, L]$, we first need a way to encode all the functions $\cos\left(\frac{n\pi}{L}x\right)$. We do this in the function **SetUpCosines**.

Listing 9.7: SetUpCosines

```
function f = SetUpCosines(L,N)
%
% Setup function handles
%
5 f = cell(N+1,1);
  for i=1:N+1
    f{i} = @(x) cos( (i-1)*pi*x/L );
  end
```

This generates handles to the functions 1, $\cos\left(\frac{\pi}{L}x\right)$, $\cos\left(\frac{2\pi}{L}x\right)$ and so forth ending with $\cos\left(\frac{N\pi}{L}x\right)$. A similar function encodes the corresponding sin functions we need for the first N terms of a Fourier Sine series. The function is called **SetUpSines** with code

Listing 9.8: SetUpSines

```
function f = SetUpSines(L,N)
%
% Setup function handles
%
5 f = cell(N,1);
  for i=1:N
    f{i} = @(x) sin( i*pi*x/L );
  end
```

This generates handles to the functions $\sin\left(\frac{\pi}{L}x\right)$, $\sin\left(\frac{2\pi}{L}x\right)$ and so forth ending with $\sin\left(\frac{N\pi}{L}x\right)$. Let's check how accurate our innerproduct calculations are on the sin and cos terms. First, we modify the functions which set up the sin and cos functions to return functions of length one on the interval $[0, L]$. This is done in the functions **SetUpOrthogCos** and **SetUpOrthogSin**.

Listing 9.9: SetUpOrthogCos

```
   function  f  =  SetUpOrthogCos (L,N)
   %
   % Setup  function  handles
   %
 5 f  =  cell (N+1,1);
   f{1}  =  @(x)  sqrt(1/L);
   for  i=2:N+1
      f{i}  =  @(x)  sqrt(2/L)*cos(  (i-1)*pi*x/L  );
   end
```

and

Listing 9.10: SetUpOrthogSin

```
 1 function  f  =  SetUpOrthogSin (L,N)
   %
   % Setup  function  handles
   %
   f  =  cell (N,1);
 6 for  i=1:N
      f{i}  =  @(x)  sqrt(2/L)*sin(  (i-1)*pi*x/L  );
   end
```

Then, we can check the accuracy of the inner product calculations by computing the matrix $d = (<f_i, f_j>)$. We do this with the new function **CheckOrtho**. We input the function **f** and interval $[a, b]$ endpoints **a** and **b** and the number of terms to use in the Riemann sum approximation of the inner product, **NIP**.

Listing 9.11: Checking orthogonality of the GSO: CheckOrtho

```
   function  CheckOrtho(f,a,b,NIP)
   %
   % Perform  Graham  -  Schmidt  Orthogonalization
   % on  a  set  of  functions  f
 5 %
   %Setup  function  handles
   N  =  length(f);

   for  i=1:N
10    for  j=1:N
         d(i,j)  =  innerproduct(f{i},f{j},a,b,NIP);
      end
   end
   d
15 end
```

Let's try it with the cos functions. We use a small number of terms for the Riemann sums, **NIP = 50** and compute $< g_i, g_j >$ for the first 7 functions.

Listing 9.12: The code to check cos orthogonality

```
f = SetUpOrthogCos(5,7);
g = CheckOrtho(f,0,5,50);
```

We do not get the identity matrix as we expect some of the off diagonal values are too large.

Listing 9.13: Checking Orthogonality for cosine functions: NIP $= 50$

```
  d =

3   1.0e+00    2.8e-02   -5.5e-17    2.8e-02   -1.2e-16    2.8e-02   -4.5e-17
       2.8e-02
     2.8e-02    1.0e+00    4.0e-02   -1.2e-16    4.0e-02    4.8e-17    4.0e-02
       1.1e-16
    -5.5e-17    4.0e-02    1.0e+00    4.0e-02   -6.2e-17    4.0e-02   -2.0e-16
       4.0e-02
     2.8e-02   -1.2e-16    4.0e-02    1.0e+00    4.0e-02   -5.5e-17    4.0e-02
      -6.9e-18
    -1.1e-16    4.0e-02   -6.2e-17    4.0e-02    1.0e+00    4.0e-02   -9.7e-17
       4.0e-02
8    2.8e-02    4.8e-17    4.0e-02   -5.5e-17    4.0e-02    1.0e+00    4.0e-02
       2.8e-17
    -4.5e-17    4.0e-02   -2.0e-16    4.0e-02   -9.7e-17    4.0e-02    1.0e+00
       4.0e-02
     2.8e-02    1.1e-16    4.0e-02   -6.9e-18    4.0e-02    2.8e-17    4.0e-02
       1.0e+00
```

Resetting **NIP = 200**, we find a better result.

Listing 9.14: Checking Orthogonality for cosine functions: NIP $= 200$

```
   d =

     1.0e+00    7.1e-03   -6.1e-17    7.1e-03   -7.1e-17    7.1e-03   -3.9e-17
       7.1e-03
     7.1e-03    1.0e+00    1.0e-02   -5.5e-17    1.0e-02   -1.1e-16    1.0e-02
      -3.8e-17
5    6.1e-17    1.0e-02    1.0e+00    1.0e-02   -7.8e-17    1.0e-02   -1.6e-16
       1.0e-02
     7.1e-03   -5.5e-17    1.0e-02    1.0e+00    1.0e-02   -2.7e-16    1.0e-02
      -3.3e-17
     7.1e-17    1.0e-02   -7.8e-17    1.0e-02    1.0e+00    1.0e-02   -1.0e-16
       1.0e-02
     7.1e-03   -1.2e-16    1.0e-02   -2.6e-16    1.0e-02    1.0e+00    1.0e-02
       2.1e-17
     3.9e-17    1.0e-02   -1.6e-16    1.0e-02   -1.0e-16    1.0e-02    1.0e+00
       1.0e-02
10   7.1e-03   -3.8e-17    1.0e-02   -3.3e-17    1.0e-02    2.1e-17    1.0e-02
       1.0e+00
```

We can do even better by increasing **NIP**, of course. We encourage you to do these experiments yourself. We did not show the results for the sin terms, but they will be similar.

9.9.1.1 Homework

Exercise 9.9.1 *Find the GSO of the functions $f_1(t) = t^2$, $f_2(t) = \cos(2t)$ and $f_3(t) = 4t^3$ on the interval $[-1, 2]$ using the implementations given in this section for various values of **NIP**. Check to see the matrix d is the diagonal.*

Exercise 9.9.2 *Find the GSO of the functions $f_1(t) = 1$, $f_2(t) = t$ and $f_3(t) = t^2$ on the interval $[-1, 1]$ using the implementations given in this section for various values of **NIP**. Check to see the matrix d is the diagonal.*

Exercise 9.9.3 *Find the GSO of the functions $f_1(t) = 1$, $f_2(t) = t$ and $f_3(t) = t^2$ on the interval $[-1, 4]$ using the implementations given in this section for various values of **NIP**. Check to see the matrix d is the diagonal.*

9.9.2 Fourier Cosine and Sine Approximation

We can then calculate the first $N + 1$ terms of a Fourier series or the first N terms of a Fourier Sine series using the functions **FourierCosineApprox** and **FourierSineApprox**, respectively. Let's look at the Fourier Cosine approximation first. This function returns a handle to the function **p** which is the approximation

$$p(x) = \sum_{n=0}^{N} <f, g_i> \; g_i(x)$$

where $g_i(x) = \cos\left(\frac{i\pi}{L}x\right)$. It also returns the Fourier cosine coefficients $A_1 = \frac{1}{L} < f, g_0 >$ through $A_{N+1} = \frac{2}{L} <f, g_N >$. We must choose how many points to use in our Riemann sum inner product estimates – this is the input variable **N**. The other inputs are the function handle **f** and the length **L**. We also plot our approximation and the original function together so we can see how we did.

Listing 9.15: FourierCosineApprox

```
function [A,p] = FourierCosineApprox(f,L,M,N)
%
% p is the Nth Fourier Approximation
% f is the original function
% M is the number of Riemann sum terms in the inner product
% N is the number of terms in the approximation
% L is the interval length

% get the first N+1 Fourier cos approximations
g = SetUpCosines(L,N);

% get Fourier Cosine Coefficients
A = zeros(N+1,1);
A(1) = innerproduct(f,g{1},0,L,M)/L;
for i=2:N+1
    A(i) = 2*innerproduct(f,g{i},0,L,M)/L;
end

% get Nth Fourier Cosine Approximation
p = @(x) 0;
for i=1:N+1
    p = @(x) (p(x) + A(i)*g{i}(x));
end

x = linspace(0,L,101);
for i=1:101
    y(i) = f(x(i));
end
yp = p(x);

figure
s = [' Fourier Cosine Approximation with ',int2str(N+1),' term(s)'];
plot(x,y,x,yp);
xlabel('x axis');
ylabel('y axis');
title(s);

end
```

We will test our approximations on two standard functions: the sawtooth curve
and the square wave. To define these functions, we will use an auxiliary function
splitfunc which defines a new function z on the interval $[0, L]$ as follows

$$z(x) = \begin{cases} f(x) & 0 \leq x < \frac{L}{2}, \\ g(x) & \frac{L}{2} \leq x \leq L \end{cases}$$

The arguments to **splitfunc** are the functions **f** and **g**, the value of **x** for we wish
to find the output **z** and the value of **L**.

Listing 9.16: Splitfunc

```
function z = splitfunc(x,f,g,L)
%
if x < L/2
    z = f(x);
else
    z = g(x);
end
```

It is easy then to define a sawtooth and a square wave with the following code. The square wave, Sq has value H on $[0, \frac{L}{2})$ and value 0 on $[\frac{L}{2}, L]$. In general, the sawtooth curve, Saw, is the straight line connecting the point $(0, 0)$ to $(\frac{L}{2}, H)$ on the interval $[0, \frac{L}{2}]$ and the line connecting $(\frac{L}{2}, H)$ to $(L, 0)$ on the interval $(\frac{L}{2}, L)$. Thus,

$$Sq(x) = \begin{cases} H & 0 \leq x < \frac{L}{2}, \\ 0 & \frac{L}{2} \leq x \leq L \end{cases}$$

and

$$Saw(x) = \begin{cases} \frac{2}{L}Hx & 0 \leq x < \frac{L}{2}, \\ 2H - \frac{2}{L}Hx & \frac{L}{2} \leq x \leq L \end{cases}$$

We implement the sawtooth in the function **sawtooth**.

Listing 9.17: Sawtooth

```
   function  f = sawtooth(L,H)
   %
   %
   % fleft  = H/(L/2)  x
 5 %             = 2Hx/L
   % check  fleft(0) = 0
   %           fleft(L/2) = 2HL/(2L) = H
   %
   % fright  = H + (H - 0)/(L/2 - L)(x - L/2)
10 %             = H - (2H/L) *x + (2H/L)*(L/2)
   %             = H - 2H/L * x + H
   %             = 2H - 2Hx/L
   % check  fright(L/2) = 2H - 2HL/(2L) = 2H -H = H
   %           fright(L)   = 2H - 2HL/L = 0
15 %
   fleft   = @(x) 2*x*H/L;
   fright  = @(x) 2*H - 2*H*x/L;
   f       = @(x) splitfunc(x,fleft,fright,L);

20 end
```

As an example, we build a square wave and sawtooth of height 10 on the interval $[0, 10]$. It is easy to plot these functions as well, although we won't show the plots yet. We will wait until we can compare the functions to their Fourier series approximations. However, the plotting code is listed below for your convenience. Note the plot must be set up in a **for loop** as the inequality checks in the function **splitfunc** do not handle a vector argument such as the **x** from a **linspace** command correctly.

Listing 9.18: Build a sawtooth and square wave

```
 f = @(x)  10;
 g = @(x)  0;
 Saw = sawtooth(10,10);
 Sq = @(x)  splitfunc(x,f,g,10);
5 x = linspace(0,10,101);
 for i=1:101
 > ySq(i) = Sq(x(i));
 > ySaw(i) = Saw(x(i));
 > end
```

9.9.2.1 Homework

Exercise 9.9.4 *Write the functions needed to generate a periodic square wave on the intervals* $[0, L] \cup [L, 2L] \cup [2L, 3L] \cup \ldots \cup [N - 1)L, NL]$ *of height H. Generate the need graphs also.*

Exercise 9.9.5 *Write the functions needed to generate a periodic square wave on the intervals* $[0, L] \cup [L, 2L] \cup [2L, 3L] \cup \ldots \cup [N - 1)L, NL]$ *with the high and low value reversed; i.e., the square wave is 0 on the front part and H on the back part of each chunk of length L. Generate the need graphs also.*

Exercise 9.9.6 *Write the functions needed to generate a periodic sawtooth wave on the intervals* $[0, L] \cup [L, 2L] \cup [2L, 3L] \cup \ldots \cup [N - 1)L, NL]$. *Generate the need graphs also.*

Exercise 9.9.7 *Write the functions needed to generate a periodic sawtooth - square wave on multiple intervals of length* $[0, 2L]$ *where the first* $[0, L]$ *is the sawtooth and the second* $[L, 2L]$ *is the square. Generate the need graphs also.*

9.9.3 Testing the Approximations

We can then test the Fourier Cosine Approximation code. This code generates the plot you see in Fig. 9.1.

Listing 9.19: The Fourier cosine approximation to a sawtooth

```
1 f = @(x)  10;
 g = @(x)  0;
 Saw = sawtooth(10,10);
 Sq = @(x)  splitfunc(x,f,g,10);
 [Acos,pcos] = FourierCosineApprox(Saw,10,100,5);
```

We can then do the same thing and approximate the square wave with the command

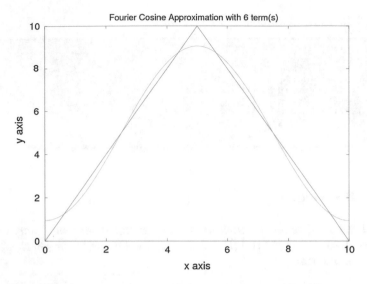

Fig. 9.1 Six term Fourier cosine series approximation to a sawtooth: NIP = 100

Listing 9.20: The Fourier cosine approximation to a square wave with 6 terms

```
[Acos,pcos] = FourierCosineApprox(Sq,10,100,5);
```

Note these approximations are done using only 100 terms in the Riemann sum approximations to the inner product. This generates the relatively poor approximation shown in Fig. 9.2.

We do better if we increase the number of terms to 11.

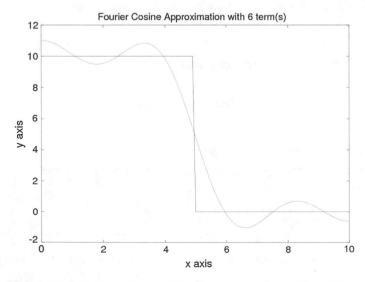

Fig. 9.2 Six term Fourier cosine series approximation to a square wave: NIP = 100

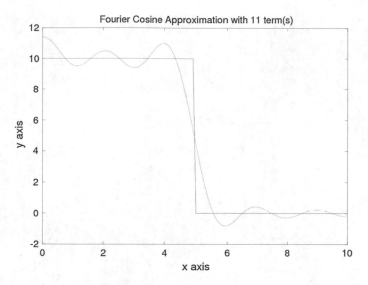

Fig. 9.3 Eleven term Fourier cosine series approximation to a square wave: NIP = 100

Listing 9.21: Fourier cosine approximation to the square wave with 11 terms

```
[Acos,pcos] = FourierCosineApprox(Sq,10,100,10);
```

This generates the improvement we see in Fig. 9.3. The code for **FourierSine-Approx** is next. It is quite similar to the cosine approximation code and so we will say little about it. It returns the Fourier sine coefficients as **A** with $A_1 = \frac{2}{L} < f, g_1 >$ through $A_N = \frac{2}{L} < f, g_N >$ where $g_i(x) = \sin\left(\frac{i\pi}{L}x\right)$. It also returns the handle to the Fourier sine approximation

$$p(x) = \sum_{n=1}^{N} <f, g_i> \ g_i(x).$$

Listing 9.22: FourierSineApprox

```
function [A,p] = FourierSineApprox(f,L,M,N)
%
% p is the Nth Fourier Approximation
% f is the original function
% M is the number of Riemann sum terms in the inner product
% N is the number of terms in the approximation
% L is the interval length
```

```
    % get the first N Fourier sine approximations
10  g = SetUpSines(L,N);

    % get Fourier Sine Coefficients
    A = zeros(N,1);
    for i=1:N
15    A(i) = 2*innerproduct(f,g{i},0,L,M)/L;
    end

    % get Nth Fourier Sine Approximation
    p = @(x) 0;
20  for i=1:N
      p = @(x) (p(x) + A(i)*g{i}(x));
    end

    x = linspace(0,L,101);
25  for i=1:101
      y(i) = f(x(i));
    end
    yp = p(x);

30  figure
    s = [' Fourier Sine Approximation with ',int2str(N),' term(s)'];
    plot(x,y,x,yp);
    xlabel('x axis');
    ylabel('y axis');
35  title(s);
    print -dpng 'FourierSineApprox.png';
    end
```

Let's test the approximation code on the square wave. using 22 terms. The sin approximations are not going to like the starting value at 10 as you can see in Fig. 9.4. The approximation is generated with the command

Listing 9.23: Fourier sine approximation to a square wave with 22 terms

```
[Asin,psin] = FourierSinApprox(Sq,10,100,22);
```

We could do similar experiments with the sawtooth function. Note, we are not using many terms in the inner product calculations. If we boost the number of terms, **NIP**, to 500, we would obtain potentially better approximations. We will leave that up to you.

9.9.3.1   Homework

Exercise 9.9.8 *Generate the Fourier Cosine Approximation with varying number of terms and different values of* **NIP** *for the periodic square wave. Draw figures as needed.*

Exercise 9.9.9 *Generate the Fourier Sine Approximation with varying number of terms and different values of* **NIP** *for the periodic square wave. Draw figures as needed.*

Exercise 9.9.10 *Generate the Fourier Cosine Approximation with varying number of terms and different values of* **NIP** *for the periodic sawtooth wave. Draw figures as needed.*

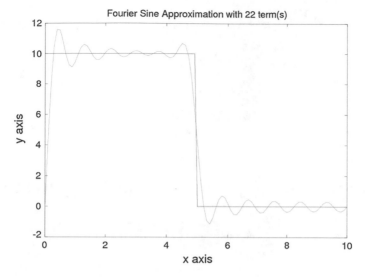

Fig. 9.4 Twenty two term Fourier sine series approximation to a square wave: NIP = 100

Exercise 9.9.11 *Generate the Fourier Sine Approximation with varying number of terms and different values of **NIP** for the periodic sawtooth wave. Draw figures as needed.*

Exercise 9.9.12 *Generate the Fourier Cosine Approximation with varying number of terms and different values of **NIP** for the periodic sawtooth-square wave. Draw figures as needed.*

Exercise 9.9.13 *Generate the Fourier Sine Approximation with varying number of terms and different values of **NIP** for the periodic sawtooth-square wave. Draw figures as needed.*

Exercise 9.9.14 *Generate the Fourier Sine Approximation with varying number of terms and different values of **NIP** for pulses applied at various locations. Draw figures as needed.*

Exercise 9.9.15 *Generate the Fourier Cosine Approximation with varying number of terms and different values of **NIP** for pulses applied at various locations. Draw figures as needed.*

9.9.4 Cable Membrane Approximations

We can use our code to approximate Fourier cosine and sine series to approximate the separation of variables solution to the cable equation with Neumann conditions. Recall the cable model is

$$\beta^2 \frac{\partial^2 \Phi}{\partial x^2} - \Phi - \alpha \frac{\partial \Phi}{\partial t} = 0, \text{ for } 0 \le x \le L, \ t \ge 0,$$

$$\frac{\partial \Phi}{\partial x}(0, t) = 0,$$

$$\frac{\partial \Phi}{\partial x}(L, t) = 0,$$

$$\Phi(x, 0) = f(x).$$

with solution

$$\Phi(x, t) = A_0 \phi_0(x, t) + \sum_{n=1}^{\infty} A_n \phi_n(x, t) = A_0 u_0(x) w_0(t) + \sum_{n=1}^{\infty} A_n u_n(x) \, w_n(t)$$

$$= A_0 e^{-\frac{1}{\alpha} t} + \sum_{n=1}^{N} A_n \cos\left(\frac{n\pi}{L} x\right) e^{-\frac{1+n^2 \pi^2 \beta^2}{\alpha L^2} t}.$$

where the coefficients A_n satisfy

$$A_0 = \frac{1}{L} \int_0^L f(x),$$

$$A_n = \frac{2}{L} \int_0^L f(x) \cos\left(\frac{n\pi}{L} x\right) dx.$$

We can now write code to find an approximation to this infinite series solution. The function will be called **CableNeumannApprox** as we have Neumann boundary conditions here. We can't solve the problem over infinite time, so we will develop the approximation for a fixed time interval $[0, T]$ for some positive T. We need to input the values of α and β, the data function **f**, the number of terms to use in our Riemann sum approximations to inner products, **M**, and finally, the number of terms of the infinite series to use, **N**. The full definition of the function is then

Listing 9.24: CableNeumannApprox arguments

```
function p = CableNeumannApprox(alpha, beta, f,M,N,L,T)
```

We then get the Fourier Cosine approximations we need and setup the inline functions as usual which store the various cos functions we need in a cell data structure.

Listing 9.25: Get Fourier cosine approximation to f

```
% Get Fourier Cosine Approximation to f for N terms
[A, fN] = FourierCosineApprox(f,L,M,N);

4 % Get Cosine Functions
u = SetUpCosines(L,N);
```

Next, we define the exponential functions we need and store in a cell.

Listing 9.26: Define the needed exponential functions

```
% Get the exponential functions
v = cell(N+1,1);
v{1} = @(t) exp(-t/alpha);
for i = 2:N+1
    c = (1 + i^2*pi^2*beta^2)/(alpha*L^2);
    v{i} = @(t) exp(-c*t);
end
```

We define an inline function to store our approximation to the series solution as follows:

Listing 9.27: Generate the approximation to the series solution

```
% Construct the approximate solution
% get Nth Fourier Sine Approximation
p = @(x,y) 0;
for i=1:N
    p = @(x,t) (p(x,t) + A(i)*u{i}(x).*v{i}(t));
end
```

We want to visualize this solution so we set up a **meshgrid** for our discretization of the space and time domain $[0, L] \times [0, T]$.

Listing 9.28: Setup the time and space domain

```
%
% draw surface for grid [0,L] x [0,T]
% set up x and y stuff
x = linspace(0,L,101);
time = linspace(0,T,101);
% set up grid of x and y pairs (x(i),t(j))
[X,Time] = meshgrid(x,time);
% set up surface
Z = p(X,Time);
```

Then, we generate the plot and end the function.

Listing 9.29: Generate the plot and end

```
%plot surface
figure
mesh(X,Time,Z,'EdgeColor','black');
xlabel('x axis');
ylabel('t axis');
zlabel('Cable Solution');
title('Cable Dirichlet Solution on Rectangle');
print -dpng 'CableDirichlet.png';

end
```

For convenience, we show the full function below.

Listing 9.30: Cable Neumann Data Fourier Series Approximation

```
    function p = CableNeumannApprox(alpha,beta,f,M,N,L,T)
    %
    % Solve beta^2 u_xx - u - alpha u_t = 0 on [0,L] x [0,T]
    %
 5  %          (0,T)---------------------(L,T)
    %            |                       |
    %            |                       |
    % u_x(0,t) =0 |                      | u_x(L,t) = 0
    %            |                       |
10  %            |                       |
    %          (0,0)---------------------(L,0)
    %                 u(x,0) = f(x)
    %
    % Separation of variables gives solution is
15  % u(x,t) =
    % A_0 exp(-(1/alpha) t)
    % + sum_{n=1}^{infinity} A_n cos(n pi x/L) exp(-(1+n^2 pi^2 beta^2)/(
      alpha L^2) t )
    %
    % where the coefficients A_n are the Fourier Cosine
20  % coefficients of f on [0,L]
    %
    % A_n = 2*<f(y), cos(n pi x/L)>
    %
    % Get Fourier Cosine Approximation to f for N terms
25  [A,fN] = FourierCosineApprox(f,L,M,N);

    % Get Cosine Functions
    u = SetUpCosines(L,N);

30  % Get the exponential functions
    v = cell(N+1,1);
    v{1} = @(t) exp(-t/alpha);
    for i = 2:N+1
      c = (1 + i^2*pi^2*beta^2)/(alpha*L^2);
35    v{i} = @(t) exp(-c*t);
    end

    % Construct the approximate solution
    % get Nth Fourier Sine Approximation
40  p = @(x,y) 0;
    for i=1:N
      p = @(x,t) (p(x,t) + A(i)*u{i}(x).*v{i}(t));
    end

45  %
    % draw surface for grid [0,L] x [0,T]
    % set up x and y stuff
    x = linspace(0,L,101);
    time = linspace(0,T,101);
50  % set up grid of x and y pairs (x(i),t(j))
    [X,Time] = meshgrid(x,time);
    % set up surface
    Z = p(X,Time);

55  %plot surface
    figure
    mesh(X,Time,Z,'EdgeColor','black');
    xlabel('x axis');
    ylabel('t axis');
60  zlabel('Cable Solution');
    title('Cable Dirichlet Solution on Rectangle');
    print -dpng 'CableDirichlet.png';

    end
```

It is interesting to look at solutions to various pulses that are applied on the cable. A rectangular pulse centered at x_0 of width r and height H on the interval $[0, L]$ has the form

$$J(x) = \begin{cases} 0, & 0 \leq x \leq x_0 - r, \\ H, & x_0 - r < x < x_0 + r, \\ 0, & x_0 + r \leq x \leq L \end{cases}$$

We code this in the function **pulsefunc** shown here.

Listing 9.31: pulsefunc

```
function z = pulsefunc(x,x0,r,H)
%
if x > x0-r && x < x0+r
   z = H;
else
   z = 0;
end
```

It is easy to approximate the solution to the cable equation with a data pulse of height 100 at location 1 of width 0.2. We'll use 200 terms in our Riemann sum inner product approximations, $\alpha = 0.5$ and $\beta = 0.025$. Hence, we are using 20 terms here.

Listing 9.32: A cable solution for a pulse

```
L = 5;
T = 30;
pulse = @(x) pulsefunc(x,2,.2,100);
p = CableNeumannApprox(0.5,.025,pulse,200,20,5,30);
```

The returned function $p(x, t)$ is shown in Fig. 9.5.

9.9.4.1 Homework

Exercise 9.9.16 *Approximate the solution to the cable equation with Neumann data on $[0, 6] \times [0, T]$ for an appropriate value of T for $\alpha = 0.3$ and $\beta = 0.8$ for a variety of pulses applied to the cable at time 0.*

Exercise 9.9.17 *Approximate the solution to the cable equation with Neumann data on $[0, 4] \times [0, T]$ for an appropriate value of T for $\alpha = 1.3$ and $\beta = 0.08$ for a variety of pulses applied to the cable at time 0.*

Exercise 9.9.18 *Approximate the solution to the cable equation with Neumann data on $[0, 10] \times [0, T]$ for an appropriate value of T for $\alpha = 0.03$ and $\beta = 0.04$ for a variety of pulses applied to the cable at time 0.*

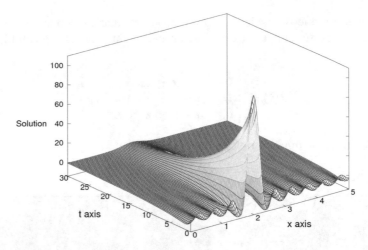

Fig. 9.5 Twenty term Fourier series approximation to the cable equation with pulse data and Neumann boundary conditions

Now it is straightforward to write a similar function for the approximation of the cable equation solution with Dirichlet boundary conditions. We leave that to you. Now redo the problems above with Dirichlet data.

Exercise 9.9.19 *Approximate the solution to the cable equation with Dirichlet data on* $[0, 6] \times [0, T]$ *for an appropriate value of T for* $\alpha = 0.3$ *and* $\beta = 0.8$ *for a variety of pulses applied to the cable at time* 0.

Exercise 9.9.20 *Approximate the solution to the cable equation with Dirichlet data on* $[0, 4] \times [0, T]$ *for an appropriate value of T for* $\alpha = 1.3$ *and* $\beta = 0.08$ *for a variety of pulses applied to the cable at time* 0.

Exercise 9.9.21 *Approximate the solution to the cable equation with Dirichlet data on* $[0, 10] \times [0, T]$ *for an appropriate value of T for* $\alpha = 0.03$ *and* $\beta = 0.04$ *for a variety of pulses applied to the cable at time* 0.

9.10   Convergence Analysis for the Cable Equation

Our general cable model is

$$\beta^2 \frac{\partial^2 \Phi}{\partial x^2} - \Phi - \alpha \frac{\partial \Phi}{\partial t} = 0, \text{ for } 0 \le x \le L, \ t \ge 0,$$

$$\frac{\partial \Phi}{\partial x}(0, t) = 0,$$

$$\frac{\partial \Phi}{\partial x}(L, t) = 0,$$

$$\Phi(x, 0) = f(x)$$

and using separation of variables, we found a series solution of the form

$$\Phi(x, t) = B_0 e^{-\frac{1}{\alpha}t} + \sum_{n=1}^{\infty} B_n \cos\left(\frac{n\pi}{L}x\right) e^{-\frac{L^2+n^2\pi^2\beta^2}{\alpha L^2}t}.$$

where since the Fourier series for the boundary data function f is given by

$$f(x) = B_0 + \sum_{n=1}^{\infty} B_n \cos\left(\frac{n\pi}{L}x\right)$$

we have

$$B_0 = \frac{1}{L}\int_0^L f(x),$$

$$B_n = \frac{2}{L}\int_0^L f(x) \cos\left(\frac{n\pi}{L}x\right)dx.$$

We have found this series solution by essentially assuming all of our series manipulations are justified. We also assume the data function f does have a Fourier series expansion. However, we haven't shown the series solution actually satisfies the cable equation and we haven't been clear about the properties the data function f should have to make this work. So in this chapter, we are going to work our way through the material we need to answer these questions carefully. Now be warned, there is a lot of reasoning in here. Many of you might prefer to just do the manipulations and call it a day, but we assure you that your understanding of what you are doing and your ability to create and solve models will be enhanced by this next journey. So roll up your sleeves and let's get started!

9.10.1   Uniform Convergence Ideas

So far we have talked about series of functions that converge pointwise to another function. A stronger type of convergence is called **uniform convergence** and a weaker one is called **least squares** or $\mathscr{L}^2$ convergence. Let's state the definition of pointwise convergence more precisely.

Definition 9.10.1 (*Pointwise Convergence of a Sequence and a Series of Functions*)
Let $\{u_n\}$ be a sequence of functions defined on the interval $[a, b]$. Let $\{S_n\}$ be the usual sequence of partial sums that this sequence determines.

We say u_n converges pointwise at a point t in $[a, b]$ if $\lim_{n\to\infty} u_n(t)$ exists. At each point t where this limit exists, we define the limit function U to have the value

$U(t) = \lim_{n\to\infty} u_n(t)$. In more precise terms, given a positive ϵ, there is a positive integer N (depending on t and ϵ) so that

$$n > N \implies |u_n(t) - U(t)| < \epsilon.$$

We say S_n converges pointwise at a point t in $[a, b]$ if $\lim_{n\to\infty} S_n(t)$ exists. At each point t where this limit exists, we define the limit function S to have the value $S(t) = \lim_{n\to\infty} S_n(t)$. In more precise terms, given a positive ϵ, there is a positive integer N (depending on t and ϵ) so that

$$n > N \implies |S_n(t) - S(t)| < \epsilon.$$

Example 9.10.1 The sequence $u_n(t) = t^n$ on the interval $[0, 1]$, converges pointwise to

$$U(t) = \begin{cases} 0 & 0 \le t < 1 \\ 1 & t = 1. \end{cases}$$

Example 9.10.2 The series built from the sequence $u_n(t) = t^n$ on the interval $(-1, 1)$, has the partial sums $S_n(t) = 1 + t + t^2 + \ldots + t^n$. A well know identity tells us that $S_n(t) = \frac{1-t^{n+1}}{1-t}$. Hence, it is easy to show that S_n converges pointwise to $S(t) = \frac{1}{1-t}$ on $(-1, 1)$.

In contrast to the idea of pointwise convergence, there is the notion of uniform convergence. Here, the N we use in the definition of pointwise convergence, depends only on the value of ϵ.

Definition 9.10.2 (*Uniform Convergence of a Sequence and a Series of Functions*) Let $\{u_n\}$ be a sequence of functions defined on the interval $[a, b]$ and assume u_n converges pointwise to U on $[a, b]$. We say u_n converges uniformly to the function U on $[a, b]$ if given a positive ϵ, there is a positive integer N (depending on ϵ only) so that

$$n > N \implies |u_n(t) - U(t)| < \epsilon, \quad \text{for all } t \in [a, b].$$

Let $\{u_n\}$ be a sequence of functions defined on the interval $[a, b]$ and let $\{S_n\}$ be the usual sequence of partial sums that this sequence determines. We say S_n converges uniformly to the function S on $[a, b]$ if given a positive ϵ, there is a positive integer N (depending on ϵ only) so that

$$n > N \implies |S_n(t) - S(t)| < \epsilon, \quad \text{for all } t \in [a, b].$$

Example 9.10.3 The sequence $u_n(t) = \frac{\sin(t)}{n}$ converges uniformly to the function $U(t) = 0$ for all t.

Definition 9.10.3 (*Least Squares Convergence of a Series of Functions*)
Let $\{u_n\}$ be a sequence of integrable functions defined on the interval $[a, b]$ and let
$\{S_n\}$ be the usual sequence of partial sums that this sequence determines. We say S_n
converges in least squares to the function S on $[a, b]$ if given a positive ϵ, there is a
positive integer N (depending on ϵ only) so that

$$n > N \Longrightarrow \int_a^b |S_n(t) - S(t)|^2 < \epsilon.$$

Comment 9.10.1 *We often denote these convergences as* $S_n \xrightarrow{ptws} S$ *(pointwise),*
$S_n \xrightarrow{unif} S$ *(uniform) and* $S_n \xrightarrow{\mathscr{L}^2} S$ *(Least Squares),*

A final result we will need is that if a sequence of continuous functions converges
uniformly on an interval, then the limit function must also be continuous.

Theorem 9.10.1 (Uniform Convergence of Continuous Functions Implies a Continuous Limit)
Let $\{u_n\}$ *be a sequence of continuous functions defined on the interval* $[a, b]$ *and
assume* u_n *converges uniformly to* U *on* $[a, b]$. *Then* U *is continuous on* $[a, b]$.
    Let $\{u_n\}$ *be a sequence of continuous functions defined on the interval* $[a, b]$ *and
let* $\{S_n\}$ *be the usual sequence of partial sums that this sequence determines. Assume
S_n *converges uniformly to the function* S *on* $[a, b]$. *Then* S *is also continuous.*

Proof Let $\epsilon > 0$ be given. Then since u_n converges uniformly to U on $[a, b]$, there
is a positive integer N so that

$$n > N \Longrightarrow |u_n(t) - U(t)| < \frac{\epsilon}{3}, \quad \text{for all } t \in [a, b].$$

Pick any two points t and consider for another s the difference $U(t) - U(s)$. We see

$$
\begin{aligned}
|U(t) - U(s)| &= |U(t) - u_n(t) + u_n(t) - u_n(s) + u_n(s) - U(s)| \\
&\leq |U(t) - u_n(t)| + |u_n(s) - u_n(s)| + |u_n(s) - U(s)|.
\end{aligned}
$$

The first and third terms are less than $\frac{\epsilon}{3}$ when $n > N$. So we have

$$|U(t) - U(s)| < \frac{2\epsilon}{3} + |u_n(s) - u_n(s)|.$$

To finish the argument, note we can fix n to be any integer larger than N and this
inequality holds. For this n, since u_n is continuous, there is a $\delta > 0$ so that $|u_n(s) -
u_n(t)| < \frac{\epsilon}{3}$ when $|t - s| < \delta$. Using this we have

$$|t - s| < \delta \implies |U(t) - U(s)| < \epsilon$$

This says U is continuous at t and since t was arbitrarily chosen, we have U is continuous on $[a, b]$. A similar argument works for the sequence S_n and shows S is continuous.   ∎

Note that **uniform** and **least squares** convergence gives us a way to decide how many terms we want to use so that we are *close* to the limit function S in some way on the entire interval $[a, b]$ but **pointwise** convergence does not let us get that sort of information. Hence, it would be nice if we could understand when our Fourier series do more than converge pointwise.

9.10.2   A Fourier Series Example

Let's do an example. Assume we have a function f extended periodically on the interval $[0, 2L]$ to $\hat{f}$ as usual with Fourier series

$$S(x) = \frac{1}{L} <f, \mathbf{1}>$$
$$+ \sum_{i=1}^{\infty} \left(\frac{2}{L}\left\langle f(x), \sin\left(\frac{i\pi}{L}x\right)\right\rangle \sin\left(\frac{i\pi}{L}x\right) + \frac{2}{L}\left\langle f(x), \cos\left(\frac{i\pi}{L}x\right)\right\rangle \cos\left(\frac{i\pi}{L}x\right) \right).$$

and using $b_0 = \frac{1}{L} <f, \mathbf{1}>$, $b_n = \frac{2}{L}\left\langle f(x), \cos\left(\frac{i\pi}{L}x\right)\right\rangle$ and $a_n = \frac{2}{L}\left\langle f(x), \sin\left(\frac{i\pi}{L}x\right)\right\rangle$ as usual, we can write

$$S(x) = b_0 + \sum_{i=1}^{\infty} \left(a_i \sin\left(\frac{i\pi}{L}x\right) + b_i \cos\left(\frac{i\pi}{L}x\right) \right).$$

Now if we assume f' exists in $[0, 2L]$, we know the Fourier series of f converges to f pointwise at each point and we have

$$f(x) = b_0 + \sum_{i=1}^{\infty} \left(a_i \sin\left(\frac{i\pi}{L}x\right) + b_i \cos\left(\frac{i\pi}{L}x\right) \right).$$

9.10.2.1   Bounded Coefficients

let S_n be the nth partial sum of the series above. Then assuming $\int_0^{2L} (f')^2(x)dx$ is finite, we have

$$0 \le \: < f - b_0 - \sum_{i=1}^{n} \left(a_i \sin\left(\frac{i\pi}{L}x\right) + b_i \cos\left(\frac{i\pi}{L}x\right) \right),$$

$$f - b_0 - \sum_{j=1}^{n} \left(a_j \sin\left(\frac{i\pi}{L}x\right) + b_j \cos\left(\frac{j\pi}{L}x\right) \right) >$$

As usual, we let $u_i(x) = \sin\left(\frac{i\pi}{L}x\right)$ and $v_i(x) = \cos\left(\frac{i\pi}{L}x\right)$ with $v_0(x) = 1$. Then, we can rewrite this as

$$0 \le \: < f - b_0 v_0(x) - \sum_{i=1}^{n} (a_i u_i(x) + b_i v_i), f - b_0 v_0(x) - \sum_{j=1}^{n} (a_j u_j(x) + b_j v_j(x)) >$$

Thus, we find, after a lot of manipulation

$$0 \le \| f - b_0 v_0(x) - \sum_{i=1}^{n} (a_i u_i(x) + b_i v_i(x)) \|^2 = < f', f' > -2b_0 < f, v_0 > + b_0^2 < v_0, v_0 >$$

$$-2 \sum_{i=1}^{n} a_i < f, u_i > -2 \sum_{i=0}^{n} b_i < f, v_i > + \sum_{i=1}^{n} a_i^2 < u_i, u_i > + \sum_{i=1}^{n} b_i^2 < v_i, v_i >$$

because all the *cross terms* such as $< v_0, u_i > = 0$, $< v_0, v_i > = \delta_0^j < u_i, u_j > = \delta_i^j$ and $< v_i, v_j > = \delta_i^j$. Hence, since $< f, v_i > = b_i$ and $< f, u_i > = a_i$ we have

$$0 \le \| f - b_0 v_0(x) - \sum_{i=1}^{n} (a_i u_i(x) + b_i v_i) \|^2 = < f, f > -b_0^2 - \sum_{i=1}^{n} a_i^2 - \sum_{i=1}^{n} b_i^2$$

We conclude that

$$b_0^2 + \sum_{i=1}^{n} (a_i^2 + b_i^2) \le \|f\|^2$$

This tells us that the series of positive terms, $b_0^2 + \sum_{i=1}^{\infty} (a_i^2 + b_i^2)$ converges.

9.10.2.2   The Derivative Series

Let's assume f is differentiable on $[0, 2L]$ and we extend f' periodically as well. We can calculate the Fourier series of f' like usual. Now if we wanted to ensure this series converged to f', we could assume f'' exists on $[0, 2L]$ except at a finite number of points where the right and left hand derivatives are both finite. Then we know the Fourier series of f' exists also and converges pointwise to $f'(x)$ at each point

x where $f'(x)$ is continuous. However, we can calculate the Fourier series without those assumptions. Let the Fourier series of f' be $T(x)$. Then

$$T(x) = \frac{1}{2L} <f', \mathbf{1}>$$
$$+ \sum_{i=1}^{\infty} \left(\frac{1}{L}\left\langle f'(x), \sin\left(\frac{i\pi}{L}x\right)\right\rangle \sin\left(\frac{i\pi}{L}x\right) + \frac{1}{L}\left\langle f'(x), \cos\left(\frac{i\pi}{L}x\right)\right\rangle \cos\left(\frac{i\pi}{L}x\right)\right).$$

The first coefficient is

$$\frac{1}{2L} <f', \mathbf{1}> = \frac{1}{2L}(f(2L) - f(0))$$

Now let's consider the other Fourier coefficients carefully. We can rewrite each coefficient using integration by parts to find,

$$\frac{1}{L}\left\langle f'(x), \cos\left(\frac{i\pi}{L}x\right)\right\rangle = \frac{1}{L}\int_0^{2L} f'(x), \cos\left(\frac{i\pi}{L}x\right)dx$$
$$= \frac{1}{L}(f(x)\cos\left(\frac{i\pi}{L}x\right)\Big|_0^{2L} + \frac{1}{L}\frac{i\pi}{L}\int_0^{2L} f(x)\sin\left(\frac{i\pi}{L}x\right)dx$$

$$\frac{1}{L}\left\langle f'(x), \sin\left(\frac{i\pi}{L}x\right)\right\rangle = \frac{1}{L}\int_0^{2L} f'(x), \sin\left(\frac{i\pi}{L}x\right)dx$$
$$= \frac{1}{L}(f(x)\sin\left(\frac{i\pi}{L}x\right)\Big|_0^{2L} - \frac{1}{L}\frac{i\pi}{L}\int_0^{2L} f(x)\cos\left(\frac{i\pi}{L}x\right)dx$$

A little thought shows we can rewrite this as

$$\frac{1}{L}\left\langle f'(x), \cos\left(\frac{i\pi}{L}x\right)\right\rangle = \frac{1}{L}\left(f(2L)\cos\left(\frac{2Li\pi}{L}\right) - f(0)\cos\left(\frac{0i\pi}{L}\right)\right) + \frac{i\pi}{L}a_i$$
$$= \frac{1}{L}\left(f(2L) - f(0)\right) + \frac{i\pi}{L}a_i$$
$$\frac{1}{L}\left\langle f'(x), \sin\left(\frac{i\pi}{L}x\right)\right\rangle = \frac{1}{L}\left(f(2L)\sin\left(\frac{2Li\pi}{L}\right) - f(0)\sin\left(\frac{0\pi}{L}\right)\right) - \frac{i\pi}{L}b_i$$

Now if we assume $f(2L) = f(0)$, these reduce to

$$\frac{1}{2L} <f', \mathbf{1}> = \frac{1}{2L}(f(2L) - f(0)) = 0$$
$$\frac{1}{L}\left\langle f'(x), \cos\left(\frac{i\pi}{L}x\right)\right\rangle = \frac{1}{L}\left(f(2L) - f(0)\right) + \frac{i\pi}{L}a_i = \frac{i\pi}{L}a_i$$
$$\frac{1}{L}\left\langle f'(x), \sin\left(\frac{i\pi}{L}x\right)\right\rangle = -\frac{i\pi}{L}b_i$$

Hence, if f is periodic, we find

$$T(x) = \sum_{i=1}^{\infty} \left(-\frac{\pi i}{L} b_i \sin\left(\frac{i\pi}{L}x\right) + \frac{\pi i}{L} a_i \cos\left(\frac{i\pi}{L}x\right) \right).$$

This is the same result we would have found if we differentiated the Fourier series for f term by term. So we conclude that the Fourier series of f' can be found be differentiating the Fourier series for f term by term.

9.10.2.3 The Derivative Coefficient Bounds

We can apply the derivation we did above for f to the series expansion for f' we have just found. We find

$$\hat{b}_0^2 + \sum_{i=1}^{n} (\hat{a}_i^2 + \hat{b}_i^2) \leq \|f'\|^2$$

where here $\hat{b}_0 = 0$, $\hat{a}_i = -\frac{\pi i}{L} b_i$ and $\hat{b}_i = \frac{\pi i}{L} a_i$. Hence, we have

$$\frac{\pi^2}{L^2} \sum_{i=1}^{n} i^2 \, (a_i^2 + b_i^2) \leq \|f'\|^2$$

9.10.3 Uniform Convergence for f

We are almost at the point where we can see circumstances where the Fourier series expansion of f converges uniformly to f on the interval $[0, 2L]$. We need a few more technical tools. From earlier calculus courses, we know what the Cauchy–Schwartz inequality is for vectors in $\Re^2$. If the vectors X and Y had components $[a_1, a_2]'$ and $[b_1, b_2]'$, respectively, we have

$$\sum_{i=1}^{2} |a_i||b_i| \leq \sqrt{\sum_{i=1}^{2} a_i^2} \sqrt{\sum_{i=1}^{2} b_i^2}$$

where we have written the dot product and the vector lengths in summation notation. We suspect we can prove this result for vectors in $\Re^n$ and also for Riemann integrable functions. Here is the result for sequences.

Theorem 9.10.2 (Cauchy–Schwartz Inequalites)
If X and Y in $\Re^n$ have components $[a_1, \ldots, a_n]'$ and $[b_1, \ldots, b_n]'$, respectively, we have

$$\sum_{i=1}^{N} |a_i||b_i| \leq \sqrt{\sum_{i=1}^{N} a_i^2} \sqrt{\sum_{i=1}^{N} b_i^2}$$

Further, if the series $\sum_{i=1}^{\infty} a_i^2$ and $\sum_{i=1}^{\infty} b_i^2$ both converge, we can say

$$\sum_{i=1}^{\infty} |a_i||b_i| \leq \sqrt{\sum_{i=1}^{\infty} a_i^2} \sqrt{\sum_{i=1}^{\infty} b_i^2}$$

Proof These statements are fairly easy to prove. We know that for any two numbers α and β that $(|\alpha| - |\beta|)^2 \geq 0$. Hence, we can say

$$|\alpha||\beta| \leq \frac{1}{2}\alpha^2 + \frac{1}{2}\beta^2.$$

Now let's consider two vectors X and Y in $\Re^N$ with components labeled as usual. Let $u = \sqrt{\sum_{i=1}^{N} a_i^2}$ and $v = \sqrt{\sum_{i=1}^{N} b_i^2}$. Now if $\alpha = \frac{a_i}{u}$ and $\beta = \frac{b_i}{v}$, we see for any index i, we have

$$|\frac{a_i}{u}||\frac{b_i}{v}| \leq \frac{1}{2u^2}a_i^2 + \frac{1}{2v^2}b_i^2.$$

Now sum this inequality over all the components to get

$$\frac{1}{uv}\sum_{i=1}^{N} |a_i||b_i| \leq \frac{1}{2u^2}\sum_{i=1}^{N} a_i^2 + \frac{1}{2v^2}\sum_{i=1}^{N} b_i^2.$$

But since $u = \sqrt{\sum_{i=1}^{N} a_i^2}$ and $v = \sqrt{\sum_{i=1}^{N} b_i^2}$, this becomes

$$\frac{1}{uv}\sum_{i=1}^{N} |a_i||b_i| \leq \frac{1}{2u^2}u^2 + \frac{1}{2v^2}v^2 = 1$$

Hence, multiplying both sides by uv, we have the inequality we want

$$\sum_{i=1}^{N} |a_i||b_i| \leq uv = \sqrt{\sum_{i=1}^{N} a_i^2} \sqrt{\sum_{i=1}^{N} b_i^2}.$$

Now if the series $\sum_{i=1}^{\infty} a_i^2$ and $\sum_{i=1}^{\infty} b_i^2$ converge, using the same argument on the first N terms of the series, we have

$$\sum_{i=1}^{N} |a_i||b_i| \leq uv = \sqrt{\sum_{i=1}^{N} a_i^2} \sqrt{\sum_{i=1}^{N} b_i^2}.$$

where now we let $u = \sqrt{\sum_{i=1}^{\infty} a_i^2}$ and $v = \sqrt{\sum_{i=1}^{\infty} b_i^2}$. The partial sums are less the the sum of the series in each case, so we have

$$\sum_{i=1}^{N} |a_i||b_i| \leq uv = \sqrt{\sum_{i=1}^{\infty} a_i^2} \sqrt{\sum_{i=1}^{\infty} b_i^2}.$$

But this says the partial sums of the series of nonnegative terms $\sum_{i=1}^{\infty} |a_i||b_i|$ are bounded above by the bound $\sqrt{\sum_{i=1}^{\infty} a_i^2} \sqrt{\sum_{i=1}^{\infty} b_i^2}$. Hence the supremum on these partial sums which is the sum of this series also satisfies this bound. We conclude

$$\sum_{i=1}^{\infty} |a_i||b_i| \leq uv = \sqrt{\sum_{i=1}^{\infty} a_i^2} \sqrt{\sum_{i=1}^{\infty} b_i^2}. \qquad \blacksquare$$

Now let's look at the result for Riemann integrable functions.

Theorem 9.10.3 (Cauchy–Schwartz Integration Inequalites)
If f and g are two Riemann integrable functions on the interval $[a, b]$, it follows that

$$\int_a^b |f(x)g(x)| \, dx \leq \sqrt{\int_a^b f^2(x) \, dx} \sqrt{\int_a^b g^2(x) \, dx}$$

Proof Again, this statements is fairly easy to prove. We know that for any two numbers α and β that

$$|\alpha||\beta| \leq \frac{1}{2}\alpha^2 + \frac{1}{2}\beta^2.$$

Now if f and g are continuous on the interval $[a, b]$, we let $u = \sqrt{\int_a^b f^2(x)dx}$ and $v = \sqrt{\int_a^b g^2(x)dx}$. Then, letting $\alpha = \frac{|f(x)|}{u}$ and $\beta = \frac{|g(x)|}{v}$, we have

$$\frac{|f(x)|}{u} \frac{|g(x)|}{v} \leq \frac{1}{2u^2}f^2(x) \frac{1}{2v^2}g^2(x).$$

or

$$|f(x)| \, |g(x)| \leq \frac{uv}{2u^2}f^2(x) \frac{uv}{2v^2}g^2(x).$$

Now integrate both sides from a to b to obtain

$$\left| \int_a^b f(x)| \, |g(x)| \le \frac{v}{2u} \int_a^b f^2(x) \, dx \; \frac{u}{2v} g^2(x) \, dx = \frac{u^2 v}{2u} + \frac{uv^2}{2v} = uv.$$

which gives the desired inequality:

$$\int_a^b |f(x)g(x)| \, dx \le \sqrt{\int_a^b f^2(x) \, dx} \sqrt{\int_a^b g^2(x) \, dx} \qquad \blacksquare$$

We also need some tests to determine if our Fourier series converge uniformly.

Theorem 9.10.4 (Tests for Uniform Convergence)
Assume we have a sequence of functions (f_i) on the interval $[a, b]$.

1. *If there is a sequence of constants (K_i) so that $\sum_{i=1}^{\infty} K_i$ converges and $|f_i(x)| \le K_i$ on $[a, b]$ for each index i, then $\sum_{i=1}^{\infty} f_i(x)$ converges uniformly to a limit function $S(x)$ on $[a, b]$.*
2. *Assume the series $\sum_{i=1}^{\infty} f_i(x)$ converges pointwise to $S(x)$ on $[a, b]$. Further, assume there is a sequence of constants (K_i) so that $\sum_{i=1}^{\infty} K_i^2$ converges. Then if*

$$m > n \implies |S_m(x) - S_n(x)| \le \sqrt{\sum_{i=n+1}^{m} K_i^2}$$

then $\sum_{i=1}^{\infty} f_i(x)$ converges uniformly to a limit function $S(x)$ on $[a, b]$.

Proof There are two separate arguments here.

1. Since we always know

$$\sum_{i=1}^{\infty} f_i(x) \le \sum_{i=1}^{\infty} K_i,$$

we know the series converges pointwise at each x in $[a, b]$. Let the sum of this series be the function $S(x)$. Consider the difference of two partial sums, $S_n(x)$ and $S_m(x)$ where $m > n$. Then, we have

$$|S_m(x) - S_n(x)| = \left| \sum_{i=n+1}^{m} f_i(x) \right|$$

$$\le \sum_{i=n+1}^{m} |f_i(x)|$$

$$\le \sum_{i=n+1}^{m} K_i,$$

Since we know the series $\sum_{i=1}^{\infty} K_i$ converges to some value K, given a positive tolerance ϵ we know there is a positive integer N so that

$$n > N \implies \left| K - \sum_{i=1}^{n} K_i \right| < \frac{1}{4}\epsilon.$$

Hence, for $m > n > N$, we have

$$\left| \sum_{i=n+1}^{m} K_i \right| = \left| \sum_{i=1}^{m} K_i - \sum_{i=1}^{n} K_i \right|$$

$$= \left| \left(K - \sum_{i=1}^{n} K_i \right) - \left(K - \sum_{i=1}^{m} K_i \right) \right|$$

$$< \frac{1}{4}\epsilon + \frac{1}{4}\epsilon = \frac{1}{2}\epsilon.$$

Thus, going back to our original inequality for $|S_m(x) - S_n(x)|$, we see there is a positive integer N so that if m and n exceed N, we have for all x in $[a, b]$, that

$$|S_m(x) - S_n(x)| = \left| \sum_{i=n+1}^{m} f_i(x) \right|$$

$$\leq \sum_{i=n+1}^{m} |f_i(x)|$$

$$\leq \sum_{i=n+1}^{m} K_i$$

$$< \frac{1}{2}\epsilon$$

Since this is true for all $m > N$, take the limit on m on both sides. Since all the terms here are continuous, we can say $\lim_m |S_m(x) - S_n(x)| = |\lim_m S_m(x) - S_n(x)|$. But we know $\lim_m S_m(x) = S(x)$ and so we can say for all x that

$$\lim_{m \to \infty} |S_m(x) - S_n(x)| = \left| S(x) - S_n(x) \right|$$

$$\leq \frac{1}{2}\epsilon < \epsilon$$

But this says S_n converges uniformly to S on $[a, b]$.

2. In the second case, we assume the series converges pointwise to S. We know that

$$m > n \implies |S_m(x) - S_n(x)| \le \sqrt{\sum_{i=n+1}^{m} K_i^2}.$$

Since we also assume that the series $\sum_{i=1}^{\infty} K_i^2$ converges, by the same sort of argument we used in the first part, we know that for any $\epsilon > 0$, there is a positive integer N so that

$$m > n \implies \sum_{i=n+1}^{n} K_i^2 < \frac{\epsilon^2}{4}$$

Combining, we see

$$m > n \implies |S_m(x) - S_n(x)| \le \sqrt{\sum_{i=n+1}^{n} K_i^2} < \frac{\epsilon}{2}$$

Now, let $m \to \infty$ to see that for all x in $[a, b]$, that

$$|S(x) - S_n(x)| \le \frac{\epsilon}{2} < \epsilon$$

This tells us the convergence is uniform. ∎

9.10.3.1 Uniform Convergence of Fourier Series

Now let's get back to the question of the uniform convergence of the Fourier series for f. We assume f is continuous and periodic on $[0, 2L]$ and that f' exists at all but finitely many points. Hence, we know the Fourier series of f exists and converges to $f(x)$. From our earlier arguments, if f has Fourier series

$$f(x) = b_0 + \sum_{i=1}^{\infty} \left(a_i \sin\left(\frac{i\pi}{L}x\right) + b_i \cos\left(\frac{i\pi}{L}x\right) \right).$$

then we know for all n

$$\frac{\pi^2}{L^2} \sum_{i=1}^{n} i^2 (a_i^2 + b_i^2) \le \|f'\|^2.$$

Let's look at the partial sums of absolute values.

$$b_0\, 1 + \sum_{i=1}^{n} (a_i \sin\left(\frac{i\pi}{L}x\right) + \sum_{i=1}^{n} (b_i \cos\left(\frac{i\pi}{L}x\right) = \sum_{i=1}^{n} (a_i \sin\left(\frac{i\pi}{L}x\right) + \sum_{i=0}^{n} (b_i \cos\left(\frac{i\pi}{L}x\right)$$

$$= \sum_{i=1}^{n} (i\, a_i\, \frac{1}{i} \sin\left(\frac{i\pi}{L}x\right) + \sum_{i=0}^{n} (i\, b_i\, \frac{1}{i} \cos\left(\frac{i\pi}{L}x\right).$$

Let T_n denote the nth partial sum here. Then, the difference of the nth and mth partial sum for $m > n$ gives

$$T_m(x) - T_n(x) = \sum_{i=n1}^{m} (i\, a_i\, \frac{1}{i} \sin\left(\frac{i\pi}{L}x\right) + \sum_{i=n+1}^{m} (i\, b_i\, \frac{1}{i} \cos\left(\frac{i\pi}{L}x\right).$$

Now apply our analogue of the Cauchy–Schwartz inequality for series.

$$T_m(x) - T_n(x) \leq \sum_{i=n+1}^{m} (i\, |a_i|\, \frac{1}{i} \left|\sin\left(\frac{i\pi}{L}x\right)\right| + \sum_{i=n+1}^{m} (i\, b_i\, \frac{1}{i} \left|\cos\left(\frac{i\pi}{L}x\right)\right|$$

$$\leq \sqrt{\sum_{i=n+1}^{m} i^2\, |a_i|^2} \sqrt{\sum_{i=n+1}^{m} \frac{1}{i^2} \left|\sin\left(\frac{i\pi}{L}x\right)\right|^2}$$

$$+ \sqrt{\sum_{i=n+1}^{m} i^2\, |b_i|^2} \sqrt{\sum_{i=n+1}^{m} \frac{1}{i^2} \left|\cos\left(\frac{i\pi}{L}x\right)\right|^2}$$

$$\leq \sqrt{\sum_{i=n+1}^{m} i^2\, |a_i|^2} \sqrt{\sum_{i=n+1}^{m} \frac{1}{i^2}} + \sqrt{\sum_{i=n+1}^{m} i^2\, |b_i|^2} \sqrt{\sum_{i=n+1}^{m} \frac{1}{i^2}}$$

Now each of the front pieces satisfy

$$\sum_{i=1}^{\infty} i^2\, a_i^2 \leq \frac{L^2}{\pi^2}\, \|f'\|^2$$

$$\sum_{i=1}^{\infty} i^2\, b_i^2 \leq \frac{L^2}{\pi^2}\, \|f'\|^2$$

Hence,

$$\left|T_m(x) - T_n(x)\right| \leq \sum_{i=n+1}^{m} (a_i \sin\left(\frac{i\pi}{L}x\right) + \sum_{i=n+1}^{m} (b_i \cos\left(\frac{i\pi}{L}x\right) \leq \frac{2L\|f'\|}{\pi} \sqrt{\sum_{i=n+1}^{m} \frac{1}{i^2}}.$$

We can then apply Theorem 9.10.4 to conclude the Fourier series of f converges uniformly to f under these circumstances! Let's summarize this result.

Theorem 9.10.5 (Uniform Convergence of Fourier Series)
Given f on $[0, 2L]$, *assume f is continuous with* $f(0) = f(2L)$ *and* f' *exists at all but a finite number of points on* $[0, 2L]$ *with finite right and left hand derivatives at these points. Then the Fourier series of f converges uniformly to f on* $[0, 2L]$.

9.10.4 Extending to Fourier Sin and Fourier Cosine Series

The arguments above also apply to the Fourier sin and Fourier cosine series we have discussed. In those cases, we have a function f defined only on $[0, L]$. If we extend f as an odd function periodically to $[0, 2L]$, the Fourier series for this extension is given by

$$\sum_{i=1}^{\infty} A_i \sin\left(\frac{i\pi}{L}x\right)$$

where

$$A_i = \frac{2}{L} \int_0^L f(x) \sin\left(\frac{i\pi}{L}x\right) dx.$$

If we extend f periodically as an even function, then the extension satisfies $f(0) = f(2L)$ and its Fourier series is

$$\sum_{i=1}^{\infty} B_i \cos\left(\frac{i\pi}{L}x\right)$$

where

$$B_0 = \frac{1}{L} \int_0^L f(x) dx$$

$$B_i = \frac{2}{L} \int_0^L f(x) \cos\left(\frac{i\pi}{L}x\right) dx.$$

9.10.4.1 The Fourier Sine Series

From our previous discussions about uniform convergence, we know that is $f(0) - f(2L)$ and f' exists on $[0, 2L]$ with at most a finite number of points where f' does not exist but the right and left hand derivatives do, then the Fourier series for f on $[0, 2L]$ converges uniformly to f. For the odd extension to f to $[0, 2L]$, if we require f be continuous on $[0, L]$, then since at $x = 0$ and $x = L$, the Fourier sine series gives

$$\sum_{i=1}^{\infty} A_i \sin\left(\frac{i\pi}{L}0\right) = 0$$

$$\sum_{i=1}^{\infty} A_i \sin\left(\frac{i\pi}{L}L\right) = 0$$

we see $f(0) = f(L) = 0$. Hence, f satisfies the periodic requirement that $f(0) = f(L)$. Hence, we suspect that the derived series

$$\sum_{i=1}^{\infty} \frac{i\pi}{L} A_i \cos\left(\frac{i\pi}{L}x\right)$$

is actually the same as f'. Let's check this out by direct calculation. We have (remember the inner products here are on the interval $[0, L]$ now)

$$\left\langle f' - \sum_{i=1}^{n} \frac{i\pi}{L} A_i \cos\left(\frac{i\pi}{L}x\right), < f' - \sum_{j=1}^{n} \frac{j\pi}{L} A_j \cos\left(\frac{j\pi}{L}x\right)\right\rangle$$

$$= <f', f'> -2\sum_{i=1}^{n} \frac{i\pi}{L} A_i \left\langle f', \cos\left(\frac{i\pi}{L}x\right)\right\rangle$$

$$+ \sum_{i=1}^{n}\sum_{j=1}^{n} \frac{i\pi}{L}\frac{j\pi}{L} A_i A_j \left\langle \cos\left(\frac{i\pi}{L}x\right), \cos\left(\frac{j\pi}{L}x\right)\right\rangle$$

$$= -2\sum_{i=1}^{n} \frac{i\pi}{L} A_i \left\langle f', \cos\left(\frac{i\pi}{L}x\right)\right\rangle$$

$$+ \sum_{i=1}^{n} \frac{i^2\pi^2}{L^2} A_i^2 \left\langle \cos\left(\frac{i\pi}{L}x\right), \cos\left(\frac{i\pi}{L}x\right)\right\rangle.$$

Integrating by parts, we find

$$\left\langle f', \cos\left(\frac{i\pi}{L}x\right)\right\rangle = f(x)\cos\left(\frac{i\pi}{L}x\right)\Big|_0^L + \frac{i\pi}{L}\int_0^L f(x)\sin\left(\frac{i\pi}{L}x\right)dx$$

$$= \frac{i\pi}{L}\int_0^L f(x)\sin\left(\frac{i\pi}{L}x\right)dx$$

$$= \frac{i\pi}{L}\frac{L}{2} A_i = \frac{i\pi}{2} A_i.$$

since $f(0) = f(L)$. Further, we know

$$l < \cos\left(\frac{i\pi}{L}x\right), \cos\left(\frac{i\pi}{L}x\right)\right\rangle = \int_0^L \cos^2\left(\frac{i\pi}{L}x\right)dx = \frac{L}{2}.$$

Thus,

$$\left\langle f' - \sum_{i=1}^{n} \frac{i\pi}{L} A_i \cos\left(\frac{i\pi}{L}x\right), < f' - \sum_{j=1}^{n} \frac{j\pi}{L} A_j \cos\left(\frac{j\pi}{L}x\right) \right\rangle = -2 \sum_{i=1}^{n} \frac{i^2\pi^2}{2L} A_i^2$$

$$+ \sum_{i=1}^{n} \frac{i^2\pi^2}{L^2} A_i^2 \frac{L}{2}$$

$$= -\sum_{i=1}^{n} \frac{i^2\pi^2}{2L} A_i^2.$$

We conclude that for all n,

$$\sum_{i=1}^{n} \frac{i^2\pi^2}{2L} A_i^2 \leq \|f'\|^2.$$

This implies that the series $\sum_{i=1}^{\infty} \frac{i^2\pi^2}{2L} A_i^2$ converges. The same argument we used before then tells us that the original Fourier sine series for f converges uniformly to $f(x)$ on $[0, L]$.

9.10.4.2   The Fourier Cosine Series

For the even extension to f to $[0, 2L]$, we will require that $f(x)$ be continuous at $x = 0$ and $x = L$. Hence, we must have the Fourier cosine series gives

$$B_0 + \sum_{i=1}^{\infty} B_i \cos\left(\frac{i\pi}{L}0\right) = B_0 + \sum_{i=1}^{\infty} B_i = f(0)$$

$$B_0 + \sum_{i=1}^{\infty} B_i \cos\left(\frac{i\pi}{L}L\right) = B_0 + \sum_{i=1}^{\infty} B_i(-1)^n = f(L).$$

We still suspect that the derived series

$$-\sum_{i=1}^{\infty} \frac{i\pi}{L} B_i \sin\left(\frac{i\pi}{L}x\right)$$

is actually the same as f'. Let's calculate this directly. We have

$$\left\langle f' + \sum_{i=1}^{n} \frac{i\pi}{L} B_i \sin\left(\frac{i\pi}{L}x\right), < f' + \sum_{j=1}^{n} \frac{j\pi}{L} B_j \sin\left(\frac{j\pi}{L}x\right) \right\rangle$$

$$= <f', f'> + 2\sum_{i=1}^{n} \frac{i\pi}{L} B_i \left\langle f', \sin\left(\frac{i\pi}{L}x\right)\right\rangle$$

$$+ \sum_{i=1}^{n}\sum_{j=1}^{n} \frac{i\pi}{L}\frac{j\pi}{L} A_i A_j \left\langle \sin\left(\frac{i\pi}{L}x\right), \sin\left(\frac{j\pi}{L}x\right)\right\rangle$$

$$= 2\sum_{i=1}^{n} \frac{i\pi}{L} A_i \left\langle f', \sin\left(\frac{i\pi}{L}x\right)\right\rangle$$

$$+ \sum_{i=1}^{n} \frac{i^2\pi^2}{L^2} A_i^2 \left\langle \sin\left(\frac{i\pi}{L}x\right), \sin\left(\frac{i\pi}{L}x\right)\right\rangle.$$

Integrating by parts, we find

$$\left\langle f', \sin\left(\frac{i\pi}{L}x\right)\right\rangle = f(x)\sin\left(\frac{i\pi}{L}x\right)\Big|_0^L - \frac{i\pi}{L}\int_0^L f(x)\cos\left(\frac{i\pi}{L}x\right)dx$$

$$= -\frac{i\pi}{L}\int_0^L f(x)\cos\left(\frac{i\pi}{L}x\right)dx$$

$$= -\frac{i\pi}{L}\frac{L}{2}B_i = -\frac{i\pi}{2}B_i.$$

Note, we no longer need to know that $f(0) = f(L)$. Further, we know

$$l < \sin\left(\frac{i\pi}{L}x\right), \sin\left(\frac{i\pi}{L}x\right)\right\rangle = \frac{L}{2}.$$

Thus,

$$\left\langle f' + \sum_{i=1}^{n} \frac{i\pi}{L} B_i \cos\left(\frac{i\pi}{L}x\right), <f' + \sum_{j=1}^{n} \frac{j\pi}{L} B_j \cos\left(\frac{j\pi}{L}x\right)\right\rangle = -2\sum_{i=1}^{n} \frac{i^2\pi^2}{2L} B_i^2$$

$$+ \sum_{i=1}^{n} \frac{i^2\pi^2}{L^2} B_i^2 \frac{L}{2}$$

$$= -\sum_{i=1}^{n} \frac{i^2\pi^2}{2L} B_i^2.$$

We conclude that for all n,

$$\sum_{i=1}^{n} \frac{i^2\pi^2}{2L} B_i^2 \le \|f'\|^2.$$

This implies that the series $\sum_{i=1}^{\infty} \frac{i^2\pi^2}{2L} B_i^2$ converges. The same argument we used before then tells us that the original Fourier cosine series for f converges uniformly to $f(x)$ on $[0, L]$.

9.10.5   Back to Convergence of the Cable Solution

Again, recall our general cable model is

$$\beta^2 \frac{\partial^2 \Phi}{\partial x^2} - \Phi - \alpha \frac{\partial \Phi}{\partial t} = 0, \text{ for } 0 \leq x \leq L, \; t \geq 0,$$

$$\frac{\partial \Phi}{\partial x}(0, t) = 0,$$

$$\frac{\partial \Phi}{\partial x}(L, t) = 0,$$

$$\Phi(x, 0) = f(x).$$

has a series solution of the form

$$\Phi(x, t) = B_0 e^{-\frac{1}{\alpha}t} + \sum_{n=1}^{\infty} B_n \cos\left(\frac{n\pi}{L}x\right) e^{-\frac{L^2+n^2\pi^2\beta^2}{\alpha L^2}t}.$$

where since the Fourier series for the boundary data function f is given by

$$f(x) = B_0 + \sum_{n=1}^{\infty} B_n \cos\left(\frac{n\pi}{L}x\right)$$

we have

$$B_0 = \frac{1}{L} \int_0^L f(x),$$

$$B_n = \frac{2}{L} \int_0^L f(x) \cos\left(\frac{n\pi}{L}x\right) dx.$$

First, let's look at some common series we will see.

Theorem 9.10.6   (Common Convergent Series)

- *For any integer n_0, The series $\sum_{n=n_0}^{\infty} \frac{1}{n^2}$ converges.*
- *For any integer n_0 and positive constant c, the series $\sum_{n=n_0}^{\infty} e^{-cn}$ converges.*
- *For any integer n_0 and positive constant c, the series $\sum_{n=n_0}^{\infty} ne^{-cn}$ converges.*
- *For any integer n_0 and positive constant c, the series $\sum_{n=n_0}^{\infty} n^2 e^{-cn}$ converges.*

Proof The arguments to show these results are valid are straightforward.

- For convenience, let's assume $n_0 = 1$ and $c = 1$. Our arguments will not change much if these constants are different from those values. Pick a positive integer N. Let P_N be the partition of the interval $[1, N]$ into pieces of length 1. Then the Riemann sum for evaluation points which are chosen to be the right hand endpoints of each of the subintervals created by the partition underestimates the area under the curve $y = \frac{1}{t^2}$ on the interval $[1, N]$. We have

$$\sum_{i=2}^{N} \frac{1}{i^2} \times 1 < \int_1^N \frac{1}{t^2} dt = 1 - \frac{1}{N} \leq 1.$$

Hence, this series is always bounded by 1 and so it must converge.

- For convenience, again assume $n_0 = 1$ and $c = 1$. Choose P_N as before. Then the Riemann sum for right hand endpoints underestimates the area under the curve $y = e^{-t}$ on the interval $[1, N]$. We have

$$\sum_{i=2}^{N} e^{-i} \times 1 < \int_0^N e^{-t} dt = 1 - e^{-N} \leq 1.$$

Hence, this series is always bounded by 1 and so it must converge.

- We assume the usual partition P_N. The function ne^{-n} has a maximum at $n = 1$ and after that it decays to 0. So from $n = 1$ on, it is a decreasing function. Then the Riemann sum for evaluation points which are chosen to be the right hand endpoints of each of the subintervals created by the partition underestimates the area under the curve $y = te^{-t}$ on the interval $[1, N]$. We have

$$\sum_{i=2}^{N} ie^{-i} \times 1 < \int_0^N te^{-t} dt.$$

The integral $\int_0^N te^{-t} dt$ can be integrated by parts to show its value is always bounded by a constant C. Hence, this series is always bounded by C and so it must converge.

- We assume the usual partition P_N. The function $n^2 e^{-n}$ has a maximum at $n = 2$ and after that it decays to 0. So from $n = 2$ on, it is a decreasing function. Then the Riemann sum for evaluation points which are chosen to be the right hand endpoints of each of the subintervals created by the partition underestimates the area under the curve $y = te^{-t}$ on the interval $[1, N]$. We have

$$\sum_{i=2}^{N} i^2 e^{-i} \times 1 < \int_0^N t^2 e^{-t} dt.$$

The integral $\int_0^N t^2 e^{-t} dt$ can also be integrated by parts to show its value is always bounded by a constant D. Hence, this series is always bounded by D and so it must converge.   ∎

Comment 9.10.2 *We can also show that series of the form* $\sum_n e^{-n^2}$, $\sum_n n e^{-n^2}$, *and* $\sum_n n^2 e^{-n^2}$ *etc. converge although it is easier to use a test we have not discussed called the* Ratio Test *to do this.*

We now need some additional tests.

Theorem 9.10.7 (Tests for Derivative Uniform Convergence)
Assume we have a sequence of continuously differentiable functions (f_i) *on the interval* $[a, b]$. *If the series* $\sum_{i=1}^{\infty} f_i(x)$ *converges pointwise to* $S(x)$ *on* $[a, b]$ *and and the series* $\sum_{i=1}^{\infty} f_i'(x)$ *converges uniformly to* $v(x)$ *on* $[a, b]$, $S'(x) = v(x)$ *on* $[a, b]$.

Proof We know $f_i(x) = f_i(a) + \int_a^x f_i'(t) dt$. Thus,

$$\sum_{i=1}^{n} f_i(x) = \sum_{i=1}^{n} f_i(a) + \int_a^x \left(\sum_{i=1}^{n} f_i'(t) \right) dt$$

Let's look at the term involving the derivatives. Consider

$$\left| \int_a^x v(t) dt - \int_a^x \left(\sum_{i=1}^{n} f_i'(t) \right) dt \right| \leq \int_a^x \left| v(t) dt - \left(\sum_{i=1}^{n} f_i'(t) \right) \right| dt$$

$$\leq \int_a^b \left| v(t) - \left(\sum_{i=1}^{n} f_i'(t) \right) \right| dt.$$

Since we assume $\sum_{i=1}^{n} f_i'(t)$ converges uniformly to $v(t)$ on $[a, b]$, for a given $\epsilon > 0$, there is a positive integer N so that

$$n > N \implies \left| v(t) - \left(\sum_{i=1}^{n} f_i'(t) \right) \right| < \frac{\epsilon}{2(b - a)}.$$

Using this, we see

$$\left| \int_a^x v(t) dt - \int_a^x \left(\sum_{i=1}^{n} f_i'(t) \right) dt \right| \leq \int_a^b \left| v(t) - \left(\sum_{i=1}^{n} f_i'(t) \right) \right| dt < \epsilon.$$

So we can conclude that $\int_a^x \left(\sum_{i=1}^{n} f_i'(t) \right) dt$ converges to $\int_a^x v(t) dt$ for all x. Letting n go to infinity, we therefore have

$$\sum_{i=1}^{n} f_i(x) = \sum_{i=1}^{n} f_i(a) + \int_a^x \left(\sum_{i=1}^{n} f_i'(t) \right) dt \implies S(x) = S(a) + \int_a^x v(t) dt$$

Applying the Fundamental Theorem of Calculus, we see $S'(x) = v(x)$ for all x. ∎

9.10.5.1 Convergence of the Cable Solution: Simplified Analysis

The formal series solution is

$$\Phi(x, t) = B_0 e^{-\frac{1}{\alpha} t} + \sum_{n=1}^{\infty} B_n \cos\left(\frac{n\pi}{L} x\right) e^{-\frac{L^2 + n^2 \pi^2 \beta^2}{\alpha L^2} t}.$$

Before we get our hands dirty with the ugly details, let's set all the parameters here to the value of 1 to make life easier. Hence, we have $L = 1$, $\alpha = 1$ and $\pi\beta = 1$. The solution is then

$$\Phi(x, t) = B_0 e^{-t} + \sum_{n=1}^{\infty} B_n \cos(n\pi x) e^{-(1+n^2)t}.$$

Now, we can estimate the B_n coefficients as

$$|B_0| \leq \frac{1}{L} \int_0^L |f(x)| dx \leq \frac{1}{L} \int_0^L |f(x)| dx,$$

$$|B_n| \leq \frac{2}{L} \int_0^L |f(x)| \left| \cos\left(\frac{n\pi}{L} x\right) \right| dx \leq \frac{2}{L} \int_0^L |f(x)| dx.$$

Letting $C = \frac{2}{L} \int_0^L |f(x)| dx$, we see Hence, $|B_n| \leq C$ for all $n \geq 0$. Note, we can get a nice constant C no matter what value L has in our solution! Further, we have

$$e^{-(1+n^2)t} \leq e^{-n^2 t}.$$

Thus, for any fixed $t > 0$, we have the estimate

$$\sum_{n=1}^{\infty} |B_n| |\cos(n\pi x)| e^{-(1+n^2)t} \leq \sum_{n=1}^{\infty} C e^{-n^2 t}.$$

Using Theorem 9.10.6, we note the series on the right hand side converges. We can then apply Theorem 9.10.4 to see the series on the left hand side converges uniformly to $\Phi(x, t)$ on the interval $[0, 1]$ as long as $t > 0$. Since each of the individual functions in the series on the left hand side is continuous and convergence is uniform, we can use Theorem 9.10.1 to see the limit function for the left hand side series is continuous on $[0, 1]$ for each $t > 0$. Hence, the function $\Phi(x, t)$ is continuous on the domain $[0, 1] \times [t, \infty)$ for any positive t. At $t = 0$, the series for $\Phi(x, t)$ becomes

$$B_0 + \sum_{n=1}^{\infty} B_n \cos(n\pi x).$$

which is the Fourier cosine series for f on $[0, 1]$. Now from our previous work with the Fourier cosine series, we know this series converges uniformly to $f(x)$ as long as f is continuous on $[0, 1]$ with a derivative that exists for all but a finite number of points. Hence, we see

$$\lim_{t \to 0} |\Phi(x, t)| = \lim_{t \to 0} \left(B_0 e^{-t} + \sum_{n=1}^{\infty} B_n \cos(n\pi x)\, e^{-(1+n^2)t} \right)$$

$$= f(x) = \Phi(x, 0).$$

So we know $\Phi(x, t)$ is continuous on $[0, L] \times [0, \infty)$. Now let's look at the partial derivatives. If we take the partial derivative with respect to t of the series for $\Phi(x, t)$ term by term, we find the function $D_t(x, t)$ given by

$$D_t(x, t) = -B_0 e^{-t} - \sum_{n=1}^{\infty} B_n(1 + n^2) \cos(n\pi x)\, e^{-(1+n^2)t}$$

The series portion for $D_t(x, t)$ satisfies the estimate

$$\sum_{n=1}^{\infty} |B_n|(1 + n^2)|\cos(n\pi x)|\, e^{-(1+n^2)t} \le \sum_{n=1}^{\infty} C(1 + n^2)\, e^{-n^2 t}.$$

Applying Theorem 9.10.6, we see the series on the right hand side converges for any $t > 0$. Then, using Theorem 9.10.4, we see the series for $D_t(x, t)$ converges uniformly. Since it is built from continuous functions, we know limit function is continuous for $t > 0$ because the convergence is uniform. Finally, applying Theorem 9.10.7, we see the series for $\frac{\partial \Phi}{\partial t}$ is the same as the series $D_t(x, t)$ we found by differentiating term by term. The formal partial derivatives of the cable series solution with respect to x are then

$$D_x(x, t) = -\sum_{n=1}^{\infty} B_n n\pi \sin(n\pi x)\, e^{-(1+n^2)t}$$

$$D_{xx}(x, t) = -\sum_{n=1}^{\infty} B_n n^2 \pi^2 \cos(n\pi x)\, e^{-(1+n^2)t}$$

Now we can estimate the B_n coefficients as usual with the constant C as before. Hence, $|B_n| \le C$ for all $n \ge 0$. Also, again, we have

$$e^{-(1+n^2)t} \le e^{-n^2 t}.$$

The series for the two partial derivatives with respect to x satisfies

$$\sum_{n=1}^{\infty} |B_n| n\pi | \sin(n\pi x)| \, e^{-(1+n^2)t} \le \sum_{n=1}^{\infty} Cn\pi \, e^{-n^2 t}$$

and

$$\sum_{n=1}^{\infty} |B_n| n^2 \pi^2 | \sin(n\pi x)| \, e^{-(1+n^2)t} \le \sum_{n=1}^{\infty} Cn^2 \pi^2 \, e^{-n^2 t}$$

Applying Theorem 9.10.6, we see both of the series on the right hand side converge for any $t > 0$. Then, using Theorem 9.10.4, we see the series for $D_x(x, t)$ and $D_{xx}(x, t)$ converge uniformly. Since these series are built from continuous functions, we then know the limit function is continuous for $t > 0$ since the convergence is uniform. Finally, applying Theorem 9.10.7, we see the series for $\frac{\partial \Phi}{\partial x}$ is the same as the series $D_x(x, t)$ we find by differentiating term by term. Further, the series for $\frac{\partial^2 \Phi}{\partial x^2}$ is the same as the series $D_{xx}(x, t)$ we also find by differentiating term by term. Hence, on $[0, 1] \times (0, \infty)$ we can compute

$$\frac{\partial^2 \Phi}{\partial x^2} - \Phi - \frac{\partial \Phi}{\partial t} = -\sum_{n=1}^{\infty} B_n n^2 \pi^2 \cos(n\pi x) \, e^{-(1+n^2)t} - B_0 e^{-t} - \sum_{n-1}^{\infty} B_n \cos(n\pi x) \, e^{-(1+n^2)t}$$

$$+ B_0 e^{-t} + \sum_{n=1}^{\infty} B_n (1 + n^2 \pi^2) \cos(n\pi x) \, e^{-(1+n^2)t}$$

$$= 0.$$

So that we see our series solution satisfies the partial differential equation. To check the boundary conditions, because of continuity, we can see

$$\frac{\partial \Phi}{\partial x}(0, t) = \left(-\sum_{n=1}^{\infty} B_n n\pi \sin(n\pi x) \, e^{-(1+n^2)t} \right)\bigg|_{x=0} = 0$$

$$\frac{\partial \Phi}{\partial x}(1, t) = \left(-\sum_{n=1}^{\infty} B_n n\pi \sin(n\pi x) \, e^{-(1+n^2)t} \right)\bigg|_{x=1} = 0.$$

and finally, the data boundary condition gives

$$\Phi(x, 0) = f(x).$$

Note, we can show that higher order partial derivatives of Φ are also continuous even though we have only limited smoothness in the data function f. Hence, the solutions of the cable equation smooth irregularities in the data.

9.10.5.2 Convergence of the General Cable Solution: Detailed Analysis

The formal series solution in the most general case is then

$$\Phi(x, t) = B_0 e^{-\frac{1}{\alpha}t} + \sum_{n=1}^{\infty} B_n \cos\left(\frac{n\pi}{L}x\right) e^{-\frac{L^2+n^2\pi^2\beta^2}{\alpha L^2}t}.$$

Now we can estimate the B_n coefficients as before by the same constant C. Hence, $|B_n| \leq C$ for all $n \geq 0$. Further, we now have

$$e^{-\frac{L^2+n^2\pi^2\beta^2}{\alpha L^2}t} \leq e^{-\frac{n^2\pi^2\beta^2}{\alpha L^2}t}.$$

Thus, for any fixed $t > 0$, we have the estimate

$$\sum_{n=1}^{\infty} |B_n| \left|\cos\left(\frac{n\pi}{L}x\right)\right| e^{-\frac{L^2+n^2\pi^2\beta^2}{\alpha L^2}t} \leq \sum_{n=1}^{\infty} C e^{-\frac{n^2\pi^2\beta^2}{\alpha L^2}t}.$$

Using Theorem 9.10.6, we note the series on the right hand side converges. We can then apply Theorem 9.10.4 to see the series on the left hand side converges uniformly to $\Phi(x, t)$ on the interval $[0, L]$ as long as $t > 0$. Since each of the individual functions in the series on the left hand side is continuous and convergence is uniform, we can use Theorem 9.10.1 to see the limit function for the left hand side series is continuous on $[0, L]$ for each $t > 0$. Hence, the function $\Phi(x, t)$ is continuous on the domain $[0, L] \times [t, \infty)$ for any positive t. At $t = 0$, the series for $\Phi(x, t)$ becomes

$$B_0 + \sum_{n=1}^{\infty} B_n \cos\left(\frac{n\pi}{L}x\right).$$

which is the Fourier cosine series for f on $[0, L]$. Now from our previous work with the Fourier cosine series, we know this series converges uniformly to $f(x)$ as long as f is continuous on $[0, L]$ with a derivative that exists for all but a finite number of points. Hence, we see

$$\lim_{t \to 0} |\Phi(x, t)| = \lim_{t \to 0} \left(B_0 e^{-\frac{1}{\alpha}t} + \sum_{n=1}^{\infty} B_n \cos\left(\frac{n\pi}{L}x\right) e^{-\frac{L^2+n^2\pi^2\beta^2}{\alpha L^2}t}\right)$$

$$= f(x) = \Phi(x, 0).$$

So we know $\Phi(x, t)$ is continuous on $[0, L] \times [0, \infty)$. Now let's look at the partial derivatives. If we take the partial derivative with respect to x of the series for $\Phi(x, t)$ term by term, we find the function $D_t(x, t)$ given by

$$D_t(x, t) = -B_0 \frac{1}{\alpha} e^{-\frac{1}{\alpha}t} - \sum_{n=1}^{\infty} B_n \frac{L^2 + n^2\pi^2\beta^2}{\alpha L^2} \cos\left(\frac{n\pi}{L}x\right) e^{-\frac{L^2 + n^2\pi^2\beta^2}{\alpha L^2}t}$$

The series portion for $D_t(x, t)$ then satisfies the estimate

$$\sum_{n=1}^{\infty} |B_n| \left|\frac{L^2 + n^2\pi^2\beta^2}{\alpha L^2}\right| \left|\cos\left(\frac{n\pi}{L}x\right)\right| e^{-\frac{L^2 + n^2\pi^2\beta^2}{\alpha L^2}t} \leq \sum_{n=1}^{\infty} C \left|\frac{L^2 + n^2\pi^2\beta^2}{\alpha L^2}\right| e^{-\frac{n^2\pi^2\beta^2}{\alpha L^2}t}.$$

Applying Theorem 9.10.6, we again see the series on the right hand side converges for any $t > 0$. Then, using Theorem 9.10.4, we see the series for $D_1(x, t)$ converges uniformly. Since it is built from continuous functions, we know limit function is continuous for $t > 0$ because the convergence is uniform. Finally, applying Theorem 9.10.7, we see the series for $\frac{\partial \Phi}{\partial t}$ is the same as the series $D_t(x, t)$ we found by differentiating term by term.

The formal partial derivatives of the cable series solution with respect to x are analyzed in a similar manner. We have

$$D_x(x, t) = -\sum_{n=1}^{\infty} B_n \frac{n\pi}{L} \sin\left(\frac{n\pi}{L}x\right) e^{-\frac{L^2 + n^2\pi^2\beta^2}{\alpha L^2}t}$$

$$D_{xx}(x, t) = -\sum_{n=1}^{\infty} B_n \frac{n^2\pi^2}{L^2} \cos\left(\frac{n\pi}{L}x\right) e^{-\frac{L^2 + n^2\pi^2\beta^2}{\alpha L^2}t}$$

Now we can estimate the B_n coefficients by C and the exponential term again as

$$e^{-\frac{L^2 + n^2\pi^2\beta^2}{\alpha L^2}t} \leq e^{-\frac{n^2\pi^2\beta^2}{\alpha L^2}t}.$$

Then series terms for the two partial derivatives with respect to x satisfy

$$\sum_{n=1}^{\infty} |B_n| \left|\frac{n\pi}{L}\right| \left|\sin\left(\frac{n\pi}{L}x\right)\right| e^{-\frac{L^2 + n^2\pi^2\beta^2}{\alpha L^2}t} \leq \sum_{n=1}^{\infty} C \left|\frac{n\pi}{L}\right| e^{-\frac{n^2\pi^2\beta^2}{\alpha L^2}t}$$

and

$$\sum_{n=1}^{\infty} |B_n| \left|\frac{n^2\pi^2}{L^2}\right| \left|\cos\left(\frac{n\pi}{L}x\right)\right| e^{-\frac{L^2 + n^2\pi^2\beta^2}{\alpha L^2}t} \leq \sum_{n=1}^{\infty} C \left|\frac{n\pi}{L}\right| e^{-\frac{n^2\pi^2\beta^2}{\alpha L^2}t}.$$

Applying Theorems 9.10.4 and 9.10.6, we see the series for $D_x(x, t)$ and $D_{xx}(x, t)$ converge uniformly for $t > 0$. Since these series are built from continuous functions, we then know the limit function is continuous for $t > 0$ Since the convergence is uniform, these limit functions are also continuous. Finally, applying Theorem 9.10.7, we see the series for $\frac{\partial \Phi}{\partial x}$ and the series for $\frac{\partial^2 \Phi}{\partial x^2}$ are the same as the series $D_x(x, t)$ and

$D_{xx}(x, t)$ that we find by differentiating term by term. It is then an easy calculation to see that the series solution satisfies the cable equation with the given boundary conditions.

Comment 9.10.3 *The arguments we presented in this chapter are quite similar to the ones we would use to analyze the smoothness qualities of the solutions to other linear partial differential equations with boundary conditions. We didn't want to go through all of them in this text as we don't think there is a lot of value in repeating these arguments over and over. Just remember that to get solutions with this kind of smoothness requires that the series we formally compute converge uniformly. If we did not have that information, our series solutions would not have this amount of smoothness. But that is another story!*

Reference

J. Peterson, in *Calculus for Cognitive Scientists: Higher Order Models and Their Analysis*. Springer Series on Cognitive Science and Technology (Springer Science+Business Media Singapore Pte Ltd., Singapore, 2015 in press)

Chapter 10
Linear Partial Differential Equations

The cable equation is an example of a linear partial differential equation and we have already seen how to solve it using separation of variables. There are several other models and we will illustrate how these models look on simple domains for convenience.

The Wave Equation: We seek functions $\Phi(x, t)$ so that

$$\frac{\partial^2 \Phi}{\partial t^2} - c^2 \frac{\partial^2 \Phi}{\partial x^2} = 0$$

$$\Phi(x, 0) = f(x), \text{ for } 0 \leq x \leq L$$

$$\frac{\partial \Phi}{\partial t}(x, 0) = g(x), \text{ for } 0 \leq x \leq L$$

$$\Phi(0, t) = 0, \text{ for } 0 \leq t$$

$$\Phi(L, t) = 0, \text{ for } 0 \leq t$$

for some positive constant c. The solution of this equation approximates the motion of a nice string with no external forces applied. The domain here is the infinite rectangle $[0, L] \times [0, \infty)$.

Laplace's Equation: The solution $\Phi(x, y)$ of this equation is a time independent solution to a problem such as the distribution of heat on a membrane stretch over the domain given that various heat sources are applied to the boundary. Here, the domain is the finite square $[0, L] \times [0, L]$. In the problem below, three of the edges of the square are clamped to 0 and the remaining one must follow the heat profile given by the function $f(x)$.

$$\frac{\partial^2 \Phi}{\partial x^2} + \frac{\partial^2 \Phi}{\partial y^2} = 0$$

$$\frac{\partial \Phi}{\partial x}(0, y) = 0, \text{ for } 0 \leq y \leq L$$

© Springer Science+Business Media Singapore 2016
J.K. Peterson, *Calculus for Cognitive Scientists*, Cognitive Science
and Technology, DOI 10.1007/978-981-287-880-9_10

$$\frac{\partial \Phi}{\partial x}(L, y) = 0, \text{ for } 0 \le y \le L$$
$$\Phi(x, L) = 0, \text{ for } 0 \le x \le L$$
$$\Phi(x, 0) = f(x), \text{ for } 0 \le x \le L$$

The Heat/Diffusion Equation: The solution of this equation, $\Phi(x, t)$, is the time dependent value of heat or temperature of a one dimensional bar which is having a heat source applied to it initially. It can also model a substance moving through a domain using diffusion with diffusion constant D as discussed in Chap. 5. The domain is again half infinite: $[0, L] \times [0, \infty)$.

$$\frac{\partial \Phi}{\partial t} - D \frac{\partial^2 \Phi}{\partial x^2} = 0$$
$$\Phi(0, t) = 0, \text{ for } 0 < t$$
$$\Phi(L, t) = 0, \text{ for } 0 < t$$
$$\Phi(x, 0) = f(x), \text{ for } 0 < x < L$$

This equation is very relevant to our needs. Indeed, in Peterson (2015), we derive this equation using a random walk model and learn how to interpret the diffusion constant D in terms of the space constant λ_C and the time constant τ_M. However, we will not discuss that here.

10.1   The Separation of Variables Method Revisited

As we know from Chap. 9, one way to find the solution to these models is to assume we can separate the variables and write $\Phi(x, t) = u(x)w(t)$ or $\Phi(x, y) = u(x)w(y)$. If we make this separation assumption, we find solutions that must be written as infinite series and to solve the boundary conditions, we must express boundary functions as series expansions.

10.1.1   The Wave Equation

We will solve this wave equation model.

$$\frac{\partial^2 \Phi}{\partial t^2} - c^2 \frac{\partial^2 \Phi}{\partial x^2} = 0$$
$$\Phi(x, 0) = f(x), \text{ for } 0 \le x \le L$$
$$\frac{\partial \Phi}{\partial t}(x, 0) = g(x), \text{ for } 0 \le x \le L$$

$$\Phi(0, t) = 0, \text{ for } 0 \leq t$$
$$\Phi(L, t) = 0, \text{ for } 0 \leq t$$

for some positive constant c. We assume a solution of the form $\Phi(x, t) = u(x)\, w(t)$ and compute the needed partials for Wave Equation. This leads to a the new equation

$$u(x) \frac{d^2 w}{dt^2} - c^2 \frac{d^2 u}{dx^2} w(t) = 0.$$

Rewriting, we find for all x and t, we must have

$$u(x) \frac{d^2 w}{dt^2} = c^2 \frac{d^2 u}{dx^2} w(t).$$

Now put all the u quantities on the left hand side and all the w ones on the right hand side. This gives

$$\frac{\frac{d^2 w}{dt^2}}{w(t)} = \frac{c^2 \frac{d^2 u}{dx^2}}{u(x)}, \quad 0 \leq x \leq L, \ t \geq 0.$$

The only way this can be true is if both the left and right hand side are always a constant. This common constant value we will denote by Θ and it is called the **separation of variables constant** for this model. This leads to the decoupled equations 10.1 and 10.2.

$$\frac{d^2 w}{dt^2} = \Theta\, w(t), \quad t \geq 0, \tag{10.1}$$

$$\frac{c^2 d^2 u}{dx^2} = \Theta\, u(x), \quad 0 \leq x \leq I_{.} \tag{10.2}$$

We also have boundary conditions. Our assumption leads to the following boundary conditions in x:

$$u(0)\, w(t) = 0, \quad t \geq 0,$$
$$u(L)\, w(t) = 0, \quad t \geq 0.$$

Since these equations must hold for all t, this forces

$$u(0) = 0, \tag{10.3}$$
$$u(L) = 0. \tag{10.4}$$

Equations 10.1–10.4 give us the boundary value problem in $u(x)$ we need to solve. We also have to solve a system for the variable $w(t)$.

10.1.1.1   Determining the Separation Constant

Let's look at the model in x and for convenience, let u'' denote the second derivative with respect to x. The model is then

$$c^2 u'' - \Theta u = 0$$
$$u(0) = 0,$$
$$u(L) = 0.$$

We are looking for nonzero solutions, so any choice of separation constant Θ that leads to a zero solution will be rejected. Hence, our job is to find the values of Θ that give rise to nonzero solutions.

Case I: $\Theta = \omega^2$, $\omega \neq 0$: The model to solve is

$$u'' - \frac{\omega^2}{c^2} u = 0$$
$$u(0) = 0,$$
$$u(L) = 0.$$

with characteristic equation $r^2 - \frac{\omega^2}{c^2} = 0$ with the real roots $\pm \frac{\omega}{c}$. The general solution of this second order model is given by

$$u(x) = A \cosh\left(\frac{\omega}{c}x\right) + B \sinh\left(\frac{\omega}{c}x\right)$$

Next, apply the boundary conditions, $u(0) = 0$ and $u(L) = 0$. Hence,

$$u(0) = 0 = A$$
$$u(L) = 0 = B \sinh\left(L\frac{\omega}{c}\right)$$

Hence, $A = 0$ and $B \sinh\left(L\frac{\omega}{c}\right) = 0$. Since sinh is never zero when ω is not zero, we see $B = 0$ also. Hence, the only u solution is the trivial one and we can reject this case. There is no need to look at the corresponding w solutions as we construct the solution to the wave equation from the products $u(x)w(t)$. Hence, if $u(x) = 0$ always, the product is also zero.

Case II: $\Theta = 0$: The model to solve is now

$$u'' = 0$$
$$u(0) = 0,$$
$$u(L) = 0.$$

with characteristic equation $r^2 = 0$ with the double root $r = 0$. Hence, the general solution is now

$$u(x) = A + Bx$$

Applying the boundary conditions, $u(0) = 0$ and $u(L) = 0$. Hence,

$$u(0) = 0 = A$$
$$u(L) = 0 = BL$$

Hence, $A = 0$ and $B = 0$ and the u solutions are trivial. We reject this case as well.

Case III: $\Theta = -\omega^2, \omega \neq 0$:

$$c^2 u'' + \frac{\omega^2}{c^2} u = 0$$
$$u(0) = 0,$$
$$u(L) = 0.$$

The general solution is given by

$$u(x) = A \, \cos\left(\frac{\omega}{c}x\right) + B \, \sin\left(\frac{\omega}{c}x\right)$$

Next, apply the boundary conditions, $u(0) = 0$ and $u(L) = 0$. Hence,

$$u(0) = 0 = A$$
$$u(L) = 0 = B \sin\left(L\frac{\omega}{c}\right)$$

Hence, $A = 0$ and $B \sin\left(L\frac{\omega}{c}\right) = 0$. We now see an interesting result. We can determine a unique value of B only if $\sin\left(L\frac{\omega}{c}\right) \neq 0$. Since $\omega \neq 0$ by assumption, if $\frac{\omega}{c}L = n\pi$ for any integer $n \neq 0$, we find the value of B can not be determined as we have the equation $0 = B \times 0$. If $\omega \neq \frac{n\pi c}{L}$, we can solve for B and find $B = 0$. So the only solutions are the trivial or zero solutions unless $\omega L = n\pi c$. Letting $\omega_n = \frac{n\pi c}{L}$, we find a a non zero solution for each nonzero value of B of the form

$$u_n(x) = B \sin\left(\frac{\omega_n}{c}x\right) = B \sin\left(\frac{n\pi}{L}x\right).$$

It is only necessary to use all integers $n \geq 1$ as sin is an odd function and so the solutions for negative integers are just the negative of the solutions for positive integers. For convenience, let's choose all the constants $B = 1$. Then we have an infinite family of nonzero solutions $u_n(x) = \sin\left(\frac{n\pi}{L}x\right)$ and an infinite family of

separation constants $\Theta_n = -\omega_n^2 = -\frac{n^2\pi^2 c^2}{L^2}$.
  We can then solve the w equation. We must solve

$$\frac{d^2 w}{dt^2} + \omega_n^2\, w(t) = 0, \ \ t \geq 0.$$

The general solution is

$$w_n(t) = C\, \cos(\omega_n t) + D\, \sin(\omega_n t)$$
$$= C\, \cos\left(\frac{n\pi c}{L}t\right) + D\, \sin\left(\frac{n\pi c}{L}t\right)$$

as we will get a different time solution for each choice of ω_n. Hence, any product

$$\phi_n(x, t) = u_n(x)\, w_n(t)$$

will solve the model with the x boundary conditions. In fact, any finite sum of the form, for arbitrary constants C_n and D_n

$$\Phi_N(x, t) = \sum_{n=1}^{N} \phi_n(x, t) = \sum_{n=1}^{N} u_n(x)\, w_n(t)$$
$$= \sum_{n=1}^{N} \sin\left(\frac{n\pi}{L}x\right)\left(C_n\, \cos\left(\frac{n\pi c}{L}t\right) + D_n\, \sin\left(\frac{n\pi c}{L}t\right)\right)$$

will solve the model with x boundary conditions as well.

10.1.1.2   The Infinite Series Solution

Using what we know about Fourier series and series in general from Chap. 9, we can show the series of the form

$$\sum_{n=1}^{\infty} \sin\left(\frac{n\pi}{L}x\right)\left(A_n \cos\left(\frac{n\pi c}{L}t\right) + B_n \sin\left(\frac{n\pi c}{L}t\right)\right)$$

converge pointwise for x in $[0, L]$ and all t. We haven't gone through all the details of this argument yet but we will. Hence, we will form the solution $\Phi(x, t)$ given by

$$\Phi(x, t) = \sum_{n=1}^{\infty} \phi_n(x, t) = \sum_{n=1}^{\infty} u_n(x)\, w_n(t)$$

$$= \sum_{n=1}^{\infty} \sin\left(\frac{n\pi}{L}x\right)\left(C_n \cos\left(\frac{n\pi c}{L}t\right) + D_n \sin\left(\frac{n\pi c}{L}t\right)\right)$$

for the arbitrary sequences of constants (C_n) and (D_n). We can show we can find the partial derivatives of this series pointwise by differentiating term by term. Hence,

$$\frac{\partial \Phi}{\partial t}(x, t) = \sum_{n=1}^{\infty} u_n(x) \frac{\partial w}{\partial t}$$

$$= \sum_{n=1}^{\infty} \sin\left(\frac{n\pi}{L}x\right)\left(-C_n \frac{n\pi c}{L} \sin\left(\frac{n\pi c}{L}t\right) + D_n \frac{n\pi c}{L} \cos\left(\frac{n\pi c}{L}t\right)\right).$$

which we can then use to solve the boundary conditions with data functions.

10.1.1.3 Handling the Other Boundary Conditions

We can easily find that

$$\Phi(x, 0) = \sum_{n=1}^{\infty} \sin\left(\frac{n\pi}{L}x\right) C_n,$$

and

$$\frac{\partial \Phi}{\partial t}(x, 0) = \sum_{n=1}^{\infty} \sin\left(\frac{n\pi}{L}x\right) D_n \frac{n\pi c}{L}.$$

The remaining boundary conditions are

$$\Phi(x, 0) = f(x), \text{ for } 0 \le x \le L$$
$$\frac{\partial \Phi}{\partial t}(x, 0) = g(x), \text{ for } 0 \le x \le L$$

These are rewritten in terms of the series solution, for $0 \le x \le L$, as

$$\sum_{n=1}^{\infty} C_n \sin\left(\frac{n\pi}{L}x\right) = f(x),$$

$$\sum_{n=1}^{\infty} D_n \frac{n\pi c}{L} \sin\left(\frac{n\pi}{L}x\right) = g(x).$$

The theory of Fourier series tells us that any piecewise differentiable function h can be written as the series

$$h(x) = \sum_{n=1}^{\infty} A_n \sin\left(\frac{n\pi}{L}x\right)$$

where the coefficients are given by the integral

$$A_n = \frac{2}{L} \int_0^L h(x) \sin\left(\frac{n\pi}{L}x\right) dx.$$

Since this is a Fourier series, the series converges pointwise. For now, apply these ideas to the functions f and g to write

$$f(x) = \sum_{n=1}^{\infty} A_n \sin\left(\frac{n\pi}{L}x\right)$$

$$g(x) = \sum_{n=1}^{\infty} B_n \sin\left(\frac{n\pi}{L}x\right)$$

with

$$A_n = \frac{2}{L} \int_0^L f(x) \sin\left(\frac{n\pi}{L}x\right) dx$$

$$B_n = \frac{2}{L} \int_0^L g(x) \sin\left(\frac{n\pi}{L}x\right) dx$$

Then, setting these series equal, we find

$$\sum_{n=1}^{\infty} C_n \sin\left(\frac{n\pi}{L}x\right) = \sum_{n=1}^{\infty} A_n \sin\left(\frac{n\pi}{L}x\right),$$

$$\sum_{n=1}^{\infty} D_n \frac{n\pi c}{L} \sin\left(\frac{n\pi}{L}x\right) = \sum_{n=1}^{\infty} B_n \sin\left(\frac{n\pi}{L}x\right).$$

Writing these series on the left hand side only we find after some term by term manipulations

$$\sum_{n=1}^{\infty} \left(C_n - A_n\right) \sin\left(\frac{n\pi}{L}x\right) = 0$$

$$\sum_{n=1}^{\infty} \left(D_n \frac{n\pi c}{L} - B_n\right) \sin\left(\frac{n\pi}{L}x\right) = 0$$

Since these series converge pointwise, the only way they can equal 0 for all x is if the individual terms equal zero. Hence, for each n, we must have

$$\left(C_n - A_n\right) \sin\left(\frac{n\pi}{L}x\right) = 0$$

$$\left(D_n\, n\pi c - B_n\right) \sin\left(\frac{n\pi}{L}x\right) = 0.$$

Since this is true for all x, the coefficients must equal zero and so we find that the solution is given by $C_n = A_n$ and $D_n = \frac{B_n L}{n\pi c}$ for all $n \geq 1$. The ideas here are really quite general so we can apply them quite easily to the other models. The technique we use above is called *coefficient matching*.

10.1.1.4 Summary

Let's summarize the approach.

- Assume the solution is in separated form, $\Phi(x, t) = u(x)w(t)$.
- Convert the model into two ordinary differential equation (ODE) models. The two models are connected by a separation constant Θ.
- The $u(x)$ ODE problem is a boundary value problem (BVP) with parameter Θ. We must find nontrivial, i.e. non zero, solutions to it. We find there are infinitely many such solutions for a sequence of Θ values. This requires we study this model for the cases $\Theta = \omega^2$, $\Theta = 0$ and $\Theta = -\omega^2$ for any $\omega \neq 0$. This requires knowledge of how to solve such ODE models.
- Once the BVP is solved, it tells us the sequence of separation constants we can usc to solve the other ODE model. This gives us one nontrivial solution family for each of these separation constant values, $\phi_n(x, t) = u_n(x)w_n(t)$.
- We form the family of series solutions $\sum_{n=1}^{\infty} \phi_n(x, t)$ and express the boundary function data given by f and g as Fourier series.
- Then equate the series solution at the boundaries which have function data to the series for the boundary data and match coefficients to find the correct solution to the model.

10.1.1.5 Homework

Solve the following problems using separation of variables.

Exercise 10.1.1

$$\frac{\partial^2 \Phi}{\partial t^2} - 25\frac{\partial^2 \Phi}{\partial x^2} = 0$$

$$\Phi(x, 0) = 0.1x^2,\ for\ 0 \leq x \leq 10$$

$$\frac{\partial \Phi}{\partial t}(x, 0) = 0,\ for\ 0 \leq x \leq 10$$

$$\Phi(0, t) = 0, \ for \ 0 \leq t$$
$$\Phi(10, t) = 0, \ for \ 0 \leq t$$

Exercise 10.1.2

$$\frac{\partial^2 \Phi}{\partial t^2} - 4\frac{\partial^2 \Phi}{\partial x^2} = 0$$
$$\Phi(x, 0) = 0, \ for \ 0 \leq x \leq 5$$
$$\frac{\partial \Phi}{\partial t}(x, 0) = 0.05x^2, \ for \ 0 \leq x \leq 5$$
$$\Phi(0, t) = 0, \ for \ 0 \leq t$$
$$\Phi(5, t) = 0, \ for \ 0 \leq t$$

10.1.2   Laplace's Equation

Next, let's focus on solving Laplace's equation.

$$\frac{\partial^2 \Phi}{\partial x^2} + \frac{\partial^2 \Phi}{\partial y^2} = 0$$
$$\frac{\partial \Phi}{\partial x}(0, y) = 0, \ for \ 0 \leq y \leq L$$
$$\frac{\partial \Phi}{\partial x}(L, y) = 0, \ for \ 0 \leq y \leq L$$
$$\Phi(x, L) = 0, \ for \ 0 \leq x \leq L$$
$$\Phi(x, 0) = f(x), \ for \ 0 \leq x \leq L$$

We assume a solution of the form $\Phi(x, y) = u(x) \, w(y)$ and compute the needed partials for Laplace's Equation. This leads to a the new equation

$$u(x) \frac{d^2 w}{dy^2} + \frac{d^2 u}{dx^2} \, w(y) = 0.$$

Rewriting, we find for all x and y, we must have

$$u(x) \frac{d^2 w}{dy^2} = -\frac{d^2 u}{dx^2} \, w(y).$$

Now put all the u quantities on the left hand side and all the w ones on the right hand side. This gives

$$-\frac{\frac{d^2 w}{dy^2}}{w(y)} = \frac{\frac{d^2 u}{dx^2}}{u(x)}, \ \ 0 \leq x, y \leq L.$$

The only way this can be true is if both the left and right hand side are always a constant. This common constant value Θ is called the **separation of variables constant** for this model. This leads to the decoupled equations 10.5 and 10.6.

$$\frac{d^2w}{dy^2} = -\Theta\, w(y), \quad 0 \le y \le L, \tag{10.5}$$

$$\frac{d^2u}{dx^2} = \Theta\, u(x), \quad 0 \le x \le L. \tag{10.6}$$

We also have boundary conditions. Our assumption leads to the following boundary conditions in x:

$$\frac{du}{dx}(0)\, w(y) = 0, \quad \text{for } 0 \le y \le L,$$

$$\frac{du}{dx}(L)\, w(y) = 0, \quad \text{for } 0 \le y \le L,$$

$$u(x)\, w(L) = 0, \quad \text{for } 0 \le x \le L.$$

Since these equations must hold for all y for the first two and all x for the last, this forces

$$\frac{du}{dx}(0) = 0, \tag{10.7}$$

$$\frac{du}{dx}(L) = 0, \tag{10.8}$$

$$w(L) = 0. \tag{10.9}$$

Equations 10.5–10.8 give us the boundary value problem in $u(x)$ we need to solve. We also have to solve a system for the variable $w(y)$ with condition 10.9.

10.1.2.1 Determining the Separation Constant

Let's look at the model in x:

$$u'' - \Theta u = 0$$
$$u'(0) = 0,$$
$$u'(L) = 0.$$

We are looking for nonzero solutions, so any choice of separation constant Θ that leads to a zero solution will be rejected. Hence, our job is to find the values of Θ that give rise to nonzero solutions.

Case I: $\Theta = \omega^2, \omega \neq 0$: The model to solve is

$$u'' - \omega^2 u = 0$$
$$u'(0) = 0,$$
$$u'(L) = 0.$$

with characteristic equation $r^2 - \omega^2 = 0$ with the real roots $\pm\omega$. The general solution of this second order model is given by

$$u(x) = A \cosh(\omega x) + B \sinh(\omega x)$$

with

$$u'(x) = A \omega \sinh(\omega x) + B \omega \cosh(\omega x)$$

Next, apply the boundary conditions, $u'(0) = 0$ and $u'(L) = 0$. Hence,

$$u'(0) = 0 = B$$
$$u'(L) = 0 = A\omega \sinh(L\omega)$$

Hence, $B = 0$ and $A \sinh(L\omega) = 0$. Since sinh is never zero when ω is not zero, we see $A = 0$ also. Hence, the only u solution is the trivial one and we can reject this case. Again, there is no need to look at the corresponding w solutions as we construct the solution to Laplace's equation from the products $u(x)w(y)$. Hence, if $u(x) = 0$ always, the product is also zero.

Case II: $\Theta = 0$: The model to solve is now

$$u'' = 0$$
$$u'(0) = 0,$$
$$u'(L) = 0.$$

with characteristic equation $r^2 = 0$ with the double root $r = 0$. Hence, the general solution is now

$$u(x) = A + B x$$

with $u'(x) = B$. Applying the boundary conditions, $u'(0) = 0$ and $u'(L) = 0$. Hence,

$$u'(0) = 0 = B$$
$$u'(L) = 0 = B$$

Hence, $B = 0$ and there is no constraint on A. Hence, any arbitrary constant which is not zero is a valid non zero solution. Choosing $A = 1$, let $u_0(x) = 1$ be our chosen nonzero solution for this case. We now need to solve for w in this case. The model to solve is

$$\frac{d^2 w}{dy^2} = 0, \quad 0 \le y \le L,$$
$$w(L) = 0.$$

The general solution is $w(y) = A + By$ and applying the boundary condition, we find $w(L) = A + BL = 0$. Hence, solving for B, we find $B = -\frac{A}{L}$. The corresponding family of solutions for w in this case is

$$w(y) = A - \frac{A}{L} y = A \left(1 - \frac{y}{L} \right)$$
$$= A \frac{L - y}{L}$$

Since A is arbitrary, we can choose the arbitrary constant $\frac{A}{L} = 1$ and set

$$w_0(y) = \frac{L - y}{L}.$$

Hence, the product $\Phi_0(x, y) = u_0(x) w_0(y)$ solves the boundary conditions. That is

$$\Phi_0(x, y) = \frac{L - y}{L}$$

is a solution.

Case III: $\Theta = -\omega^2, \omega \ne 0$:

$$u'' + \omega^2 u = 0$$
$$u'(0) = 0,$$
$$u'(L) = 0.$$

The general solution is given by

$$u(x) = A \cos(\omega x) + B \sin(\omega x)$$

with

$$u'(x) = -A \omega \sin(\omega x) + B \omega \cos(\omega x)$$

Next, apply the boundary conditions, $u'(0) = 0$ and $u'(L) = 0$. Hence,

$$u'(0) = 0 = B$$
$$u'(L) = 0 = A \sin(L\omega)$$

Hence, $B = 0$ and $A \sin(L\omega) = 0$. We can determine a unique value of A only if $\sin(L\omega) \neq 0$. Since $\omega \neq 0$ by assumption, if $\omega L = n\pi$ for any integer $n \neq 0$, we find the value of A can not be determined. Of course, if $\omega \neq \frac{n\pi c}{L}$, we can solve for A and find $A = 0$. So the only solutions are the trivial or zero solutions unless $\omega L = n\pi$. Letting $\omega_n = \frac{n\pi}{L}$, we find a a non zero solution for each nonzero value of A of the form

$$u_n(x) = A \cos(\omega_n x) = A \cos\left(\frac{n\pi}{L}x\right).$$

It is only necessary to use all integers $n \geq 1$ as cos is an even function and so the solutions for negative integers are the same of the solutions for positive integers. For convenience, let's choose all the constants $A = 1$. Then we have an infinite family of nonzero solutions $u_n(x) = \cos(\frac{n\pi}{L}x)$ and an infinite family of separation constants $\Theta_n = -\omega_n^2 = -\frac{n^2\pi^2}{L^2}$. We can then solve the w equation. We must solve

$$\frac{d^2 w}{dy^2} - \omega_n^2 \, w(y) = 0, \quad 0 \geq y \geq L,$$

$$w_n(L) = 0.$$

The general solution is

$$w_n(y) = C \, \cosh(\omega_n y) + D \, \sinh(\omega_n y)$$
$$= C \, \cosh\left(\frac{n\pi}{L}y\right) + D \, \sinh\left(\frac{n\pi}{L}y\right)$$

Applying the boundary condition, we have

$$w_n(L) = C \, \cosh(\omega_n L) + D \, \sinh(\omega_n L) = 0.$$

Neither cosh or sinh are zero here, so we can solve for D in terms of C to find

$$D = -C \, \frac{\cosh(\omega_n L)}{\sinh(\omega_n L)}.$$

Hence, we find a family of nontrivial solutions of the form

$$w_n(y) = C \, \cosh(\omega_n y) - C \, \frac{\cosh(\omega_n L)}{\sinh(\omega_n L)} \, \sinh(\omega_n y)$$

$$= C \left(\frac{\cosh(\omega_n y)\ \sinh(\omega_n L) - \cosh(\omega_n L)\ \sinh(\omega_n y)}{\sinh(\omega_n L)} \right)$$

Using the hyperbolic function identity $\sinh(u - v) = \sin(u)\cosh(v) - \sinh(u)\cosh(v)$, we can rewrite this as

$$w_n(y) = \frac{C}{\sinh(\omega_n L)}\ \sinh\Big(\omega_n(L - y)\Big).$$

Finally, the term $\frac{C}{\sinh(\omega_n L)}$ is arbitrary as C is arbitrary. Choosing the arbitrary constant to be 1 always, we obtain a family of solutions of the form

$$w_n(y) = \sinh\Big(\omega_n(L - y)\Big)$$

for all integers $n \geq 1$. Hence, any product

$$\phi_n(x, t) = u_n(x)\ w_n(y)$$

will solve the model with the x boundary conditions. In fact, any finite sum of the form, for arbitrary constants A_n

$$\Phi_N(x, y) = \sum_{n=1}^{N} A_n \phi_n(x, y) = \sum_{n=1}^{N} A_n u_n(x)\ w_n(y)$$

$$= \sum_{n=1}^{N} A_n \cos\left(\frac{n\pi}{L}x\right) \sinh\left(\frac{n\pi}{L}(L - y)\right)$$

will solve the model with our boundary conditions as well.

10.1.2.2 The Infinite Series Solution

Again, we can show series of the form

$$\sum_{n=0}^{\infty} \cos\left(\frac{n\pi}{L}x\right) \sinh\left(\frac{n\pi}{L}(L - y)\right)$$

converge pointwise for x, y in $[0, L]$. Hence, we will form the solution $\Phi(x, y)$ given by

$$\Phi(x, y) = \sum_{n=0}^{\infty} A_n\ \phi_n(x, y) = \sum_{n=0}^{\infty} A_n\ u_n(x)\ w_n(y)$$

$$= A_0 \frac{L-y}{L} + \sum_{n=1}^{\infty} A_n \cos\left(\frac{n\pi}{L}x\right) \sinh\left(\frac{n\pi}{L}(L-y)\right)$$

for the arbitrary sequence of constants (A_n).

10.1.2.3   Handling the Other Boundary Conditions

We can easily find that

$$\Phi(x,0) = A_0 + \sum_{n=1}^{\infty} A_n \cos\left(\frac{n\pi}{L}x\right) \sinh(n\pi),$$

The remaining boundary condition is

$$\Phi(x,0) = f(x), \text{ for } 0 \le x \le L$$

We know we can write

$$f(x) = B_0 + \sum_{n=1}^{\infty} B_n \cos\left(\frac{n\pi}{L}x\right)$$

with

$$B_0 = \frac{1}{L} \int_0^L f(x)dx$$

$$B_n = \frac{2}{L} \int_0^L f(x) \cos\left(\frac{n\pi}{L}x\right)dx, \ \ n \ge 1$$

Then, setting the series equal, we find

$$A_0 + \sum_{n=1}^{\infty} A_n \cos\left(\frac{n\pi}{L}x\right) \sinh(n\pi) = B_0 + \sum_{n=1}^{\infty} B_n \cos\left(\frac{n\pi}{L}x\right),$$

Equating coefficients, we have $A_0 = B_0$ and $A_n = \frac{B_n}{\sinh(n\pi)}$.

10.1.2.4   Summary

Let's summarize this approach again.

- Assume the solution is in separated form $\Phi(x,y) = u(x)w(y)$.

- Convert the model into two ordinary differential equation (ODE) models. These two models are connected by a separation constant Θ.
- The $u(x)$ problems is a boundary value problem (BVP) with parameter Θ. We must find nontrivial, i.e. non zero, solutions to it. We find there are infinitely many such solutions for a sequence of Θ values. This requires we study this model for the cases $\Theta = \omega^2$, $\Theta = 0$ and $\Theta = -\omega^2$ for any $\omega \neq 0$. This requires knowledge of how to solve such ODE models.
- Once the BVP is solved, it tells us the sequence of separation constants we can use to solve the other ODE model. This gives us one nontrivial solution family for each of these separation constant values, ϕ_n.
- We form the family of series solutions $\sum_{n=1}^{\infty} \phi_n$ and express the boundary function data given by f as a Fourier series.
- Then equate the series solution at the boundary which has function data to the series for the boundary data and match coefficients to find the correct solution to the model.

10.1.2.5 Homework

Exercise 10.1.3

$$\frac{\partial^2 \Phi}{\partial x^2} + \frac{\partial^2 \Phi}{\partial y^2} = 0$$

$$\frac{\partial \Phi}{\partial x}(0, y) = 0, \; for \; 0 \leq y \leq 10$$

$$\frac{\partial \Phi}{\partial x}(7, y) = 0, \; for \; 0 \leq y \leq 10$$

$$\Phi(x, 10) = 0, \; for \; 0 \leq x \leq 7$$

$$\Phi(x, 0) = 1.2x^4, \; for \; 0 \leq x \leq 7.$$

Exercise 10.1.4

$$\frac{\partial^2 \Phi}{\partial x^2} + \frac{\partial^2 \Phi}{\partial y^2} = 0$$

$$\frac{\partial \Phi}{\partial x}(0, y) = 0, \; for \; 0 \leq y \leq 2$$

$$\frac{\partial \Phi}{\partial x}(5, y) = 0, \; for \; 0 \leq y \leq 2$$

$$\Phi(x, 2) = 0, \; for \; 0 \leq x \leq 5$$

$$\Phi(x, 0) = 1.2x^2, \; for \; 0 \leq x \leq 5.$$

10.1.3   The Heat/Diffusion Equation

We will solve this heat equation model.

$$\frac{\partial \Phi}{\partial t} - D \frac{\partial^2 \Phi}{\partial x^2} = 0$$

$$\Phi(0, t) = 0, \ \text{for } 0 < t$$
$$\Phi(L, t) = 0, \ \text{for } 0 < t$$
$$\Phi(x, 0) = f(x), \ \text{for } 0 < x < L$$

for some positive constant D which in the case of diffusion is called the diffusion constant. We again assume a solution of the form $\Phi(x, t) = u(x)\, w(t)$ and compute the needed partials. This leads to a the new equation

$$u(x) \frac{dw}{dt} - D \frac{d^2 u}{dx^2} w(t) = 0.$$

Rewriting, we find for all x and t, we must have

$$u(x) \frac{dw}{dt} = D \frac{d^2 u}{dx^2} w(t).$$

Rewriting, we have

$$\frac{\frac{dw}{dt}}{w(t)} = \frac{D \frac{d^2 u}{dx^2}}{u(x)}, \ \ 0 \leq x << L, \ t > 0.$$

As usual, the only way this is true is if both the left and right hand side equal the separation constant Θ for this model. This leads to the decoupled equations 10.10 and 10.11.

$$\frac{dw}{dt} = \Theta\, w(t), \ \ t > 0, \tag{10.10}$$

$$D \frac{d^2 u}{dx^2} = \Theta\, u(x), \ \ 0 \leq x \leq L, \tag{10.11}$$

We also have boundary conditions. Our assumption leads to the following boundary conditions in x:

$$u(0)\, w(t) = 0, \ \ t > 0,$$
$$u(L)\, w(t) = 0, \ \ t > 0.$$

Since these equations must hold for all t, this forces

$$u(0) = 0, \tag{10.12}$$
$$u(L) = 0. \tag{10.13}$$

Equations 10.10–10.13 give us the boundary value problem in $u(x)$ we need to solve. Once we have the u solution, we can solve the system for the variable $w(t)$.

10.1.3.1 Determining the Separation Constant

The model is then

$$D u'' - \Theta u = 0$$
$$u(0) = 0,$$
$$u(L) = 0.$$

We are looking for nonzero solutions, so any choice of separation constant Θ that leads to a zero solution will be rejected. Hence, our job is to find the values of Θ that give rise to nonzero solutions.

Case I: $\Theta = w^2$, $w \neq 0$: The model to solve is

$$u'' - \frac{w^2}{D} u = 0$$
$$u(0) = 0,$$
$$u(L) = 0.$$

with characteristic equation $r^2 - \frac{w^2}{D} = 0$ with the real roots $\pm \frac{w}{\sqrt{D}}$. The general solution of this second order model is given by

$$u(x) = A \, \cosh\left(\frac{w}{\sqrt{D}}x\right) + B \, \sinh\left(\frac{w}{\sqrt{D}}x\right)$$

Next, apply the boundary conditions, $u(0) = 0$ and $u(L) = 0$. Hence,

$$u(0) = 0 = A$$
$$u(L) = 0 = B \sinh\left(L\frac{w}{\sqrt{D}}\right)$$

Hence, $A = 0$ and $B \sinh\left(L\frac{w}{\sqrt{D}}\right) = 0$. Since sinh is never zero when w is not zero, we see $B = 0$ also. Hence, the only u solution is the trivial one and we can reject this case.

Case II: $\Theta = 0$: The model to solve is now

$$u'' = 0$$
$$u(0) = 0,$$
$$u(L) = 0.$$

with characteristic equation $r^2 = 0$ with the double root $r = 0$. Hence, the general solution is now

$$u(x) = A + Bx$$

Applying the boundary conditions, $u(0) = 0$ and $u(L) = 0$. Hence,

$$u(0) = 0 = A$$
$$u(L) = 0 = BL$$

Hence, $A = 0$ and $B = 0$ and the u solutions are trivial. We reject this case as well.

Case III: $\Theta = -\omega^2, \omega \neq 0$:

$$u'' + \frac{\omega^2}{D}u = 0$$
$$u(0) = 0,$$
$$u(L) = 0.$$

The general solution is given by

$$u(x) = A \cos\left(\frac{\omega}{\sqrt{D}}x\right) + B \sin\left(\frac{\omega}{\sqrt{D}}x\right)$$

Next, apply the boundary conditions, $u(0) = 0$ and $u(L) = 0$. Hence,

$$u(0) = 0 = A$$
$$u(L) = 0 = B\sin\left(L\frac{\omega}{\sqrt{D}}\right)$$

Hence, $A = 0$ and $B\sin\left(L\frac{\omega}{\sqrt{D}}\right) = 0$. Thus, we can determine a unique value of B only if $\sin\left(L\frac{\omega}{\sqrt{D}}\right) \neq 0$. If $\omega \neq \frac{n\pi c}{L}$, we can solve for B and find $B = 0$, but otherwise, B can't be determined. So the only solutions are the trivial or zero solutions unless $\omega L = n\pi\sqrt{D}$. Letting $\omega_n = \frac{n\pi\sqrt{D}}{L}$, we find a a non zero solution for each nonzero value of B of the form

$$u_n(x) = B \sin\left(\frac{\omega_n}{\sqrt{D}}x\right) = B \sin\left(\frac{n\pi}{L}x\right).$$

For convenience, let's choose all the constants $B = 1$. Then we have an infinite family of nonzero solutions $u_n(x) = \sin\left(\frac{n\pi}{L}x\right)$ and an infinite family of separation constants $\Theta_n = -\omega_n^2 = -\frac{n^2\pi^2 D}{L^2}$. We can then solve the w equation. We must solve

$$\frac{dw}{dt} = -\omega_n^2\, w(t), \quad t \geq 0.$$

The general solution is

$$w(t) = B_n\, e^{-\omega_n^2 t} = B_n\, e^{-\frac{n^2\pi^2 D}{L^2}t}$$

as we will get a different time solution for each choice of ω_n. Choosing the constants $B_n = 1$, we obtain the w_n functions

$$w_n(t) = e^{-\frac{n^2\pi^2 D}{L^2}t}$$

Hence, any product

$$\phi_n(x, t) = u_n(x)\, w_n(t)$$

will solve the model with the x boundary conditions and any finite sum of the form, for arbitrary constants A_n

$$\Phi_N(x, t) = \sum_{n=1}^{N} \phi_n(x, t) = \sum_{n=1}^{N} u_n(x)\, w_n(t)$$

$$= \sum_{n=1}^{N} \sin\left(\frac{n\pi}{L}x\right) e^{-\frac{n^2\pi^2 D}{L^2}t}$$

will solve the model with x boundary conditions as well.

10.1.3.2   The Infinite Series Solution

We know series of the form

$$\sum_{n=1}^{\infty} \sin\left(\frac{n\pi}{L}x\right) e^{-\frac{n^2\pi^2 D}{L^2}t}$$

converge pointwise for x in $[0, L]$ and all t. Hence, form the solution $\Phi(x, t)$ as

$$\Phi(x, t) = \sum_{n=1}^{\infty} \phi_n(x, t) = \sum_{n=1}^{\infty} u_n(x)\, w_n(t)$$

$$= \sum_{n=1}^{\infty} \sin\left(\frac{n\pi}{L}x\right) e^{-\frac{n^2\pi^2 D}{L^2}t}$$

for the arbitrary sequence of constants (A_n).

10.1.3.3   Handling the Other Boundary Condition

We can easily find that

$$\Phi(x, 0) = \sum_{n=1}^{\infty} A_n \sin\left(\frac{n\pi}{L}x\right)$$

The remaining boundary condition is

$$\Phi(x, 0) = f(x), \text{ for } 0 \leq x \leq L$$

Rewriting in terms of the series solution, for $0 \leq x \leq L$, we find

$$\sum_{n=1}^{\infty} A_n \sin\left(\frac{n\pi}{L}x\right) = f(x)$$

The Fourier series for f is given by

$$f(x) = \sum_{n=1}^{\infty} A_n \sin\left(\frac{n\pi}{L}x\right)$$

with

$$B_n = \frac{2}{L} \int_0^L f(x) \sin\left(\frac{n\pi}{L}x\right) dx$$

Then, setting these series equal, we find that the solution is given by $A_n = B_n$ for all $n \geq 1$.

10.1.3.4   Homework

Exercise 10.1.5

$$\frac{\partial \Phi}{\partial t} - 10 \frac{\partial^2 \Phi}{\partial x^2} = 0$$

$$\Phi(0, t) = 0, \; for \; 0 < t$$
$$\Phi(8, t) = 0, \; for \; 0 < t$$
$$\Phi(x, 0) = 10e^{-0.002t}, \; for \; 0 < x < 8.$$

Exercise 10.1.6

$$\frac{\partial \Phi}{\partial t} - 4 \frac{\partial^2 \Phi}{\partial x^2} = 0$$

$$\Phi(0, t) = 0, \; for \; 0 < t$$
$$\Phi(12, t) = 0, \; for \; 0 < t$$
$$\Phi(x, 0) = 0.05 \cos(t^2), \; for \; 0 < x < 12.$$

10.2   Fourier Series Approximate Methods

Using the Fourier approximation tools we built in Chap. 9, we can now write application code that will approximate the separation of variables solutions to linear PDE. We will show code for one particular set of boundary conditions for each problem. We will write functions to solve models with *Dirichlet* boundary conditions (i.e. the solution is set to 0 on two boundaries) and those with *Dirichlet* boundary conditions (a partial of the solution is set to 0 on two boundaries).

10.2.1   The Wave Equation

In order to write the approximation code for the wave equation solutions, we first must solve the model using separation of variables.

10.2.1.1   Dirichlet Boundary Conditions

A typical Wave equation with *Dirichlet* boundary conditions is given below.

$$\frac{\partial^2 \Phi}{\partial t^2} - c^2 \frac{\partial^2 \Phi}{\partial x^2} = 0$$

$$\Phi(x, 0) = f(x), \; for \; 0 \le x \le L$$

$$\frac{\partial \Phi}{\partial t}(x, 0) = 0, \text{ for } 0 \leq x \leq L$$

$$\Phi(0, t) = 0, \text{ for } 0 \leq t$$

$$\Phi(L, t) = 0, \text{ for } 0 \leq t$$

Applying separation of variables, we let $\Phi(x, t) = u(x)w(t)$ and find

$$c^2 u'' - \Theta_n u = 0, \text{ for } 0 \leq x \leq L$$

$$u(0) = 0,$$

$$u(L) = 0.$$

$$w'' - \Theta_n w = 0, \text{ for } t \geq 0,$$

$$w'(0) = 0.$$

where the separation constants are $\Theta_n = -\left(\frac{n\pi c}{L}\right)^2$ for $n = 1, 2, \ldots$. The boundary conditions on x lead to the solutions $u_n(x) = \sin\left(\frac{n\pi}{L}x\right)$ and $w_n(t) = \cos\left(\frac{n\pi c}{L}t\right)$. The infinite series solution is then

$$\Phi(x, t) = \sum_{n=1}^{\infty} A_n \sin\left(\frac{n\pi}{L}x\right) \cos\left(\frac{n\pi c}{L}t\right).$$

Hence, to satisfy the last boundary condition, we must have

$$\Phi(x, 0) = \sum_{n=1}^{\infty} A_n \sin\left(\frac{n\pi}{L}x\right)$$

$$= f(x) = \sum_{n=1}^{\infty} B_n \sin\left(\frac{n\pi}{L}x\right),$$

where $B_n = \frac{2}{L} \int_0^L f(s) \sin\left(\frac{n\pi}{L}s\right) ds$. This tells us we choose the constants (A_n) so that $A_n = B_n$. The code to compute the approximation for N terms is given in **WaveDirichletApprox**. The code approximates the solution to the Dirichlet wave model on the rectangle $[0, L1] \times [0, L2]$. The parameter M is the number of terms we want to use in the Riemann sum approximations to our needed inner products $\int_0^L f(s) \sin\left(\frac{n\pi}{L}s\right) ds$. Finally, we draw the surface which corresponds to the approximate solution $\Phi_N(x, t)$ using **nx** points for the x interval and **nt** points for the t interval. If we use reasonable small values of nx and nt, we will get a surface we can rotate easily with our mouse to see in different perspectives. However, if these values are large, the Octave to gnuplot interface is flawed and the resulting surface will still rotate, but with a significant time delay. In this function, we approximate the series as follows: First, we get the first N Fourier sine series coefficients.

Listing 10.1: Get Fourier sine coefficients

```
% Get Fourier Sine Approximation to f for N terms
[A,fN] = FourierSineApprox(f,L1,M,N);
```

Then, we set up function handles for all of the needed sin functions.

Listing 10.2: Setup sine functions

```
% Get Sine Functions
u = SetUpSines(L1,N);
```

We then construct the cos function handles we will use in the approximation.

Listing 10.3: Setup cosine functions

```
  % Get the cos functions
  v = cell(N,1);
3 for i = 1:N
    v{i} = @(t) cos( i*pi*c*t/L1);
  end
```

Finally, we define the surface function $p(x, t)$ as the partial sum of the first N terms.

Listing 10.4: Construct the approximate solution

```
  % Construct the approximate solution
  % get Nth Fourier Sine Approximation
  p = @(x,t) 0;
  for i=1:N
5   p = @(x,t) (p(x,t) + A(i)*u{i}(x).*v{i}(t));
  end
```

Once $p(x, t)$ is defined, we can plot the approximate surface. We set up a mesh of (x, t) points and calculate all the needed surface values for all the mesh points using the code below.

Listing 10.5: Setup plot domain

```
  %
  % draw surface for grid [0,L1] x [0,L2]
  % set up x and y stuff
4 x = linspace(0,L1,nx);
  t = linspace(0,L2,nt);
  % set up grid of x and y pairs (x(i),t(j))
  [X,T] = meshgrid(x,t);
  % set up surface
9 Z = p(X,T);
```

We then plot the surface using the **mesh** command setting labels and so forth as usual.

Listing 10.6: Plot approximate solution surface

```
%plot surface
figure
mesh(X,T,Z,'EdgeColor','black');
xlabel('x axis');
ylabel('t axis');
zlabel('Wave Solution');
title('Wave Dirichlet Solution on Rectangle');
print -dpng 'WaveDirichlet.png';
```

The full code then looks like this:

Listing 10.7: WaveDirichletApprox

```
function p = WaveDirichletApprox(f,c,M,N,nx,nt,L1,L2)
% f is the boundary data
% M is the number of terms in the Riemann sum approximations
%   to the inner products
% N is the number of terms in the series approximation to the
%   solution
% nx is the number of points to use in the surface plot for x
% nt is the number of points to use in the surface plot for t
% L1 is the length of the x interval
% L2 is the length of the t interval
%
% Solve u_tt - c^2 U_xx u = 0 on [0,L] x [0,L]
%
%         (0,L2)--------------------(L1,L2)
%               |                    |
%               |                    |
% u(0,t) =0     |                    | u(L1,t) = 0
%               |                    |
%               |                    |
%         (0,0)---------------------(L1,0)
%                       u(x,0)   = f(x)
%                       u_t(x,0) = 0
%
% Separation of variables gives solution is
% u(x,t) =
% sum_{n=1}^{infinity} A_n sin(n pi x/L1) cos( n pi c t/L1)
%
% where the coefficients A_n are the Fourier Sine
% coefficients of f on [0,L]
%
% A_n = (2/L1) <f, sin(n pi x/L1>
%
%   Get Fourier Sine Approximation to f for N terms
[A,fN] = FourierSineApprox(f,L1,M,N);

% Get Sine Functions
u = SetUpSines(L1,N);

% Get the cos functions
v = cell(N,1);
for i = 1:N
    v{i} = @(t) cos( i*pi*c*t/L1);
end

% Construct the approximate solution
% get Nth Fourier Sine Approximation
p = @(x,t) 0;
for i=1:N
    p = @(x,t) (p(x,t) + A(i)*u{i}(x).*v{i}(t));
end
%
```

```
52 % draw surface for grid [0,L1] x [0,L2]
   % set up x and y stuff
   x = linspace(0,L1,nx);
   t = linspace(0,L2,nt);
   % set up grid of x and y pairs (x(i),t(j))
57 [X,T] = meshgrid(x,t);
   % set up surface
   Z = p(X,T);

   %plot surface
62 figure
   mesh(X,T,Z,'EdgeColor','black');
   xlabel('x axis');
   ylabel('y axis');
   zlabel('Wave Solution');
67 title('Wave Dirichlet Solution on Rectangle');
   print -dpng 'WaveDirichlet.png';

   end
```

Here is a typical session with a sawtooth curve as the data f.

Listing 10.8: Sample Wave Equation Approximate solution

```
  L = 5;
  H = 10;
  saw = sawtooth(L,H);
  c = 1;
5 p = WaveDirichletApprox(saw,c,500,5,21,41,5,20);
```

The surface approximation can be seen in Fig. 10.1. Note even though the number of points in our mesh is low, the surface plot is reasonably good.

We can check to see if the boundary conditions are satisfied by adding a few lines of code. To the function **WaveDirichletApprox**, we can calculate the

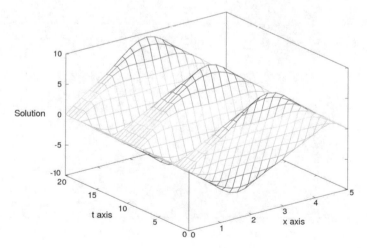

Fig. 10.1 5 term Fourier series approximation to the wave equation with sawtooth wave data and Dirichlet boundary conditions

approximations to $\frac{\partial \Phi}{\partial x}$ and $\frac{\partial \Phi}{\partial t}$ and return them as the functions **pd1** and **pd2**. The first line of the function becomes

Listing 10.9: Altering WaveDirichletApprox to return partials

```
function  [p,pd1,pd2]  =  WaveDirichletApprox(f,c,M,N,nx,nt,L1,L2)
```

and we add lines to handle the partial derivative approximations.

Listing 10.10: Setting up partial derivative approximations

```
   % Get Sine Functions
   % returns sin( i*pi*x/L )
   u = SetUpSines(L1,N);
4
   % Get the sin derivative functions
   up = cell(N,1);
   for i = 1:N
       up{i} = @(x) ((i*pi*c)/L1)*cos( i*pi*c*x/L1);
9  end

   % Get the cos functions
   v = cell(N,1);
   for i = 1:N
14     v{i} = @(t) cos( i*pi*c*t/L1);
   end

   % Get the cos derivative functions
   vp = cell(N,1);
19 for i = 1:N
       vp{i} = @(t) -((i*pi*c)/L1)*sin( i*pi*c*t/L1);
   end

   % Construct the approximate solution
24 p = @(x,t) 0;
   for i=1:N
       p = @(x,t) (p(x,t) + A(i)*u{i}(x).*v{i}(t));
   end

29 % Construct the approximate solution for pd1:
   % partial of p wrt argument 1
   pd1 = @(x,t) 0;
   for i=1:N
       pd1 = @(x,t) (pd1(x,t) + A(i)*up{i}(x).*v{i}(t));
34 end

   % Construct the approximate solution for pd2:
   % partial of p wrt argument 2
   pd2 = @(x,t) 0;
39 for i=1:N
       pd2 = @(x,t) (pd2(x,t) + A(i)*u{i}(x).*vp{i}(t));
   end
```

The rest of the code is the same and so we do not show it. You can alter the code to add these checks easily if you want. When we run the code using the sawtooth function like before, we use

Listing 10.11: The wave solution with dirichlet BC using saw data

```
[p,pd1,pd2] = WaveDirichletApprox(saw,c,500,5,21,41,5,20);
```

and we test the boundary conditions by constructing functions which we can plot to see if we obtain the requested boundary behavior. We define functions **bottom**, **bottomprime**, **right** and **left** for these checks. We will leave it to you to verify that we are indeed satisfying the boundary conditions. Note since the approximations are built from functions that satisfy the x boundary conditions, the **left** and **right** functions are guaranteed to work.

Listing 10.12: Testing the boundary conditions

```
   bottom = @(x) p(x,0);
   bottomprime = @(x) pd2(x,0);
   left = @(t) p(0,t);
   right = @(t) p(5,t);
 5 x = linspace(0,5,21);
   time = linspace(0,10,41);
   for i=1:21
   > ysaw(i) = saw(x(i));
   > end
10 plot(x,ysaw,x,bottom(x));
   plot(x,bottomprime(x));
   plot(time,right(time));
   plot(time,left(time));
```

10.2.1.2 Neumann Boundary Conditions

A Wave equation with *Neumann* boundary conditions has the following form:

$$\frac{\partial^2 \Phi}{\partial t^2} - c^2 \frac{\partial^2 \Phi}{\partial x^2} = 0$$

$$\Phi(x,0) = f(x), \text{ for } 0 \le x \le L$$

$$\frac{\partial \Phi}{\partial t}(x,0) = 0, \text{ for } 0 \le x \le L$$

$$\frac{\partial \Phi}{\partial x}(0,t) = 0, \text{ for } 0 \le t$$

$$\frac{\partial \Phi}{\partial x}(L,t) = 0, \text{ for } 0 \le t$$

Applying separation of variables, we let $\Phi(x,t) = u(x)w(t)$ and find

$$c^2 u'' - \Theta_n u = 0, \text{ for } 0 \le x \le L$$

$$u'(0) = 0,$$
$$u'(L) = 0.$$
$$w'' - \Theta_n w = 0, \text{ for } t \geq 0,$$
$$w'(0) = 0.$$

where the separation constants are again $\Theta_n = -\left(\frac{n\pi c}{L}\right)^2$ for $n = 0, 1, \ldots$. The boundary conditions on x lead to the solutions $u_n(x) = \cos\left(\frac{n\pi}{L}x\right)$ and $w_n(t) = \cos\left(\frac{n\pi c}{L}t\right)$. The infinite series solution is then

$$\Phi(x, t) = A_0 + \sum_{n=1}^{\infty} A_n \cos\left(\frac{n\pi}{L}x\right) \cos\left(\frac{n\pi c}{L}t\right).$$

Hence, to satisfy the last boundary condition, we must have

$$\Phi(x, 0) = A_0 + \sum_{n=1}^{\infty} A_n \cos\left(\frac{n\pi}{L}x\right)$$
$$= f(x) = B_0 + \sum_{n=1}^{\infty} B_n \cos\left(\frac{n\pi}{L}x\right),$$

where $B_0 = \frac{1}{L} \int_0^L f(s)\, ds$n and for $n > 1$, $B_n = \frac{2}{L} \int_0^L f(s) \cos\left(\frac{n\pi}{L}s\right) ds$. This tells us we choose the constants (A_n) so that $A_n = B_n$. The code to compute this approximation is quite similar to what we did in the code for **WaveDirichletApprox**. This code is named **WaveNeumannApprox**. Since we are using Neumann boundary conditions, the series expansions are in terms of cos functions this time. There are a few changes in the calculations. We return $N + 1$ coefficients when we call **FourierCosineApprox**.

Listing 10.13: Get fourier cosine approximations

```
%   Get Fourier Cosine Approximation to f for N terms
%  returns  A(1)  =  (1/L) < f(x),  1>
%           A(2)  =  (2/L) < f(x),  cos(pi x/L>
[A,fN]  =  FourierCosineApprox(f,L1,M,N);
```

The call to **SetUpCosines** also returns $N + 1$ cos functions rather than N handles. We also set up $N + 1$ cos functions of time as well.

Listing 10.14: Setup cosine functions

```
% Get Cosine Functions
% returns u{1} = 1, u{2} = cos(pi x/L) etc
u = SetUpCosines(L1,N);

5 % Get the cos functions
v = cell(N+1,1);
v{1} = @(t) 1;
for i = 2:N+1
    v{i} = @(t) cos( (i-1)*pi*c*t/L1);
10 end
```

Finally, we set up the approximation as we did before.

Listing 10.15: Setup approximation solution

```
% Construct the approximate solution
p = @(x,t) 0;
for i=1:N+1
    p = @(x,t) (p(x,t) + A(1)*u{1}(x).*v{1}(t));
5 end
```

The full listing is given below.

Listing 10.16: WaveNeumannApprox

```
function p = WaveNeumannApprox(f,c,M,N,nx,nt,L1,L2)
% f is the boundary data
% M is the number of terms in the Riemann sum approximations
%   to the inner products
5 % N is the number of terms in the series approximation to the
%   solution
% nx is the number of points to use in the surface plot for x
% nt is the number of points to use in the surface plot for t
% L1 is the length of the x interval
10 % L2 is the length of the t interval
%
% Solve u_tt - c^2 U_xx u = 0 on [0,L] x [0,L]
%
%       (0,L2)--------------------(L1,L2)
15 %          |                    |
%          |                    |
% u_x(0,t) =0 |                  | u_x(L1,t) = 0
%          |                    |
%          |                    |
20 %       (0,0)-----------------(L1,0)
%              u(x,0)   = f(x)
%              u_t(x,0) = 0
%
% Separation of variables gives solution is
25 % u(x,t) = A_0 +
% sum_{n=1}^{infinity} A_n cos(n pi x/L1) cos( n pi c t/L1)
%
% where the coefficients A_n are the Fourier Sine
% coefficients of f on [0,L]
30 %
% A_0 = (1/L1) <f, 1>
% A_n = (2/L1) <f, sin(n pi x/L1>
%
% Get Fourier Cosine Approximation to f for N terms
35 % returns A(1) = (1/L) < f(x), 1>
%          A(2) = (2/L) < f(x), cos(pi x/L>
[A,fN] = FourierCosineApprox(f,L1,M,N);

% Get Cosine Functions
```

```
40 % returns u{1} = 1, u{2} = cos(pi x/L) etc
   u = SetUpCosines(L1,N);

   % Get the cos functions
   v = cell(N+1,1);
45 v{1} = @(t) 1;
   for i = 2:N+1
     v{i} = @(t) cos( (i-1)*pi*c*t/L1);
   end

50 % Construct the approximate solution
   p = @(x,t) 0;
   for i=1:N+1
     p = @(x,t) (p(x,t) + A(i)*u{i}(x).*v{i}(t));
   end
55 %
   % draw surface for grid [0,L1] x [0,L2]
   % set up x and y stuff
     x = linspace(0,L1,nx);
60   t = linspace(0,L2,nt);
   % set up grid of x and y pairs (x(i),t(j))
     [X,T] = meshgrid(x,t);
   % set up surface
     Z = p(X,T);
65
   %plot surface
   figure
   mesh(X,T,Z,'EdgeColor','black');
   xlabel('x axis');
70 ylabel('t axis');
   zlabel('Wave Solution');
   title('Wave Neumann Solution on Rectangle');
   print -dpng 'WaveNeumann.png';

75 end
```

Using a sawtooth curve as the data f, we generate the approximation solution with a few lines of code.

Listing 10.17: Testing on saw data

```
   L = 5;
   H = 10;
   saw = sawtooth(L,H);
   c = 1;
 5 p = WaveNeumannApprox(saw,c,500,5,21,41,5,20);
```

The surface approximation can be seen in Fig. 10.2.

We can again check to see if the boundary conditions are satisfied by adding a few lines of code. To the function **WaveNeumannApprox**, we can calculate the approximations to $\frac{\partial \Phi}{\partial x}$ and $\frac{\partial \Phi}{\partial t}$ as before and return them as the functions **pd1** and **pd2**. The first line of the function becomes

Listing 10.18: Altering WaveNeumannApprox to return partial information

```
   function [p,pd1,pd2] = WaveNeumannApprox(f,c,M,N,nx,nt,L1,L2)
```

and we add lines similar to what we did before to handle the partial derivative approximations.

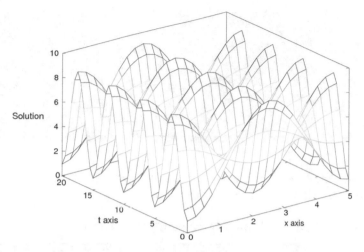

Fig. 10.2 5 term Fourier series approximation to the wave equation with sawtooth wave data and Neumann boundary conditions

Listing 10.19: Setting up the partial derivative approximations

```
% Get Cosine Functions
% returns u{1} = 1, u{2} = cos(pi x/L) etc
u = SetUpCosines(L1,N);
4
% Get the cos derivative functions
up - cell(N+1,1);
up{1} = @(x) 0;
for i = 2:N+1
9    up{i} = @(x) -(((i-1)*pi*c)/L1)*sin( (i-1)*pi*c*x/L1);
end

% Get the cos functions
v = cell(N+1,1);
14 v{1} = @(t) 1;
for i = 2:N+1
    v{i} = @(t) cos( (i-1)*pi*c*t/L1);
end

19 % Get the cos derivative functions
vp = cell(N+1,1);
vp{1} = @(t) 0;
for i = 2:N+1
    vp{i} = @(t) -((i*pi*c)/L1)*sin( i*pi*c*t/L1);
24 end

% Construct the approximate solution
p = @(x,t) 0;
for i=1:N+1
29    p = @(x,t) (p(x,t) + A(i)*u{i}(x).*v{i}(t));
end

% Construct the approximate solution for pd1:
% partial of p wrt argument 1
```

```
34 pd1 = @(x,t)  0;
   for  i=1:N+1
     pd1 = @(x,t)  (pd1(x,t) + A(i)*up{i}(x).*v{i}(t));
   end

39 % Construct the approximate solution for pd2:
   % partial of p wrt argument 2
   pd2 = @(x,t)  0;
   for  i=1:N+1
     pd2 = @(x,t)  (pd2(x,t) + A(i)*u{i}(x).*vp{i}(t));
44 end
```

You can alter the code to add these checks easily if you want. When we run the code using the sawtooth function like before, we use

Listing 10.20: Solving and returning the partial information

```
1 p,pd1,pd2]  = WaveNeumannApprox(saw,c,500,5,21,41,5,20);
```

and we test the boundary conditions by constructing functions which we can plot to see if we obtain the requested boundary behavior. We define the functions **bottom**, **bottomprime**, **right** and **left** for these checks as we did for the Dirichlet data and perform the check as usual.

10.2.2   Laplace's Equation

Now let's look at approximations to Laplace's equation on the rectangle $[0, L1] \times [0, L2]$.

10.2.2.1   Dirichlet Boundary Conditions

We solve a problem of this form

$$\frac{\partial^2 \Phi}{\partial x^2} + \frac{\partial^2 \Phi}{\partial y^2} = 0$$

$$\Phi(0, y) = 0, \text{ for } 0 \le y \le L1$$

$$\Phi(L1, y) = 0, \text{ for } 0 \le y \le L2$$

$$\Phi(x, L2) = 0, \text{ for } 0 \le x \le L1$$

$$\Phi(x, 0) = f(x), \text{ for } 0 \le x \le L1$$

We assume a solution of the form $\Phi(x, y) = u(x)\, w(y)$. This leads to

$$u'' - \Theta_n u = 0, \text{ for } 0 \leq x \leq L1,$$
$$u(0) = 0,$$
$$u(L1) = 0,$$
$$w'' + \Theta_n w = 0, \text{ for } 0 \leq y \leq L2,$$
$$w(L2) = 0.$$

where the separation constants are $\Theta_n = -\left(\frac{n\pi}{L1}\right)^2$ for $n = 1, 2, \ldots$. The boundary conditions on x lead to the solutions $u_n(x) = \sin\left(\frac{n\pi}{L1}x\right)$ and $w_n(y) = \sinh\left(\frac{n\pi}{L1}(L2 - y)\right)$. The infinite series solution is then

$$\Phi(x, y) = \sum_{n=1}^{\infty} A_n \sin\left(\frac{n\pi}{L1}x\right)\, \sinh\left(\frac{n\pi}{L1}(L2 - y)\right).$$

Hence, to satisfy the last boundary condition, we must have

$$\Phi(x, 0) = \sum_{n=1}^{\infty} A_n \sin\left(\frac{n\pi}{L1}x\right)\, \sinh(n\pi)$$

$$= f(x) = \sum_{n=1}^{\infty} B_n \sin\left(\frac{n\pi}{L1}x\right),$$

where $B_n = \frac{2}{L1} \int_0^{L1} f(s) \sin\left(\frac{n\pi}{L1}s\right) ds$. This tells us we choose the constants (A_n) so that $A_n = \dfrac{B_n}{\sinh\left(\frac{n\pi L2}{L1}\right)}$. The function we will use to approximate this solution is **LaplaceDirichletApprox** whose calling signature is similar to the ones we used in the approximations to the Wave equation.

Listing 10.21: LaplaceDirichletApprox arguments

```
function p = LaplaceDirichletApprox(f,M,N,nx,ny,L1,L2)
```

We have added the arguments **nx** and **ny** that are the number of x and y points to use in the surface plot of the approximate solution. The code to build the approximation is similar. First, we get the Fourier sine coefficients of the boundary data function **f** and then we create the handles to the component sin and sinh functions. Finally, when we construct the function **p** which is the approximate solution, we scale the individual terms by the needed $\sinh\left(\frac{n\pi L2}{L1}\right)$ terms.

Listing 10.22: Build approximation solution

```
   %  Get  Fourier  Sine  Approximation  to  f  for  N  terms
   %  returns  (2/L1)  <f,  u_n>
   [A,fN]  =  FourierSineApprox(f,L1,M,N);

 5 %  Get  Sine  Functions
   u  =  SetUpSines(L1,N);

   %  Get  the  sinh  functions
   v  =  cell(N,1);
10 for  i  =  1:N
       v{i}  =  @(y)  sinh(  i*pi*(L2-y)/L1);
   end

   %  Construct  the  approximate  solution
15 %  get  Nth  Fourier  Sine  Approximation
   p  =  @(x,y)  0;
   for  i=1:N
       p  =  @(x,y)  (p(x,y)  +  (A(i)/sinh(i*pi*L2/L1))*u{i}(x).*v{i}(y))
       ;
   end
```

The full code is listed below.

Listing 10.23: LaplaceDirichletApprox

```
 1 function  p  =  LaplaceDirichletApprox(f,M,N,nx,ny,L1,L2)
   %
   %  Solve  grad^2  u  =  0  on  [0,L1]  x  [0,L2]
   %                    u(x,L2)  =  0
   %       (0,L2)------------------------(L1,L2
 6 %             |                    |
   %             |                    |
   %  u(0,y)  =0  |                    |  u(L1,y)  =  0
   %             |                    |
   %             |                    |
11 %       (0,0)------------------------(L1,0)
   %                    u(x,0)  =  f(x)
   %
   %  Separation  of  variables  gives  solution  is
   %  u(x,y)  =
16 %  sum_{n=1}^{infinity}  A_n  sin(n  pi  x/L1)  sinh(  n  pi  (L2-y)/L1)
   %
```

```
    % where the coefficients A_n are the Fourier Sine
    % coefficients of f on [0,L1] divided by sinh(n pi)
    %
21  % A_n = 2*<f(y), sin(n pi y/L1)> / (L1 sinh(n pi L2/L1))
    %
    %  Get Fourier Sine Approximation to f for N terms
    % returns (2/L1) <f, u_n>
    [A,fN] = FourierSineApprox(f,L1,M,N);
26
    % Get Sine Functions
    u = SetUpSines(L1,N);

    % Get the sinh functions
31  v = cell(N,1);
    for i = 1:N
      v{i} = @(y) sinh( i*pi*(L2-y)/L1);
    end

36  % Construct the approximate solution
    % get Nth Fourier Sine Approximation
    p = @(x,y) 0;
    for i=1:N
      p = @(x,y) (p(x,y) + (A(i)/sinh(i*pi*L2/L1))*u{i}(x).*v{i}(y));
41  end

    %
    % draw surface for grid [0,L1] x [0,L2]
    % set up x and y stuff
46  x = linspace(0,L1,nx);
    y = linspace(0,L2,ny);
    % set up grid of x and y pairs (x(i),y(j))
    [X,Y] = meshgrid(x,y);
    % set up surface
51  Z = p(X,Y);

    %plot surface
    figure
    mesh(X,Y,Z,'EdgeColor','black');
56  xlabel('x axis');
    ylabel('y axis');
    zlabel('Laplace Solution');
    title('Laplace Dirichlet Solution on Rectangle');
    print -dpng 'LaplaceDirichlet.png';
61
    end
```

Let's look at how we use this function. We construct a square wave data function and then solve Laplace's equation on the rectangle.

Listing 10.24: A sample solution with 10 terms

```
    L1 = 5;
    L2 = 10;
3   f = @(x) 10;
    g = @(x) 0;
    Sq = @(x) splitfunc(x,f,g,L1);
    p = LaplaceDirichletApprox(Sq,500,10,41,51,L1,L2);
```

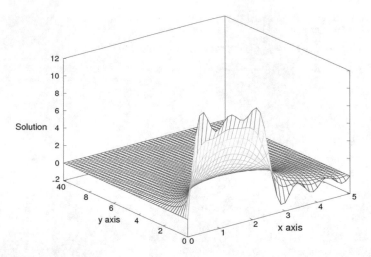

Fig. 10.3 10 term Fourier series approximation to Laplace's equation with square wave data and Dirichlet boundary conditions

The returned function $p(x, y)$ is the approximation to our model with the generated surface shown in Fig. 10.3.

We can do a better job by using more terms as is shown next. This time we use 20 terms. The resulting surface plot is better as is seen in Fig. 10.4.

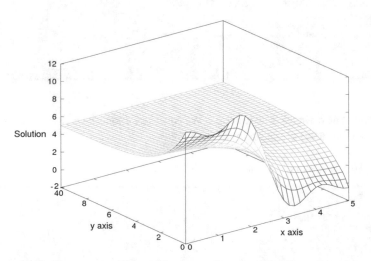

Fig. 10.4 20 term Fourier series approximation to Laplace's equation with square wave data and Dirichlet boundary conditions

Listing 10.25: A sample solution with 20 terms

```
p = LaplaceDirichletApprox(Sq,500,41,51,20,L1,L2);
```

10.2.2.2 Neumann Boundary Conditions

We next solve a Laplace model with Neumann conditions.

$$\frac{\partial^2 \Phi}{\partial x^2} + \frac{\partial^2 \Phi}{\partial y^2} = 0$$

$$\frac{\partial \Phi}{\partial x}(0, y) = 0, \text{ for } 0 \le y \le L$$

$$\frac{\partial \Phi}{\partial x}(L, y) = 0, \text{ for } 0 \le y \le L$$

$$\Phi(x, L) = 0, \text{ for } 0 \le x \le L$$

$$\Phi(x, 0) = f(x), \text{ for } 0 \le x \le L$$

We assume a solution of the form $\Phi(x, y) = u(x) w(y)$. This leads to

$$u'' - \Theta_n u = 0, \text{ for } 0 \le x \le L1,$$

$$u'(0) = 0,$$

$$u'(L1) = 0,$$

$$w'' + \Theta_n w = 0, \text{ for } 0 \le y \le L2,$$

$$w(L2) = 0.$$

where the separation constants are $\Theta_n = -\left(\frac{n\pi}{L1}\right)^2$ for $n = 0, 1, 2, \ldots$. The boundary conditions on x lead to the solutions $u_n(x) = \cos\left(\frac{n\pi}{L1}x\right)$ and $w_n(y) = \sinh\left(\frac{n\pi}{L1}(L2 - y)\right)$. The infinite series solution is then

$$\Phi(x, y) = A_0 \left(\frac{L2 - y}{L2}\right) + \sum_{n=1}^{\infty} A_n \cos\left(\frac{n\pi}{L1}x\right) \sinh\left(\frac{n\pi}{L1}(L2 - y)\right).$$

Hence, to satisfy the last boundary condition, we must have

$$\Phi(x, 0) = A_0 + \sum_{n=1}^{\infty} A_n \sin\left(\frac{n\pi}{L1}x\right) \sinh\left(\frac{n\pi L2}{L1}\right)$$

$$= f(x) = B_0 + \sum_{n=1}^{\infty} B_n \cos\left(\frac{n\pi}{L1}x\right),$$

where $B_0 = \frac{1}{L1}$ and for $n \geq 1$, $B_n = \frac{2}{L1}\int_0^{L1} f(s)\cos\left(\frac{n\pi}{L1}s\right) ds$. This tells us we choose the constants $A_0 = B_0$ and for $n \geq 1$, (A_n) so that $A_n = \dfrac{B_n}{\sinh\left(\frac{n\pi L2}{L1}\right)}$. The function we will use to approximate this solution is **LaplaceNeumannApprox**.

Listing 10.26: LaplaceNeumannApprox arguments

```
function p = LaplaceNeumannApprox(f,M,N,nx,ny,L1,L2)
```

The full source code is shown below.

Listing 10.27: LaplaceNeumannApprox

```
function p = LaplaceNeumannApprox(f,M,N,nx,ny,L1,L2)
%
% Solve grad^2 u = 0 on [0,L1] x [0,L2]
%                 u(x,L2) = 0
%         (0,L2)-------------------(L1,L2)
%              |                   |
%              |                   |
%   u_x(0,y) =0|                   | u_x(L1,y) = 0
%              |                   |
%              |                   |
%         (0,0)-------------------(L1,0)
%                 u(x,0) = f(x)
%
% Separation of variables gives solution is
% u(x,y) = A_0 (1-y/L2)+
%    sum_{n=1}^{infinity} A_n cos(n pi x/L1) sinh( (n pi/L1) (L2- y))
%
% where the coefficients A_n are the Fourier Sine
% coefficients of f on [0,L1] divided by sinh(n pi L2/L1)
%
% A_0 = <f,1>/L1
% A_n = 2*<f, cos(n pi x/L1)> / (L1 sinh(n pi L2/L1)
%
%   Get Fourier Cosine Approximation to f for N+1 terms
  [A,fN] = FourierCosineApprox(f,L1,M,N);

  % Get Cosine Functions
  u = SetUpCosines(L1,N);

  % Get the y functions
  v = cell(N+1,1);
  v{1} = @(y) 1 - y/L2;
  for i = 2:N+1
    v{i} = @(y) sinh( (i-1)*pi*(L2-y)/L1);
  end

  % Construct the approximate solution
  % get (N+1)th Approximation to PDE solution
```

```
39 c = A(1);
   p = @(x,y) c;
   for i=2:N+1
      d = sinh( (i-1)*pi*L2/L1 );
      p = @(x,y) (p(x,y) + (A(i)/ d )*u{i}(x).*v{i}(y));
44 end

   %
   % draw surface for grid [0,L1] x [0,L2]
   % set up x and y stuff
49 x = linspace(0,L1,nx);
   y = linspace(0,L2,ny);
   % set up grid of x and y pairs (x(i),y(j))
   [X,Y] = meshgrid(x,y);
   % set up surface
54 Z = p(X,Y);

   %plot surface
   figure
   mesh(X,Y,Z,'EdgeColor','black');
59 xlabel('x axis');
   ylabel('y axis');
   zlabel('Laplace Solution');
   title('Laplace Neumann Solution on Square');
   print -dpng 'LaplaceNeumann.png';
64
   end
```

Next, we approximate the solution to the Laplace Neumann model for square wave boundary data using 5 terms.

Listing 10.28: Sample solution with 5 terms

```
p = LaplaceNeumannApprox(Saw,500,5,31,41,5,10);
```

In this approximation, we generate the surface shown in Fig. 10.5.

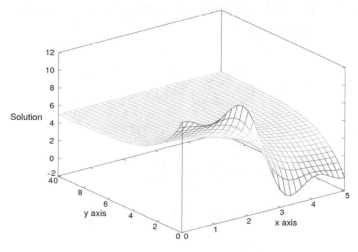

Fig. 10.5 5 term Fourier series approximation to Laplace's equation with square wave data and Neumann boundary conditions

10.2.3   The Heat Equation

Next, we consider heat equations on a rectangle.

10.2.3.1   Dirichlet Boundary Conditions

The problem we want to solve is

$$\frac{\partial \Phi}{\partial t} - D \frac{\partial^2 \Phi}{\partial x^2} = 0$$

$$\Phi(0, t) = 0, \text{ for } 0 < t$$
$$\Phi(L, t) = 0, \text{ for } 0 < t$$
$$\Phi(x, 0) = f(x), \text{ for } 0 < x < L1$$

We assume a solution of the form $\Phi(x, y) = u(x)\, w(t)$. This leads to

$$D\, u'' - \Theta_n u = 0, \text{ for } 0 \le x \le L1,$$
$$u(0) = 0,$$
$$u(L1) = 0,$$
$$w' - \Theta_n w = 0, \text{ for } 0 \ge t,$$

where the separation constants are $\Theta_n = -D\left(\frac{n\pi}{L1}\right)^2$ for $n = 1, 2, \ldots$. The boundary conditions on x lead to the solutions $u_n(x) = \sin\left(\frac{n\pi}{L1}x\right)$ and $w_n(t) = e^{-\Theta_n t}$. The infinite series solution is then

$$\Phi(x, t) = \sum_{n=1}^{\infty} A_n \sin\left(\frac{n\pi}{L1}x\right) e^{-D\left(\frac{n\pi}{L1}\right)^2 t}.$$

Hence, to satisfy the last boundary condition, we must have

$$\Phi(x, 0) = \sum_{n=1}^{\infty} A_n \sin\left(\frac{n\pi}{L1}x\right)$$

$$= f(x) = \sum_{n=1}^{\infty} B_n \sin\left(\frac{n\pi}{L1}x\right),$$

where $B_n = \frac{2}{L1} \int_0^{L1} f(s) \sin\left(\frac{n\pi}{L1}s\right) ds$. This tells us we choose the constants (A_n) so that $A_n = B_n$. The function we will use to approximate this solution is **HeatDirichletApprox** whose calling signature is similar to the ones we have used before in the earlier models. The full code is listed below.

Listing 10.29: HeatDirichletApprox

```
   function p = HeatDirichletApprox(f,M,N,nx,nt,D,L1,L2)
   %
   % Solve  u_t -D u_{xx} = 0  on  [0,L1]  x  [0,L2]
   %
 5 %        (0,t)--------------------(L1,t)
   %              |                  |
   %              |                  |
   %  u(0,t) =0   |                  |   u(L,t) - 0
   %              |                  |
10 %              |                  |
   %        (0,0)--------------------(L1,0)
   %                  u(x,0) = f(x)
   %
   % Separation of variables gives solution is
15 % u(x,y) =
   % sum_{n=1}^{infinity} A_n sin(n pi x/L1) exp( -D n^2 pi^2 t/L^2)
   %
   % where the coefficients A_n are the Fourier Sine
   % coefficients of f on [0,L1]
20 %
   % A_n = 2*<f(y), sin(n pi y/L1)>
   %
   %  Get Fourier Sine Approximation to f for N terms
   [A,fN] = FourierSineApprox(f,L1,M,N);
25
   % Get Sine Functions
   u = SetUpSines(L1,N);

   % Get the exp functions
30 v = cell(N,1);
   for i = 1:N
     v{i} = @(t) exp( -D*i^2*pi^2*t/L1^2);
   end

35 % Construct the approximate solution
   % get Nth Fourier Sine Approximation
   p = @(x,t)  0;
   for i=1:N
     p = @(x,t) (p(x,t) + A(i)*u{i}(x).*v{i}(t));
40 end

   %
   % draw surface for grid [0,L1] x [0,L2]
   % set up x and y stuff
45 x = linspace(0,L1,nx);
   t = linspace(0,L2,nt);
   % set up grid of x and y pairs (x(i),t(j))
   [X,T] = meshgrid(x,t);
   % set up surface
50 Z = p(X,T);

   %plot surface
   figure
   mesh(X,T,Z,'EdgeColor','black');
```

```
55 xlabel('x axis');
   ylabel('t axis');
   zlabel('Heat Solution');
   title('Heat Dirichlet Solution on Square');
   print -dpng 'HeatDirichlet.png';
60
   end
```

Let's look a quick test of this function to solve the Heat equation using square wave data. We set the diffusion constant to $D = 0.02$ and solve the model over 80 time units. Note we use a relatively coarse gridding strategy for the plot; $nx = 41$ and $nt = 121$. The decay of the solution is quite slow and unless we plot it over a long time interval, we don't see much change.

Listing 10.30: Sample solution with 10 terms

```
   L1 = 5;
   f = @(x) 10;
   g = @(x) 0;
 4 Sq = @(x) splitfunc(x,f,g,L1);
   p = HeatDirichletApprox(Sq,500,10,41,121,.02,L1,80);
```

The returned function $p(x, t)$ is the approximation to our model with the generated surface shown in Fig. 10.6.

We can also solve the heat equation using pulse data. Recall, a rectangular pulse centered at x_0 of width r and height H on the interval $[0, L]$ has the form

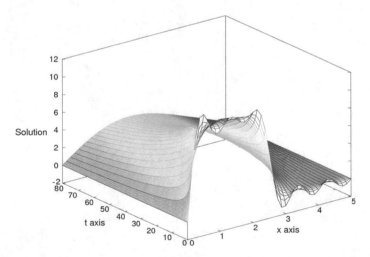

Fig. 10.6 10 term Fourier series approximation to the heat equation with square wave data and Dirichlet boundary conditions

$$J(x) = \begin{cases} 0, & 0 \le x \le x_0 - r, \\ H, & x_0 - r < x < x_0 + r, \\ 0, & x_0 + r \le x \le L \end{cases}$$

which we coded in the function **pulsefunc**.

Listing 10.31: pulsefunc

```
function  z  =  pulsefunc(x,x0,r,H)
%
if  x  >  x0−r  &&  x  <  x0+r
    z  =  H;
else
    z  =  0;
end
```

It is then easy to approximate the solution to the heat equation with a data pulse of height 100 at location 1 of width 0.2. As you would expect, we need more terms in the Fourier series to approximate this pulse reasonably well. Hence, we are using 40 terms here.

Listing 10.32: Sample solution with 40 terms

```
pulse  =  @(x)  pulsefunc(x,2,.2,100);
p  =  HeatDirichletApprox(pulse,500,40,31,221,.0002,L1,800);
```

The returned function $p(x, t)$ is shown in Fig. 10.7.

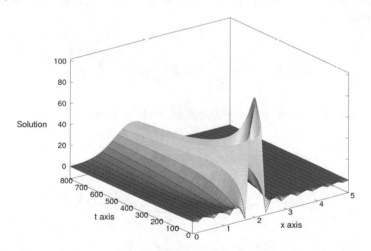

Fig. 10.7 40 term Fourier series approximation to the heat equation with pulse data and Dirichlet boundary conditions

10.2.3.2   Neumann Boundary Conditions

The problem we want to solve is

$$\frac{\partial \Phi}{\partial t} - D \frac{\partial^2 \Phi}{\partial x^2} = 0$$

$$\frac{\partial \Phi}{\partial x}(0, t) = 0, \text{ for } 0 < t$$

$$\frac{\partial \Phi}{\partial x}(L, t) = 0, \text{ for } 0 < t$$

$$\Phi(x, 0) = f(x), \text{ for } 0 < x < L1$$

We assume a solution of the form $\Phi(x, y) = u(x)\, w(t)$. This leads to

$$D\, u'' - \Theta_n u = 0, \text{ for } 0 \le x \le L1,$$

$$u'(0) = 0,$$

$$u'(L1) = 0,$$

$$w' - \Theta_n w = 0, \text{ for } 0 \ge t,$$

where the separation constants are $\Theta_n = -D \left(\frac{n\pi}{L1} \right)^2$ for $n = 0, 1, 2, \ldots$. The boundary conditions on x lead to the solutions $u_0(x) = 1$ and $u_n(x) = \cos\left(\frac{n\pi}{L1} x \right)$ for $n \ge 1$. The solutions for t are then $w_0(t) = 1$ and $w_n(t) = e^{-\Theta_n t}$ for $n \ge 1$. The infinite series solution is then

$$\Phi(x, t) = A_0 + \sum_{n=1}^{\infty} A_n \cos\left(\frac{n\pi}{L1} x \right) e^{-D\left(\frac{n\pi}{L1}\right)^2 t}.$$

Hence, to satisfy the last boundary condition, we must have

$$\Phi(x, 0) = A_0 + \sum_{n=1}^{\infty} A_n \cos\left(\frac{n\pi}{L1} x \right)$$

$$= f(x) = B_0 + \sum_{n=1}^{\infty} B_n \cos\left(\frac{n\pi}{L1} x \right),$$

where $B_0 = \frac{1}{L}$ and $B_n = \frac{2}{L1} \int_0^{L1} f(s) \cos\left(\frac{n\pi}{L1} s \right) ds$ for $n \ge 1$; hence, (A_n) so that $A_n = B_n$. The function we will use to approximate this solution is

HeatNeumannApprox whose calling signature is similar to the ones we have used before in the earlier models. The full code is listed below.

Listing 10.33: HeatNeumannApprox

```
  function p = HeatNeumannApprox(f,M,N,nx,nt,D,L1,L2)
  %
  % Solve u_t −D u_{xx} = 0 on [0,L1] x [0,L2]
  %
5 %          (0,t)--------------------(L1,t)
  %            |                    |
  %            |                    |
  % u_x(0,t) =0 |                    |  u_x(L,t) = 0
  %            |                    |
10 %            |                    |
  %          (0,0)--------------------(L1,0)
  %              u(x,0) = f(x)
  %
  % Separation of variables gives solution is
15 % u(x,y) = A_0 +
  %    sum_{n=1}^{infinity} A_n cos(n pi x/L) exp( −D n^2 pi^2 t/L^2)
  %
  % where the coefficients A_n are the Fourier Cosine
  % coefficients of f on [0,L]
20 %
  % A_0 = <f,1>/L
  % A_n = 2*<f, cos(n pi x/L)>
  %
  %  Get Fourier Cosine Approximation to f for N+1 terms
25 [A,fN] = FourierCosineApprox(f,L1,M,N);

  % Get Cosine Functions
  u − SetUpCosines(L1,N);

30 % Get the exp functions
  v = cell(N+1,1);
  v{1} = @(t) 1;
  for i = 2:N+1
     v{i} = @(t) exp( −D*i^2*pi^2*t/L1^2);
35 end

  % Construct the approximate solution
  % get (N+1)th Approximation to PDE solution
  p = @(x,t) 0;
40 for i=1:N+1
     p − @(x,t) (p(x,t) + A(i)*u{i}(x).*v{i}(t));
  end

  %
45 % draw surface for grid [0,L1] x [0,L2]
  % set up x and y stuff
    x = linspace(0,L1,nx);
    t = linspace(0,L2,nt);
  % set up grid of x and y pairs (x(i),t(j))
50   [X,T] = meshgrid(x,t);
  % set up surface
  Z = p(X,T);

  %plot surface
55 figure
  mesh(X,T,Z,'EdgeColor','black');
  xlabel('x axis');
  ylabel('t axis');
  zlabel('Laplace Solution');
60 title('Laplace Neumann Solution on Square');
  print −dpng 'HeatNeumann.png';

  end
```

As an example, let's solve the heat equation using multiple pulse data. The function **pulsefunc(x,x0,r,H)** gives us a rectangular pulse centered at x_0 of width r and height H on any interval. For convenience, denote such a pulse of height 1 by the function $P(x_0, r)$. Then a pulse of height H is simply $HP(x_0, r)$. We now apply two pulses, $100P(2, 0.3)$ and $200P(4, 0.1)$ as boundary data. Since this boundary data is more complicated, let's look at it Fourier cosine approximations first. We generate this pulse with the code

Listing 10.34: Two pulses

```
p1 = @(x)  pulsefunc(x,2,.3,100);
p2 = @(x)  pulsefunc(x,4,.1,200);
pulse = @(x)  p1(x) + p2(x);
```

and we compute the Fourier cosine approximations as usual with these lines. Recall the line **FourierCosineApprox(pulse,5,100,5);** is setting the number of terms in the Riemann sum approximations to the inner product calculations to 100, the interval to $[0, 5]$ and the number of Fourier terms to 6. Hence, we generate approximations all the way to 1400 Riemann sum terms and 61 Fourier cosine coefficients.

Listing 10.35: Fourier cosine approximation to the double pulse

```
    [A,p]  =  FourierCosineApprox(pulse,5,100,5);
    [A,p]  =  FourierCosineApprox(pulse,5,100,10);
    [A,p]  =  FourierCosineApprox(pulse,5,100,20);
    [A,p]  =  FourierCosineApprox(pulse,5,100,40);
  5 [A,p]  =  FourierCosineApprox(pulse,5,100,60);
    [A,p]  =  FourierCosineApprox(pulse,5,400,60);
    [A,p]  =  FourierCosineApprox(pulse,5,800,60);
    [A,p]  =  FourierCosineApprox(pulse,5,1200,60);
    [A,p]  =  FourierCosineApprox(pulse,5,1400,60);
```

We show the approximation we will use in our approximation of the heat model in Fig. 10.8 which uses 41 Fourier coefficients and 500 terms in the Riemann sum.

Then, we can approximate the solution to the heat model for a diffusion coefficient $D = 0.0002$ over 300 time units.

Linear Partial Differential Equations!the Heat equation with Neumann BC: a sample solution with 40 terms for two pulses

Listing 10.36: Sample solution with 40 terms

```
 1 pulse = @(x)  pulsefunc(x,2,.2,100);
   L1 = 5;
   p = HeatNeumannApprox(pulse,500,40,31,121,.0002,L1,300);
```

The returned function $p(x, t)$ is shown in Fig. 10.9.

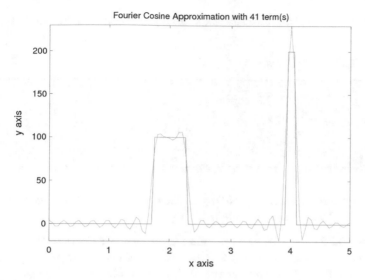

Fig. 10.8 41 term Fourier series approximation for dual pulse data

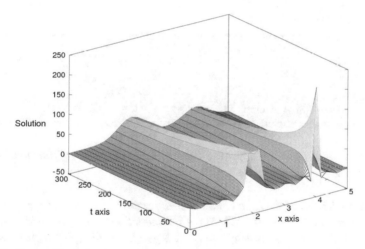

Fig. 10.9 41 term Fourier series approximation to the heat equation with dual pulse data and Neumann boundary conditions

10.2.3.3 Homework

Exercise 10.2.1 *Solve the Heat equation using square wave data (L1 = 4) with the diffusion constant to D = 0.0004. Solve the model over an appropriate length of*

time using a the gridding strategy you think is good. Use the split function as coded below.

Listing 10.37: Sample session

```
   L1 = 4;
 2 f = @(x) 20;
   g = @(x) 0;
   Sq = @(x) splitfunc(x,f,g,L1);
   % Here we use N = 500, nx = 10 and nt = 41 but experiment with
   % different choices
 7 p = HeatDirichletApprox(Sq,500,10,41,121,.02,L1,80);
```

Experiment with a variety of gridding strategies and the length of time until you get a good plot of the decay of the solution.

Exercise 10.2.2 *Solve the Heat equation using square wave data (L1 = 14) with the diffusion constant to D = 0.4. Solve the model over an appropriate length of time using a the gridding strategy you think is good. Use the split function as coded below.*

Listing 10.38: Sample session

```
   L1 = 4;
   f = @(x) 80;
 3 g = @(x) 0;
   Sq = @(x) splitfunc(x,f,g,L1);
   % Here we use N = 500, nx = 10 and nt = 41 but experiment with
   % different choices
   p = HeatDirichletApprox(Sq,500,10,41,121,.02,L1,80);
```

Experiment with a variety of gridding strategies and the length of time until you get a good plot of the decay of the solution.

Exercise 10.2.3 *Solve the Heat equation using square wave data (L1 = 4) with the diffusion constant to D = 0.0004. Solve the model over an appropriate length of time using a the gridding strategy you think is good. Use the split function as coded below.*

Listing 10.39: Sample session

```
   L1 = 9;
   f = @(x) 0;
 3 g = @(x) 30;
   Sq = @(x) splitfunc(x,f,g,L1);
   % Here we use N = 500, nx = 10 and nt = 41 but experiment with
   % different choices
   p = HeatDirichletApprox(Sq,500,10,41,121,.02,L1,80);
```

Experiment with a variety of gridding strategies and the length of time until you get a good plot of the decay of the solution.

Reference

J. Peterson, *BioInformation Processing: A Primer on Computational Cognitive Science*. Springer
Series on Cognitive Science and Technology (Springer Science+Business Media Singapore Pte
Ltd., Singapore, 2015 in press)

Chapter 11
Simplified Dendrite—Soma—Axon Information Processing

Let's review the basics of information processing in a typical neuron. There are many first sources for this material; some of them are *Introduction to Neurobiology* (Hall 1992), *Ionic Channels of Excitable Membranes* (Hille 1992), *Foundations of Cellular Neurophysiology* (Johnston and Wu 1995)], *Rall's review of cable theory in the 1977 Handbook of Physiology* (Rall 1977) and *Cellular Biophysics: Transport and Electrical Properties* (Weiss 1996a, b).

Our basic model consists of the following structural elements: A neuron which consists of a *dendritic tree* (which collects sensory stimuli and sums this information in a temporally and spatially dependent way), a cell body (called the *soma*) and an output fiber (called the *axon*). Individual dendrites of the dendritic tree and the axon are all modeled as cylinders of some radius a whose length ℓ is very long compared to this radius and whose walls are made of a bilipid membrane. The inside of each cylinder consists of an intracellular fluid and we think of the cylinder as lying in a bath of extracellular fluid. So for many practical reasons, we can model a dendritic or axonal fiber as two concentric cylinders; an inner one of radius a (this is the actual dendrite or axon) and an outer one with the extracellular fluid contained in the space between the inner and outer membranes.

The potential difference across the inner membrane is essentially due to a balance between the electromotive force generated by charge imbalance, the driving force generated by charge concentration differences in various ions and osmotic pressures that arise from concentration differences in water molecules on either side of the membrane. Roughly speaking, the ions of importance in our simplified model are the potassium K^+, sodium Na^+ and chloride Cl^- ions. The equilibrium potential across the inner membrane is about -70 millivolts and when the membrane potential is driven above this rest value, we say the membrane is *depolarized* and when it is driven below the rest potential, we say the membrane is *hyperpolarized*. The axon of one neuron interacts with the dendrite of another neuron via a site called a *synapse*. The synapse is physically separated into two parts: the *presynaptic* side (the side the axon is on) and the *postsynaptic* side (the side the dendrite is on). There is an actual physical gap, the *synaptic cleft*, between the two parts of the synapse. This cleft is filled with extracellular fluid.

© Springer Science+Business Media Singapore 2016
J.K. Peterson, *Calculus for Cognitive Scientists*, Cognitive Science
and Technology, DOI 10.1007/978-981-287-880-9_11

If there is a rapid depolarization of the presynaptic site, a chain of events is initialized which culminates in the release of specialized molecules called *neurotransmitters* into the synaptic cleft. There are pores embedded in the postsynaptic membrane whose opening and closing are dependent on the potential across the membrane that are called *voltage-dependent gates*. In addition, the gates generally allow the passage of a specific ion; so for example, there are sodium, potassium and chloride gates. The released neurotransmitters bind with the sites specific for the Na^+ ion. Such sites are called *receptors*. Once bound, Na^+ ions begin to flow across the membrane into the fiber at a greater rate than before. This influx of positive ions begins to drive the membrane potential above the rest value; that is, the membrane begins to depolarize. The flow of ions across the membrane is measured in gross terms by what are called *conductances*. Conductance has the units of reciprocal ohms; hence, high conductance implies high current flow per unit voltage. Thus the conductance of a gate is a good way to measure its flow. We can say that as the membrane begins to depolarize, the sodium conductance, g_{Na}, begins to increase. This further depolarizes the membrane. However, the depolarization is self-limited as the depolarization of the membrane also triggers the activation of voltage-dependent gates for the potassium ion, K^+, which allow potassium ions to flow through the membrane out of the cell. So the increase in the sodium conductance, g_{Na} triggers a delayed increase in potassium conductance, g_K (there are also conductance effects due to chloride ions which we will not mention here). The net effect of these opposite driving forces is the generation of a potential pulse that is fairly localized in both time and space. It is generated at the site of the synaptic contact and then begins to propagate down the dendritic fiber toward the soma. As it propagates, it attenuates in both time and space. We call these voltage pulses *Post Synaptic Pulses* or PSPs.

We model the soma itself as a small isopotential sphere, small in surface area compared to the surface area of the dendritic system. The possibly attenuated values of the PSPs generated in the dendritic system at various times and places are assumed to propagate without change from any point on the soma body to the initial segment of the axon which is called the *axon hillock*. This is a specialized piece of membrane which generates a large output voltage pulse in the axon by a coordinated rapid increase in g_{Na} and g_K once the axon hillock membrane depolarizes above a critical trigger value. The axon itself is constructed in such a way that this output pulse, called the *action potential*, travels without change throughout the entire axonal fiber. Hence, the initial depolarizing voltage impulse that arrives at a given presynaptic site is due to the action potential generated in the presynaptic neuron by its own dendritic system.

The salient features of the ball stick model are thus:

- Axonal and dendritic fibers are modeled as two concentric membrane cylinders.
- The axon carries action potentials which propagate without change along the fiber once they are generated. Thus if an axon makes 100 synaptic contacts, we assume that the depolarizations of each presynaptic membrane are the same.
- Each synaptic contact on the dendritic tree generates a time and space localized depolarization of the postsynaptic membrane which is attenuated in space as the

pulse travels along the fiber from the injection site and which decrease in magnitude the longer the time is since the pulse was generated.

- The effect of a synaptic contact is very dependent on the position along the dendritic fiber that the contact is made—in particular, how far was the contact from the axon hillock (i.e., in our model, how far from the soma)? Contacts made in essentially the same space locality have a high probability of reinforcing each other and thereby possibly generating a depolarization high enough to trigger an action potential.
- The effect of a synaptic contact is very dependent on the time at which the contact is made. Contacts made in essentially the same time frame have a high probability of reinforcing each other and thereby possibly generating a depolarization high enough to trigger an action potential.

11.1 A Simple Model of a Dendrite: The Core Conductor Model

We can model the above dendrite fiber reasonably accurately by using what is called the core conductor model (see Fig. 11.1). We have gone over this before, but let's repeat it here so we can easily refer to it for this discussion. We assume the following:

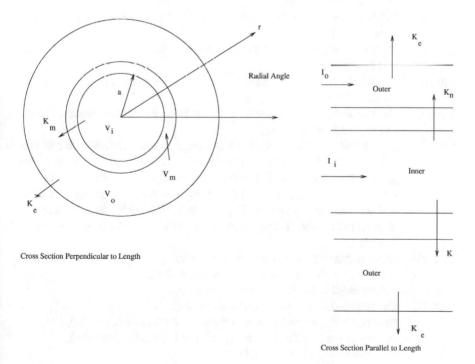

Fig. 11.1 The dendrite fiber model

- The dendrite is made up of two concentric cylinders. Both cylinders are bilayer lipid membranes with the same electrical characteristics. There is conducting fluid between both the inner and outer cylinder (extracellular solution) and inside the inner cylinder (intracellular solution). These solutions are assumed homogeneous and isotropic; in addition, Ohm's law is valid within them.
- All electrical variables are assumed to have cylindrical symmetry; hence, all variables are independent of the traditional polar angle θ. In particular, currents in the inner and outer fluids are longitudinal only (that is, up and down the dendritic fiber). Finally, current through the membrane is always normal to the membrane (that is the membrane current is only radial).
- A circuit theory description of currents and voltages is adequate for our model. At a given position along the cylinder, the inner and outer conductors are at the same potential (if you slice the cylinder at some point along its length perpendicular to its length, all voltage measurements in the inner and outer solutions will have the same voltage). The only radial voltage variation occurs in the membrane itself.

There are many variables associated with this model; although the model is very simplified from the actual biological complexity, it is still formidably detailed. These variables are described below:

z: the position of along the cable measured from some reference zero (m),

t: the current time (sec),

$I_o(z, t)$:   the total longitudinal current flowing in the $+z$ direction in the outer conductor (amps),

$I_i(z, t)$:   the total longitudinal current flowing in the $+z$ direction in the inner conductor (amps),

$J_m(z, t)$:   the membrane current density in the inner conductor to the outer conductor (amp/m^2),

$K_m(z, t)$:   the membrane current per unit length from the inner conductor to the outer conductor (amp/m),

$K_e(z, t)$:   the current per unit length due to external sources applied in a cylindrically symmetric manner. So it we think of a presynaptic neuron's axon generating a postsynaptic pulse in a given postsynaptic neuron, we can envision this synaptic contact occurring at some point z along the cable and the resulting postsynaptic pulse as a Dirac delta function impulse applied to the cable as a high current K_e which lasts for a very short time (amp/m),

$V_m(z, t)$:   the membrane potential which is consider $+$ when the inner membrane is positive with respect to the outer one (volts),

$V_i(z, t)$:   the potential in the inner conductor (volts),

$V_o(z, t)$:   the potential in the outer conductor (volts),

r_o: the resistance per unit length in the outer conductor (ohms/m),

r_i: the resistance per unit length in the inner conductor (ohms/m),

a: the radius of the inner cylinder (m).

Careful reasoning using Ohm's law and Kirchhoff's laws for current and voltage balance lead to the well-known steady state equations:

$$\frac{\partial I_i}{\partial z} = -K_m(z, t), \quad \frac{\partial I_o}{\partial z} = K_m(z, t) - K_e(z, t) \tag{11.1}$$

$$\frac{\partial V_i}{\partial z} = -r_i I_i(z, t), \quad \frac{\partial V_o}{\partial z} = -r_o I_o(z, t), \quad V_m = V_i - V_o \tag{11.2}$$

These equations look at what is happening in the concentric cylinder model at equilibrium; hence, the change in the potential across the inner membrane is due entirely to the longitudinal variable z. From Eq. 11.2, we see

$$\frac{\partial V_m}{\partial z} = \frac{\partial V_i}{\partial z} - \frac{\partial V_o}{\partial z} = r_o I_o(z, t) - r_i I_i(z, t)$$

implying

$$\frac{\partial^2 V_m}{\partial z^2} = r_o \frac{\partial I_o}{\partial z} - r_i \frac{\partial I_i}{\partial z} = (r_i + r_o) K_m(z, t) - r_o K_e(z, t).$$

Using Eq. 11.2, we then obtain the **core conductor** equation

$$\frac{\partial^2 V_m}{\partial z^2} = (r_i + r_o) K_m(z, t) - r_o K_e(z, t). \tag{11.3}$$

It is much more useful to look at this model in terms of transient variables which are perturbations from rest values. We define

$$V_m(z, t) = V_m^0 + v_m(z, t), \quad K_m(z, t) = K_m^0 + k_m(z, t), \quad K_e(z, t) = K_e^0 + k_e(z, t) \tag{11.4}$$

where the rest values are respectively V_m^0 (membrane rest voltage), K_m^0 (membrane current per length base value) and K_e^0 (injected current per length base value). With the introduction of these transient variables, we are able to model the flow of current across the inner membrane more precisely. We introduce the conductance per length g_m (Siemens/cm or 1/(ohms cm)) and capacitance per length c_m (fahrads/cm) of the membrane and note that we can think of a patch of membrane as as simple RC circuit. This leads to the **transient cable** equation

$$\frac{\partial^2 v_m}{\partial z^2} = (r_i + r_o) g_m v_m(z, t) + (r_i + r_o) c_m \frac{\partial v_m}{\partial t} - r_o k_e(z, t). \tag{11.5}$$

If we write the transient cable equation into an appropriate scaled form, we gain great insight into how membrane voltages propagate in time and space relative to what may be called fundamental scales. Define

$$\tau_M = \frac{c_m}{g_m} \tag{11.6}$$

$$\lambda_c = \frac{1}{\sqrt{(r_i + r_o)\, g_m}} \tag{11.7}$$

Note that τ_M is independent of the geometry of the cable and depends only on dendritic fiber characteristics. We will call τ_M the fundamental time constant (that is, this constant determines how quickly a membrane potential decays to one half on its initial value) of the solution for reasons we will see shortly. On the other hand, the constant λ_c is dependent on the geometry of the cable fiber and we will find it is the fundamental space constant of the system (that is, the membrane potential decays to one half of its value within this distance along the cable fiber).

If we let C_M and G_M denote the capacitance and conductance per square cm of the membrane, the circumference of the cable is $2\pi a$ and hence the capacitance and conductance per length are given by

$$c_m = 2\pi a C_M \tag{11.8}$$

$$g_m = 2\pi a G_M \tag{11.9}$$

and we see clearly that the ratio τ_M is simply $\frac{C_M}{G_M}$. This clearly has no dependence on the cable radius showing yet again that this constant is geometry independent. We note that the units of τ_M are in seconds. The space constant can be shown to have units of cm and defining ρ_i to be the resistivity (ohm-cm) of the inner conductor, we can show that

$$r_i = \frac{\rho_i}{\pi\, a^2}. \tag{11.10}$$

From this it follows that

$$\lambda_C = \frac{1}{\sqrt{2\rho_i G_M}}\, a^{\frac{1}{2}}. \tag{11.11}$$

Clearly, the space constant is proportional to the square root of the fiber radius and the proportionality constant is geometry independent. Another important constant is the Thévenin equivalent conductance, G_∞, which is defined to be

$$G_\infty = \lambda_C\, g_m = \sqrt{\frac{g_m}{r_i + r_o}}. \tag{11.12}$$

For most biologically plausible situations, the outer conductor resistance per unit length is very small in comparison to the inner conductor's resistance per unit length. Hence, $r_i \gg r_o$ and Eq. 11.12 can be rewritten using Eqs. 11.9 and 11.10 to have the form

$$G_\infty = \sqrt{\frac{g_m}{r_i}} = \pi \sqrt{\frac{2G_M}{\rho_i}} \, a^{\frac{3}{2}}. \tag{11.13}$$

which shows that G_∞ is proportional to the three halves power of the fiber radius a with a proportionality constant which is geometry independent. With all this said, we note that $r_i + r_o = \frac{1}{\lambda_C^2 g_m}$ and hence, we can rewrite the transient cable equation as

$$\lambda_C^2 \frac{\partial^2 v_m}{\partial z^2} = v_m + \tau_M \frac{\partial v_m}{\partial t} - r_o \lambda_C^2 k_e. \tag{11.14}$$

11.2 Time Independent Solutions to the Cable Model

There are three important classes of solution to the properly scaled cable equation: for the cases of an infinite, semi-infinite and finite length cable respectively; further, we are interested in both time independent and dependent solutions, which we will discuss in the next section. The three types of solution here are for the *infinite, semi-infinite* and *finite cable* models. We model the applied current as a current impulse of the form $k_e I_e \, \delta(z - 0)$, where I_e is the magnitude of the impulse which is applied at position $z = 0$ using the Dirac delta function $\delta(z - 0)$. The resulting steady state equation is given by

$$\lambda_C^2 \frac{d^2 v_m}{dz^2} = v_m - r_o \lambda_C^2 I_e \delta(z - 0).$$

which has solution

$$v_m(z) = \frac{\lambda_C \, r_o \, I_e}{2} e^{\frac{-|z|}{\lambda_C}} \tag{11.15}$$

The noticeable characteristics of this solution are that its spatial decay is completely determined by the space constant λ_C and the decay is symmetric across the injection site at $z = 0$. Within one space constant, the potential drops by $\frac{1}{e}$.

In the *semi-infinite cable* case, we assume the cable begins at $z = 0$ and extends out to infinity to the right. We assume a current I_e is applied at $z = 0$ and we can show the appropriate differential equation to solve is

$$\lambda_C^2 \frac{d^2 v_m}{dz^2} = v_m, \quad i_i(0) = I_e$$

where i_i denotes the transient inner conductor current. This system has solution

$$v_m(z) = \frac{I_e}{g_m \, \lambda_C} \, e^{-\frac{z}{\lambda_C}}$$

$$= \sqrt{\frac{r_i + r_o}{g_m}} \, I_e \, e^{-\frac{z}{\lambda_C}}$$

which we note is defined only for $z \geq 0$. Note that the ratio $\frac{i_i(0)}{v_m(0)}$ reduces to the Thévenin constant G_∞. This ratio is current to voltage, so it has units of conductance. Therefore, it is very useful to think of this ratio as telling us what *resistance* we see looking into the *mouth* of the semi-infinite cable. While heuristic in nature, this will give us a powerful way to judge the capabilities of dendritic fibers for information transmission.

Finally, in the *finite cable* version, we consider a finite length, $0 \leq z \leq \ell$, of cable with current I_e pumped in at $z =$ and a conductance load of G_e applied at the far end $z = \ell$. The cable can be thought of as a length of fiber whose two ends are capped with membrane that is identical to the membrane that makes up the cable. The load conductance G_e represents the conductance of the membrane capping the cable or the conductance of another cell body attached at that point. The system to solve is now

$$\lambda_C^2 \frac{d^2 v_m}{dz^2} = v_m, \quad i_i(0) = I_e, \quad i_i(\ell) = G_e \, v_m(\ell).$$

This has solution

$$v_m(z) = \frac{I_e}{G_\infty} \frac{\cosh\left(\frac{\ell-z}{\lambda_C}\right) + \frac{G_e}{G_\infty} \sinh\left(\frac{\ell-z}{\lambda_C}\right)}{\sinh\left(\frac{\ell}{\lambda_C}\right) + \frac{G_e}{G_{infty}} \cosh\left(\frac{\ell}{\lambda_C}\right)}, \quad 0 \leq z \leq \ell. \qquad (11.16)$$

with Thévenin equivalent conductance $\frac{i_i(0)}{v_m(0)}$ given by

$$G_T(\ell) = G_\infty \frac{\sinh\left(\frac{\ell}{\lambda_C}\right) + \frac{G_e}{G_\infty} \cosh\left(\frac{\ell}{\lambda_C}\right)}{\cosh\left(\frac{\ell}{\lambda_C}\right) + \frac{G_e}{G_\infty} \sinh\left(\frac{\ell}{\lambda_C}\right)}, \quad 0 \leq z \leq \ell. \qquad (11.17)$$

Now if the end caps of the cable are patches of membrane whose specific conductance is the same as the rest of the fiber, the surface area of the cap is πa^2 which is much smaller than the surface of the cylindrical fiber, $2\pi a \ell$. Hence, the conductance of the cylinder without caps is $2\pi a \ell G_M$ which is very large compared to the conductance of a cap, $\pi a^2 G_M$. So little current will flow through the caps and we can approximate this situation by thinking of the cap as an open circuit (thereby setting $G_e = 0$). We can think of this as the *open-circuit* case. Using Eqs. 11.16 and 11.17, we find

$$v_m(z) = \frac{I_e}{G_\infty} \frac{\cosh\left(\frac{\ell - z}{\lambda_C}\right)}{\sinh\left(\frac{\ell}{\lambda_C}\right)}, \quad 0 \le z \le \ell. \tag{11.18}$$

$$G_T(\ell) = G_\infty \tanh\left(\frac{\ell}{\lambda_C}\right), \quad 0 \le z \le \ell. \tag{11.19}$$

11.3 The Dendritic Voltage Model

The cable equation for the normalized membrane voltage is a nice model of the voltage on the excitable nerve cell dendrite which results from a voltage source being applied to the dendritic cable. For **zero-rate end cap** conditions, the model, as discussed and solved in Sect. 9.4, is stated below: which we have derived using ideas from biology and physics.

$$\lambda_C^2 \frac{\partial^2 v_m(z, t)}{\partial z^2} = v_m(z, t) + \tau_M \frac{\partial v_m}{\partial t}, \quad 0 \le z \le \ell, \ t \ge 0,$$
$$\frac{\partial v_m(0, t)}{\partial z} = 0,$$
$$\frac{\partial v_m(\ell, t)}{\partial z} = 0,$$
$$v_m(z, 0) = f(z), \quad 0 \le z \le \ell.$$

This is the general cable equation with $\beta = \lambda_C$ and $\alpha = \tau$. We then normalized the variables using the transformations

$$\tau = \frac{t}{\tau_M}, \quad \lambda = \frac{z}{\lambda_C}, \quad L = \frac{\ell}{\lambda_C},$$
$$\hat{v}_m = v_m(\lambda_C \lambda, \tau_M \tau).$$

This gave the system

$$\frac{\partial^2 \hat{v}_m}{\partial \lambda^2} = \hat{v}_m + \frac{\partial \hat{v}_m}{\partial \tau}, \quad 0 \le \lambda \le L, \ \tau \ge 0$$
$$\frac{\partial \hat{v}_m(0, \tau)}{\partial \lambda} = 0$$
$$\frac{\partial \hat{v}_m(L, \tau)}{\partial \lambda} = 0,$$
$$\hat{v}_m(\lambda, 0) = \hat{f}(z), \quad 0 \le \lambda \le L.$$

where $\hat{f} = f(\lambda_C \lambda)$. Now denote the applied voltage at time zero by $V(\lambda)$. Since, this is the general cable equation with $\beta = 1$ and $\alpha = 1$, the solution is

$$\hat{v}_m(\lambda, \tau) = A_0 e^{-\tau} + \sum_{n=1}^{\infty} A_n \cos\left(\frac{n\pi}{L}\lambda\right) e^{-\frac{1+n^2\pi^2}{L^2}\tau}.$$

with $A - n = B_n$ with

$$B_0 = \frac{1}{L} \int_0^L f(x)dx,$$

$$B_n = \frac{2}{L} \int_0^L f(x) \cos\left(\frac{n\pi}{L}x\right) dx.$$

The functions $u_n(\lambda) = \cos(\omega_n \lambda)$ and the scalars ω_n are known as the **eigenvalues** and **eigenfunctions** of the differential equation operator—a fancy name for the model – u'' with zero rate endcap boundary conditions. We haven't called them this before in our previous chapters on the separation of variables technique, but in a sense the work we do here is very analogous to the usual eigenvalue/eigenvector analysis we do for matrices. We can say a lot more about this in more advanced courses; in particular we discuss this carefully in Peterson (2015a). However, that is beyond our scope here. We wanted to mention this type of phrasing because we will need to mention it in a bit when we actually try to solve the Ball Stick model. Anyway, for this problem, the spatial eigenvalues are then $\beta_0 = 0$ with eigenfunction $u_0(\lambda) = 1$ and $\beta_n = -\omega_n^2 = -\frac{n^2\pi^2}{L^2}$ for any nonzero integer n with associated eigenfunctions $u_n(\lambda) = \cos(\omega_n \lambda)$. It is also easy to redo our original analysis to shown that we can use $u_n(\lambda) = \cos(\omega_n(L - \lambda))$ as the eigenfunctions. Still in either case, there are an infinite number of eigenvalues satisfying $|\beta_n| \to \infty$. Further, these eigenfunctions are mutually orthogonal. The normalized eigenfunctions are then

$$\hat{u}_0(\lambda) = \frac{1}{\sqrt{L}},$$

$$\hat{u}_n(\lambda) = \sqrt{\frac{2}{L}} \cos(\omega_n \lambda), \quad n \geq 1$$

or

$$\hat{u}_0(\lambda) = \frac{1}{\sqrt{L}},$$

$$\hat{u}_n(\lambda) = \sqrt{\frac{2}{L}} \cos(\omega_n(L - \lambda)), \quad n \geq 1$$

It is easy to see that with our usual inner product, $< \hat{u}_n, \hat{u}_n > = 1$ for all integers $n \geq 0$. It is possible to show—again, the details are in other texts such as Peterson (2015a)—that any continuous function f has the representation

$$f(\lambda) = \sum_{n=0}^{\infty} <f, \hat{u}_n(\lambda) > \hat{u}_n(\lambda$$

which converges in the least squares sense. Recall we talked about these ideas in Chap. 9 so you can go back there and review if you need to. When this is true, we say these functions form a complete orthonormal sequence.

11.4 The Ball and Stick Model

We can now extend our simple dendritic cable model to what is called the *ball and stick* neuron model. This consists of an isopotential sphere to model the cell body or soma coupled to a single dendritic fiber input line. We will model the soma as a simple parallel resistance/ capacitance network and the dendrite as a finite length cable as previously discussed (see Fig. 11.2). In Fig. 11.2, you see the terms I_0, the input current at the soma/dendrite junction starting at $\tau = 0$; I_D, the portion of the input current that enters the dendrite (effectively determined by the input conductance to the finite cable, G_D); I_S, the portion of the input current that enters the soma (effectively determined by the soma conductance G_S); and C_S, the soma membrane capacitance. We assume that the electrical properties of the soma and dendrite membrane are the same; this implies that the fundamental time and space constants of the soma and dendrite are given by the same constant (we will use our standard notation τ_M and λ_C as usual). It takes a bit of work, but it is possible to show that with a reasonable zero-rate left end cap condition the appropriate boundary condition at $\lambda = 0$ is given by

$$\rho \frac{\partial \hat{v}_m}{\partial \lambda}(0, \tau) = \tanh(L) \left[\hat{v}_m(0, \tau) + \frac{\partial \hat{v}_m}{\partial \tau}(0, \tau) \right], \tag{11.20}$$

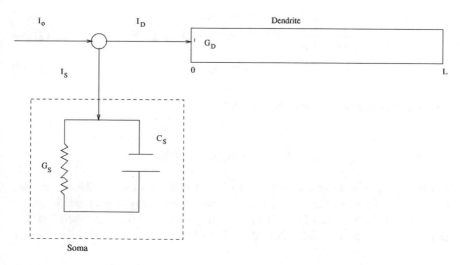

Fig. 11.2 The ball and stick model

where we introduce the fundamental ratio $\rho = \frac{G_D}{G_S}$, the ratio of the dendritic conductance to soma conductance. For more discussion of the ball and stick model boundary conditions we use here, you can look at the treatment in Rall (1977, Chap. 7, Sect. 2). The full system to solve is therefore:

$$\frac{\partial^2 \hat{v}_m}{\partial \lambda^2} = \hat{v}_m + \frac{\partial \hat{v}_m}{\partial \tau}, \quad 0 \leq \lambda \leq L, \quad \tau \geq 0. \tag{11.21}$$

$$\frac{\partial \hat{v}_m}{\partial \lambda}(L, \tau) = 0, \tag{11.22}$$

$$\rho \frac{\partial \hat{v}_m}{\partial \lambda}(0, \tau) = \tanh(L)\left[\hat{v}_m(0, \tau) + \frac{\partial \hat{v}_m}{\partial \tau}(0, \tau) \right]. \tag{11.23}$$

Applying the technique of separation of variables, $\hat{v}_m(\lambda, \tau) = u(\lambda)w(\tau)$, leads to the system:

$$u''(\lambda)w(\tau) = u(\lambda)w(\tau) + u(\lambda)w'(\tau)$$
$$\rho u'(0)w(\tau) = \tanh(L)\left(u(0)w(\tau) + u(0)w'(\tau) \right)$$
$$u'(L)w(\tau) = 0$$

This leads again to the ratio equation

$$\frac{u''(\lambda) - u(\lambda)}{u(\lambda)} = \frac{w'(\tau)}{w(\tau)}.$$

Since these ratios hold for all τ and λ, they must equal a common constant Θ. Thus, we have

$$\frac{d^2 u}{d\lambda^2} = (1 + \Theta)u, \quad 0 \leq \lambda \leq L, \tag{11.24}$$

$$\frac{dw}{d\tau} = \Theta w, \quad \tau \geq 0. \tag{11.25}$$

The boundary conditions then become

$$u'(L) = 0$$
$$\rho u'(0) = (1 + \Theta) \tanh(L)u(0).$$

The only case where we can have non trivial solutions to Eq. 11.24 occur when $1 + \Theta = -\alpha^2$ for some constant α. There are several cases to consider, but the details are quite similar to those of the zero rate endcap case and so are not included. The general solution to Eq. 11.24 is then of the form $A \cos(\alpha(L - \lambda)) + B \sin(\alpha(L - \lambda))$. Since

$$u'(\lambda) = \alpha A \sin(\alpha(L - \lambda)) - \alpha B \cos(\alpha(L - \lambda))$$

we see

$$u'(L) = -\alpha B = 0,$$

and so $B = 0$. Then,

$$u(0) = A \cos(\alpha L)$$
$$u'(0) = \alpha A \sin(\alpha L)$$

to satisfy the last boundary conditions, we find, since $1 + \Theta = -\alpha^2$,

$$\alpha A \left(\rho \sin(\alpha L) + \alpha \tanh(L) \cos(\alpha L) \right) = 0$$

A non trivial solution for A then requires α must satisfy the transcendental equation

$$\tan(\alpha L) = -\alpha \, \frac{\tanh(L)}{\rho} = -\kappa(\alpha L), \qquad (11.26)$$

where $\kappa = \frac{\tanh(L)}{\rho L}$. The values of α that satisfy Eq. 11.26 give us the eigenvalue of our original problem, $\Theta = -1 - \alpha^2$. The eigenvalues of our system can be determined by the solution of the transcendental Eq. 11.26. This is easy to do graphically as you can see in Fig. 11.3). It can be shown that the eigenvalues form a monotonically increasing sequence starting with $\alpha_0 = 0$ and with the values α_n approaching asymptotically the values $\frac{2n-1}{2} \pi$. Hence, there are a countable number of eigenvalues of the form $\Theta_n = -1 - \alpha_n^2$ leading to a general solution of the form

$$\hat{v}_m^n(\lambda, \tau) = A_n \, \cos(\alpha_n \lambda) \, e^{-(1+\alpha_n^2)\tau}. \qquad (11.27)$$

Hence, this system uses the pairs α_n (the solution to the transcendental Eq. 11.26) and $\cos[\alpha_n (L - \lambda)]$. In fact, we can show by direct integration that for $n \neq m$,

$$\int_0^L \cos(\alpha_n(L - \lambda)) \cos(\alpha_m(L - \lambda)) \, d\lambda = \frac{\sin((\alpha_n + \alpha_m)L)}{2(\alpha_n + \alpha_m)} + \frac{\sin((\alpha_n - \alpha_m)L)}{2(\alpha_n - \alpha_m)} \neq 0.$$

Since $\lim \left(\alpha_n - \frac{(2n-1)\pi}{2} \right) = 0$, we see there is an integer Q so that

$$\int_0^L \cos(\alpha_n(L - \lambda)) \cos(\alpha_m(L - \lambda)) \approx 0$$

if n and m exceed Q. We could say these functions are approximately orthogonal. The separation of variables technique gives rise to general solutions of the form

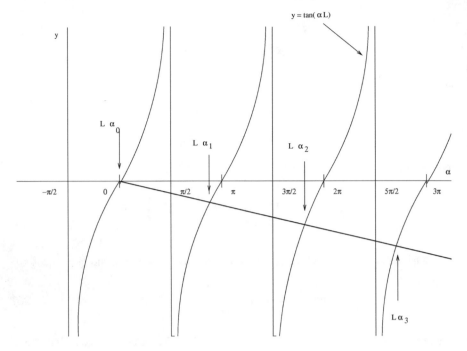

Fig. 11.3 The ball and stick eigenvalue problem

$$\hat{v}_m(\lambda, \tau) = A_0 e^{-\tau} + \sum_{n=1}^{\infty} A_n \cos(\alpha_n(L - \lambda))e^{-(1+\alpha_n^2)\tau}.$$

When $\tau = 0$, this means we want to know if the series

$$\hat{v}_m(\lambda, 0) = A_0 + \sum_{n=1}^{\infty} A_n \cos(\alpha_n(L - \lambda))$$

converges for the appropriate numbers A_n. To solve this model, for an applied continuous voltage pulse V to the cable we will want to expand V in terms of the eigenfunctions and equate coefficients. That is, we would write

$$V(\lambda) = V_0 + \sum_{n=1}^{\infty} V_n \cos(\alpha_n(L - \lambda))$$

and match coefficients from the expansion of $\hat{v}_m(\lambda, 0)$ given above. However, to do that we need to know if any V can be written like this and to know that the expansion of $\hat{v}_m$ at $\tau = 0$ is valid. Remember, that a family of functions is complete if we can do this. Now our family of functions here is not orthogonal so we certainly can't get a complete orthonormal sequence. However, as discussed very carefully in Peterson (2015a), it is possible to show that expansions like this are indeed permissible using this family of functions. Also, note for $\tau > 0$, the series expansions are eventually so strongly damped by the decay exponentials, that convergence is guaranteed so it is only a potential problem at 0. Further, note since the numbers α_n are not integers this is not a Fourier series expansion as we have discussed previously and so none of that analysis applies. Hence, suffice it to say, there is a need to dig into this question carefully even though doing so in this book is not appropriate. We will not present these details here, therefore, and ask you go to a source like Peterson (2015a) for the details. We conclude this discussion by noting these expansions are permitted.

Thus, as usual, we expect the most general solution is given by

$$\hat{v}_m(\lambda, \tau) = A_0 \, e^{-\tau} + \sum_{n=1}^{Q} A_n \, \cos(\alpha_n(L - \lambda)) e^{-(1+\alpha_n^2)\tau} \qquad (11.28)$$

$$+ \sum_{n=Q+1}^{\infty} A_n \, \cos(\alpha_n(L - \lambda)) e^{-(1+\alpha_n^2)\tau} \qquad (11.29)$$

Since the spatial functions are approximately orthogonal, the computation of the coefficients A_n for $n > Q$ can be handled with a straightforward inner product calculation. The calculation of the first Q coefficients must be handled as a linear algebra problem.

11.5 Ball-Stick MatLab Solutions

There are several ways to solve the ball-stick cable equation with a time dependency. One way to find a solution is to transform the cable equation into what is called a diffusion equation. We then use the properties of the solution to the diffusion equation and some variable transformations to obtain the cable solution. This is done in Peterson (2015c). We will use the *Separation of Variables* technique as discussed above here.

Let's assume we are applying a spatially varying voltage $V(\lambda)$ at $\tau = 0$ to our model. This means our full problem is

$$\frac{\partial^2 \hat{v}_m}{\partial \lambda^2} = \hat{v}_m + \frac{\partial \hat{v}_m}{\partial \tau}, \ 0 \leq \lambda \leq L, \ t \geq 0.$$

$$\hat{v}_m(\lambda, 0) = V(\lambda), \ \tau \geq 0$$

$$\frac{\partial \hat{v}_m}{\partial \lambda}(L, \tau) = 0,$$

$$\rho \frac{\partial \hat{v}_m}{\partial \lambda}(0, \tau) = \tanh(L) \left[\hat{v}_m(0, \tau) + \frac{\partial \hat{v}_m}{\partial \tau}(0, \tau) \right].$$

We have solved this problem and found

$$\hat{v}_m(\lambda, \tau) = A_0 e^{-\tau} + \sum_{n=1}^{\infty} A_n \cos\left(\frac{n\pi}{L}\lambda\right) e^{-\frac{1+n^2\pi^2}{L^2}\tau}.$$

We know we can approximate the solution $\hat{v}_m$ as

$$\hat{v}_m(\lambda, \tau) = A_0 \, e^{-\tau} + \sum_{n=1}^{Q} A_n \, \cos(\alpha_n(L - \lambda)) e^{-(1+\alpha_n^2)\tau}$$

for any choice of integer Q we wish. At $\tau = 0$, to satisfy the desired applied voltage condition, we must have

$$\hat{v}_m(\lambda, 0) = A_0 + \sum_{n=1}^{Q} A_n \, \cos(\alpha_n(L - \lambda)).$$

We can approximately solve this problem then by finding constants A_0 to A_Q so that

$$V(\lambda) = A_0 + \sum_{n=1}^{Q} A_n \, \cos(\alpha_n(L - \lambda)).$$

11.5.1 Approximating the Solution

For convenience, let $\phi_i(\lambda)$ denote the function $\cos(\alpha_n(L - \lambda))$, where $\phi_0(\lambda) = 1$ always. Then, note for each integer $i, 0 \le i \le Q$, we have

$$\frac{2}{L} \int_0^L V(s) \, \phi_i(s) \, ds = A_0 \frac{2}{L} \int_0^L \phi_0(s) \, \phi_i(s) \, ds$$

$$+ A_n \sum_{n=1}^{Q} \frac{2}{L} \int_0^L \phi_n(s) \, \phi_i(s) \, ds$$

To make this easier to write out, we again use "inner product" $< f, g >$ defined by

$$< f, g > \, = \frac{2}{L} \int_0^L f(s)g(s) \, ds.$$

Using this new notation, we see we are trying to find constants A_0 to A_Q so that

$$
\begin{bmatrix}
<\phi_0,\phi_0> & <\phi_1,\phi_0> & \cdots & <\phi_Q,\phi_0> \\
<\phi_0,\phi_1> & <\phi_1,\phi_1> & \cdots & <\phi_Q,\phi_1> \\
\vdots & \vdots & \vdots & \vdots \\
<\phi_0,\phi_i> & <\phi_1,\phi_i> & \cdots & <\phi_Q,\phi_i> \\
\vdots & \vdots & \vdots & \vdots \\
<\phi_0,\phi_Q> & <\phi_1,\phi_Q> & \cdots & <\phi_Q,\phi_Q>
\end{bmatrix}
\begin{bmatrix}
A_0 \\ A_1 \\ \vdots \\ A_i \\ \vdots \\ A_Q
\end{bmatrix}
=
\begin{bmatrix}
<V,\phi_0> \\ <V,\phi_1> \\ \vdots \\ <V,\phi_i> \\ \vdots \\ <V,\phi_Q>
\end{bmatrix}
$$

For convenience, call this coefficient matrix M and define the vectors B and D by

$$
B = \begin{bmatrix} A_0 \\ A_1 \\ \vdots \\ A_i \\ \vdots \\ A_Q \end{bmatrix}, \quad
D = \begin{bmatrix} <V,\phi_0> \\ <V,\phi_1> \\ \vdots \\ <V,\phi_i> \\ \vdots \\ <V,\phi_Q> \end{bmatrix}
$$

Then we find the desired constants A_0 to A_Q by solving

$$
M B = D.
$$

If M is invertible, this is an easy calculation. We can simplify this a bit further. We know

1. **The case $n \neq m$ with n and m not equal to** 1: Here, we have

$$
\int_0^L \cos(\alpha_n(L - \lambda)) \cos(\alpha_m(L - \lambda)) \, d\lambda = \frac{\sin((\alpha_n + \alpha_m)L)}{2(\alpha_n + \alpha_m)} + \frac{\sin((\alpha_n - \alpha_m)L)}{2(\alpha_n - \alpha_m)}
$$

However, we also know

$$
\sin(\alpha_n L) = -\kappa \, (\alpha_n L) \, \cos(\alpha_n L)
$$

Standard trigonometric formulae tell us

$$
\sin(\alpha_n L + \alpha_m L) = \sin(\alpha_n L) \, \cos(\alpha_m L) + \sin(\alpha_m L) \, \cos(\alpha_n L)
$$
$$
\sin(\alpha_n L - \alpha_m L) = \sin(\alpha_n L) \, \cos(\alpha_m L) - \sin(\alpha_m L) \, \cos(\alpha_n L)
$$

So, the pieces of the "inner products" become

$$
\frac{\sin((\alpha_n + \alpha_m)L)}{2(\alpha_n + \alpha_m)} = \frac{\sin(\alpha_n L) \, \cos(\alpha_m L) + \sin(\alpha_m L) \, \cos(\alpha_n L)}{2(\alpha_n + \alpha_m)}
$$

$$= \frac{-\kappa \, \alpha_n L \, \cos(\alpha_m L) \, \cos(\alpha_n L) - \kappa \, \alpha_m L \, \cos(\alpha_m L) \, \cos(\alpha_n L)}{2(\alpha_n + \alpha_m)}$$

$$= -\kappa \frac{L}{2} \, \cos(\alpha_m L) \, \cos(\alpha_n L)$$

and

$$\frac{\sin((\alpha_n - \alpha_m)L)}{2(\alpha_n - \alpha_m)} = \frac{\sin(\alpha_n L) \, \cos(\alpha_m L) + \sin(\alpha_m L) \, \cos(\alpha_n L)}{2(\alpha_n + \alpha_m)}$$

$$= \frac{-\kappa \, (\alpha_n L) \, \cos(\alpha_m L) \, \cos(\alpha_n L) + \kappa \, (\alpha_m L) \, \cos(\alpha_m L) \, \cos(\alpha_n L)}{2(\alpha_n + \alpha_m)}$$

$$= -\kappa \frac{L}{2} \, \cos(\alpha_m L) \, \cos(\alpha_n L).$$

and hence,

$$\frac{2}{L} \int_0^L \cos(\alpha_n(L - \lambda)) \, \cos(\alpha_m(L - \lambda)) \, d\lambda = -2\kappa \, \cos(\alpha_m L) \, \cos(\alpha_n L).$$

2. **If $n = 0$ with $m > 0$:** Since $n = 0$, the "inner product" integration is a bit simpler. We have

$$\frac{2}{L} \int_0^L \cos(\alpha_m(L - \lambda)) \, d\lambda = 2 \frac{\sin((\alpha_m)L)}{\alpha_m L}$$

However, of course, we also know

$$\sin(\alpha_m L) = -\kappa \, (\alpha_m L) \, \cos(\alpha_m L)$$

So, the "inner product" becomes

$$2 \frac{\sin(\alpha_m L)}{\alpha_m L} = -2\kappa \, \cos(\alpha_m L)$$

Hence, we conclude

$$\frac{2}{L} \int_0^L \cos(\alpha_m(L - \lambda)) = -2\kappa \, \cos(\alpha_m L).$$

3. $n = m, n > 0$: Here, a direct integration gives

$$\frac{2}{L} \int_0^L \cos^2(\alpha_n(L - \lambda)) \, d\lambda = \frac{2}{L} \int_0^L \frac{1 + \cos(\alpha_n(L - \lambda))}{2} \, d\lambda$$

$$= 1 + \frac{1}{2\alpha_n L} \, \sin(\alpha_n L).$$

4. $n = m = 0$: Finally, we have

$$\frac{2}{L} \int_0^L d\lambda = 2.$$

Plugging this into our "inner products", we see we need to solve

$$
\begin{bmatrix}
-\frac{1}{2\kappa} & \frac{1}{2}\cos(\beta_1) & \cdots & \frac{1}{2}\cos(\beta_Q) \\
\frac{1}{2}\cos(\beta_1) & -\frac{1}{4\kappa}\gamma_1 & \cdots & \cos(\beta_1)\cos(\beta_Q) \\
\vdots & \vdots & \vdots & \vdots \\
\frac{1}{2}\cos(\beta_i) & \cos(\beta_1)\cos(\beta_i) & \cdots & \cos(\beta_i)\cos(\beta_Q) \\
\vdots & \vdots & \vdots & \vdots \\
\frac{1}{2}\cos(\beta_Q) & \cos(\beta_1)\cos(\beta_Q) & \cdots & -\frac{1}{4\kappa}\gamma_Q
\end{bmatrix}
\begin{bmatrix}
A_0 \\ A_1 \\ \vdots \\ A_i \\ \vdots \\ A_Q
\end{bmatrix}
= -\frac{1}{4\kappa}
\begin{bmatrix}
<V,\phi_0> \\ <V,\phi_1> \\ \vdots \\ <V,\phi_i> \\ \vdots \\ <V,\phi_Q>
\end{bmatrix}
$$

where we use β_i to denote $\alpha_i L$ and the γ_i to replace the term $1 + \frac{1}{2\alpha_i L}\sin(2\alpha_i L)$. For example, if $L = 5$, $\rho = 10$ and $Q = 5$, we find with an easy MatLab calculation that M is

Listing 11.1: M Matrix

```
   -25.0023      0.4991      0.4962     -0.4917      0.4855     -0.4778
    -0.4991     12.2521      0.9906      0.9815     -0.9691      0.9538
3    0.4962     -0.9906    -12.2549     -0.9760      0.9637     -0.9485
    -0.4917      0.9815     -0.9760    -12.2594     -0.9548      0.9397
     0.4855     -0.9691      0.9637     -0.9548    -12.2655     -0.9279
    -0.4778      0.9538     -0.9485      0.9397     -0.9279    -12.2728
```

which has a non zero determinant so that we can solve the required system.

11.5.2 Applied Voltage Pulses

If we specialize to an applied voltage pulse $V(\lambda) = V^*\delta(\lambda - \lambda_0)$, note this is interpreted as a sequence of pulses k_e^C satisfying for all positive C

$$\int_{\lambda_0 - C}^{\lambda_0 + C} k_e^C(s)\, ds = V^*.$$

Hence, for all C, we have

$$
\begin{aligned}
<k_e^C, \phi_0> &= \frac{2}{L}\int_{\lambda_0 - C}^{\lambda_0 + C} k_e^C(s)\, ds \\
&= \frac{2}{L}V^*
\end{aligned}
$$

Thus, we use

$$< V^* \delta(\lambda - \lambda_0), \phi_0 > = \frac{2V^*}{L}.$$

Following previous discussions, we then find for any i larger than 0, that

$$< V^* \delta(\lambda - \lambda_0), \phi_i > = \frac{2V^*}{L} \cos(\alpha_i(L - \lambda_0)).$$

Therefore, for an impulse voltage applied to the cable at position λ_0 of size V^*, we find the constants we need by solving $M B = D$ for

$$D = -\frac{V^*}{2kL} \begin{bmatrix} 1 \\ \cos(\alpha_1(L - \lambda_0)) \\ \vdots \\ \cos(\alpha_i(L - \lambda_0)) \\ \vdots \\ \cos(\alpha_Q(L - \lambda_0)) \end{bmatrix}.$$

If we had two impulses applied, one at λ_0 of size V_0^* and one at λ_1 of size V_1^*, we can find the desired solution by summing the solution to each of the respective parts. In this case, we would find a vector solution B_0 and B_1 following the procedure above. The solution at $\lambda = 0$, the axon hillock, would then be

$$\hat{v}_m(0, \tau) = (B_{00} + B_{10}) e^{-\tau} + \sum_{n=1}^{Q} (B_{0n} + B_{1n}) \cos(\alpha_n L) e^{-(1+\alpha_n^2)\tau}$$

11.6   Ball and Stick Numerical Solutions

To see how to handle the Ball and Stick model numerically with MatLab, we will write scripts that do several things. We will try to explain, as best we can, the general principles of our numerical modeling using home grown code rather than built in MatLab functions. We need

1. A way to find Q eigenvalues numerically.
2. A routine to construct the matrix M we need for our approximations.
3. A way to solve the resulting matrix system for the coefficients we will use to build our approximate voltage model.

To solve for the eigenvalues numerically, we will use root finding techniques. to solve for the roots of

$$\tan(\alpha L) + \frac{tanh(L)}{\rho L} \ (\alpha L) = 0.$$

for various choices of L and ρ. We know the roots we seek are always in the intervals $[\frac{\pi}{2}, \pi], [\frac{3\pi}{2}, 2\pi]$ and so on, so it is easy for us to find useful lower and upper bounds for the root finding method. Since tan is undefined at the odd multiples of $\frac{\pi}{2}$, numerically, it is much easier to solve instead

$$\sin(\alpha L) + \frac{tanh(L)}{\rho L} \ (\alpha L) \ \cos(\alpha L) = 0.$$

To do this, we write the function **f2** as

Listing 11.2: Ball and Stick Eigenvalues

```
function  y = f2(x,L,rho)
%
% L = length  of  the  dendrite  cable  in  spatial  lengths
% rho = ratio  of  dendrite  to  soma  conductance ,  G_D/G_S
5 %
%
kappa = tanh(L)/(rho*L);
u = x*L;
y = sin(u) + kappa*u*cos(u);
```

11.6.1 Root Finding and Simple Optimization

The original finite difference Newton Method was for functions of one argument and was discussed in Sect. 11.6.1. Hence, we will have to modify the MatLab code for our purposes now. Our function **f2(x,L,rho)** depends on three parameters, so we modify our **feval** statements to reflect this. The new code is pretty much the same with each **feval** statement changed to add **,L,rho** to the argument list.

Listing 11.3: Modified Finite Difference Newton Method

```
function  [x,fx ,nEvals ,aF ,bF] = ...
          GlobalNewtonFDTwo(fName,L,rho ,a,b, tolx , tolf ,nEvalsMax)
%
% fName          a  string  that  is  the  name  of  the  function  f(x)
5 % a,b            we  look  for  the  root  in  the  interval  [a,b]
% tolx           tolerance  on  the  size  of  the  interval
% tolf           tolerance  of  f(current  approximation  to  root)
% nEvalsMax       Maximum  Number  of  derivative  Evaluations
%
10 % x              Approximate  zero  of  f
% fx             The  value  of  f  at  the  approximate  zero
% nEvals         The  Number  of  Derivative  Evaluations  Needed
% aF,  bF        the  final  interval  the  approximate  root  lies  in ,
%                [aF,bF]
```

```
15 %
   % Termination    Interval [a,b] has size < tolx
   %                |f(approximate root)| < tolf
   %                Have exceeded nEvalsMax derivative Evaluations
   %
20 fa = feval(fName,a,L,rho);
   fb = feval(fName,b,L,rho);
   x = a;
   fx = feval(fName,x,L,rho);
   delta = sqrt(eps)* abs(x);
25 fpval = feval(fName,x+delta,L,rho);
   fpx = (fpval-fx)/delta;

   nEvals = 1;
   k = 1;
30 disp(' ')
   disp(' Step         |     k     |      a(k)     |      x(k)     |      b(
        k) ')
   disp(sprintf('Start     | %6d | %12.7f | %12.7f | %12.7f',k,a,x,b
        ));
   while  (abs(a-b)>tolx) && (abs(fx)> tolf) && (nEvals<nEvalsMax) || (
        nEvals==1)
      %[a,b] brackets a root and x=a or x=b
35    check = StepIsIn(x,fx,fpx,a,b);
      if check
         %Take Newton Step
         x = x - fx/fpx;
      else
40       %Take a Bisection Step:
         x = (a+b)/2;
      end
      fx = feval(fName,x,L,rho);
      fpval = feval(fName,x+delta,L,rho);
45    fpx = (fpval-fx)/delta;
      nEvals = nEvals+1;
      if fa*fx<=0
         %there is a root in [a,x]. Use right endpoint.
         b = x;
50       fb = fx;
      else
         %there is a root in [x,b]. Bring in left endpoint.
         a = x;
         fa = fx;
55    end
      k = k+1;
      if(check)
         disp(sprintf('Newton     | %6d | %12.7f | %12.7f | %12.7f',k,a
              ,x,b));
      else
60       disp(sprintf('Bisection | %6d | %12.7f | %12.7f | %12.7f',k,a
              ,x,b));
      end
   end
   aF = a;
   bF = b;
```

We can then write the script **FindRoots**.

Listing 11.4: Finding Ball and Stick Eigenvalues

```
   function z = FindRoots(Q,L,rho,lbtol,ubtol)
   %
   % Q       = the number of eigenvalues we want to find
   % L       = the length of the dendrite in space constants
 5 % rho     = the ratio of dendrite to soma conductance, G_D/G_S
   % lbtol   = a small tolerance to add to the lower bounds we use
   % uptol   = a small tolerance to add to the upper bounds we use
   %
   z = zeros(Q,1);
10 LB = zeros(Q,1);
   UB = zeros(Q,1);
```

```
      kappa = tanh(L)/(rho*L);
      for n=1:Q
          LB(n) = ((2*n-1)/(2*L))*pi*lbtol;
15        UB(n) = n*pi/L*ubtol;
          [z(n),fx,nEvals,aF,bF] = GlobalNewtonFDTwo('f2'  ,L,rho,LB(n),MB(n
              ),10^-6,10^-8,500);
          w = z*L;
          lb = L*LB;
          disp(sprintf('n = %3d  EV = %12.7f   LB =  %12.7f   Error = %12.7f
              ',n,w(n),lb(n),w(n)-lb(n)));
20        disp(sprintf('n %3d function value = %12.7f original function
              value = %12.7f ',n,sin(w(n)) + kappa*w(n)*cos(w(n)),tan(w(n))
              + kappa*w(n)));

      end
```

11.6.1.1 Sample Runs

Let's try running this for the first two eigenvalues. Here is the sesssion:

Listing 11.5: Finding Two Eigenvalues

```
     FindRoots(2,5,10,1.0,1.0);

3    Step          |     k   |      a(k)      |      x(k)      |       b(k
              )
     Start         |     1   |   0.3141593   |   0.3141593   |
         0.6283185
     Bisection |       2   |   0.4712389   |   0.4712389   |
         0.6283185
     Bisection |       3   |   0.5497787   |   0.5497787   |
         0.6283185
     Newton    |       4   |   0.5497787   |   0.6186800   |
         0.6186800
8    Newton    |       5   |   0.6160148   |   0.6160148   |
         0.6186800
     Newton    |       6   |   0.6160148   |   0.6160149   |
         0.6160149
     n =    1  EV =    3.0800745  LB =      1.5707963  Error =
         1.5092782
     n   1  function value =   -0.0000000
          original function value =    0.0000000
13
     Step          |     k   |      a(k)      |      x(k)      |       b(k
              )
     Start         |     1   |   0.9424778   |   0.9424778   |
         1.2566371
     Bisection |       2   |   1.0995574   |   1.0995574   |
         1.2566371
     Bisection |       3   |   1.1780972   |   1.1780972   |
         1.2566371
18   Newton    |       4   |   1.1780972   |   1.2335644   |
         1.2335644
     Newton    |       5   |   1.2321204   |   1.2321204   |
         1.2335644
     n =    2  EV =    6.1606022  LB =      4.7123890  Error =
         1.4482132
     n   2  function value =   -0.0000000
          original function value =    -0.0000000
```

Now we will run this for 20 eigenvalues for $L = 5$ and $\rho = 10$ but we have commented out the intermediate prints in **GlobalNewtonFDTwo** to save space! We keep printing out the actual $f2$ values just to make sure our numerical routines are working like we want.

Listing 11.6: 20 Eigenvalues

```
  FindRoots(20,5,10,1.0,1.0);
  n =    1  EV =      3.0800745  LB =        1.5707963   Error =
     1.5092782
  n    1  function  value =     −0.0000000
         original  function  value =      0.0000000
5 n =    2  EV =      6.1606022  LB =        4.7123890   Error =
     1.4482132
  n    2  function  value =     −0.0000000
         original  function  value =     −0.0000000
  n =    3  EV =      9.2420168  LB =        7.8539816   Error =
     1.3880352
  n    3  function  value =     −0.0000000
10         original  function  value =      0.0000000
  n =    4  EV =     12.3247152  LB =       10.9955743   Error =
     1.3291410
  n    4  function  value =      0.0000000
         original  function  value =      0.0000000
  ...
15 n =   20  EV =     61.9402349  LB =       61.2610567   Error =
     0.6791782
  n   20  function  value =      0.0000000
         original  function  value =      0.0000000
```

11.6.2 The Ball and Stick Matrix Construction

To construct the matrix M, we need, we will use the following function.

Listing 11.7: Finding the Ball and Stick coefficient Matrix

```
  function M = FindM(Q,L,rho,z)
  %
3 % Q      = the number of eigenvalues we want to find
  % L      = the length of the dendrite in space constants
  % rho    = the ratio of dendrite to soma conductance, G_D/G_S
  % M      = matrix of coefficients for our approximate voltage
  %            model
8 %          In MatLab the root numbering is off by 1; so
  %          the roots we find start at 1 and end at Q.
  %          So M is Q+1 by Q+1.
  kappa = tanh(L)/(rho*L);
  w = zeros(1+Q,1);
13 M = zeros(1+Q,1+Q);
  w(1) = 0;
  % set first diagonal position
  M(1,1) = −1/(2*kappa);
  for n=1:Q
18   w(n+1) = z(n)*L;
  end
```

```
   % set first column
   for n = 2:Q+1
     M(1,n) = 0.5*cos(w(n));
23 end
   % set first row
   for n = 2:Q+1
     M(n,1) = 0.5*cos(w(n));
   end
28 % set main block
   for m=2:Q+1
     for n = 2:Q+1
       if m ~= n
         M(m,n) = cos(w(m))*cos(w(n));
33     end
       if m == n
         M(m,m) = -1/(4*kappa)*(1 + sin(2*w(m))/(2*w(m)));
       end
     end
38 end
```

This is pretty easy to use. For example, for $Q = 6$, we have

Listing 11.8: $Q = 6$

```
   Q = 6;
 2 L = 5;
   rho = 10;
   z = FindRoots(Q,L,rho,1.0,1.0);
   n =    1  EV =     3.0800745  LB =       1.5707963  Error =
      1.5092782
   n    1 function value =    -0.0000000
 7      original function value =   0.0000000
   n =    2  EV =     6.1606022  LB =       4.7123890  Error =
      1.4482132
   n    2 function value =    -0.0000000
        original function value =   -0.0000000
   n =    3  EV =     9.2420168  LB =       7.8539816  Error =
      1.3880352
12 n    3 function value =    -0.0000000
        original function value =   0.0000000
   n =    4  EV =    12.3247152  LB =      10.9955743  Error =
      1.3291410
   n    4 function value =     0.0000000
        original function value =   0.0000000
17 n =    5  EV =    15.4090436  LB =      14.1371669  Error =
      1.2718767
   n    5 function value =    -0.0000000
        original function value =   0.0000000
   n =    6  EV =    18.4952884  LB =      17.2787596  Error =
      1.2165288
   n    6 function value =     0.0000000
22      original function value =   0.0000000
   M = FindM(Q,L,rho,z);
```

The resulting M matrix is

Listing 11.9: M Matrix

```
M =
2
     -25.0023      -0.4991       0.4962      -0.4917       0.4855      -0.4778
       0.4690
      -0.4991     -12.2521      -0.9906       0.9815      -0.9691       0.9538
      -0.9361
       0.4962      -0.9906     -12.2549      -0.9760       0.9637      -0.9485
       0.9309
      -0.4917       0.9815      -0.9760     -12.2594      -0.9548       0.9397
      -0.9223
7      0.4855      -0.9691       0.9637      -0.9548     -12.2655      -0.9279
       0.9106
      -0.4778       0.9538      -0.9485       0.9397      -0.9279     -12.2728
      -0.8963
       0.4690      -0.9361       0.9309      -0.9223       0.9106      -0.8963
     -12.2812
```

Since we know that eventually, the roots or eigenvalues will become close to an odd multiple of $\frac{\pi}{2}$, we see that the axon voltage which is $\hat{v}_m(\lambda, 0)$ is

$$\hat{v}_m(\lambda, 0) = A_0 + \sum_{n=1}^{Q} A_n \cos(\alpha_n(L - \lambda)) + \sum_{n=Q+1}^{\infty} A_n \cos(\alpha_n(L - \lambda))$$

where $Q + 1$ is the value where the $\cos(\alpha_n L) = 0$ within our tolerance choice. At the Axon Hillock, $\lambda = 0$, we have

$$\hat{v}_m(0, 0) = A_0 + \sum_{n=1}^{Q} A_n \cos(\alpha_n L) + \sum_{n=Q+1}^{\infty} A_n \cos(\alpha_n L)$$

However, the terms past $Q + 1$ are zero and so our approximate solution at the axon hillock at time 0 is just

$$\hat{v}_m(0, 0) = A_0 + \sum_{n=1}^{Q} A_n \cos(\alpha_n L).$$

However, as time increases, the voltage wave initiated by impulse travels towards the axon hillock. It dampens or attenuates as it goes. The magnitude of the depolarization or hyperpolarization seen at the axon hillock depends on the time we wish to look at. Recall, the solution to our impulse at the axon hillock, $\lambda = 0$, is given by

$$\hat{v}_m(0, \tau) = A_0 \, e^{-\tau} + \sum_{n=1}^{Q} A_n \, \cos(\alpha_n L) \, e^{-(1+\alpha_n^2)\tau} \qquad (11.30)$$

$$+ \sum_{n=Q+1}^{\infty} A_n \, \cos(\alpha_n L) \, e^{-(1+\alpha_n^2)\tau}. \qquad (11.31)$$

The strong damping given by $e^{-(1+\alpha_n^2)\tau}$ for $n > Q$ tells us these last terms are very negligible. So a good approximation to the transient membrane voltage at the axon hillock as a function of time is

$$\hat{v}_m(0, \tau) = A_0 \, e^{-\tau} + \sum_{n=1}^{Q} A_n \, \cos(\alpha_n L) \, e^{-(1+\alpha_n^2)\tau}. \qquad (11.32)$$

The membrane voltage at the axon hillock is then the sum of the equilibrium voltage V_R (typically -70 mV) and $\hat{v}_m(0, \tau)$. An interesting question is how do we estimate Q? We did it with a short MatLab script.

Listing 11.10: Finding Q

```
 1 function N = FindRootSize(EndValue,L,rho,lbtol,ubtol)
   %
   % EndValue = how far we want to search to find
   %              where approximation can stop
   % Q        = the number of eigenvalues we want to find
 6 % L        = the length of the dendrite in space constants
   % rho      = the ratio of dendrite to soma conductance, G_D/G_S
   % lbtol    = a small tolerance to add to the lower bounds we use
   % uptol    = a small tolerance to add to the upper bounds we use
   %
11 % set Q
   Q = EndValue;
   z = zeros(Q,1);
   LB = zeros(Q,1);
   UB = zeros(Q,1);
16 kappa = tanh(L)/(rho*L);
   N = Q;
   for n=1:Q
         LB(n) = ((2*n-1)/(2*L))*pi*lbtol;
         UB(n) = n*pi/L*ubtol;
21       [z(n),fx,nEvals,aF,bF] = GlobalNewtonFDTwo('f2',L,rho,LB(n),UB(n
              ),10^-6,10^-8,500);
         w = z*L;
         testvalue = abs(cos(w(n)));
         if (abs(testvalue) < 10^-3)
            disp(sprintf('n = %3d  root = %12.7f testvalue = %12.7f',n,w(n
                 ),testvalue));
26          N = n;
            break;
         end
   end
   disp(sprintf('n = %3d  root = %12.7f testvalue = %12.7f',n,w(n),
        testvalue));
```

We can find Q for given values of L and ρ as follows:

Listing 11.11: Finding the Minimal Q For a Resonable Orthogonality

```
L = 5;
rho = 10;
N = FindRootSize(20000,L,rho,1.0,1.0);
n = 15918   root = 50006.3020635 testvalue =      0.0010000
```

Of course, we can't really use this information! If you try to solve our approximation problem using $Q = 15918$, you will probably do as we did and cause your laptop to die in horrible ways. The resulting matrix M we need is 15918×15918 in size and most of us don't have enough memory for that! There are ways around this, but for our purposes, we don't really need to use a large value of Q.

11.6.3 Solving for Impulse Voltage Sources

First, we need to set up the data vector. We do this with the script

Listing 11.12: Finding the Ball and Stick Data Vector

```
   function D = FindData(Q,L,rho,z,Vmax,lambda0)
   %
   % Q        = the number of eigenvalues we want to find
   % L        = the length of the dendrite in space constants
 5 % rho      = the ratio of dendrite to soma conductance, G_D/G_S
   % z        = eigenvalue vector
   % D        = data vector
   % Vmax     = size of voltage impulse
   % lambda0  =
10 % M        = matrix of coefficients for our approximate voltage
   %            model
   %
   kappa = tanh(L)/(rho*L);
   w = zeros(1+Q,1);
15 D = zeros(1+Q,1);
   multiplier = -Vmax/(2*kappa*L);
   %
   w(1) = 0;
   for n=1:Q
20    w(n+1) = z(n);
   end
   D(1) = multiplier;
   for n = 2:Q+1
      D(n) = multiplier*cos(w(n)*(L - lambda0));
25 end
```

Next, we can use the linear system solvers we have discussed in Peterson (2015b). To solve our system $MB = D$, we then use the LU decomposition and the lower and upper triangular solvers as follows:

1. We find the LU decomposition of M.
2. Then, we compute apply Gaussian elimination to transform the system into lower triangular form.
3. Finally, we use our upper triangular solver tool to find the solution.

A typical MatLab session would have this form for our previous $L = 5$, $Q = 6$ (remember, this gives us 7 values as the value 0 is also used in our solution), $\rho - 10$ for a pulse of size 200 administered at location $\lambda = 2.5$.

Listing 11.13: Finding the Approximate Solution

```
[L,U,piv] = GePiv(M);
y = LTriSol(L,D(piv));
B = UTriSol(U,y);
```

where B is the vector of coefficients we need to form the solution

$$V(\lambda, \tau) = B_0 e^{-\tau} + \sum_{n=1}^{Q} B_n \cos(\alpha_n(L - \lambda)) \, e^{-(1+\alpha_n^2)\tau}.$$

The axon hillock voltage is then $V(0, \tau)$. We compute the voltage with the script

Listing 11.14: Finding the Axon Hillock Voltage

```
     function [V,AH] = FindVoltage(Q,L,rho,z,M,D)
2    %
     % Q       = the number of eigenvalues we want to find
     % L       = the length of the dendrite in space constants
     % rho     = the ratio of dendrite to soma conductance, G_D/G_S
     % z       = eigenvalue vector
7    % D       = data vector
     % M       = matrix of coefficients for our approximate voltage
     %           model
     %
     % set time scale for dendrite model
12   TD = 5.0;
     %
     [Lower,Upper,piv] = GePiv(M);
     y = LTriSol(Lower,D(piv));
     B = UTriSol(Upper,y);
17   Error - Lower*Upper*B - D(piv);
     Diff = M*B-D;
     e = norm(Error);
     e2 = norm(Diff);
     disp(sprintf(' norm of LU residual = %12.7f norm of MB-D = %12.7f',e,
         e2));
22   % set spatial and time bounds
     % divide L into 300 parts
     lambda = linspace(0,L,301);
     %divide time into 100 parts
     tau    = linspace(0,TD,101);
27   V = zeros(301,101);
     % find voltage at space point lambda(s) and time point tau(t)
     for s = 1:301
         for t = 1:101
             V(s,t) = B(1)*exp(-tau(t));
32           for n=1:Q
                 V(s,t) = V(s,t) + B(n+1)*cos(z(n)*(L - lambda(s))) ...
                     *exp(-(1+z(n)^2)*tau(t));
             end
         end
37   end
     AH = V(1,:);
```

We can compute a representative axon hillock voltage as follows:

Listing 11.15: Axon Hillock Voltage Computation

```
  Q = 40;
2 L = 5;
  rho = 10;
  z = FindRoots(Q,L,rho ,1.0 ,1.0);
  M = FindM(Q,L,rho ,z);
  Vmax = 100;
7 location = 2.0;
  D = FindData(Q,L,rho ,z ,Vmax, location );
  [V,AH]  = FindVoltage (Q,L,rho ,z ,M,D);
  t = linspace (0 ,5 ,101);
  x = linspace (0 ,5 ,301);
12 plot (t ,AH);
  xlabel ('Time  mS');
  ylabel ('Voltage  mV');
  title ('Transient  Axon  Hillock  Response  to  Pulse  100  at  2.0');
```

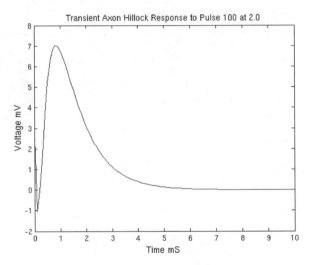

Fig. 11.4 Initial axon hillock voltage for pulse of size 100 at location 2.0. This uses the $t = 0$ values which possess least square solution artifacts

This generates the plot of Fig. 11.4. We are plotting the transient axon hillock voltage versus time. Note we are using the computed axon hillock voltage over the entire time interval [0, 5]. The least squares solution used at $t = 0$ is not very representative of the true voltage. A better approximation is obtained by starting the plot one time value up from zero: here, the index 1 in the t vector corresponds to $5/100$ milliseconds. This small amount of time allows some attenuation from the damped exponentials of time to remove the least squares artifacts. We show this in Fig. 11.5. We can also plot voltage versus distance. V evaluated at time 0 tells us how we have approximated our impulse at location 2.0 of magnitude 100 as a function of distance along the

Fig. 11.5 Initial axon hillock voltage for pulse of size 100 at location 2.0. This starts at $t = 0.05$ to lessen the least square solution artifacts

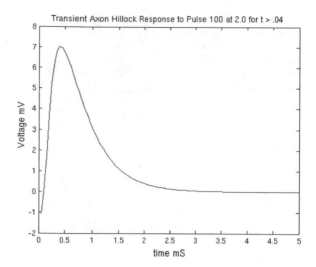

Fig. 11.6 Model approximation to an impulse of size 100 at location 2.0: This starts with the voltage at $t = 0$ so there are least squares solution artifacts

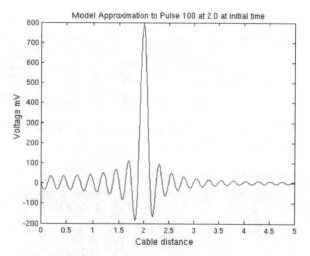

dendritic cable. Of course, if we use the voltage values starting at $t = 0$, we will see least squares artifacts. We show this in Fig. 11.6 where we obtain a typical least squares approximation to the pulse. If we only count voltage values past the first time tick in our time vector (i.e. $t \geq 0.05$), we get a better measure of the axon hillock voltage. We show this in Fig. 11.7.

Fig. 11.7 Model
approximation to an impulse
of size 100 at location 2.0.
This starts at $t = 0.05$ to
lessen the least squares
solution artifacts

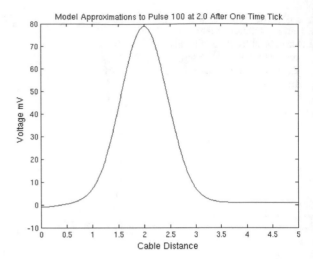

Listing 11.16: Plot Of Model Approximation To Pulse

```
plot(x,V(:,2));
xlabel('Cable distance');
ylabel('Voltage mV');
title('Model Approximation to Pulse 100 at 2.0 at initial time')
  ;
```

We can combine these MatLab scripts into one function: **GetAxonHillock2**
which is shown below.

Listing 11.17: GetAxonHillock2

```
   function [tau,lambda,V,AH,Input] = GetAxonHillock2(Q,L,TD,rho,Vmax,
       location)
   % Inputs:
   % Q              the number of eigenvalues we want to find
   % L              the length of the dendrite in space constants
 5 % rho            the ratio of dendrite to soma conductance, G_D/G_S
   % TD             length of time interval
   % Vmax           size of voltage impulse
   % location       location of pulse
   %
10 % Outputs
   % tau            dendritic time in units of time constants
   % lambda         dendritic position in units of space constants
   % V              the solution at (z,t)
   % AH             the solution at (1,t)
15 % Input          the solution as (z,0) to see if match to
   %                input voltage is reasonable
   %
   % Variables used in the code
   % z              eigenvalue vector
20 % D              data vector
   % M              matrix of coefficients for our approximate voltage
   %                model
```

```
     %
 25  % get eigenvalue vector z
     z = FindRoots(Q,L,rho,1.0,1.0);
     % get coefficient matrix M
     M = FindM(Q,L,rho,z);
     % compute data vector for impulse
 30  D = FindData(Q,L,rho,z,Vmax,location);
     % Solve MB = D system
     [Lower,Upper,piv] = GePiv(M);
     y = LTriSol(Lower,D(piv));
     B = UTriSol(Upper,y);
 35  % check errors
     Error = Lower*Upper*B - D(piv);
     Diff = M*B-D;
     e = norm(Error);
     e2 = norm(Diff);
 40  disp(sprintf(' norm of LU residual = %12.7f norm of MB-D = %12.7f',e,
         e2));
     % set dendritic spatial and time bounds
     % divide dendritic time into 100 parts
     tau     = linspace(0,TD,101);
     % divide dendritic space L into 300 parts
 45  lambda = linspace(0,L,301);
     %
     V = zeros(301,101);
     A = zeros(Q,1);
     for n = 1:Q
 50    A(n) = B(n+1);
     end
     % find voltage at space point lambda(s) and time point tau(t)
     for s = 1:301
       for t = 1:101
 55      w = z*(L-lambda(s));
         u = -(1+z.*z)*tau(t);
         V(s,t) = B(1)*exp(-tau(t))+ dot(A,cos(w).*exp(u));
       end
     end
 60  % axon hillock is at lambda = 1
     AH = V(1,:);
     % pulse is modeled at initial time by
     Input = V(:,1);
```

Now, let's see what happens if we increase Q.

Listing 11.18: Increasing Q

```
 1  [tau,lambda,V40,AH40,Input40] = GetAxonHillock2(40,L,10,rho,Vmax
      ,location);
    norm of LU residual =        0.0000000 norm of MB-D =        0.0000000
    [tau,lambda,V80,AH80,Input80] = GetAxonHillock2(80,L,10,rho,Vmax
      ,location);
    norm of LU residual =        0.0000000 norm of MB-D =        0.0000000
    [tau,lambda,V120,AH120,Input120] = GetAxonHillock2(120,L,10,rho,
      Vmax,location);
 6  norm of LU residual =        0.0000000 norm of MB-D =        0.0000000
    plot(t,AH40,'-',t,AH80,'--',t,AH120,'-.');
    legend('Q = 40','Q = 80','Q = 120');
```

The new plot in Fig. 11.8 shows that the how the axon hillock voltage approximation is altered by the use of more terms in the expansion. In this plot, we plot the voltages starting at $t = 0$ so that all exhibit the least squares solution artifacts. We could

Fig. 11.8 Axon hillock
response to an impulse of
size 100 at location 2.0;
Q = 40, 80 and 120: time
starts at 0

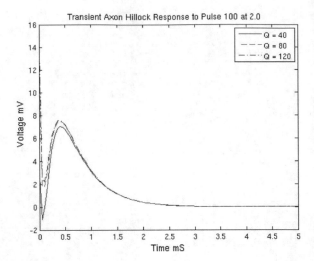

Fig. 11.8 Axon hillock
response to an impulse of
size 100 at location 2.0;
Q = 40, 80 and 120: time
starts at 0

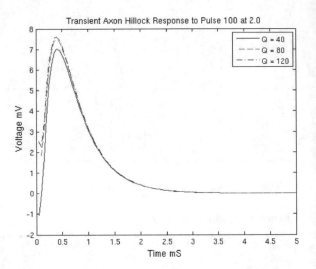

Fig. 11.9 Axon hillock
response to an impulse of
size 100 at location 2.0; Q =
40, 80 and 120: t starts at
0.05

eliminate them by starting the plots after one time tick as we do in Fig. 11.9. Higher
Q values are, of course, better although they cost more to compute.

Listing 11.19: Simple Plot

```
plot(t(2:101),AH40(2:101),'-',t(2:101),AH80(2:101),'--',t(2:101)
    ,AH120(2:101),'-.');
```

We can also look at how the approximation to the pulse changes with Q in
Fig. 11.10 starting at $t = 0$ and after one time tick in Fig. 11.11.

Fig. 11.10 Model approximation to an impulse of size 100 at location 2.0; Q = 40, 80 and 120 starting at time zero

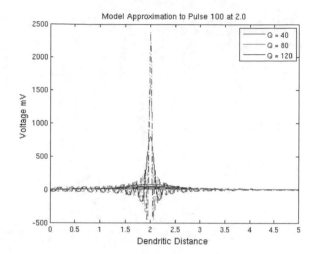

Fig. 11.11 Model approximation to an impulse of size 100 at location 2.0; Q = 40, 80 and 120 starting at time 0.05

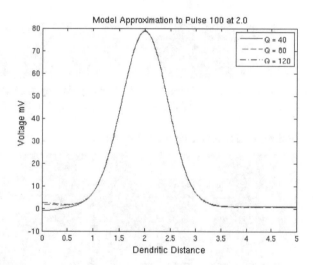

Now let's look at what happens with several pulses. We have a pulse of 100 at 2.0, one of 300 at 4.5. The code is as follows:

Listing 11.20: Two Pulses

```
  Q = 40;
  L = 5;
  rho = 10;
4 TD = 5.0;
  Vmax1 = 100;
  location1 = 2.0;
  [tau, lambda, V1, AH1, Input1] = GetAxonHillock2(Q, L, TD, rho, Vmax1,
     location1);
```

Fig. 11.12 Transient axon
hillock voltage for pulse of
size 300 at location 4.5

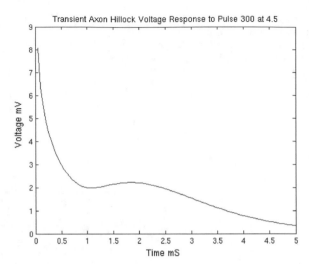

```
   Vmax2 = 350;
 9 location2 = 4.5;
   [tau,lambda,V2,AH2,Input2] = GetAxonHillock2(Q,L,TD,rho,Vmax2,
      location2);
```

The axon hillock voltage for the second pulse starting with time at one time tick is
seen in Fig. 11.12. We also show the pulse itself in Fig. 11.13.

Listing 11.21: Axon Hillock Responses

```
   plot(tau(2:101),AH2(2:101));
   xlabel('Time mS');
   ylabel('Voltage mV');
   title('Transient Axon Hillock Voltage Response to Pulse 300 at
      4.5');
 5 plot(lambda,V2(:,2));
   xlabel('Dendritic Distance');
   ylabel('Voltage mV');
   title('Model Approximation to Pulse 300 at 4.5');
```

Finally, we add the responses together in Fig. 11.14 and the inputs in Fig. 11.15.

Listing 11.22: Adding Responses

```
   plot(tau(2:101),AH1(2:101),'-',tau(2:101),AH2(2:101),'--',tau
      (2:101),AH1(2:101)+AH2(2:101),'-.');
 2 legend('AH Pulse 1','AH Pulse 2','Summed Pulse');
   xlabel('Time mS');
   ylabel('Voltage mV');
   title('Summed Transient Axon Hillock Response: Pulse 100 at 2.0,
      300 at 4.5');
   plot(lambda,V1(:,2),'-',lambda,V2(:,2),'--',lambda,V1(:,2)+V2
      (:,2),'-.');
 7 legend('Pulse 1 at 2.0','Pulse 2 at 4.5','Summed Pulse','
      location','best');
```

Fig. 11.13 Model approximation to an impulse of size 300 at location 4.5

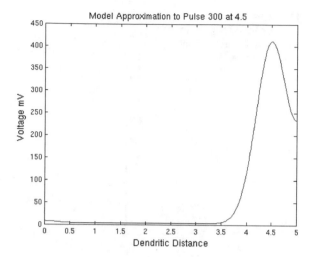

Fig. 11.14 Axon hillock voltage for pulse of size 300 at location 4.5 and Size 100 at location 2.0

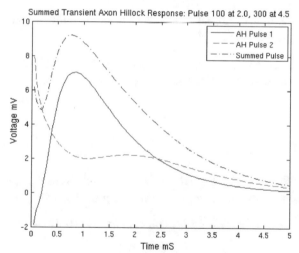

11.7 Ball Stick Project

Let's compute the axon—hillock voltage given above for a number of possible choices. It is clear that the solution $\hat{v}_m$ we obtain depends on

1. The length of the cable L.
2. The ratio of the dendritic to soma conductance, ρ.
3. The space constant λ_c as it depends on the cable radius a.
4. The truncation constant Q; we clearly make error by using only a finite number of terms.
5. The values of α_1 to α_Q. These have to be determined numerically as the solution of a transcendental equation as discussed in the text.

Fig. 11.15 Model
approximation for pulse of
size 300 at location 4.5 and
size 100 at location 2.0

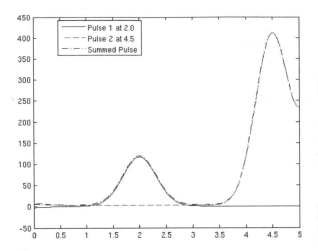

With this said, let's calculate the axon—hillock voltage for various given applied
voltages to the cable.

Exercise 11.7.1 *Compute this voltage for sums of impulses voltages applied at integer spatial constant distances:* $\lambda = 0$ *to* $\lambda = L$:

$$V(\lambda) = \sum_{i=0}^{L} V_i^* \delta(\lambda - i).$$

*Use the following voltage profiles and use a variety of values of Q to do the approx-
imations. Do these for the choices* $\rho \in \{0.1, 1.0, 10.0\}$.

1. *$L = 1$: Here V_0^* is applied at location 0.25 and V_1^* is applied at location 0.75.*

   (a) *Use $V_0^* = 10$ and $V_1^* = 0$.*
   (b) *Use $V_0^* = 0$ and $V_1^* = 10$.*
   (c) *Use $V_0^* = 100$ and $V_1^* = 0$.*
   (d) *Use $V_0^* = 0$ and $V_1^* = 100$.*

2. *$L = 5$: Here V_0^* is applied at location 0.25 and V_5^* is applied at location 4.75.
   The others are applied at integer locations.*

   (a) *Use $V_0^* = 10$, $V_1^* = 0$, $V_2^* = 0$, $V_3^* = 0$, $V_4^* = 0$ and $V_5^* = 10$*
   (b) *Use $V_0^* = 0$, $V_1^* = 0$, $V_2^* = 10$, $V_3^* = 0$, $V_4^* = 0$ and $V_5^* = 0$*
   (c) *Use $V_0^* = 100$, $V_1^* = 0$, $V_2^* = 0$, $V_3^* = 0$, $V_4^* = 0$ and $V_5^* = 100$*
   (d) *Use $V_0^* = 0$, $V_1^* = 0$, $V_2^* = 100$, $V_3^* = 0$, $V_4^* = 0$ and $V_5^* = 0$*

3. *$L = 10$: Since the cable is so long, we will assume all V_i^* are 0 except the ones
   listed. Here V_0^* is applied at location 0.25 and V_{10}^* is applied at location 9.75.
   The others are applied at integer locations.*

(a)   Use $V_0^* = 10$ and $V_{10}^* = 10$
(b)   Use $V_5^* = 10$
(c)   Use $V_0^* = 100$ and $V_{10}^* = 100$
(d)   Use $V_5^* = 100$

- *Use your judgment to provide reasonable plots of the information you have gathered above.*
- *Can you draw any conclusions about how to choose Q, L and ρ for our modeling efforts?*

References

Z. Hall, *An Introduction to Molecular Neurobiology* (Sinauer Associates Inc., Sunderland, 1992)

B. Hille, *Ionic Channels of Excitable Membranes* (Sinauer Associates Inc., Sunderland, 1992)

D. Johnston, S. Miao-Sin Wu, *Foundations of Cellular Neurophysiology* (MIT Press, Cambridge, 1995)

J. Peterson, *Boundary Value Problems and Fourier Series* (Lulu Press, Raleigh, 2015a). Unpublished notes available from author

J. Peterson *Calculus for Cognitive Scientists: Higher Order Models and Their Analysis* (Springer Series on Cognitive Science and Technology, Singapore, 2015b in press)

J. Peterson, *BioInformation Processing: A Primer On Computational Cognitive Science* (Springer Series on Cognitive Science and Technology, Singapore, 2015c in press)

W. Rall. Core Conductor Theory and Cable Properties of Neurons, chapter 3, pages 39–67. Unknown, 1977

T. Weiss, *Cellular Biophysics*, vol. 1, Transport (MIT Press, Cambridge, 1996a)

T. Weiss, *Cellular Biophysics*, vol. 2, Electrical properties (MIT Press, Cambridge, 1996b)

Chapter 12
The Basic Hodgkin–Huxley Model

We now discuss the Hodgkin–Huxley model for the generation of the action potential of an excitable nerve cell. We use the standard Ball and Stick model of Chap. 11. As previously mentioned, there are many variables are needed to describe what is happening inside and outside the cellular membrane, and to some extent, inside the membrane for a standard cable model. There are many of them, so it doesn't hurt to keep repeating them! Standard references for this material are Johnston and Wu (1995) and Weiss (1996a, b). These variables, their meanings and their units are given in Table 12.1.

The salient variables needed to describe what is happening inside and outside the cellular membrane and to some extent, inside the membrane are then

- V_m^0 is the rest value of the membrane potential.
- K_m^0 is the rest value of the membrane current per length density.
- K_e^0 is the rest value of the externally applied current per length density.
- I_i^0 is the rest value of the inner current.
- I_o^0 is the rest value of the outer current.
- V_i^0 is the rest value of the inner voltage.
- V_o^0 is the rest value of the outer voltage.
- r_i is the resistance of the inner fluid of the cable.
- r_o is the resistance of the outer fluid surrounding the cable.
- g_m is the membrane conductance per unit length.
- c_m is the membrane capacitance per unit length.

The membrane voltage can be shown to satisfy the partial differential equation 12.1 which dimensional analysis shows has units of volts/(cm^2).

$$\frac{\partial^2 V_m}{\partial z^2} = (r_i + r_o)K_m - r_o K_e \tag{12.1}$$

© Springer Science+Business Media Singapore 2016
J.K. Peterson, *Calculus for Cognitive Scientists*, Cognitive Science
and Technology, DOI 10.1007/978-981-287-880-9_12

Table 12.1 Hodgkin–Huxley variable units

Variable	Meaning	Units
V_m	Membrane potential	mV
K_m	Membrane current per length	nA/cm
K_e	Externally applied current	nA/cm
I_i	Inner current	nA
I_o	Outer current	nA
I_e	External current	nA
I_m	Membrane current	nA
V_i	Inner voltage	mV
V_o	Outer voltage	mV
r_i	Resistance inner fluid per length	μohms/cm
r_o	Resistance outer fluid per length	μohms/cm
g_m	Membrane conductance per length	μSiemens/cm
c_m	Membrane capacitance per length	nano Fahrads/cm
G_M	Membrane conductance	μSiemens/cm
C_M	Membrane capacitance	nano Fahrads/cm

In the standard core conductor model, the membrane is not modeled at all, but we now need to be more careful. A realistic description of how the membrane activity contributes to the membrane voltage must use models of ion flow which are controlled by gates in the membrane. A simple model of this sort is based on work that Hodgkin and Huxley (1952), Hodgkin (1952, 1954) performed in the 1950s. We start by expanding the membrane model to handle potassium, sodium and an all purpose current, called leakage current, using a modification of our original simple electrical circuit model of the membrane. We will think of a gate in the membrane as having an intrinsic resistance and the cell membrane itself as having an intrinsic capacitance as shown in Fig. 12.1. This is a picture of an idealized cell with a small portion of the membrane blown up into an idealized circuit: we see a small piece of the lipid membrane with an inserted gate. Thus, we expand the single branch of our old circuit model to multiple branches—one for each ion flow we wish to model. The ionic current consists of the portions due to potassium, K_K, sodium, K_{Na} and leakage K_L. The leakage current is due to all other sources of ion flow across the membrane which are not being explicitly modeled. This would include ion pumps; gates for other ions such as Calcium, Chlorine; neurotransmitter activated gates and so forth. We will assume that the leakage current is chosen so that there is no excitable neural activity at equilibrium. We are thus led to the model shown in Fig. 12.2. We will replace our empty membrane box by a parallel circuit model. Now this box is

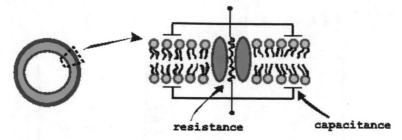

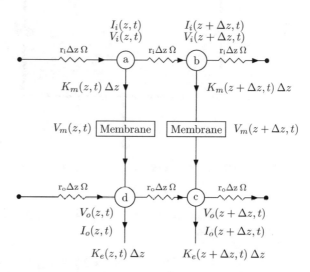

Fig. 12.1 The membrane and gate circuit model

Fig. 12.2 The equivalent electrical network model

really a chunk of membrane that is Δz wide. In the first cable model, we assume our membrane has a constant resistance and capacitance. We know that conductance is reciprocal resistance, so our model will consist to a two branch circuit: one branch is contains a capacitor and the other, the conductance element. We will let c_m denote the membrane capacitance density per unit length (measured in $\frac{\text{fahrad}}{\text{cm}}$). Hence, in our membrane box which is Δz wide, we see the value of capacitance should be $c_m \Delta z$. Similarly, we let g_m be the conductance per unit length (measured in $\frac{1}{\text{ohm}-\text{cm}}$) for the membrane. The amount of conductance for the box element is thus $g_m \Delta z$. In Fig. 12.3, we illustrate our new membrane model. Since this is a resistance-capacitance parallel circuit, it is traditional to call this an RC membrane model.

In Fig. 12.3, the current going into the element is $K_m(z, t)\Delta z$ and we draw the rest voltage for the membrane as a battery of value V_m^0. We know that the membrane current, K_m, satisfies Eq. 12.2:

$$K_m(z, t) = g_m \, V_m(z, t) + c_m \frac{\partial V_m}{\partial t} \tag{12.2}$$

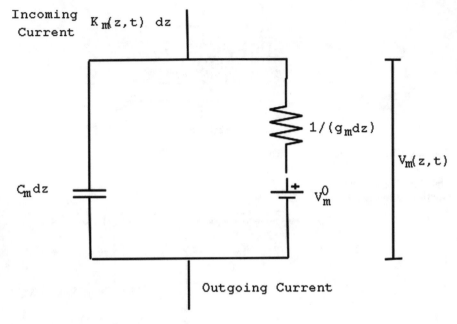

Fig. 12.3 The RC membrane model

In terms of membrane current densities, all of the above details come from modeling the simple equation

$$K_m = K_c + K_{ion}$$

where K_m is the membrane current density, K_c is the current through the capacitative side of the circuit and K_{ion} is the current that flows through the side of the circuit that is modeled by the conductance term, g_m. We see that in this model

$$K_c = c_m \frac{\partial V_m}{\partial t}, \quad K_{ion} = g_m V_m$$

However, we can come up with a more realistic model of how the membrane activity contributes to the membrane voltage by adding models of ion flow controlled by gates in the membrane. We have not done this before. In this chapter, and subsequent ones, we will study intensely increasingly more sophisticated versions that model the membrane gates as well as diffusion processes inside the cell that contribute to ion flow across the membrane. Our models will initially be based on work that Hodgkin and Huxley performed in the 1950s.

12.1 Adding Ion Membrane Currents

The standard Hodgkin–Huxley of an excitatory neuron consists of the equation for the total membrane current, K_M, obtained from Ohm's law

$$K_m = c_m \frac{\partial V_m}{\partial t} + K_K + K_{Na} + K_L, \quad (12.3)$$

where we have expanded the K_{ion} term to include the contributions from the sodium and potassium currents and the leakage current. The new equation for the membrane voltage is thus

$$\frac{\partial^2 V_m}{\partial z^2} = (r_i + r_o)K_m - r_o K_e$$

$$= (r_i + r_o) c_m \frac{\partial V_m}{\partial t} + (r_i + r_o) K_K + (r_i + r_o) K_{Na} + (r_i + r_o) K_L - r_o K_e$$

which can be simplified to what is seen in Eq. 12.5:

$$\frac{1}{r_i + r_o} \frac{\partial^2 V_m}{\partial z^2} = K_m - \frac{r_o}{r_i + r_o} K_e \quad (12.4)$$

$$= c_m \frac{\partial V_m}{\partial t} + K_K + K_{Na} + K_L - \frac{r_o}{r_i + r_o} K_e \quad (12.5)$$

12.2 The Hodgkin–Huxley Gates

In Fig. 12.4, we show an idealized cell with a small portion of the membrane blown up into an idealized circuit. We see a small piece of the lipid membrane with an inserted gate. We think of the gate as having some intrinsic resistance and capacitance. Now for our simple Hodgkin–Huxley model here, we want to model a sodium and potassium gate as well as the cell capacitance. So we will have a resistance for both the sodium and potassium. In addition, we know that other ions move across the membrane due to pumps, other gates and so forth. We will temporarily model this additional ion current as a **leakage** current with its own resistance. We also know that each ion has its own equilibrium potential which is determined by applying the Nernst equation. The **driving electromotive force** or **driving emf** is the difference between the ion equilibrium potential and the voltage across the membrane itself. Hence, if E_c is the equilibrium potential due to ion c and V_m is the membrane potential, the driving force is $V_c - V_m$. In Fig. 12.4, we see an electric schematic that summarizes what we have just said. We model the membrane as a parallel circuit with a branch for the sodium and potassium ion, a branch for the leakage current and a branch for the membrane capacitance. From circuit theory, we know that the charge q across a capacitor is $q = C E$, where C is the capacitance and E is the voltage across the

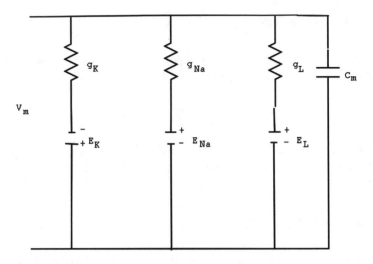

Fig. 12.4 The simple Hodgkin–Huxley membrane circuit model

capacitor. Hence, if the capacitance C is a constant, we see that the current through the capacitor is given by the time rate of change of the charge

$$\frac{dq}{dt} = C\frac{dE}{dt}$$

If the voltage E was also space dependent, then we would write $E(z, t)$ to indicate its dependence on both a space variable z and the time t. Then the capacitative current would be

$$\frac{dq}{dt} = C\frac{\partial E}{\partial t}$$

From Ohm's law, we know that voltage is current times resistance; hence for each ion c, we can say

$$V_c = I_c\, R_c$$

where we label the voltage, current and resistance due to this ion with the subscript c. This implies

$$I_c = \frac{1}{R_c}\, V_c = G_c\, V_c$$

where G_c is the reciprocal resistance or conductance of ion c. Hence, we can model all of our ionic currents using a conductance equation of the form above. Of course, the potassium and sodium conductances are nonlinear functions of the membrane voltage V and time t. This reflects the fact that the amount of current that flows through the membrane for these ions is dependent on the voltage differential across the membrane which in turn is also time dependent. The general functional form for an ion c is thus

$$I_c = G_c(V, t)(V(t) - E_c(t))$$

where as we mentioned previously, the driving force, $V - E_c$, is the difference between the voltage across the membrane and the equilibrium value for the ion in question, E_c. Note, the ion battery voltage E_c itself might also change in time (for example, extracellular potassium concentration changes over time). Hence, the driving force is time dependent. The conductance is modeled as the product of a activation, m, and an inactivation, h, term that are essentially sigmoid nonlinearities. The activation and inactivation are functions of V and t also. The conductance is assumed to have the form

$$G_c(V, t) = G_0 m^p(V, t) \, h^q(V, t)$$

where appropriate powers of p and q are found to match known data for a given ion conductance. We model the leakage current, I_L, as

$$I_L = g_L \, (V(t) - E_L)$$

where the leakage battery voltage, E_L, and the conductance g_L are constants that are data driven. Hence, in terms of current densities, letting g_K, g_{Na} and g_L respectively denote the ion conductances per length, our full model would be

$$K_K = g_K(V - E_K)$$
$$K_{NA} = g_{Na}(V - E_{Na})$$
$$K_L = g_L(V - E_L)$$

We know the membrane voltage satisfies:

$$\frac{1}{r_i + r_o} \frac{\partial^2 V_m}{\partial z^2} = c_m \frac{\partial V_m}{\partial t} + K_K + K_{Na} + K_L - \frac{r_o}{r_i + r_o} K_e$$

We can rewrite this as

$$\frac{1}{r_i + r_o} \frac{\partial^2 V_m}{\partial z^2} = c_m \frac{\partial V_m}{\partial t} + g_K(V_m - E_K) + g_{Na}(V_m - E_{Na})$$
$$+ g_L(V_m - E_L) - \frac{r_o}{r_i + r_o} K_e$$

12.2.1   Activation and Inactivation Variables

We assume that the voltage dependence of our activation and inactivation has been fitted from data. Hodgkin and Huxley modeled the time dependence of these variables using first order kinetics. They assumed a typical variable of this type, say Φ, satisfies for each value of voltage, V:

$$\frac{d\Phi(V)}{dt} = \alpha_\Phi(V) \, (1 - \Phi(V)) - \beta_\Phi(V) \, \Phi(V)$$
$$\Phi(0) = \Phi_0(V)$$

For convenience of exposition, we usually drop the functional dependence of Φ on V and just write:

$$\frac{d\Phi}{dt} = \alpha_\Phi \, (1 - \Phi) - \beta_\Phi \, \Phi$$
$$\Phi(0) = \Phi_0$$

Rewriting, we see

$$\frac{1}{\alpha_\Phi + \beta_\Phi} \frac{d\Phi}{dt} = \frac{\alpha_\Phi}{\alpha_\Phi + \beta_\Phi} - \Phi$$
$$\Phi(0) = \Phi_0$$

We let

$$\tau_\Phi = \frac{1}{\alpha_\Phi + \beta_\Phi}$$
$$\Phi_\infty = \frac{\alpha_\Phi}{\alpha_\Phi + \beta_\Phi}$$

allowing us to rewrite our rate equation as

$$\tau_\Phi \frac{d\Phi}{dt} = \Phi_\infty - \Phi$$
$$\Phi(0) = \Phi_0$$

12.3   The Voltage Clamped Protocol

Under certain experimental conditions, we can force the membrane voltage to be independent of the spacial variable z. In this case, we find

$$\frac{\partial^2 V_m}{\partial z^2} = 0$$

which allows us to rewrite Eq. 12.6 as follows

$$c_m \frac{dV_m}{dt} + K_K + K_{Na} + K_L - \frac{r_o}{r_i + r_o} K_e = 0 \tag{12.6}$$

The replacement of the partial derivatives with a normal derivative reflects the fact that in the voltage clamped protocol, the membrane voltage depends only on the one variable, time t. Since, c_m is capacitance per unit length, the above equation can also be interpreted in terms of capacitance, C_m, and currents, I_K, I_{Na}, I_L and an external type current I_e. This leads to Eq. 12.7

$$C_m \frac{dV_m}{dt} + I_K + I_{Na} + I_L - \frac{r_o}{r_i + r_o} I_e = 0 \tag{12.7}$$

Finally, if we label as external current, I_e, the term

$$I_E = \frac{r_o}{r_i + r_o} I_e,$$

the equation we need to solve under the voltage clamped protocol becomes Eq. 12.8.

$$\frac{dV_m}{dt} = \frac{1}{C_M} (-I_K - I_{Na} - I_L + I_E) \tag{12.8}$$

12.4 The Hodgkin–Huxley Sodium and Potassium Model

Hodgkin and Huxley modeled the sodium and potassium gates as

$$g_{Na}(V) = g_{Na}^{Max} \mathcal{M}_{NA}{}^3(V) \, \mathcal{H}_{NA}(V)$$
$$g_K(V) = g_K^{Max} \mathcal{M}_K{}^4(V)$$

where the two activation variables, $\mathcal{M}_{NA}$ and $\mathcal{M}_K$, and the one inactivation variable, $\mathcal{H}_{NA}$, all satisfy the first order Φ kinetics as we have discussed. Hence, we know

$$\tau_{\mathcal{M}_{NA}} \, \mathcal{M}'_{NA}(t) = (\mathcal{M}_{NA})_\infty - \mathcal{M}_{NA}$$
$$\tau_{\mathcal{H}_{NA}} \, \mathcal{H}'_{NA}(t) = (\mathcal{H}_{NA})_\infty - \mathcal{H}_{NA}$$
$$\tau_{\mathcal{M}_K} \, \mathcal{M}'_K(t) = (\mathcal{M}_K)_\infty - \mathcal{M}_K$$

with

$$\tau_{\mathcal{M}_{NA}} = \frac{1}{\alpha_{\mathcal{M}_{NA}} + \beta_{\mathcal{M}_{NA}}}$$
$$(\mathcal{M}_{NA})_\infty = \frac{\alpha_{\mathcal{M}_{NA}}}{\alpha_{\mathcal{M}_{NA}} + \beta_{\mathcal{M}_{NA}}}$$

$$\tau_{\mathcal{H}_{NA}} = \frac{1}{\alpha_{\mathcal{H}_{NA}} + \beta_{\mathcal{H}_{NA}}}$$

$$(\mathcal{H}_{NA})_\infty = \frac{\alpha_{\mathcal{H}_{NA}}}{\alpha_{\mathcal{H}_{NA}} + \beta_{\mathcal{H}_{NA}}}$$

$$\tau_{\mathcal{M}_K} = \frac{1}{\alpha_{\mathcal{M}_K} + \beta_{\mathcal{M}_K}}$$

$$(\mathcal{M}_K)_\infty = \frac{\alpha_{\mathcal{M}_K}}{\alpha_{\mathcal{M}_K} + \beta_{\mathcal{M}_K}}$$

Further, the coefficient functions, α and β for each variable required data fits as functions of voltage. These were determined to be

$$\alpha_{\mathcal{M}_{NA}} = -0.10 \, \frac{V + 35.0}{e^{-0.1\,(V+35.0)} - 1.0}$$

$$\beta_{\mathcal{M}_{NA}} = 4.0 \, e^{\frac{-(V+60.0)}{18.0}}$$

$$\alpha_{\mathcal{H}_{NA}} = 0.07 \, e^{-0.05\,(V+60.0)}$$

$$\beta_{\mathcal{H}_{NA}} = \frac{1.0}{(1.0 + e^{-0.1\,(V+30.0)})} \tag{12.9}$$

$$\alpha_{\mathcal{M}_K} = -\frac{0.01 * (V + 50.0)}{(e^{-0.1(V+50.0)} - 1.0)}$$

$$\beta_{\mathcal{M}_K} = 0.125 \, e^{-0.0125\,(V+60.0)}$$

Of course these data fits were obtained at a certain temperature and assumed values for all the other constants needed. These other parameters are given in the units below

voltage	mV	milli volts	10^{-3} Volts
current	na	nano amps	10^{-9} Amps
time	ms	milli seconds	10^{-3} Seconds
concentration	mM	milli moles	10^{-3} Moles
conductance	μ S	micro Siemens	10^{-6} ohms^{-1}
capacitance	nF	nano Fahrads	10^{-9} Fahrads

The Hodgkin–Huxley gates model of the membrane dynamics here thus consists of the following differential equations:

$$\tau_{\mathcal{M}_{NA}} \frac{d\mathcal{M}_{NA}}{dt} = (\mathcal{M}_{NA})_\infty - \mathcal{M}_{NA}$$

$$\tau_{\mathcal{H}_{NA}} \frac{d\mathcal{H}_{NA}}{dt} = (\mathcal{H}_{NA})_\infty - \mathcal{H}_{NA}$$

$$\tau_{\mathcal{M}_K} \frac{d\mathcal{M}_K}{dt} = (\mathcal{M}_K)_\infty - \mathcal{M}_K$$

$$\frac{dV}{dt} = \frac{I_M - I_K - I_{Na} - I_L}{C_M}$$

with initial conditions

$$\mathcal{M}_{NA}(0) = (\mathcal{M}_{NA})_\infty(V_0, 0)$$

$$\mathcal{H}_{NA}(0) = (\mathcal{H}_{NA})_\infty(V_0, 0)$$

$$\mathcal{M}_K(0) = (\mathcal{M}_K)_\infty(V_0, 0)$$

$$V(0) = V_0$$

We note that at equilibrium there is no current across the membrane. Hence, the sodium and potassium currents are zero and the activation and inactivation variables should achieve their steady state values which would be m_∞, h_∞ and n_∞ computed at the equilibrium membrane potential which is here denoted by V_0.

12.4.1 Encoding the Dynamics

Now these dynamics are more difficult to solve than you might think. The sequence of steps is this:

- Given the time t and voltage V compute

$$\alpha_{\mathcal{M}_{NA}} = -0.10 \, \frac{V + 35.0}{e^{-0.1\,(V+35.0)} - 1.0}$$

$$\beta_{\mathcal{M}_{NA}} = 4.0 \, e^{\frac{-(V+60.0)}{18.0}}$$

$$\alpha_{\mathcal{H}_{NA}} = 0.07 \, e^{-0.05\,(V+60.0)}$$

$$\beta_{\mathcal{H}_{NA}} = \frac{1.0}{(1.0 + e^{-0.1\,(V+30.0)})}$$

$$\alpha_{\mathcal{M}_K} = -\frac{0.01 * (V + 50.0)}{(e^{-0.1(V+50.0)} - 1.0)}$$

$$\beta_{\mathcal{M}_K} = 0.125 \, e^{-0.0125\,(V+60.0)}$$

- Then compute the τ and steady state activation and inactivation variables

$$\tau_{\mathcal{M}_{NA}} = \frac{1}{\alpha_{\mathcal{M}_{NA}} + \beta_{\mathcal{M}_{NA}}}$$

$$(\mathcal{M}_{NA})_\infty = \frac{\alpha_{\mathcal{M}_{NA}}}{\alpha_{\mathcal{M}_{NA}} + \beta_{\mathcal{M}_{NA}}}$$

$$\tau_{\mathcal{H}_{NA}} = \frac{1}{\alpha_{\mathcal{H}_{NA}} + \beta_{\mathcal{H}_{NA}}}$$

$$(\mathcal{H}_{NA})_\infty = \frac{\alpha_{\mathcal{H}_{NA}}}{\alpha_{\mathcal{H}_{NA}} + \beta_{\mathcal{H}_{NA}}}$$

$$\tau_{\mathcal{M}_K} = \frac{1}{\alpha_{\mathcal{M}_K} + \beta_{\mathcal{M}_K}}$$

$$(\mathcal{M}_K)_\infty = \frac{\alpha_{\mathcal{M}_K}}{\alpha_{\mathcal{M}_K} + \beta_{\mathcal{M}_K}}$$

- Then compute the sodium and potassium potentials. In this model this is easy as each is set only once since the internal and external ion concentrations always stay the same and so Nernst's equation only has to used one time. Here we use the concentrations

$$[NA]_o = 491.0$$
$$[NA]_i = 50.0$$
$$[K]_o = 20.11$$
$$[K]_i = 400.0$$

These computations are also dependent on the temperature.
- Next compute the conductances since we now know ($\mathcal{M}_{NA}(V)$, $\mathcal{H}_{NA}(V)$ and $\mathcal{M}_K(V)$).

$$g_{Na}(V) = g_{Na}^{Max} \, \mathcal{M}_{NA}{}^3(V) \, \mathcal{H}_{NA}(V)$$
$$g_K(V) = g_K^{Max} \, \mathcal{M}_K{}^4(V)$$

Now here we will need the maximum sodium and potassium conductances g_{Na}^{Max} and g_K^{Max} to finish the computation. These values must be provided as data and in this model are not time dependent. Here we use

$$g_{Na}^{Max} = 120.0$$
$$g_K^{Max} = 36.0$$

- Then compute the ionic currents:

$$I_{Na} = g_{NA}(V)(V(t) - E_{NA})$$
$$I_K = g_K(V)(V(t) - E_K)$$
$$I_L = g_L(V)(V(t) - E_L)$$

where we use suitable values for the leakage conductance and battery voltage such as

$$g_L = 0.0287$$
$$E_L = -50.0$$

We generally obtain these values with the **rest.m** computation which we will discuss later.

- Finally compute the total current

$$I_T = I_{NA} + I_K + I_L$$

- We can now compute the dynamics of our system at time t and voltage V: we let I_E denote the external current to our system which we must supply.

$$\frac{dV}{dt} = \frac{I_E - I_T}{C_M}$$
$$\frac{d\mathcal{M}_{NA}}{dt} = \frac{(\mathcal{M}_{NA})_\infty - \mathcal{M}_{NA}}{\tau_{\mathcal{M}_{NA}}}$$
$$\frac{d\mathcal{H}_{NA}}{dt} = \frac{(\mathcal{H}_{NA})_\infty - \mathcal{H}_{NA}}{\tau_{\mathcal{H}_{NA}}}$$
$$\frac{d\mathcal{M}_K}{dt} = \frac{(\mathcal{M}_K)_\infty - \mathcal{M}_K}{\tau_{\mathcal{M}_K}}$$

where we use $C_M = 1.0$.

The basic Hodgkin–Huxley model thus needs four independent variables which we place in a four dimensional vector y which components assigned as follows:

$$y = \begin{bmatrix} y[0] = V \\ y[1] = \mathcal{M}_{NA} \\ y[2] = \mathcal{H}_{NA} \\ y[3] = \mathcal{M}_K \end{bmatrix}$$

We encode the dynamics calculated above into a four dimensional vector f whose components are interpreted as follows:

$$f = \begin{bmatrix} f[0] = \frac{I_E - I_T}{C_M} \\ f[1] = \frac{(\mathcal{M}_{NA})_\infty - \mathcal{M}_{NA}}{\tau_{\mathcal{M}_{NA}}} \\ f[2] = \frac{(\mathcal{H}_{NA})_\infty - \mathcal{H}_{NA}}{\tau_{\mathcal{H}_{NA}}} \\ f[3] = \frac{(\mathcal{M}_K)_\infty - \mathcal{M}_K}{\tau_{\mathcal{M}_K}} \end{bmatrix}$$

which in terms of our vector y becomes

$$
\begin{bmatrix}
f[0] = \frac{I_E - I_T}{C_M} \\
f[1] = \frac{(\mathscr{M}_{NA})_\infty - y[1]}{\tau_{\mathscr{M}_{NA}}} \\
f[2] = \frac{(\mathscr{H}_{NA})_\infty - y[2]}{\tau_{\mathscr{H}_{NA}}} \\
f[3] = \frac{(\mathscr{M}_K)_\infty - y[3]}{\tau_{\mathscr{M}_K}}
\end{bmatrix}
$$

So to summarize, at each time t and V, we need to calculate the four dimensional dynamics vector f for use in our choice of ODE solver.

12.5   Computing the Solution Numerically

As an example of what we might do to solve this kind of a problem, we note that this system can now be written in vector form as

$$
\frac{dy}{dt} = f(t, y)
$$
$$
y(0) = y0
$$

where we have found that

$$
y =
\begin{bmatrix}
V \\
\mathscr{M}_{NA} \\
\mathscr{H}_{NA} \\
\mathscr{M}_K
\end{bmatrix},
$$

$$
f =
\begin{bmatrix}
\frac{I_M - I_T}{C_M} \\
\frac{(\mathscr{M}_{NA})_\infty - \mathscr{M}_{NA}}{\tau_{\mathscr{M}_{NA}}} \\
\frac{(\mathscr{H}_{NA})_\infty - \mathscr{H}_{NA}}{\tau_{\mathscr{H}_{NA}}} \\
\frac{(\mathscr{M}_K)_\infty - \mathscr{M}_K}{\tau_{\mathscr{M}_K}}
\end{bmatrix}
$$

with initial data

$$
y_0 =
\begin{bmatrix}
V_0 \\
(\mathscr{M}_{NA})_\infty \\
(\mathscr{H}_{NA})_\infty \\
(\mathscr{M}_K)_\infty
\end{bmatrix}
$$

The hardest part is to encode all of this in the vector form. The dynamic component $f[0]$ is actually a complicated function of $(\mathcal{M}_{NA})$, $(\mathcal{H}_{NA})$ and $(\mathcal{M}_K)$ as determined by the encoding sequence we have previously discussed. Hence, the calculation of the dynamics vector f is not as simple as defining each component of f as a functional expression. Now recall that the Runge–Kutta order 4 requires multiple vector function f evaluations per time step so that there is no way around the fact that solving this system numerically is expensive!

12.6 The Simple MatLab Implementation

This MatLab is more complicated and it is worth your while to have review the MatLab implementations we have discussed earlier. Simple code to manage our solution of the Hodgkin–Huxley equations is given below. We simply integrate the equations using a fixed time step repeatedly with a Runge–Kutta method of order 4. This is not terribly accurate but it illustrates the results nicely nevertheless. The MatLab code is organized as follows:

1. **SolveSimpleHHOrig.m** manages the action potential simulation. It calls **HHFixedRK.m** to manage the solution of the model using the Runge–Kutta approximation. We can do better than this in several ways. Right now, our dynamics function does not have the form $f(p, t, y)$ where p is a parameter vector which allows us to pass in many convenient values each time the dynamics function is called. We will fix that later. Also, we still use a fixed step size Runge–Kutta method. We will use RKF5 code later so that we can use an adjustable step size approach.
2. The code in **HHFixedRK.m** calls the implementation of Runge–Kutta order 2–4 in **HHRKstep.m**. RK code which does not have an adjustable step size algorithm can handle sudden discontinuities in current inputs because there is no error checking around the time point where the current suddenly jumps. In our later RKF5 code, we will have to use modified current pulses which ramp up quickly to their maximum value and then go down to their initial state in a continuous way so that the errors can be calculated correctly. However, that is a story for later.
3. The Runge–Kutta code must evaluate the dynamics function **simpleHH.m** many times.
4. We also need to initialize the simulation to set the leakage currents to be 0; this is done with the code in **rest.m**.
5. We also need to calculate Nernst voltages using the function **IonBattery Voltages** which uses the function **Nernst**.

So here are the code snippets we need.

12.6.1   Managing the Simulation: SolveSimpleHH.m

The code to manage the solution of the ODE is given below:

Listing 12.1: Managing The MatLab Hodgkin–Huxley Simulation One

```
    function [tvals,g_NA,g_K,V,m_NA,h_NA,n_K] = ...
             SolveSimpleHHOrig(fname,iename,tinit,tfinal,y0,h,k,g_NAmax,
             g_Kmax)

    %
5   % We use a simple Runge-Kutta scheme
    % using the Matlab function below:
    %
    % find number of interations to do
    n = ceil((tfinal-tinit)/h)+1;
10  %
    m_NA = zeros(1,n);
    h_NA = zeros(1,n);
    n_K  = zeros(1,n);
    g_Na = zeros(1,n);
15  g_K  = zeros(1,n);
    V    = zeros(1,n);

    [tvals,yvals] = HHFixedRK(fname,iename,tinit,y0,h,k,n);

20  %
    % store values in physical variables
    %
    V    = yvals(1,1:n);
    m_NA = yvals(2,1:n);
25  h_NA = yvals(3,1:n);
    n_K  = yvals(4,1:n);

    %
    %   Compute g_Na and G_K
30  %
    for i = 1:n
        u       = m_NA(i)*m_NA(i)*m_NA(i)*h_NA(i);
        g_NA(i) = g_NAmax*u;
    end
35
    for i = 1:n
        u       = n_K(i)*n_K(i)*n_K(i)*n_K(i);
        g_K(i)  = g_Kmax * u;
    end
40
    end
```

12.6.2   Gate Code

In all of our calculations, we will need the equilibrium Nernst voltages for various ions. We find these with a call to the function **IonBatteryVoltages**.

Listing 12.2: Calculating Ion Voltages: IonBatteryVoltages.m

```
function  [EK,ENA,T]  =  IonBatteryVoltages(KIn,KOut,NaIn,NaOut,TF)
%
% KIn  is  the  inside  potassium  concentration
% KOut  is  the  outside  potassium  concentration
5 % NaIN  is  the  inner  sodium  concentration
% NaOut  is  the  outer  sodium  concentration
% TF  is  the  temperature  in  Fahrenheit
%
% ===============================================
10 %   Constants  for  Equilibrium  Voltage  Calculations
% ===============================================
%

% Rydberg's  Constant
15   R = 8.31;
% Kelvin  Temperature;  convert  fahrenheit  to  Celsius
% default  is  9.3  Celsius
T = (5/9)*(TF-32);
% Faraday's  constant
20  F = 9.649e+4;
%
% Compute  Nernst  voltages  of  E_K  and  E_Na
% voltage  =  Nernst(valence,Temperature,InConc,OutConc)
%
25 % Sodium
% NA_O = 491.0;
% NA_I = 50.0;
ENA = Nernst(1,T,NaIn,NaOut);
% Potassium
30 % K_O = 20.11;
% K_I = 400.0;
EK = Nernst(1,T,KIn,KOut);
```

Now, the typical dynamics for the Hodgkin–Huxley model has the following form in MatLab. It has input arguments of time, t, the current vector of y values and the name of the function which calculates an external current, **IE**. We encode all of the sodium and potassium gate model parameters directly into the function. This makes the function one we have to edit every time we change any aspect of the Hodgkin–Huxley gate models.

Listing 12.3: simpleHH Arguments

```
function  f = simpleHH(t,y,IE)
%       Standard  Hodgkin  -  Huxley  Model
%       units
%       voltage        mV
5 %      current        na
%       time           ms
%       concentration  mM
%       conductance    micro  Siemens
%       capacitance    nF
10 %
%       t  =  time
%       y  =  dynamic  variables
%       y  input
%       y(1) = V
```

```
15 %        y(2)  =  m_NA
   %        y(3)  =  h_NA
   %        y(4)  =  m_K
   aa
   %        IE  =  external  input
20 %
```

The dynamics vector here is four dimensional, so we set its size next.

Listing 12.4: Set dimension of dynamics

```
  function  f  =  simpleHH(t,y,IE)
  % set  size  of  f
  %
  f  =  zeros(1,4);
5 %        ================================
  %        f  vector  assignments
  %        ================================
  %        f(1)  =  V  dynamics
  %        f(2)  =  m_NA  dynamics
10 %       f(3)  =  h_NA  dynamics
  %        f(4)  =  m_K  dynamics
  %
```

We compute the Nernst battery voltages and set the maximum sodium and potassium ion conductances.

Listing 12.5: Get Ion battery voltages

```
  %        ================================
  %        Constants  for  Equilibrium  Voltage  Calculations
3 %        ================================
  [E_K,E_NA]  =  IonBatteryVoltages(400.0,20.11,50.0,491.0,69);

  % max  conductance  for  NA
    g_NA_bar  =  120.0;
8
  % max  conductance  for  K
    g_K_bar  =  36.0;
```

Then, we find the sodium current.

Listing 12.6: Calculate sodium current

```
  %        ================================
  %        Fast  Sodium  Current
  %        ================================
  % activation/inactivation  parameters  for  NA
5 %
  % alpha_mNA ,  beta_mNA
  %
    sum  =  y(1)+35.0;
    if  sum  >  0
```

```
10      alpha_mNA        = −0.10*sum/(exp(−sum/10.0) − 1.0);
     else
        alpha_mNA        = −0.10*exp( sum/ 10.0)*sum/(1.0 − exp( sum/
           10.0));
     end
  %
15   sum = y(1)+60.0;
     if sum > 0
        beta_mNA         = 4.0*exp( −sum/18.0);
     else
        beta_mNA         = 4.0*exp( −sum/18.0);
20   end
  %
     m_NA_infinity = alpha_mNA/(alpha_mNA+beta_mNA);
     t_m_NA          = 1.0/(alpha_mNA+beta_mNA);
  %
25   f(2)             = (m_NA_infinity − y(2))/t_m_NA;
  %
  % activation, inactivation parameter for I_NA
  %
  % alpha_hNA , beta_mNA
30 %
     sum = (y(1)+60.0)/20.0;
     if sum < 0
        alpha_hNA        = 0.07*exp( −sum);
     else
35      alpha_hNA        = 0.07*exp( −sum);
     end
  %
     sum = y(1)+30.0;
     if sum>0
40      beta_hNA         = 1.0/(1.0 + exp(−sum/10.0));
     else
        beta_hNA         = exp( sum/10.0)/(exp( sum/10.0) + 1.0);
     end
  %
45   h_NA_infinity = alpha_hNA/(alpha_hNA+beta_hNA);
     t_h_NA          = 1.0/(alpha_hNA+beta_hNA);
     f(3)             = (h_NA_infinity − y(3))/t_h_NA;
  %
  % I_NA current
50 %
     I_NA = g_NA_bar*(y(1)−E_NA)*y(2)*y(2)*y(2)*y(3);
```

Next, we calculate the potassium current.

Listing 12.7: Calculate Potassium current

```
  %  ============================================
  %    Potassium  Current
  %  ============================================
4 %
  % activation/inactivation  parameters  for  K
  %
  % alpha_mK
  %
9   sum = y(1)+50.0;
     if sum > 0
        alpha_mK         = −0.01*sum/(exp(−sum/10.0) − 1.0);
     else
        alpha_mK         = −0.01*exp(sum/10.0)*sum/(1.0 − exp(sum/10.0))
           ;
```

```
14      end
        sum = (y(1)+60.0)*0.0125;
        if sum > 0
           beta_mK        = 0.125*exp( -sum);
        else
19         beta_mK        = 0.125*exp( -sum);
        end
        t_m_K             = 1.0/(alpha_mK+beta_mK);
        m_K_infinity      = alpha_mK/(alpha_mK+beta_mK);
        f(4)              = (m_K_infinity - y(4))/t_m_K;
24 %
   % I_K  current
   %
        I_K = g_K_bar*(y(1)-E_K)*y(4)*y(4)*y(4)*y(4);
```

Finally, we set the leakage current.

Listing 12.8: Set leakage current

```
   %  =====================================================
   %     Leakage  Current:  run  rest.m  to  find  appropriate
3  %     g_L  value
   %  =====================================================
        g_leak = .0287;
        E_leak = -50.0;

8  %
   % I_L  current
   %
        I_leak = g_leak*(y(1)-E_leak);
```

We then set the cell capacitance, find the external input at this time and find the **f(1)**
term which gives the full cell dynamics.

Listing 12.9: Set cell capacitance and cable dynamics

```
   %
   % Cell  Capacitance
   %
        C_M = 1.0;
5       s = feval(IE,t);
   %    s = 0;
        f(1) = (s - I_NA - I_K - I_leak)/C_M;
```

Let's change this, and at the same time make the code simpler and cleaner, by
introducing a more abstract model of the ion gates. The function used in the gate
models are all special cases of the of the general mapping $F(V_m, p, q)$ with $p \in \Re^4$
and $q \in \Re^2$ defined by

$$F(V_m, p, q) = \frac{p_0(V_m + q_0) + p_1}{e^{p_2(V_m + q_1)} + p_3}.$$

For ease of exposition, here we will denote $\mathcal{M}_{Na}$ by m, $\mathcal{H}_{Na}$ by h and $\mathcal{M}_K$ by n as,
historically, m, h and n have been used for these Hodgkin–Huxley models. The α

and β pairs are thus described using the generic F mapping by

$$\alpha_m = F(V_m, \ p_m^\alpha = \{-0.10, 0.0, -0.1, -1.0\},$$
$$q_m^\alpha = \{35.0, 35.0\}),$$
$$\beta_m = F(V_m, \ p_m^\beta = \{0.0, 4.0, 0.0556, 0.0\},$$
$$q_m^\beta = \{60.0, 60.0\}),$$
$$\alpha_h = F(V_m, \ p_h^\alpha = \{0.0, 0.07, 0.05, 0.0\},$$
$$q_h^\alpha = \{60.0, 60.0\}),$$
$$\beta_h = F(V_m, \ p_h^\beta = \{0.0, 1.0, -0.1, 1.0\},$$
$$q_h^\beta = \{30.0, 30.0\}),$$
$$\alpha_n = F(V_m, \ p_n^\alpha = \{-0.01, 0.0, -0.1, -1.0\},$$
$$q_n^\alpha = \{50.0, 50.0\}),$$
$$\beta_n = F(V_m, \ p_n^\beta = \{0.0, 0.125, 0.0125, 0.0\},$$
$$q_n^\beta = \{60.0, 60.0\}).$$

The p and q parameters control the shape of the action potential in a complex way. We will think of alterations in the (p, q) pair associated with a given α and/or β as a way of modeling how passage of ions through the gate are altered by the addition of various ligands. These effects may or may not be immediate. For example, the alterations to the p and q parameters may be due to the docking of ligands which are manufactured through calls to the genome in the cell's nucleus. In that case, there may be a long delay between the initiation of a second messenger signal to the genome and the migration of the ligands to the outside of the cell membrane. In addition, proteins can be made which bind to the inside of a gate and thereby alter the ion flow. We understand that this type of modeling is not attempting to explain the details of such interactions. Instead, we are exploring an approach for rapid identification and differentiation of the signals due to various toxins. For now, we assume our standard Hodgkin–Huxley model uses $g_{Na}^{Max} = 120$, $g_K^{Max} = 36.0$ and the classical α, β functions are labeled as the nominal values. Hence, all we need to do is code the information below in a vector Λ_0 given by

$$\Lambda_0 = \begin{bmatrix}
G_{Na}^0 = 120.0 & & \\
G_K^0 = 36.0 & & \\
(p_m^\alpha)^0 & \{-0.10, 0.0, -0.1, -1.0\} & (p_m^\beta)^0 \ \{0.0, 4.0, 0.0556, 0.0\} \\
(p_h^\alpha)^0 & \{0.0, 0.07, 0.05, 0.0\} & (p_h^\beta)^0 \ \{0.0, 1.0, -0.1, 1.0\} \\
(p_n^\alpha)^0 & \{-0.01, 0.0, -0.1, -1.0\} & (p_n^\beta)^0 \ \{0.0, 0.125, 0.0125, 0.0\} \\
(q_m^\alpha)^0 & \{35.0, 35.0\} & (q_m^\beta)^0 \ \{60.0, 60.0\} \\
(q_h^\alpha)^0 & \{60.0, 60.0\} & (q_h^\beta)^0 \ \{30.0, 30.0\} \\
(q_n^\alpha)^0 & \{50.0, 50.0\} & (q_n^\beta)^0 \ \{60.0, 60.0\}
\end{bmatrix}$$

First, we write MatLab code to implement the function F.

Listing 12.10: Gate Dynamics Function

```
  function  F = HHGates(V,p,q)
  %
3 % V = incoming  voltage
  % p = parameters  for  gate  model
  % q = parameters  for  gate  model
  % gate  value
  %
8 Numerator    = p(1)*(V + q(1)) + p(2);
  Denominator = exp( p(3)*(V + q(2))) + p(4);
  F = Numerator/Denominator;
```

We can then implement the classical Hodgkin–Huxley model dynamics as follows:

Listing 12.11: Classical HH Gate Model

```
   function  alpha = GatesNAK(gatesname,V)
   %
   % gatesname = name  of  gate  function
   % V         = voltage
 5 % alpha     = (alpha,beta)  pairs  in  HH  model
   %                as  many  pairs  as  there  are  ions
   %                in  the  model
   % This  is  simple  Sodium,  Potassium  model  so
   %  alpha,  beta  for  mNA = alpha(1),  alpha(2)
10 %  alpha,  beta  for  hNA = alpha(3),  alpha(4)
   %  alpha,  beta  for  mK  = alpha(5),  alpha(6)
   %
   % we  need  p,  q  parameter  vectors  for  each  of  the
   % six  alpha,  beta  pairs.   These  values  come  from
15 % the  literature
   %
   p = zeros(6,4);
   q = zeros(6,2);
   p = [-0.1,0.0,-0.1,-1.0;...
20      0.0,4.0,0.0556,0.0;...
        0.0,0.07,0.05,0.0;...
        0.0,1.0,-0.1,1.0;...
        -0.01,0.0,-0.1,-1.0;...
        0.0,0.125,0.0125,0.0];
25 %
   q = [35.0,35.0;...
        60.0,60.0;...
        60.0,60.0;...
        30.0,30.0;...
30      50.0,50.0;...
        60.0,60.0];
   %
   alpha(1) = feval(gatesname,V,p(1,:),q(1,:));
   alpha(2) = feval(gatesname,V,p(2,:),q(2,:));
35 alpha(3) = feval(gatesname,V,p(3,:),q(3,:));
   alpha(4) = feval(gatesname,V,p(4,:),q(4,:));
   alpha(5) = feval(gatesname,V,p(5,:),q(5,:));
   alpha(6) = feval(gatesname,V,p(6,:),q(6,:));
   end
```

This returns the three (α, β) pairs associated with the classical HH model as a single vector α. We write the (p, q) values we need directly into this file. However, if we wanted to implement a different gate model, we could easily either edit this file or copy it to a new file and edit that. This gives us a more flexible way to do our modeling. Our dynamics are given in the rewritten code **simpleHH.m**.

Listing 12.12: HH Dynamics: simpleHH.m

```
function f = simpleHH(t,y,IE)

%       Standard Hodgkin - Huxley Model
%       voltage         mV
5 %     current         na
%       time            ms
%       concentration   mM
%       conductance     micro Siemens
%       capacitance     nF
10 % ============================================
%       y vector assignments
% ============================================
%       y(1) = V
%       y(2) = m_NA
15 %     y(3) = h_NA
%       y(4) = m_K
%
% set size of f
%
20
f = zeros(1,4);

% ============================================
%       f vector assignments
25 % ============================================
%       y(1) = V dynamics
%       y(2) = m_NA dynamics
%       y(3) = h_NA dynamics
%       y(4) = m_K dynamics
30 % ============================================
%
%       Constants for Equilibrium Voltage Calculations
% ============================================
[E_K,E_NA,T] = IonBatteryVoltages(400.0,20.11,50.0,491.0,69);
35
% max conductance for NA
  g_NA_bar = 120.0;
% max conductance for K
  g_K_bar = 36.0;
40
alpha = GatesNAK('HHGates',y(1));
%
% activation/inactivation parameters for NA
% alpha(1), alpha(2) alpha and beta for mna
45 % alpha(3), alpha(4) alpha and beta for hna
% alpha(5), alpha(6) alpha and beta for mk
%
% ============================================
%       Fast Sodium Current
%
50 % ============================================
  m_NA_infinity   = alpha(1)/(alpha(1)+alpha(2));
  t_m_NA          = 1.0/(alpha(1)+alpha(2));
  f(2)            = (m_NA_infinity - y(2))/t_m_NA;
%
55 h_NA_infinity   = alpha(3)/(alpha(3)+alpha(4));
  t_h_NA          = 1.0/(alpha(3)+alpha(4));
  f(3)            = (h_NA_infinity - y(3))/t_h_NA;
  I_NA            = g_NA_bar*(y(1)-E_NA)*y(2)^3*y(3);
%
60 %    Potassium Current
%
%
  m_K_infinity    = alpha(5)/(alpha(5)+alpha(6));
  t_m_K           = 1.0/(alpha(5)+alpha(6));
65 f(4)            = (m_K_infinity - y(4))/t_m_K;
  I_K             = g_K_bar*(y(1)-E_K)*y(4)^4;
```

```
     %
     % leakage current: run rest.m to find appropriate g_L value
     %
70     g_leak  =  .0084170;
       E_leak  =  -50.0;
     %
     % I_L current
       I_leak  =  g_leak*(y(1)-E_leak);
75
     %
     % Cell Capacitance
     %
       C_M =  1.0;
80     s  =  feval(IE,t);
     % s  =  0;
       f(1)  =  (s - I_NA - I_K - I_leak)/C_M;
```

12.6.3   Setting Ion Currents to Zero: Rest.m

The initial conditions for the activation and inactivation variables for this model should be calculated carefully. When the excitable nerve cell is at equilibrium, with no external current applied, there is no net current flow across the membrane. At this point, the voltage across the membrane should be the applied voltage E_M. Therefore,

$$
\begin{aligned}
0 = {} & I_T \\
= {} & I_{NA} + I_{K1} + I_L \\
= {} & g_{NA}^{Max} \, (E_M - E_{NA})(\mathscr{M}_{NA})_\infty^3 \, (\mathscr{H}_{NA})_\infty + g_K^{Max} \, (E_M - E_K) \, (\mathscr{H}_{NA})_\infty^4 \\
& + g_L \, (E_M - E_L)
\end{aligned}
$$

The two parameters, g_L and E_L, are used to take into account whatever ionic currents flow across the membrane that are not explicitly modeled. We know this model does not deal with various pumps, a variety of potassium gates, calcium gates and so forth. We also know there should be no activity at equilibrium in the absence of an external current. However, it is difficult to choose these parameters. So first, solve to the leakage parameters in terms of the rest of the variables. Also, we can see that the activation and inactivation variables for each gate at equilibrium will take on the values that are calculated using the voltage E_M. We can use the formulae given in the Hodgkin–Huxley dynamics to compute the values of activation/inactivation that occur when the membrane voltage is at rest, E_M. We label these values with a superscript r for notational convenience.

$$
\begin{aligned}
\mathscr{M}_{NA}^r &\equiv (\mathscr{M}_{NA})_\infty(E_M) \\
\mathscr{H}_{NA}^r &\equiv (\mathscr{H}_{NA})_\infty(E_M) \\
\mathscr{M}_K^r &\equiv (\mathscr{M}_K)_\infty(E_M)
\end{aligned}
$$

Then, we can calculate the equilibrium currents

$$I_{NA}^r = g_{NA}^{max} (E_M - E_{NA}) (\mathcal{M}_{NA}^r)^3 \mathcal{H}_{NA}^r$$
$$I_K^r = g_K^{Max} (E_M - E_K) (\mathcal{M}_K^r)^4$$

Thus, we see we must choose g_L and E_L so that

$$g_L (E_L - E_M) = I_{NA}^r + I_K^r$$

or

$$g_L (E_L - E_M) = I_{NA}^r + I_K^r$$

If we choose to fix E_L, we can solve for the leakage conductance

$$g_L = \frac{I_{NA}^r + I_K^r}{E_L - E_M}$$

We do this in the function **rest.m**.

Listing 12.13: Initializing The Simulation: rest.m

```
      function  [g_L ,m_NA0, h_NA0 ,m_K0, E_K, E_NA]  =  rest (E_L ,V_R, g_NA_bar ,
          g_K_bar ,p ,F)
      %
   3  % inputs :
      % E_L       = leakage  voltage
      % g_NA_bar  = maximum  sodium  conductance
      % g_K_bar   = maximum  potassium  conductance
      % V_R       = rest  voltage
   8  % p         = parameter  vector
      % p [1]     = Inside  Potassium  Concentration
      % p [2]     = Outside  Potassium  Concentration
      % p [3]     = Inside  Sodium  Concentration
      % p [4]     = Outside  Sodium  Concentration
  13  % p [5]     = reference  temperature
      % F         = name  of  gates  function
      %
      % outputs :
      % g_L       = leakage  conductance
  18  % m_NA0     = initial  value  of  m_NA
      % h_NA0     = initial  value  of  h_NA
      % m_K0      = initial  value  of  m_K
      % E_K       = potassium  Nernst  voltage
      % E_NA      = sodium  Nernst  voltage
  23  %
      % =======================================================
      %     Constants  for  Ion  Equilibrium  Voltage  Calculations
      % =======================================================
      KIn   = p(1) ;
  28  KOut  = p(2) ;
      NaIn  = p(3) ;
      NaOut = p(4) ;
      TF    = p(5) ;

  33  [E_K, E_NA]  =  IonBattery Voltages (KIn, KOut, NaIn, NaOut, TF) ;
      %
      alpha  =  GatesNAK (F, V_R) ;
      %
      % activation / inactivation  parameters  for  NA
```

```
38  % alpha(1), alpha(2)  alpha and beta for mna
    % alpha(3), alpha(4)  alpha and beta for hna
    % alpha(5), alpha(6)  alpha and beta for mk
    %
      m_NA0 = alpha(1)/(alpha(1)+alpha(2));
43    h_NA0 = alpha(3)/(alpha(3)+alpha(4));
      m_K0  = alpha(5)/(alpha(5)+alpha(6));
    % I_NA current, I_K current
      I_NA = g_NA_bar*(V_R-E_NA)*m_NA0^3*h_NA0;
      I_K = g_K_bar*(V_R-E_K)*m_K0^4;
48  %
      numerator   = -I_NA - I_K;
      denominator = V_R - E_L;
    %
    % compute g_L: Note we want
53  % I_NA + I_K + g_L * (V_R - E_L) = 0
    % which gives the equation below assuming we are given E_L.
    %
      g_L = -(I_NA + I_K)/(V_R - E_L);
    end
```

12.6.4   The Runge–Kutta Manager: HHFixedRK.m

The code that manages the call to a fixed Runge–Kutta order method from an earlier
chapter has been modified to accept an argument which is the name of the function
which models the current injection.

Listing 12.14: The MatLab Fixed Runge–Kutta Method

```
    function [tvals,yvals] = HHFixedRK(fname,iename,t0,y0,h,k,n)
    %
3   %            Gives approximate solution to
    %                  y'(t) = f(t,y(t))
    %                  y(t0) = y0
    %            using a kth order RK method
    %
8   % t0          initial time
    % y0          initial state
    % h           stepsize
    % k           RK order   1<= k <= 4
    % n           Number of steps to take
13  %
    % tvals       time values of form
    %                  tvals(j) = t0 + (j-1)*h, 1 <= j <= n
    % yvals       approximate solution
    %                  yvals(:j) = approximate solution at
18  %                  tvals(j),  1 <= j <= n
    % fname       name of dynamics function
    % iename      name of current injection function
    %
      tc = t0;
23  yc = y0;
      tvals = tc;
      yvals = zeros(4,n);
      yvals(1:4,1) = transpose(yc);
      fc = feval(fname,tc,yc,iename);
28  for j=1:n-1
        [tc,yc,fc] = HHRKstep(fname,iename,tc,yc,fc,h,k);
        yvals(1:4,j+1) = transpose(yc);
        tvals = [tvals tc];
      end
```

12.6.4.1   The Runge–Kutta Implementation: HHRKstep.m

The actual Runge–Kutta code was also modified to allow the name of the current injection function to be entered as an argument.

Listing 12.15: Adding Injection Current To The Runge–Kutta Code

```
  function [tnew,ynew,fnew] = HHRKstep(fname,iename,tc,yc,fc,h,k)
  %
3 % fname        the name of the right hand side function f(t,y)
  %              t is a scalar usually called time and
  %              y is a vector of size d
  % yc           approximate solution to y'(t) = f(t,y(t)) at t=tc
  % fc           f(tc,yc)
8 % h            The time step
  % k            The order of the Runge–Kutta Method 1<= k <= 4
  %
  % tnew         tc+h
  % ynew         approximate solution at tnew
13 % fnew        f(tnew,ynew)
  %
  if k==1
     k1 = h*fc;
     ynew = yc+k1;
18 elseif k==2
     k1 = h*fc;
     k2 = h*feval(fname,tc+(h/2),yc+(k1/2),iename);
     ynew = yc + (k1+k2)/2;
  elseif k==3
23    k1 = h*fc;
     k2 = h*feval(fname,tc+(h/2),yc+(k1/2),iename);
     k3 = h*feval(fname,tc+h,yc-k1+2*k2,iename);
     ynew = yc+(k1+4*k2+k3)/6;
  elseif k==4
28    k1 = h*fc;
     k2 = h*feval(fname,tc+(h/2),yc+(k1/2),iename);
     k3 = h*feval(fname,tc+(h/2),yc+(k2/2),iename);
     k4 = h*feval(fname,tc+h,yc+k3,iename);
     ynew = yc+(k1+2*k2+2*k3+k4)/6;
33 else
     disp(sprintf('The RK method %2d order is not allowed!',k));
  end
  tnew = tc+h;
  fnew = feval(fname,tnew,ynew,iename);
```

12.6.5   A Simple External Current: IE.m

Listing 12.16: The Injected Current Pulse

```
  function ie = IE(t)
  %
3 %
  %
  ts1 =   9.95;
  te1 = 10.0;
  ts2 = 10.62;
8 te2 = 10.67;
  Imax1 = 100.0;
  ie = 0;
```

```
     if  t >= ts1 && t <= te1
        ie = (Imax1/(te1-ts1))*(t - ts1);
13   elseif  t>te1 && t < ts2
        ie = Imax1;
     elseif  t >= ts2 && t < te2
        ie = Imax1 -(Imax1/(te2-ts2))*(t-ts2);
     end
```

12.6.6   Runtime Results

For this simulation, we will use a Fahrenheit temperature of 69.0, a maximum sodium conductance of 120.0 and a maximum potassium conductance of 36.0. There are a few things we have to watch for when we run a simulation:

- The Celsius temperature is actually typed into the code files **simpleHH.m** and **rest.m**. So changing this temperature means you have to edit and change a line in both files.
- The maximum sodium and potassium conductances is hard coded into the file **simpleHH.m**. So changing these values requires editing this files.
- Once values of temperature, maximum conductances and leakage voltage are chosen, run rest.m to determine the value of g_L that ensures no current flows across the cell membrane when there is no external current. This function call also returns the initial values for the Hodgkin–Huxley dynamical system. Lines like

Listing 12.17: Finding Hodgkin–Huxley model initial conditions

```
         V_R = -70;
         E_L = -50;
3        Kin = 400.0;
         Kout = 20.11;
         Nain = 50.0;
         Naout = 491.0;
         TF = 69.0;
8        p = [Kin;Kout;Nain;Naout;TF];
         g_NA_bar = 120;
         g_K_bar = 36.0;
         F = 'HHGates';
         [g_L ,m_NA0,h_NA0,m_K0,E_K,E_NA] = ...
13            rest(E_L,V_R,g_NA_bar,g_K_bar,p,F)
         g_L =   0.0084170
         m_NA0 =   0.015385
         h_NA0 =   0.86517
         m_K0 =   0.18100
18       E_K = -75.607
         E_NA =   57.760
```

allow us to find the initial condition for the Hodgkin–Huxley dynamical system we want to solve. We find

$$\begin{bmatrix} y[1] \\ y[2] \\ y[3] \\ y[4] \end{bmatrix} = \begin{bmatrix} V\_R \\ m\_NA0 \\ h\_NA0 \\ m\_K0 \end{bmatrix}.$$

Then the value of g_L returned must then be typed into the appropriate place in the file **simpleHH.m**.

• Once the above steps have been done, we are ready to run the simulation.

We ran our simple simulation using this session (we edit a bit to shorten lines and so forth):

Listing 12.18: Sample session to generate an action potential

```
1    [g_L,m_NA0,h_NA0,m_K0,E_K,E_NA] = ...
            rest(E_L,V_R,g_NA_bar,g_K_bar,p,F)
     g_L =   0.0084170
     m_NA0 =   0.015385
     h_NA0 =   0.86517
6    m_K0 =   0.18100
     E_K = -75.607
     E_NA =   57.760
% we then edited simpleHH.m to add this g_L and our E_L
     fname = 'simpleHH';
11   iename = 'IE';
     tinit = 0;
     tfinal = 30;
     y0 = [V_R,m_NA0,h_NA0,m_K0];
     y0
16   y0 =

      -70.000000     0.015385     0.865168     0.181001
     [t_vals,g_NA,g_K,V,m_NA,h_NA,n_K] = ...
        SolveSimpleHHOrig(fname,iename,tinit,tfinal,...
21                      y0,0.01,4,g_NA_bar,g_K_bar);
```

This result uses the applied current given by the code **IE.m**. This pulse applies a constant current of 100 namps between on the time interval [10.0, 10.62]. On the interval [9.95, 10.0], the current ramps up from 0 to 100 and on the interval [10.62, 10.67], the current moves smoothly from 100 back to 0. Otherwise, it is zero. Hence, this is a continuous current pulse. In Fig. 12.5, we see the membrane voltage versus time curve that results from the application of this injected current. Note we see a nice action potential here. We show the corresponding sodium and potassium conductances during this pulse in Fig. 12.6. In Fig. 12.7 we show the activation and inactivation variables for sodium and in Fig. 12.8, we show the activation variable for potassium.

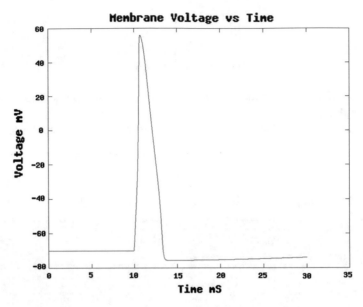

Fig. 12.5 Membrane voltage versus time: $g_L = 0.0084170$, $E_L = -50$, $g_{NA}^{max} = 120$ and $g_K^{max} = 36.0$

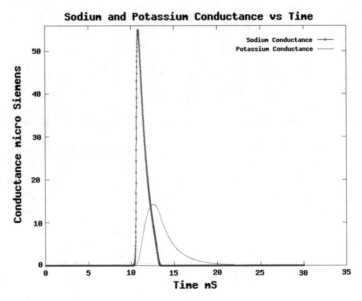

Fig. 12.6 Sodium and potassium conductances versus time for $g_{NA}^{max} = 120$ and $g_K^{max} = 36$ for impulse current

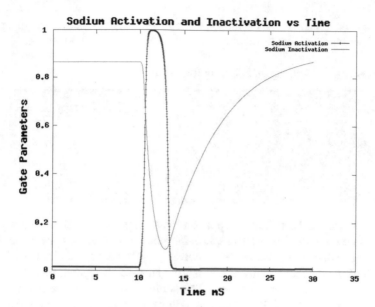

Fig. 12.7 Sodium activation and inactivation versus time

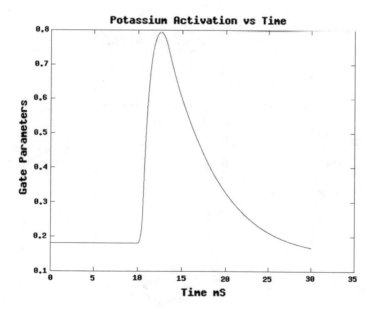

Fig. 12.8 Potassium activation versus time

It is also instructive to look at the ion currents and conductances during the pulses. Simply compute and plot the currents with the following MatLab commands

Listing 12.19: Calculating and plotting ion currents

```
  E_NA = 58.0840;
  E_K = -76.0304;
  I_NA = g_NA_bar*g_NA.*(V-E_NA);
  I_K = g_K_bar*g_K.*(V- E_K);
5 plot(t_vals ,I_NA ,tvals ,I_K);
  xlabel('Time ms');
  ylabel('Current na');
  title('Sodium and Potassium Currents vs Time');
  legend('Sodium','Potassium');
```

In Fig. 12.9, we see the sodium and potassium currents plotted simultaneously. Note that the sodium current is positive while the potassium current is negative. If you look closely, you can see the potassium current lags behind the sodium one. Finally, we can look at how the sodium and potassium gates action depends on voltage. They are voltage dependent gates after all! The sodium conductance versus membrane voltage is shown in Fig. 12.10 and the potassium conductance voltage curve is shown in Fig. 12.11.

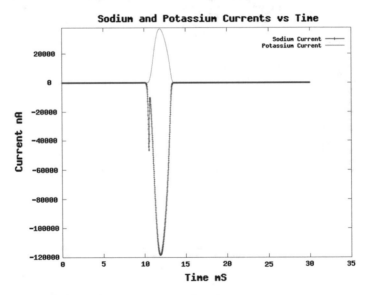

Fig. 12.9 Sodium and potassium currents versus time

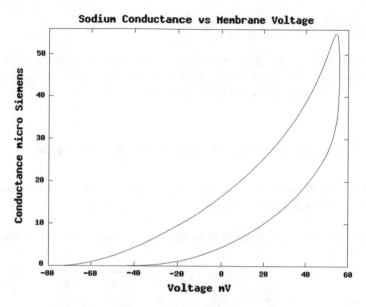

Fig. 12.10 Sodium conductance versus voltage

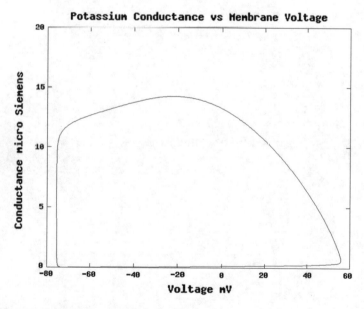

Fig. 12.11 Potassium conductance versus voltage

12.7   Printing Additional Variables

In order to plot the τ_{mNA} and m_{NA}^{∞} values and so forth for our simulation, we have to rewrite some code. We need to return all of the time constants and asymptotic values so that we can generate plots. First, we need to modify the dynamics function.

12.7.1   The New Dynamics Function: simpleHHnew.m

We add additional variables to the dynamics so we can return information on the time constants and steady state values of the α–β gate functions.

Listing 12.20: Adding new variables to the dynamics to return more information: simpleHHnew

```
1 function [f,z] = simpleHHnew(t,y,IE)
  %      Standard Hodgkin - Huxley Model
  %      voltage         mV
  %      current         na
  %      time            ms
6 %      concentration  mM
  %      conductance     micro Siemens
  %      capacitance     nF
  %
  %      y vector assignments
11 %      ================================
  %      y(1) = V
  %      y(2) = m_NA
  %      y(3) = h_NA
  %      y(4) = m_K
16 %      auxiliary variables
  %      z(1) =   t_m_NA
  %      z(2) =   m_NA_infinity
  %      z(3) =   t_h_NA
  %      z(4) =   h_NA_infinity
21 %      z(5) =   t_m_K
  %      z(6) =   m_K_infinity
  %
  % set size of f
  %
26
  f = zeros(1,4);

  %      ================================
  %      f vector assignments
31 %      ================================
  %      y(1) = V dynamics
  %      y(2) = m_NA dynamics
  %      y(3) = h_NA dynamics
  %      y(4) = m_K dynamics
36 %
  %      ================================
  %      Constants for Equilibrium Voltage Calculations
  %      ================================
  [E_K,E_NA,T] = IonBatteryVoltages(400.0,20.11,50.0,491.0,69);
41
  % max conductance for NA
    g_NA_bar = 120.0;
  % max conductance for K
    g_K_bar = 36.0;
```

```
46 %
       alpha = GatesNAK('HHGates',y(1));
   %
   % activation/inactivation  parameters  for  NA
   % alpha(1),  alpha(2)  alpha  and  beta  for  mna
51 % alpha(3),  alpha(4)  alpha  and  beta  for  hna
   % alpha(5),  alpha(6)  alpha  and  beta  for  mk
   %
   %    Fast  Sodium  Current
   %
56 %
       m_NA_infinity    = alpha(1)/(alpha(1)+alpha(2));
       t_m_NA           = 1.0/(alpha(1)+alpha(2));
       f(2)             = (m_NA_infinity - y(2))/t_m_NA;
   %
61     h_NA_infinity    = alpha(3)/(alpha(3)+alpha(4));
       t_h_NA           = 1.0/(alpha(3)+alpha(4));
       f(3)             = (h_NA_infinity - y(3))/t_h_NA;
       I_NA             = g_NA_bar*(y(1)-E_NA)*y(2)^3*y(3);
       z(1)             = m_NA_infinity;
66     z(2)             = t_m_NA;
       f(2)             = (m_NA_infinity   y(2))/t_m_NA;
       z(3)             = h_NA_infinity;
       z(4)             = t_h_NA;
       f(3)             = (h_NA_infinity - y(3))/t_h_NA;
71 %
   %    Potassium  Current
   %
   %
       m_K_infinity     = alpha(5)/(alpha(5)+alpha(6));
76     t_m_K            = 1.0/(alpha(5)+alpha(6));
       z(5)             = m_K_infinity;
       z(6)             = t_m_K;
       f(4)             = (m_K_infinity - y(4))/t_m_K;
       I_K              = g_K_bar*(y(1)-E_K)*y(4)^4;
81 %
   % leakage  current:  run  rest.m  to  find  appropriate  g_L  value
   %
       g_leak = .0084170;
       E_leak = -50.0;
86 %
   % I_L  current
       I_leak = g_leak*(y(1)-E_leak);

   %
91 % Cell  Capacitance
   %
       C_M = 1.0;

   % external  current
96     s = feval(IE,t);
   %
       f(1) = (s - I_NA - I_K - I_leak)/C_M;
```

12.7.2   The New Manager: SolveSimpleHHnew.m

We alter the **SolveSimpleHH.m** function to **SolveSimpleHHnew.m** as follows:

Listing 12.21: A new HH dynamics function: SolveSimpleHHnew

```
     function [tvals,g_NA,g_K,V,m_NA,h_NA,n_K,...
              t_m_NA,m_NA_infinity,...
              t_h_NA,h_NA_infinity,...
              t_m_K,m_K_infinity] = ...
5            SolveSimpleHHnew(fname,iename,tinit,tfinal,y0,h,k,g_NAmax,
                 g_Kmax)
     % find number of interations to do
     n = ceil((tfinal-tinit)/h)+1;
     %
     m_NA          = zeros(1,n);
10   h_NA          = zeros(1,n);
     n_K           = zeros(1,n);
     g_NA          = zeros(1,n);
     g_K           = zeros(1,n);
     V             = zeros(1,n);
15   t_m_NA        = zeros(1,n);
     m_NA_infinity = zeros(1,n);
     t_h_NA        = zeros(1,n);
     h_NA_infinity = zeros(1,n);
     t_m_K         = zeros(1,n);
20   m_K_infinity  = zeros(1,n);

     [tvals,yvals,zvals] = HHFixedRKnew(fname,iename,tinit,y0,h,k,n);

     %
25   % store values in physical variables
     %
     V             = yvals(1,1:n);
     m_NA          = yvals(2,1:n);
     h_NA          = yvals(3,1:n);
30   n_K           = yvals(4,1:n);
     t_m_NA        = zvals(1,1:n);
     m_NA_infinity = zvals(2,1:n);
     t_h_NA        = zvals(3,1:n);
     h_NA_infinity = zvals(4,1:n);
35   t_m_K         = zvals(5,1:n);
     m_K_infinity  = zvals(6,1:n);

     %
     %  Compute g_Na and g_K
40   %
     for i = 1:n
         u        = m_NA(i)*m_NA(i)*m_NA(i)*h_NA(i);
         g_NA(i) = g_NAmax*u;
     end
45
     for i = 1:n
         u        = n_K(i)*n_K(i)*n_K(i)*n_K(i);
         g_K(i)  = g_Kmax * u;
     end
```

Note we are now returning all of the time constants and asymptotic values for each time step. To do this, we have to alter **HHFixedRK.m** to the new code **HHFixedRKnew.m**.

12.7.3 The New Runge–Kutta Manager: HHFixedRKnew.m

This code is quite similar to the original as you can see.

Listing 12.22: A new Runge–Kutta manager: HHFixedRKnew

```
   function [tvals,yvals,zvals] = HHFixedRKnew(fname,iename,t0,y0,h,k,n)
   %
   %          Gives approximate solution to
   %             y'(t) = f(t,y(t))
 5 %             y(t0) = y0
   %          using a kth order RK method
   %
   % t0       initial time
   % y0       initial state
10 % h        stepsize
   % k        RK order  1<= k <= 4
   % n        Number of steps to take
   %
   % tvals    time values of form
15 %             tvals(j) = t0 + (j-1)*h,  1 <= j <= n
   % yvals    approximate solution
   %             yvals(:j) = approximate solution at
   %             tvals(j),  1 <= j <= n
   % zvals    auxiliary dynamics variables
20 %             zvals(:j) = auxiliary variables at
   %             tvals(j),  1 <= j <= n
   % fname    name of dynamics function
   % iename   name of current injection function
   %
25 tc = t0;
   yc = y0;
   tvals = tc;
   yvals = zeros(4,n);
   zvals = zeros(6,n);
30 yvals(1:4,1) = transpose(yc);
   [fc,zc] = feval(fname,tc,yc,iename);
   zvals(1:6,1) = transpose(zc);
   for j=1:n-1
     [tc,yc,fc] = HHRKstep(fname,iename,tc,yc,fc,h,k);
35   yvals(1:4,j+1) = transpose(yc);
     [fc,zc] = feval(fname,tc,yc,iename);
     zvals(1:6,j+1) = transpose(zc);
     tvals = [tvals tc];
   end
```

We leave the file **HHRKstep.m** alone as we don't need to change that.

12.7.4 Runtime Results

To generate the plots, we set up a MatLab session like this:

Listing 12.23: A sample session

```
   E_L = -50;
   V_R = -70;
   Kin = 400.0;
   Kout = 20.11;
 5 Nain = 50.0;
   Naout = 491.0;
   TF = 69.0;
   p = [Kin;Kout;Nain;Naout;TF];
   g_NA_bar = 120.0;
10 g_K_bar = 36.0;
   F = 'HHGates';
   [g_L ,m_NA0,h_NA0,m_K0] = ...
        rest(E_L,V_R,g_NA_bar,g_K_bar,p,F);
   g_L
15 g_L            =    0.0084
   [tvals ,g_NA,g_K,V,mNA,h_NA,n_K,...
               t_m_NA,m_NA_infinity ,...
               t_h_NA,h_NA_infinity ,...
               t_m_K,m_K_infinity] = ...
20 SolveSimpleHHnew('simpleHHnew','IE',0,30,...
       [V_R,m_NA0,h_NA0,m_K0] ,...
       .01,4,g_K_bar,g_NA_bar);
       plot(tvals ,t_m_NA);
```

The values of τ_m^{NA} and m_{NA}^{∞} versus time are plotted simultaneously in Fig. 12.12.

We can also plot τ_m^{NA} and m_{NA}^{∞} versus membrane voltage to see how they vary with depolarization (Fig. 12.13).

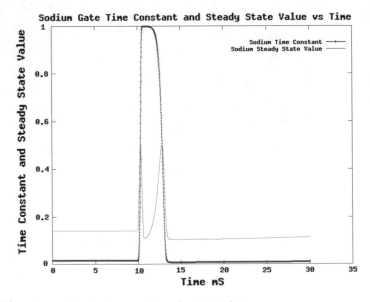

Fig. 12.12 m time constant and asymptotic value versus time

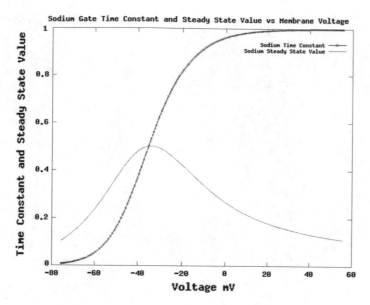

Fig. 12.13 m time constant and asymptotic value versus membrane voltage

12.8 Initiation of the Action Potential

Let's explore how much depolarization of the membrane at the axon hillock is needed to initiate an action potential. For this experiment, we set the size of the pulse generated by **IE.m** to be zero and we solve the Hodgkin–Huxley equations for a series of initial voltages that start at V_R and slowly increase. In this Chapter, we have developed a simple model of how an action potential is generated when axon hillock of the excitable nerve cell is depolarized. This is not much of a model, of course. Right now, the depolarization is accomplished by setting the initial voltage to be $V_R + \Delta$ for some depolarization Δ or by injected a current I_E into the membrane. We don't have a real model of an excitable nerve cell that consists of a dendritic cable, a soma and an axon. That is the topic we address in Chap. 11 on neural processing.

Listing 12.24: Sample session to generate action potentials

```
    E_L = -50;
    V_R = -70;
    Kin = 400.0;
    Kout = 20.11;
 5  Nain = 50.0;
    Naout = 491.0;
    TF = 69.0;
    p = [Kin; Kout; Nain; Naout; TF];
    F = 'HHGates';
10  g_NA_bar = 120.0;
    g_K_bar = 36.0;
```

```
    [g_L,m_NA_infinity,h_NA_infinity,m_K_infinity,E_K,E_NA] =...
        rest(E_L,V_R,g_NA_bar,g_K_bar,p,f);
    % now set g_L in simpleHH.m
15  % solve HH model for various axon hillock voltages
    % axon hillock = V_R = -70
    [tvals,g_NA0,g_K0,V0,m_NA0,h_NA0,n_K0] = ...
        SolveSimpleHHOrig('simpleHH','IE',0,100.0,...
        [V_R,m_NA_infinity,h_NA_infinity,m_K_infinity],...
20      0.01,4,g_NA_bar,g_K_bar);
    % axon hillock = V_R = -69
    [tvals,g_NA1,g_K1,V1,m_NA1,h_NA1,n_K1] = ...
        SolveSimpleHHOrig('simpleHH','IE',0,100.0,...
        [V_R+1,m_NA_infinity,h_NA_infinity,m_K_infinity],...
25      0.01,4,g_NA_bar,g_K_bar);
    % axon hillock = V_R = -68
    [tvals,g_NA2,g_K2,V2,m_NA2,h_NA2,n_K2] = ...
        SolveSimpleHHOrig('simpleHH','IE',0,100.0,...
        [V_R+2,m_NA_infinity,h_NA_infinity,m_K_infinity],...
30      0.01,4,g_NA_bar,g_K_bar);
    % axon hillock = V_R = -67
    ...
    % axon hillock = V_R = -61
    [tvals,g_NA9,g_K9,V9,m_NA9,h_NA9,n_K9] = ...
35      SolveSimpleHHOrig('simpleHH','IE',0,100.0,...
        [V_R+9,m_NA_infinity,h_NA_infinity,m_K_infinity],...
        0.01,4,g_NA_bar,g_K_bar);
    % axon hillock = V_R = -60
    [tvals,g_NA10,g_K10,V10,m_NA10,h_NA10,n_K10] = ...
40      SolveSimpleHHOrig('simpleHH','IE',0,100.0,...
        [V_R+10,m_NA_infinity,h_NA_infinity,m_K_infinity],...
        0.01,4,g_NA_bar,g_K_bar);
    % plot action potentials
    plot(tvals,V0,tvals,V4,...
45          tvals,V6,tvals,V8,...
            tvals,V9,tvals,V10);
    % set legend, labels and so forth
    % the legend is one line
    legend('-70 mV','-66 mV','-64 mV',
50          '-62 mV','-61 mV','-60 mV');
    xlabel('Time mS');
    ylabel('Voltage mV');
    title('Action Potential versus Time');
```

Fig. 12.14 Action potentials generated by a series of depolarizations starting from none to 10 mV

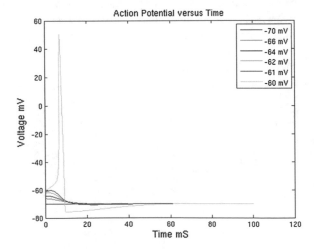

Fig. 12.15 Action potentials generated by a series of depolarizations starting from none to 11 mV

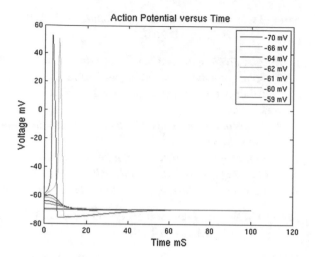

Now, in Fig. 12.14, you can see that the action potential is not initiated until the depolarization is about 10 mV.

If we depolarize one more degree, we obtain Fig. 12.15. This shows clearly that the additional degree causes the action potential to initiate much sooner.

12.9 Homework

Exercise 12.9.1 *Run the Hodgkin–Huxley simulation as has been done in the work above and write a complete report in your favorite editor which documents how an action potential is generated using the plots and evidence obtained in the following studies.*

For all of these studies, we will use the current impulse defined by the code in **IE.m** *where you define the pulse appropriately.*

Study One: *maximum sodium conductance is* 120 *and maximum potassium conductance is* 36.0:

1. *IE is as given in the text; a pulse of size* 100.
2. *The plots of voltage, m_NA, h_NA and m_K versus time for* 100 *mSec.*
3. *The plots of the time constant and the asymptotic value for m_NA, h_NA and m_K versus time.*
4. *The plots of the time constant and the asymptotic value for m_NA, h_NA and m_K versus voltage.*
5. *The plots of the sodium and potassium conductance versus time.*
6. *The plots of the sodium and potassium conductance versus voltage.*
7. *The plots of sodium and potassium current versus time.*

Study Two: *maximum sodium conductance is* 120 *and maximum potassium conductance is* 36.0:

1. *IE is as given in the text but change the pulse size to* 400.
2. *The plots of voltage, m_NA, h_NA and m_K versus time for* 100 mSec.
3. *The plots of the time constant and the asymptotic value for m_NA, h_NA and m_K versus time.*
4. *The plots of the time constant and the asymptotic value for m_NA, h_NA and m_K versus voltage.*
5. *The plots of the sodium and potassium conductance versus time.*
6. *The plots of the sodium and potassium conductance versus voltage.*
7. *The plots of sodium and potassium current versus time.*

Study Three: *maximum sodium conductance is* 120 *and maximum potassium conductance is* 16.0:

1. *IE is as given in the text but change the pulse size to* 800.
2. *The plots of voltage, m_NA, h_NA and m_K versus time for* 100 mSec.
3. *The plots of the time constant and the asymptotic value for m_NA, h_NA and m_K versus time.*
4. *The plots of the time constant and the asymptotic value for m_NA, h_NA and m_K versus voltage.*
5. *The plots of the sodium and potassium conductance versus time.*
6. *The plots of the sodium and potassium conductance versus voltage.*
7. *The plots of sodium and potassium current versus time.*

Study Four: *maximum sodium conductance is* 20.0 *and maximum potassium conductance is* 3.6:

1. *IE is as given in the text but change the pulse size to* 400.
2. *The plots of voltage, m_NA, h_NA and m_K versus time for* 100 mSec.
3. *The plots of the time constant and the asymptotic value for m_NA, h_NA and m_K versus time overlaid on the plots from Study One and Study Two.*
4. *The plots of the time constant and the asymptotic value for m_NA, h_NA and m_K versus voltage overlaid on the plots from Study One and Study Two.*
5. *The plots of the sodium and potassium conductance versus time overlaid on the plots from Study One and Study Two.*
6. *The plots of the sodium and potassium conductance versus voltage overlaid on the plots from Study One and Study Two.*
7. *The plots of sodium and potassium current versus time overlaid on the plots from Study One and Study Two.*

12.10   Another MatLab Implementation

We are now ready to integrate the ball stick model and our Hodgkin–Huxley model in Matlab. We modify the existing MatLab code to reflect the fact that we are now going to allow our dendritic model to provide the axon hillock voltage. We will need to rewrite many of our original Hodgkin–Huxley functions in order to take advantage of the ball-stick model for dendritic inputs. We also will now move to a more efficient ODE solver: RKF5 which has automatic step size adjustment algorithms.

Recall, we introduced the general mapping F to encode our gate dynamics more efficiently as defined by

$$F(V_m, p, q) = \frac{p_0(V_m + q_0) + p_1}{e^{p_2(V_m + q_1)} + p_3}.$$

under some choices of the parameter vectors p and q, these functions may be undefined. However, if the numerator and denominator are both 0, this occurs when the numerator $N = p_0(V_m + q_0) + p_1 = 0$ and the denominator $D = e^{p_2(V_m + q_1)} + p_3 = 0$. This gives two equations in the one unknown V_m. If these equations are consistent call this critical voltage value V_m^*.

To understand what to do here, we need another idea from the calculus of one variable called L'Hopital's rule. We only need the easy form but you can read up about other variations and we encourage you to do so. Let's assume we have two functions $f(x)$ and $g(x)$ both of which are differentiable and both of which vanish at the value x^*. If we are interested in knowing the value of $\lim_{x \to x^*} \frac{f(x)}{g(x)}$ how can we find that out? The naive thing to do would be to plug x^* into the fraction getting $\frac{f(x^*)}{g(x^*)} = \frac{0}{0}$ which is undefined. But we can figure this out by using the definition of differentiability with error terms for both f and g. We have

$$\lim_{x \to x^*} \frac{f(x)}{g(x)} = \lim_{x \to x^*} \frac{f(x^*) + f'(x^*)(x - x^*) + E_f(x - x^*)}{g(x^*) + g'(x^*)(x - x^*) + E_g(x - x^*)}$$

where we assume we are in the simplest case where both $f'(x^*)$ and $g'(x^*)$ are not zero. We can still figure out what to do if one or both of them is zero, but we don't need that complication here. We know $f(x^*)$ and $g(x^*)$ are zero, so we have

$$\lim_{x \to x^*} \frac{f(x)}{g(x)} = \lim_{x \to x^*} \frac{f'(x^*)(x - x^*) + E_f(x - x^*)}{g'(x^*)(x - x^*) + E_g(x - x^*)}$$

Now divide top and bottom by $x - x^*$ to get

$$\lim_{x \to x^*} \frac{f(x)}{g(x)} = \lim_{x \to x^*} \frac{f'(x^*) + \frac{E_f(x - x^*)}{(x - x^*)}}{g'(x^*) + \frac{E_g(x - x^*)}{(x - x^*)}}$$

Since f and g are differentiable, we know $\frac{E_f(x - x^*)}{(x - x^*)} \to 0$ and $\frac{E_f(x - x^*)}{(x - x^*)} \to 0$ as $x \to x^*$. Hence, we have

$$\lim_{x \to x^*} \frac{f(x)}{g(x)} = \frac{f'(x^*)}{g'(x^*)}$$

We can state this as a general theorem.

Theorem 12.10.1 (A Simple L'Hopital's Rule)
If $f(x)$ and $g(x)$ are both differentiable in a circle defined about x^* and if $f(x^*) = g(x^*) = 0$ with $f'(x^*)$ and $g'(x^*)$ both not zero, then

$$\lim_{x \to x^*} \frac{f(x)}{g(x)} = \frac{f'(x^*)}{g'(x^*)}$$

Proof We worked out the argument in the lines above!   ∎

Then, by L'Hopital's rule, the limiting value at V_m^* is

$$\lim_{V_m \to V_m^*} F(V_m, p, q) = \frac{p_0}{p_2 e^{p_2(V_m + q_1)}} = \frac{-p_0}{p_2 p_3}$$

as $e^{p_2(V_m^* + q_1)} = -p_3$. Since the axon hillock voltage is a voltage spike, the incoming voltages may cross this critical voltage value V_m^*. Hence, we will rewrite the α and β computations to be more careful to avoid division by zero. Such a division by zero would introduce a discontinuity into our numerical solution with potential loss of accuracy. Here is a more careful version of our previous **HHGates** code.

Listing 12.25: A more careful gates function: HHGates

```
   function  F = HHGates(V,p,q)
   %
   % V = incoming  voltage
   % p = parameters  for  gate  model
 5 % q = parameters  for  gate  model
   % gate  value
   %
   tol = 1.0e-6;
   Numerator    = p(1)*(V + q(1)) + p(2);
10 Denominator = exp( p(3)*(V + q(2))) + p(4);
   if  abs(Denominator) < tol
      if  abs(Numerator) < tol
         F = -p(1)/(p(3)*p(4));
      elseif  sign(Denominator) == 0 && Numerator >= tol
15       F = 100;
      elseif  sign(Denominator) == 0 && Numerator <= -tol
         F = -100;
      elseif  sign(Denominator) == 1 && Numerator >= tol
         F = 100;
20    elseif  sign(Denominator) == 1 && Numerator <= -tol
         F = -100;
      elseif  sign(Denominator) == -1 && Numerator >= tol
         F = -100;
      elseif  sign(Denominator) == -1 && Numerator <= -tol
25       F = 100;
      end
   else
      F = Numerator/Denominator;
   end
```

In this code, we handle the $\frac{0}{0}$ case carefully. If the numerator and denominator of F are both in the interval $(-\epsilon, \epsilon)$ for ϵ set equal to the tolerance **tol = 1.0e-6**, we use the L'Hopital's limiting value. Otherwise, depending on the signs of the numerator and denominator values, we set the $\frac{\pm 1}{0}$ case to be 100 or -100. Finally, if the numerator and denominator are both reasonable numbers, we simply calculate the usual ratio. An incoming axon hillock voltage spike could easily be very large and so we don't want any of the α and β calculations to be undefined at some step in the numerical process. We can then define α and β functions easily as follows.

Listing 12.26: Handling indeterminacies in the α/β calculations: HHGatesGraph

```
function F = HHGatesGraph(p,q)
%
% V = incoming voltage
% p = parameters for gate model
5 % q = parameters for gate model
%
% gate value

  F = @(V) HHGates(V,p,q);
10
  end
```

This function returns the α or β function as $F(V)$ using Matlab/Octave's ability to define a function inline. Once this is done, we can easily write a function to graph all the α and β functions for a typical simulation. This is done in the function **GatesNAKGraph** which returns a set of handles to the plots, **AlphaBetaPlot**.

Listing 12.27: Graphing the α/β functions: GatesNAKGraph

```
function AlphaBetaPlot = GatesNAKGraph()
%
% We graph the (alpha,beta) pairs in HH model
%
5 % This is a simple Sodium, Potassium model so
%   alpha, beta for mNA = F{1}, F{2}
%   alpha, beta for hNA = F{3}, F{24
%   alpha, beta for mK  = F{5}, F{6}
%
10 % we need p, q parameter vectors for each of the
%  six alpha, beta pairs.  These values come from
%  the literature.
%
  p = zeros(6,4);
15 q = zeros(6,2);
  p = [-0.1,0.0,-0.1,-1.0;...
       0.0,4.0,0.0556,0.0;...
       0.0,0.07,0.05,0.0;...
       0.0,1.0,-0.1,1.0;...
20      -0.01,0.0,-0.1,-1.0;...
       0.0,0.125,0.0125,0.0];
%
  q = [35.0,35.0;...
       60.0,60.0;...
25      60.0,60.0;...
       30.0,30.0;...
       50.0,50.0;...
       60.0,60.0];
%
```

```
30  F= {};
    message = {}'
    F{1} = HHGatesGraph(p(1,:),q(1,:));
    message{1} = 'alphaMNA';
    F{2} = HHGatesGraph(p(2,:),q(2,:));
35  message{2} = 'betaMNA';
    F{3} = HHGatesGraph(p(3,:),q(3,:));
    message{3} = 'alphaHNA';
    F{4} = HHGatesGraph(p(4,:),q(4,:));
    message{4} = 'betaHNA';
40  F{5} = HHGatesGraph(p(5,:),q(5,:));  .
    message{5} = 'alphaMK';
    F{6} = HHGatesGraph(p(6,:),q(6,:));
    message{6} = 'betaMK';

45  V = linspace(-120,120,241);
    Y = {};
    S = {};
    for  i=1:6
      %filename=[message{i},'.png'];
50    % uncomment if you want to print results to files
      figure;
      for  j = 1:241
        Y{i}(j) = F{i}(V(j));
      end
55    plot(V,Y{i});
      AlphaBetaPlot{i} = gcf();
      xlabel('Voltage mV');
      ylabel('Gate Response');
      title(message{i});
60    %print('-dpng',filename);
      %uncomment if you want to print results to files
    end
    %
    end
```

To make it easy to clean up these figures, we use the script **CleanUpFigures AlphaBeta.m**. In the previous code, the ith plot was assigned the handle **AlphaBetaPlot{i}** and so a simple **delete** call will remove the plot. This is useful when there are a lot of plots to remove.

Listing 12.28: Cleanup figures

```
    N = length(AlphaBetaPlot);
    for  i=1:N
      delete(AlphaBetaPlot{i});
    end
```

To graph and then remove all these plots, a typical session would be as follows:

Listing 12.29: Removing plots

```
1  AlphaBetaPlot = GatesNAKGraph();
   CleanUpFiguresAlphaBeta;
```

The plots for our typical Hodgkin–Huxley model are shown in Figs. 12.16a, b, 12.17a, b and 12.18a, b for a voltage range of -200 to 200 mV. In the simulations to come, we will be using these improved α–β calculations.

Fig. 12.16 $\mathscr{M}$ α and β for sodium. **a** $\alpha_{\mathscr{M}_{NA}}$. **b** $\beta_{\mathscr{M}_{NA}}$

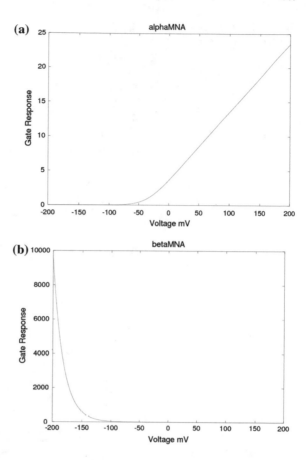

12.10.1 *The Rewritten Dynamics*

We now rewrite our previous HH dynamics function to accept a parameter vector. This makes the code cleaner and easier to follow. This code uses the function **IE2** to provide an external current input which can be shaped as we see fit. For example, a two pulse external current might be **IE2.m** given below. This has two impulses: pulse one is on maximum intensity 100 and lasts for 0.05 mS and pulse two is of size 200 and lasts for 0.05 mS as well. Since we are now using RKF5 as our model integrator, the errors due to a rapid discontinuous rise from 0 to the maximum value in the pulses cause problems in the optimal step size calculations. Hence, we now introduce a smooth linear rise from 0 to the maximum and then down again into the functions. This creates continuous pulses although they are not differentiable at two points. This allows for a more accurate error test. It would be even better to

Fig. 12.17 $\mathcal{H}$ α and β for sodium. **a** $\alpha_{\mathcal{H}_{NA}}$. **b** $\beta_{\mathcal{H}_{NA}}$

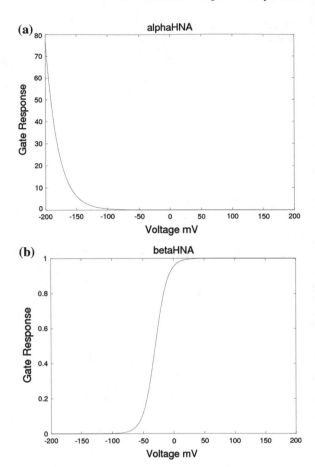

approximate the pulse with a six times differentiable function, but this suffices for now.

Listing 12.30: Continuous Pulses: IE2.m

```
function ie = IE2(t)
%
ts1 = 10.0;
ts2 = 10.05;
te1 = 10.10;
te2 = 10.15;
ts3 = 20.0;
ts4 = 20.05;
te3 = 20.10;
te4 = 20.15;
Imax1 = 100.0;
Imax2 = 300.0;
ie = 0;
if t >= ts1 & t <= ts2
```

```
15    ie = (Imax1/(ts2-ts1))*(t - ts1);
    elseif t>ts2 & t < te1
       ie = Imax1;
    elseif t >= te1 & t < te2
       ie = Imax1 -(Imax1/(te2-te1))*(t-te1);
20  elseif t >= ts3 & t <= ts4
       ie = (Imax2/(ts4-ts3))*(t - ts3);
    elseif t>ts3 & t < te3
       ie = Imax2;
    elseif t >= te3 & t < te4
25     ie = Imax2 -(Imax2/(te4-te3))*(t-te3);
    end
```

The new dynamics function is given in Listing 12.32. We are handling the parameter **p** differently here also. We now want to send in numerical data and string data. We do this as follows. We create a data structure **p** having two parts, **p.a** for the numerical data, and **p.b** for the string data. We can then access the elements of **p.a**

Fig. 12.18 $\mathscr{M}$ α and β for potassium. **a** $\alpha_{\mathscr{M}_K}$. **b** $\beta_{\mathscr{M}_K}$

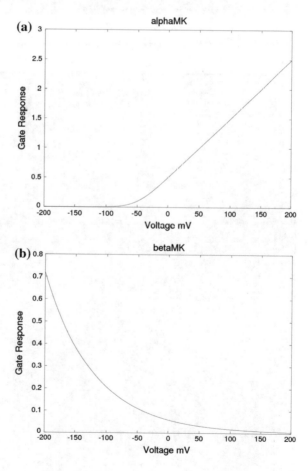

as usual, **p.a(1)**, **p.a(2)** and so forth. The syntax **p.b** will give us the string name we need.

Listing 12.31: p structure

```
p.a = [E_K;E_NA;E_L;g_L;g_NA_bar;g_K_bar];
p.b = 'HHGates';
p.a
ans =

5
      -7.5607e+01
       5.7760e+01
      -5.0000e+01
       8.4170e-03
10     1.2000e+02
       3.6000e+01


p.a(1)
ans = -75.607
```

Listing 12.32: A rewritten HH dynamic function using parameter arguments: HHdynamics.m

```
 1 function f = HHdynamics(p,t,y)
   %    Standard Hodgkin - Huxley Model
   %    voltage          mV
   %    current          na
   %    time             ms
 6 %    concentration mM
   %    conductance     micro Siemens
   %    capacitance     nF
   %
   %    arguments
11 %    p   parameter vector
   %    t   time
   %    y   state vector
   %
   %  ========================================
16 %    y vector assignments
   %  ========================================
   %    y(1) = V
   %    y(2) = m_NA
   %    y(3) = h_NA
21 %    y(4) = m_K
   %
   %    parameter assignments
   %    p comes in as a data structure.
   %
26 %    p.a is the numeric data
   %    p.a(1) = EK, the potassium battery voltage
   %    p.a(2) = ENA is the sodium battery voltage
   %    p.a(3) = EL, the leakage voltage
   %    p.a(4) = GL, the leakage conductance
31 %    p.a(5) = gNAmax, maximum sodium conductance
   %    p.a(6) = gKmax, maximum potassium conductance
   %
   %    p.b is the string data
   %    To use
36 %    p.a = [EK;ENA;EL;GL;gNAmax;gKMax];
   %    p.b = 'string';
   %

   EK          = p.a(1);
```

```
41 ENA        = p.a(2);
   % leakage current: run rest.m to find appropriate g_L value
   % default E_leak = -50.0
   E_leak     = p.a(3);
   g_leak     = p.a(4);
46 % max conductance for NA: default is 120
   g_NA_bar   = p.a(5);
   % max conductance for K: default is 36.0
   g_K_bar    = p.a(6);
   gatename   = p.b;
51
   f = zeros(4,1);

   %   =========================================
   %     f vector assignments
56 %   =========================================
   %     y(1) = V dynamics
   %     y(2) = m_NA dynamics
   %     y(3) = h_NA dynamics
   %     y(4) = m_K dynamics
61 %
   alpha = GatesNAK(gatename,y(1));
   %
   % activation/inactivation parameters for NA
   % alpha(1), alpha(2) alpha and beta for mna
66 % alpha(3), alpha(4) alpha and beta for hna
   % alpha(5), alpha(6) alpha and beta for mk
   %
     m_NA_infinity   = alpha(1)/(alpha(1)+alpha(2));
     t_m_NA          = 1.0/(alpha(1)+alpha(2));
71   f(2)            = (m_NA_infinity - y(2))/t_m_NA;
   %
     h_NA_infinity   = alpha(3)/(alpha(3)+alpha(4));
     t_h_NA          = 1.0/(alpha(3)+alpha(4));
     f(3)            = (h_NA_infinity - y(3))/t_h_NA;
76   I_NA            = g_NA_bar*(y(1)-ENA)*y(2)^3*y(3);
   %
     m_K_infinity    = alpha(5)/(alpha(5)+alpha(6));
     t_m_K           = 1.0/(alpha(5)+alpha(6));
     f(4)            = (m_K_infinity - y(4))/t_m_K;
81   I_K             = g_K_bar*(y(1)-EK)*y(4)^4;
   % I_L current
     I_leak = g_leak*(y(1)-E_leak);
   % Cell Capacitance
     C_M = 1.0;
86 % full dynamics
     s1 = IE2(t);
     f(1) = (s1 - I_NA - I_K - I_leak)/C_M;
   end
```

12.10.2   A New SolveSimpleHH.m

The code to manage the solution of the ODE is then given below in Listing 12.33.

Listing 12.33: Managing The Hodgkin–Huxley Simulation

```
 1 function  [tvals ,yvals ,fvals ,hvals ,g_NA,g_K,V,m_NA,h_NA,n_K] = ...
             SolveSimpleHH(fname ,q,p,t0 ,tf ,yinit ,hinit )

   %
   %      fname   dynamics function
 6 %      q parameter vector for RKF5 settings
   %      p parameter vector for HH dynamics
   %      t0      initial time
   %      tf      final time
   %      yinit   initial state
11 %      hinit   initial stepsize
   %
   %      We use RKF5
   %
   %      p parameter assignments
16 %
   %      p.a(1) = EK, the potassium battery voltage
   %      p.a(2) = ENA is the sodium battery voltage
   %      p.a(3) = EL, the leakage voltage
   %      p.a(4) = GL, the leakage conductance
21 %      p.a(5) = g_NA_bar , maximum sodium conductance
   %      p.a(6) = g_K_bar , maximum potassium conductance
   %
   %      To use
   %      p.a = [EK;ENA;EL;GL;g_NA_bar;g_K_bar];
26 %      p.b = 'name of the gate dynamics function ';
   %
   %      q parameter assignments
   %
   %      q(1) = errortol , step size tolerance
31 %      q(2) = steptol , step size tolerance
   %      q(3) = minstep , minimum allowable step size
   %      q(4) = maxstep , maximum allowable step size
   %
   %      To use
36 %      q = [errortol; steptol; minstep; maxstep]
   %
   %      initial y values
   %
   %      y(1) = V
41 %      y(2) = m_NA
   %      y(3) = h_NA
   %      y(4) = m_K
   %
   %      To use
46 %      yinit = [V0;m_NA0;h_NA0;m_K0]
   %
   EK       = p.a(1);
   ENA      = p.a(2);
   EL       = p.a(3);
51 GL       = p.a(4);
   g_NA_bar = p.a(5);
   g_K_bar  = p.a(6);
   %
   errortol = q(1);
56 steptol  = q(2);
   minstep  = q(3);
   maxstep  = q(4);

   [tvals ,yvals ,fvals ,hvals ] = RKF5(errortol ,steptol ,minstep ,maxstep ,
       fname ,p, hinit ,t0 ,tf ,yinit );
61 %
   % store values in physical variables
   %
   V     = yvals (1,:);
   m_NA  = yvals (2,:);
66 h_NA  = yvals (3,:);
   n_K   = yvals (4,:);

   %
   %  Compute g_Na and G_K
71 %
   g_NA = g_NA_bar*m_NA.*m_NA.*m_NA.*h_NA;
   g_K  = g_K_bar*n_K.*n_K.*n_K.*n_K;
   end
```

12.11   More Action Potentials

Let's examine how an action potential is generated with this model. We can inject
both a current using the function **IE2.m** and apply an axon hillock voltage using the
techniques from Chap. 11. For our work in this section, we will not use an external
current so we will set **IE2.m** to be always 0. Then, for a given dendritic pulse at
some location, we use the function **GetAxonHillock2** given in Listing 12.34.

Listing 12.34: Generate An Axon Hillock Voltage Curve

```
    function [tau,lambda,V,AH,Input] = GetAxonHillock2(Q,L,TD,rho,Vmax,
       location)
    %
    % Q          the number of eigenvalues we want to find
    % L          the length of the dendrite in space constants
 5  % rho        the ratio of dendrite to soma conductance, G_D/G_S
    % z          eigenvalue vector
    % D          data vector
    % Vmax       size of voltage impulse
    % location   location of pulse
10  % M          matrix of coefficients for our approximate voltage
    %            model
    % Input      the solution as (z,0) to see if match to
    %            input voltage is reasonable
    % AH         the solution at (1,t)
15  % V          the solution at (z,t)
    % tau        dendritic time
    % lambda     dendritic position
    %
    % get eigenvalue vector z
20  z = FindRoots(Q,L,rho,1.0,1.0);
    % get coefficient matrix M
    M = FindM(Q,L,rho,z);
    % compute data vector for impulse
    D = FindData(Q,L,rho,z,Vmax,location);
25  % Solve MB = D system
    [Lower,Upper,piv] = GePiv(M);
    y = LTriSol(Lower,D(piv));
    B = UTriSol(Upper,y);
    % check errors
30  Error = Lower*Upper*B - D(piv);
    Diff = M*B-D;
    e = norm(Error);
    e2 = norm(Diff);
    display(sprintf(' norm of LU residual = %12.7f norm of MB-D = %12.7f'
       ,e,e2));
35  % set dendritic spatial and time bounds
    % divide dendritic time into 100 parts
    tau    = linspace(0,TD,101);
    % divide dendritic space L into 300 parts
    lambda = linspace(0,L,301);
40  %
    V = zeros(301,101);
    A = zeros(Q,1);
    for n = 1:Q
       A(n) = B(n+1);
45  end
    % find voltage at space point lambda(s) and time point tau(t)
    for s = 1:301
       for t = 1:101
          w = z*(L-lambda(s));
```

```
50        u = -(1+z.*z)*tau(t);
          V(s,t) = B(1)*exp(-tau(t))+ dot(A,cos(w).*exp(u));
       end
    end
    % axon hillock is at lambda = 1
55  AH = V(1,:);
    % pulse is modeled at initial time by
    Input = V(:,1);
```

A typical voltage trace generated by an impulse of size 1100 at location 3.0 is seen in
Fig. 12.19. This was generated by the following lines in MatLab. First, we generate
the axon hillock voltage trace.

Listing 12.35: Generating axon hillock voltage trace

```
   Q = 40;
   L = 5;
 3 TD = 5;
   rho = 10;
   Vmax1 = 1100;
   location1 = 3;
   [tau,lambda,V,AH,Input] = GetAxonHillock2(Q,L,TD,rho,Vmax1,
       location1);
 8 norm of LU residual =      0.0000000 norm of MB-D =      0.0000000
   plot(tau(2:101),AH(2:101));
   xlabel('Dendritic Time mS');
   ylabel('Axon Hillock Voltage mV');
   title('Axon Hillock Voltage: Pulse 1100 at 3.0');
```

The transient axon hillock voltage is shown in Fig. 12.19. The axon hillock voltage
is then $V_R + AH$ which for us is $-70 + AH(\tau)$.

Then, we calculate the action potential corresponding to various axon hillock
voltages. We choose a dendritic time, τ, and use the value of the axon hillock voltage,

Fig. 12.19 A voltage pulse
1100 at 3.0

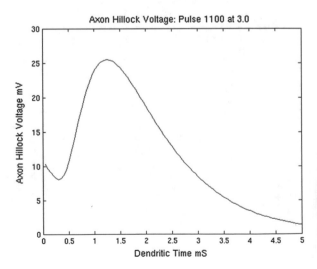

$AH(\tau)$, to be the initial value in the HH simulation. Here are the MatLab lines to do that.

Listing 12.36: Calculating an action potential

```
   E_L = -50;
   V_R = -70;
 3 g_NA_bar = 120.0;
   g_K_bar = 36.0;
   gatesname = 'HHGates';
   Kin = 400.0;
   Kout = 20.11;
 8 Nain = 50.0;
   Naout = 491.0;
   TF = 69.0;
   p = [Kin;Kout;Nain;Naout;TF];
   [g_L,m_NA0,h_NA0,m_K0,E_K,E_NA] =...
13      rest(E_L,V_R,g_NA_bar,g_K_bar,p,gatesname);
   pp.a = [E_K;E_NA;E_L;g_L;g_NA_bar;g_K_bar];
   pp.b = 'HHGates';
   errortol = 1.0e-6;
   steptol = 1.0e-8;
18 minstep = 1.0e-4;
   maxstep = 2.0;
   q = [errortol; steptol;minstep;maxstep];
   V_0 = V_R + AH(2)
   V_0 =
23
      -59.7262
   [tvals,yvals,fvals,hvals,g_NA,g_K,V,m_NA,h_NA,m_K] = ...
          SolveSimpleHH('HHdynamics',q,pp,0,30,...
                        [V_R+AH(2);m_NA0;h_NA0;m_K0],.4);
28 plot(tvals,V);
   xlabel('Time mS');
   ylabel('Voltage mV');
   title('Action Potential vs Time using AH(2): Pulse 1100 at 3.0')
      ;
```

As this current flows into the membrane, the membrane begins to depolarize. The current flow of sodium and potassium through the membrane is voltage dependent and so this increase in the voltage across the membrane causes changes in the ion gates. Using $V_0 = -70 + AH(0)$, we generate the action potential shown in Fig. 12.20. Recall that in the ion current equations the nonlinear conductance is modeled by

$$g_{Na}(V,t) = g_{Na}^{Max} \mathscr{M}_{NA}{}^3(V) \mathscr{H}_{NA}(V)$$
$$g_K(V,t) = g_K^{Max} \mathscr{M}_K{}^4(V).$$

The full current equations are then

$$I_{Na} = g_{NA}(V,t)(V(t) - E_{NA})$$
$$I_K = g_K(V,t)(V(t) - E_K)$$
$$I_L = g_L(V,t)(V(t) - E_L)$$

Fig. 12.20 Action potential: pulse 200 at 3.0

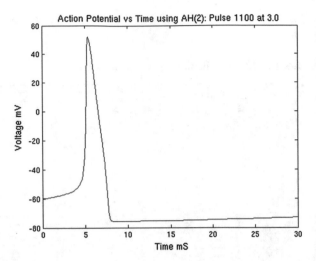

The conductance is in micro Siemens or units of 10^{-6} 1/ohms. Time is in milliseconds or 10^{-3} seconds and voltage is in millivolts or 10^{-3} volts. The currents here are thus in units of 10^{-9} amps or nano amps. We show the plots of these currents versus time in Fig. 12.21. Note the sodium current is negative and the potassium current is positive. The MatLab commands are

Fig. 12.21 Sodium and potassium current during a pulse 1100 at 3.0

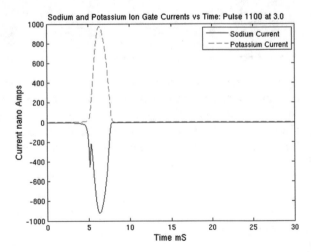

Fig. 12.22 Absolute sodium
and potassium current during
a pulse 1100 at 3.0

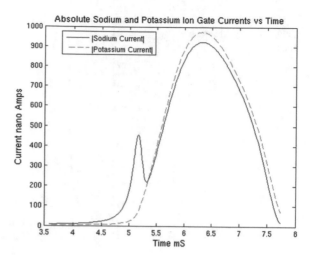

Listing 12.37: Generating currents

```
INA = g_NA.*(V-E_NA);
IK = g_K.*(V-E_K);
IL = g_L.*(V-E_L);
4 plot(tvals,INA,'-',tvals,IK,'--');
xlabel('Time mS' );
ylabel('Current nano Amps');
legend('Sodium Current','Potassium Current');
title('Sodium and Potassium Ion Gate Currents vs Time: Pulse
   1100 at 3.0');
```

To see how the sodium current activates first and *then* the potassium current, we can
plot the absolute value of I_{Na} and I_K simultaneously. This is done in Fig. 12.22. The
time interval here is setup with a linspace command dividing 30 mS into 300 equal
parts. So to see the currents better, we will plot them on the time interval [3.5, 20.0]
which is time tick 35 to 200.

Listing 12.38: Plotting currents

```
plot(tvals(35:200),abs(INA(35:200)),'-',tvals(35:200),abs(IK
   (35:200)),'--');
2 xlabel('Time mS');
ylabel('Current nano Amps');
title('Absolute Sodium and Potassium Ion Gate Currents vs Time')
   ;
legend('|Sodium Current|','|Potassium Current|');
```

We can study the activation and inactivation terms more closely using some carefully
chosen plots. We show the plot of $\mathscr{M}_{NA}$ in Fig. 12.23 and $\mathscr{H}_{NA}$ in Fig. 12.24.

Finally, the potassium activation $\mathscr{M}_K$ is shown in Fig. 12.25.

Fig. 12.23 Sodium activation $\mathcal{M}_{NA}$ during a pulse

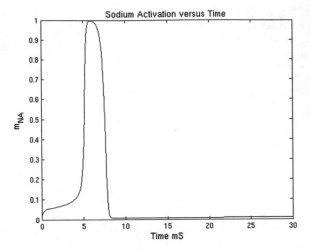

Fig. 12.24 Sodium inactivation $\mathcal{H}_{NA}$ during a pulse

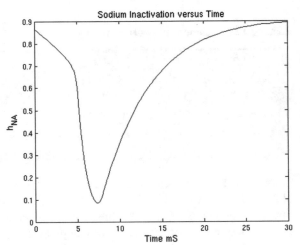

Of even more interest is the products that are proportional to the sodium and potassium current. We plot $\mathcal{M}_{NA}^{3}\,\mathcal{H}_{NA}$ versus time in Fig. 12.26 and $\mathcal{M}_{K}^{4}$'s time trace in Fig. 12.27.

We note that if we apply a voltage pulse at a location further from the axon hillock, say at location 3, we will not generate an action potential! In this chapter, we have started to put together our models: we have a simple ball stick neuron model which couples voltage pulses applied to the dendritic cable to a simple cell body. We assume the resulting membrane voltage at the junction of the dendrite and the cell body is available without change at the axon hillock which is potentially on the other side of the cell body. From Fig. 12.19, we know the voltage we seen at the axon hillock varies with time. In the simulations in this Chapter, we have chosen a time (usually time tick 2 which corresponds to a time of 0.1 time constants) and used the

Fig. 12.25 Potassium activation $\mathscr{M}_K$ during a pulse

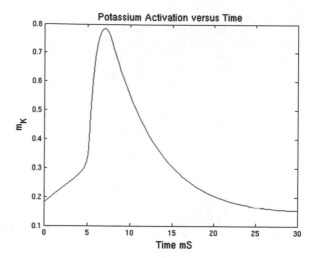

Fig. 12.26 The product $\mathscr{M}_{NA}^3 \ \mathscr{H}_{NA}$ during a pulse 1100 at 3.0 scaled by the maximum sodium and potassium conductance values

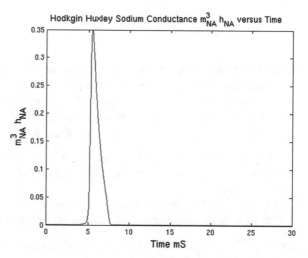

axon hillock voltage there as our depolarizing input to the Hodgkin–Huxley model. Clearly, we could use other starting values from the axon hillock voltage time trace. Each of these values may or may not trigger an action potential. The neurobiology of the axon sends a voltage wave backwards through the axon hillock at each time also. If an action potential is generated, this backward potential hyperpolarizes the dendritic cable and effectively shuts down depolarization for some time. This is the well-known refractory period for an excitable neuron. We are not ready to model these effects yet.

Fig. 12.27 The product $\mathcal{M}_K{}^4$ during a pulse 1100 at 3.0 scaled by the maximum sodium and potassium conductance values

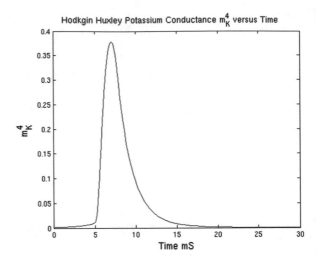

12.11.1 Exercise

Exercise 12.11.1 *Run the Hodgkin–Huxley simulation as has been done in the work above and write a complete report in your favorite editor which includes the following things: For this study, we use the maximum sodium conductance is* 120, *the maximum potassium conductance is* 36, *the leakage voltage is* −50 *with temperature* 70 *degrees Fahrenheit. We will use* I_M *is zero always. Then, find*

1. *The plots of voltage V,* $\mathcal{M}_{NA}$, $\mathcal{H}_{NA}$ *and* $\mathcal{M}_K$ *versus time for* 20 *milliseconds.*
2. *The plots of the time constant and the asymptotic value for* $\mathcal{M}_{NA}$, $\mathcal{H}_{NA}$ *and* $\mathcal{M}_K$ *versus time.*
3. *The plots of* $\mathcal{M}_{NA}{}^3 \mathcal{H}_{NA}$ *and* $\mathcal{M}_K{}^4$ *versus time.*
4. *The plots of sodium and potassium current versus time.*

for the following applied voltage pulses on a cable of length 5 *with* $\rho = 10$. *Use suitable Q for the following pulses:*

1. *Pulse of size* 300 *at* 1.5.
2. *Pulse of size* 200 *at* 2.0 *and* 100 *at* 1.0.
3. *Pulse of size* 300 *at* 4.0 *and* 200 *at* 1.0.

Exercise 12.11.2 *Run the Hodgkin–Huxley simulation as has been done in the work above and write a complete report in your favorite editor which includes the following things: For this study, we use the maximum sodium conductance is* 120, *the maximum potassium conductance is* 36, *the leakage voltage is* −50 *with temperature* 70 *degrees Fahrenheit. We will still use* I_M *is zero always. Then, find*

1. *The plots of voltage V,* $\mathcal{M}_{NA}$, $\mathcal{H}_{NA}$ *and* $\mathcal{M}_K$ *versus time for* 20 *milliseconds.*
2. *The plots of the time constant and the asymptotic value for* $\mathcal{M}_{NA}$, $\mathcal{H}_{NA}$ *and* $\mathcal{M}_K$ *versus time.*

3. The plots of $\mathcal{M}_{NA}{}^3\,\mathcal{H}_{NA}$ and $\mathcal{M}_K{}^4$ versus time.
4. The plots of sodium and potassium current versus time.

for the following applied voltage pulses on a cable of length 5 with $\rho = 1$. Use suitable Q for the following pulses:

1. Pulse of size 300 at 1.5.
2. Pulse of size 200 at 2.0 and 400 at 1.0.
3. Pulse of size 500 at 4.0 and 300 at 1.0.

Exercise 12.11.3 Run the Hodgkin–Huxley simulation as has been done in the work above and write a complete report in your favorite editor which includes the following things: For this study, we use the maximum sodium conductance is 120, the maximum potassium conductance is 36, the leakage voltage is -50 with temperature 70 degrees Fahrenheit. We will still use I_M is zero always. Then, find

1. The plots of voltage V, $\mathcal{M}_{NA}$, $\mathcal{H}_{NA}$ and $\mathcal{M}_K$ versus time for 20 milliseconds.
2. The plots of the time constant and the asymptotic value for $\mathcal{M}_{NA}$, $\mathcal{H}_{NA}$ and $\mathcal{M}_K$ versus time.
3. The plots of $\mathcal{M}_{NA}{}^3\,\mathcal{H}_{NA}$ and $\mathcal{M}_K{}^4$ versus time.
4. The plots of sodium and potassium current versus time.

for the following applied voltage pulses on a cable of length 5 with $\rho = 0.1$. Use suitable Q for the following pulses:

1. Pulse of size 300 at 1.5.
2. Pulse of size 200 at 2.0 and 100 at 1.0.
3. Pulse of size 300 at 4.0 and 200 at 1.0.

Exercise 12.11.4 Run the Hodgkin–Huxley simulation as has been done in the work above and write a complete report in your favorite editor which includes the following things: For this study, we use the maximum sodium conductance is 120, the maximum potassium conductance is 36, the leakage voltage is -50 with temperature 70 degrees Fahrenheit. We will still use I_M is zero always. Then, find

1. The plots of voltage V, $\mathcal{M}_{NA}$, $\mathcal{H}_{NA}$ and $\mathcal{M}_K$ versus time for 20 milliseconds.
2. The plots of the time constant and the asymptotic value for $\mathcal{M}_{NA}$, $\mathcal{H}_{NA}$ and $\mathcal{M}_K$ versus time.
3. The plots of $\mathcal{M}_{NA}{}^3\,\mathcal{H}_{NA}$ and $\mathcal{M}_K{}^4$ versus time.
4. The plots of sodium and potassium current versus time.

for the following applied voltage pulses on a cable of length 1 with $\rho = 10.0$. Use suitable Q for the following pulses:

1. Pulse of size 300 at 1.0.
2. Pulse of size 200 at 1.0 and 100 at 0.6.

12.12   Action Potential Dynamics

Let's redo our code to allow sequences of pulses to enter the dendritic system of an excitable nerve cell. First, we will rewrite our **GetAxonHillock** code to return our approximate solution to the ball-stick model as a function of the two variables spatial distance along the dendrite and local dendrite time. Most of the code below is standard and has been discussed already. However, in this code we do two new things: first, as mentioned, we return the solution V as a function of two variables and second, we generate a plot of this solution as a surface. The returned axon hillock function is generated as follows:

Listing 12.39: Generating the axon hillock function

```
V = @(s,t)  B(1)*exp(-t);
for  n=1:Q
    V = @(s,t)  (V(s,t) + B(n+1)*cos(z(n)*(L - s)).*exp(-(1+z(n)^2)
        *t));
end
```

The surface plot uses the **mesh** and **meshgrid** commands.

Listing 12.40: Drawing the surface

```
   %
   % draw surface for grid [0,L] x [0,5]
   % set up space and time stuff
     space = linspace(0,L,101);
 5   time  = linspace(0,5,101);
   % set up grid of x and y pairs (space(i),time(j))
     [Space,Time] = meshgrid(space,time);
   % set up surface
   Z = V(Space,Time);
10
   %plot surface
   figure
   mesh(Space,Time,Z,'EdgeColor','black');
   xlabel('Dendrite Cable axis');
15 ylabel('Time axis');
   zlabel('Voltage');
   title('Dendritic Voltage');
   print -dpng 'DendriticVoltage.png';
```

Note, we print out the graph as a portable network graphic file for inclusion in our documentation. The full code is then given below.

Listing 12.41: GetAxonHillockThree

```
   function V = GetAxonHillockThree(Q,L,rho,Vmax,location)
 2 %
   % Arguments
   %
   % Q            = the number of eigenvalues we want to find
   % L            = the length of the dendrite in space constants
 7 % rho          = the ratio of dendrite to soma conductance, G_D/G_S
   % Vmax         = size of voltage impulse
   % location     = location of pulse
   % z            = eigenvalue vector
   % D            = data vector
12 %
   % Computed Quantities
   %
   % M            = coefficient matrix for approximate voltage model
   % AxonHillock  = the solution at (0,t)
17 % V            = the solution at (z,t)
   %
   % get eigenvalue vector z
   z = FindRoots(Q,L,rho,1.0,1.0);
   % get coefficient matrix M
22 M = FindM(Q,L,rho,z);
   % compute data vector for impulse
   D = FindDataTwo(Q,L,rho,z,Vmax,location);
   % Solve MB = D system
   [Lower,Upper,piv] = GePiv(M);
27 y = LTriSol(Lower,D(piv));
   B = UTriSol(Upper,y);
   % check errors
   Error = Lower*Upper*B - D(piv);
   Diff = M*B-D;
32 e = norm(Error);
   e2 = norm(Diff);
   normmessage = sprintf(' norm of LU residual = %12.7f norm of MB-D =
       %12.7f',e,e2);
   disp(normmessage);

37 % set spatial and time bounds
   V = @(s,t) B(1)*exp(-t);
   for n=1:Q
     V = @(s,t) (V(s,t) + B(n+1)*cos(z(n)*(L - s)).*exp(-(1+z(n)^2)*t));
   end
42
   %
   % draw surface for grid [0,L] x [0,5]
   % set up space and time stuff
   space = linspace(0,L,101);
47 time  = linspace(0,5,101);
   % set up grid of x and y pairs (space(i),time(j))
   [Space,Time] = meshgrid(space,time);
   % set up surface
   Z = V(Space,Time);
52
   %plot surface
   figure
   mesh(Space,Time,Z,'EdgeColor','black');
   xlabel('Dendrite Cable axis');
57 ylabel('Time axis');
   zlabel('Voltage');
   title('Dendritic Voltage');
   print -dpng 'DendriticVoltage.png';
62 end
```

To test this code, here is a sample session.

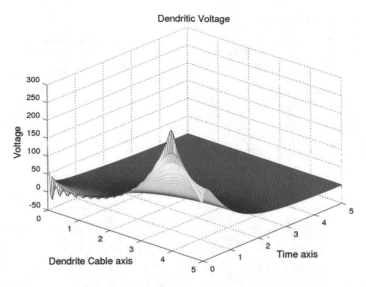

Fig. 12.28 Dendritic voltage surface approximation due to a spike of 300 mV at location 4.0 using 40 terms

Listing 12.42: GetAxonHillockThree test

```
Q = 40;
L = 5;
rho = 10;
V = GetAxonHillockThree (Q, L , rho ,300 ,4.0 );
   norm of LU residual =      0.0000000 norm of MB–D =      0.0000000
```

We have oriented the surface you see in Fig. 12.28 so you can see the voltage solution at spacial position 0; this is the axon hillock voltage as a function of space and time. Note the separation of variables technique generates a voltage which has spurious oscillations in it do the fact that we are calculating an approximate solution with $Q = 40$ terms in the expansion.

We can focus on the axon hillock voltage by defining an axon hillock function and plotting it as usual. To see the axon hillock voltage as a function of time, we define an appropriate time axis and the compute the plot.

Listing 12.43: Axon hillock voltage time dependence

```
% define the axon hillock function
AH = @( t )  V( 0 , t );
% setup plot axis
time = linspace (0 ,5 ,101);
y = AH( time );
% do the plot
plot ( time , y );
```

The approximation to the dendritic pulse shows the oscillations we expect.

Listing 12.44: Approximation to the dendritic pulse

```
  space = linspace(0,5,101);
  DV = @(x) V(x,0);
3 z = DV(space);
  plot(space,z);
```

We next modify this function a bit to make it easier to use when there are multiple pulses. This is the function **GetAxonHillockFour**.

Listing 12.45: GetAxonHillockFour arguments

```
function [V,Surface] = GetAxonHillockFour(Q,L,rho,Vmax,x0,t0,tf)
```

which returns the dendrite voltage function **V** as well as the handle of the surface we plot, **Surface**. Most of this code is similar to that of the third version but this time, we can specify the time the pulse is delivered; hence, the pulse is of magnitude **Vmax** and is applied at time **t0** and location **x0**. We also can choose how long we want to time we want to use in the plot of the resulting surface by setting the variable **tf** as we see fit. The new code for the surface plot is slightly different as we have to choose the underlying **linspace** commands to we get a reasonable plot. We choose a time and space step of .1 here as it works fairly well for many of our surfaces, but it is possible it would be too coarse in some situations. So, you might have to alter this in some simulations. We also change the voltage calculation to use absolute values by changing to **-abs(t-t0)** in the code.

Listing 12.46: Adding absolute values

```
  V = @(s,t) B(1)*exp(-abs(t-t0)) + sum( ( BB.*cos(z*(L-s)) ).*exp
    (EZ*abs(t-t0)) );
```

We setup the surface plot in the usual manner.

Listing 12.47: Setup the surface plot

```
  %
  % draw surface for grid [0,L] x [0,tf]
  % set up space and time stuff
4 sizetime = tf/.1+1;
  sizespace = L/.1+1;
  space = linspace(0,L,sizespace);
  time  = linspace(0.3,tf,sizetime);
  % set up grid of x and y pairs (space(i),time(j))
9 [Space,Time] = meshgrid(space,time);

  % rows of Space are copies of space
  % as many rows are there are time points
  % cols of Time are copies of time
```

```
14 % as  many  cols  are  there  are  space  points
   [rowspace , colspace ] = size ( Space );
   [rowstime , coltime ] = size ( Space );
   %
   % set  up  surface
19 %
   % for  this  to  work
   % rowspace = coltime
   % colspace = rowtime
   for  i=1:sizespace
24   for  j=1:sizetime
        Z(j , i ) = V( space ( i ) , time ( j ));
     end
   end
```

After the construction of the surface points to plot, **z**, is done, we do the plot itself.

Listing 12.48: Generating the plot

```
%Z = V( Space , Time );
%plot  surface
figure
mesh ( Space , Time , Z , 'EdgeColor' , 'black' );
5 Surface = gcf ();
xlabel ('Dendrite Cable axis' );
ylabel ('Time axis' );
zlabel ('Voltage' );
title ('Dendritic Voltage' );
```

The only thing new here is that after we generate the plot with the **mesh** command, we store its *handle* with the line **Surface = gcf();**. We can then refer to this plot in later code by using the name **Surface**. The full code is then as follows:

Listing 12.49: GetAxonHillockFour

```
   function  [V, Surface ] = GetAxonHillockFour (Q, L, rho , Vmax, x0 , t0 , tf )
   %
   % Arguments
   %
5  % Q             = the  number  of  eigenvalues  we  want  to  find
   % L             = the  length  of  the  dendrite  in  space  constants
   % rho           = the  ratio  of  dendrite  to  soma  conductance ,  G_D/G_S
   % Vmax          = size  of  voltage  impulse
   % x0            = cable  location  of  pulse
10 % t0            = start  time  of  pulse
   % tf            = end  time  for  axonal  pulse
   %
   % Computed  Quantities
   %
15 % z             = eigenvalue  vector
   % D             = data  vector
   % M             = coefficient  matrix  for  approximate  voltage  model
   % AxonHillock   = the  solution  at  (0,t)
   % V             = the  solution  at  (z,t)
20 %
   % get  eigenvalue  vector  z
   z = FindRoots (Q, L, rho ,1.0 ,1.0 );
   % get  coefficient  matrix  M
   M = FindM (Q, L, rho , z );
25 % compute  data  vector  for  impulse
   D = FindDataTwo (Q, L, rho , z , Vmax, x0 );
```

```matlab
      % Solve MB = D system
      [Lower,Upper,piv] = GePiv(M);
      y = LTriSol(Lower,D(piv));
30  B = UTriSol(Upper,y);
      % check errors
      Error = Lower*Upper*B - D(piv);
      Diff = M*B-D;
      e = norm(Error);
35  e2 = norm(Diff);

      EZ = zeros(Q,1);
      for n = 1:Q
         EZ(n) = -(1+z(n)^2);
40  end

      BB = zeros(Q,1);
      for n = 1:Q
         BB(n) = B(n+1);
45  end

      %
      V = @(s,t) B(1)*exp(-abs(t-t0)) + sum( ( BB.*cos(z*(L-s)) ).*exp(EZ*
         abs(t-t0)) );
      %                                        1 2        3    4   43 2      2
                       3       32 1
50
      %
      % draw surface for grid [0,L] x [0,tf]
      % set up space and time stuff
      sizetime = tf/.1+1;
55  sizespace = L/.1+1;
      space = linspace(0,L,sizespace);
      time  = linspace(0.3,tf,sizetime);
      % set up grid of x and y pairs (space(i),time(j))
      [Space,Time] = meshgrid(space,time);
60
      % rows of Space are copies of space
      % as many rows are there are time points
      % cols of Time are copies of time
      % as many cols are there are space points
65  [rowspace,colspace] = size(Space);
      [rowtime,coltime] = size(Space);
      %
      % set up surface
      %
70  % for this to work
      % rowspace = coltime
      % colspace = rowtime
      for i=1:sizespace
         for j=1:sizetime
75        Z(j,i) = V(space(i),time(j));
         end
      end

      %Z = V(Space,Time);
80  %plot surface
      figure
      mesh(Space,Time,Z,'EdgeColor','black');
      Surface = gcf();
      xlabel('Dendrite Cable axis');
85  ylabel('Time axis');
      zlabel('Voltage');
      title('Dendritic Voltage');
      %print -dpng 'DendriticVoltage.png';

90  end
```

12.12.1 Multiple Pulses

To handle multiple pulses occurring at different times, we will use the function **SetUpPulses**. The function returns the a collection of handles to the surface plots of each of the incoming pulses **s{1}**, **s{2}** and so forth; the handle of the plot of the voltage for the cumulative pulses on the dendritic cable, **DVplot**, the handle of the axon hillock voltage for the cumulative pulses, **AHplot** and finally, the summed voltage function, **Vin**.

Listing 12.50: SetUpPulses arguments

```
function [S,DVplot,AHplot,Vin] = SetUpPulses(VMax,x0,td0,scale,
    xs,ts,tf)
```

The data structure we use here in the **cell**. The voltage pulses are sent in as the cell **VMax{1}** to **VMax{N}** where **N** is the number of elements in the cell. We store **N** in the variable **NumberOfPulses** and then we setup the summed dendritic impulse function by calling the function **GetAxonHillockFour** repeatedly for all **N** incoming pulses. This gives us a cell of handles to the individual surface plots **s{}** and a cell of the individual pulse functions **v{}**. We also generate all of the individual axon hillock potentials and save them in the cell **AH{}**. Since there are multiple pulses, the locations of the pulses and their arrival times are also stored in cells; here, **x0{}** and **td0{}**, respectively. Finally, we sum the response to the incoming pulses in the function **Vin**.

Listing 12.51: Generating the response for an impulse sequence

```
% initialize pulse cell
V = {};
% initialize surface handle cell
S = {};
% initialize axon potential cell
AV = {};
W = @(t) 0;
for i = 1:NumberOfPulses
    [V{i},S{i}] = GetAxonHillockFour(Q,L,rho,VMax{i},x0{i},td0{i},
        tf);
    AV{i} = @(t) ( V{i}(.2,0.2+t)+ V{i}(.2,0.2+t+delt)+V{i
        }(.2,0.2+t+2*delt) )/3;
    W = @(t) W(t) + AV{i}(t);
end
Vin = @(t) scale*W(t);
```

Note, we use the argument **scale** here. We may or may not want to use the generated pulse voltages unscaled. Our approximation strategies can generate really high voltages at times; hence, we opt for prudence and allow ourselves a scaling choice. If we set **scale = 1.0** we use the full generated voltages. On the other hand, if we set **scale = 0.4** or anything less than 1, we artificially diminish the size of the pulse. We then compute the summed dendritic voltage, **DV**, on the dendritic cable itself as a function. Also, to avoid irregularities with the approximation algorithms at time 0 and dendritic cable location 0, we choose to start time at 0.2 and space at 0.2 rather than 0, respectively.

Listing 12.52: Constructing the summed dendritic pulse

```
DV = @(x)  0;
for  i  =  1:NumberOfPulses
    DV = @(x)  DV(x) + V{i}(x,td0{i});
end
```

Next, we generate the plots of the summed dendritic voltage **Vin** and summed axon hillock voltage and return their handles **DVPlot{ }** and **AHplot**. There are multiple handles for the dendritic voltage plots as their are multiple pulses. Here we use the variables **delx** and **delt** to try to set up reasonable linspace commands for our surface plots. These variables are set earlier in the code.

Listing 12.53: Generating plots

```
%
sizeSpace = L/delx + 1;
sizeTime = tf/delt + 1;
time = linspace(ts,tf,sizeTime);
space = linspace(xs,L,sizeSpace);
%
figure;
for  n=1:sizeTime
    AH(n) = Vin(time(n));
end
plot(time,AH);
AHplot = gcf();
xlabel('Time');
ylabel('Voltage');
title('Axon Hillock Voltage');
%
DVplot = {};
for  i  = 1:NumberOfPulses
    figure;
    for  n=1:sizeSpace
        DVY(n) = V{i}(space(n),td0{i});
    end
    plot(space,DVY);
    DVplot{i} = gcf();
    xlabel('Space');
    ylabel('Applied Input Voltage');
    message = ['Initial Voltage ',int2str(i) ' on Dendrite'];
    title(message);
end
```

Here is the full code.

Listing 12.54: SetUpPulses

```
 1 function [S,DVplot,AHplot,Vin] = SetUpPulses(VMax,x0,td0,scale,xs,ts,
      tf)
   %
   % VMax  = cell of dendritic impulse sizes
   % x0    = cell of dendritic impulse locations
   % td0   = time dendritic impulse arrives
 6 % scale = multiplier for pulse
   % ts    = start time of plot
   % tf    = end time of plot
   % xs    = start position of space in plot
   %
11 % Calculate dendritic pulse
   %
   Q = 48;
   L = 5;
   rho = 10;
16 NumberOfPulses = length(VMax);
   delt = .1;
   delx = .05;
   V = {};
   S = {};
21 AV = {};
   W = @(t) 0;
   for i = 1:NumberOfPulses
     [V{i},S{i}] = GetAxonHillockFour(Q,L,rho,VMax{i},x0{i},td0{i},tf);
     AV{i} = @(t) ( V{i}(.2,0.2+t)+ V{i}(.2,0.2+t+delt)+V{i}(.2,0.2+t+2*
        delt) )/3;
26   W = @(t) W(t) + AV{i}(t);
   end
   Vin = @(t) scale*W(t);
   %
   % Applied Voltage
31 %
   DV = @(x) 0;
   for i = 1:NumberOfPulses
     DV = @(x) DV(x) + V{i}(x,td0{i});
   end
36 %
   sizeSpace = L/delx + 1;
   sizeTime = tf/delt + 1;
   time = linspace(ts,tf,sizeTime);
   space = linspace(xs,L,sizeSpace);
41 %
   figure;
   for n=1:sizeTime
     AH(n) = Vin(time(n));
   end
46 plot(time,AH);
   AHplot = gcf();
   xlabel('Time');
   ylabel('Voltage');
   title('Axon Hillock Voltage');
51 %
   DVplot = {};
   for i = 1:NumberOfPulses
     figure;
     for n=1:sizeSpace
56     DVY(n) = V{i}(space(n),td0{i});
     end
     plot(space,DVY);
     DVplot{i} = gcf();
     xlabel('Space');
61   ylabel('Applied Input Voltage');
     message = ['Initial Voltage ',int2str(i),' on Dendrite'];
     title(message);
   end

66 end
```

Let's try out the code. We send in two pulses: one of magnitude 200 at time 5 and location 2.3 and the other of size 400 at time 10.0 and location 2.5. We setup the time axis to run from 0 to 15 and we choose to use full scaling; i.e., **scale = 1.0**.

Listing 12.55: Setting up pulses

```
[S,DVplot,AHplot,Vin] = SetUpPulses
  ({200;400},{2.3;2.5},{5.0;10.0},1.0,0,0,15.0);
```

This returns handles to our graphics and the summed axon hillock function **Vin**. We then print out our plots. First, we find the handle numbers.

Listing 12.56: Getting handles for the plots

```
      S
      S =

      {
5         [1,1] =    2
          [1,2] =    3
      }
      AHplot
      AHplot =   4
10
      DVplot
      DVplot =

      {
15        [1,1] =    5
          [1,2] =    6
      }
```

Then, we do the plots.

Listing 12.57: Generating the plots

```
print('-f2','-dpng','DendriticVoltageSurfaceOne.png');
print('-f3','-dpng','DendriticVoltageSurfaceTwo.png');
3 print('-f4','-dpng','AxonHillockVoltage.png');
print('-f5','-dpng','DendriticPulseOne.png');
print('-f6','-dpng','DendriticPulseTwo.png');
```

This gives us the two surface plots, Figs. 12.29 and 12.30.

The voltage trace for the two pulses along the cable are approximated as shown in Figs. 12.31 and 12.32.

Finally, the generated axon hillock voltage is shown in Fig. 12.33.

We clean up our generated figures with the utility function **CleanUpFigures-Two** given below. To use this, simply type **CleanUpFiguresTwo**.

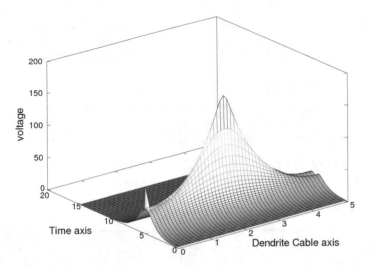

Fig. 12.29   Voltage surface for one pulse

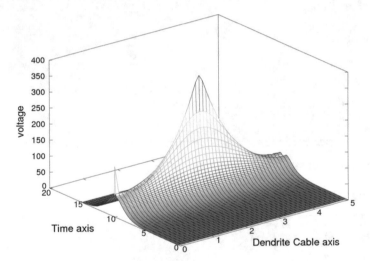

Fig. 12.30   Voltage surface for two pulses

Listing 12.58: Cleaning up the generated figures

```
%CleanUpFiguresTwo.m
N = length(S);
for i=1:N
    delete(S{i});
    delete(DVplot{i});
end

delete(AHplot);
```

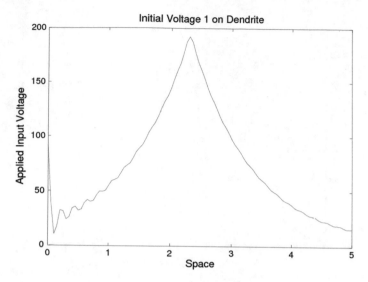

Fig. 12.31 Approximation to pulse one on the dendritic cable

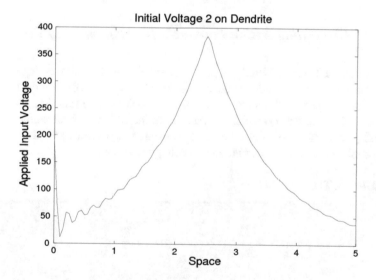

Fig. 12.32 Approximation to pulse two on the dendritic cable

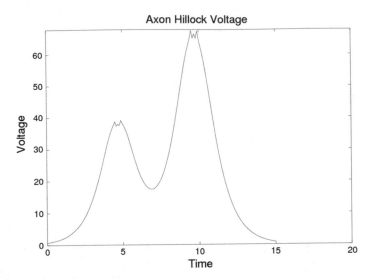

Fig. 12.33   Axon hillock voltage for two pulses

12.12.2   Generating Action Potentials for Sequences of Pulses

We must alter the Hodgkin–Huxley dynamics to handle an incoming voltage. Most of this code is similar to the code we have used previously for the dynamics. However, we now want to pass in the name of the axon hillock voltage function. To do this, we alter the data structure **p** which we pass in. In the code fragment below, note the **p.a** and **p.b** portions of **p** are the same as usual, but we have added a new field **p.c** to contain the name of our incoming voltage function.

Listing 12.59: The p.c field

```
     %    p comes in as a data structure.
     %
     %    p.a(1) = EK, the potassium battery voltage
     %    p.a(2) = ENA is the sodium battery voltage
  5  %    p.a(3) = EL, the leakage voltage
     %    p.a(4) = GL, the leakage conductance
     %    p.a(5) = gNAmax, maximum sodium conductance
     %    p.a(6) = gKmax, maximum potassium conductance
     %
  10 %    p(1,2) is the string data
     %    To use
     %    p.a = [EK;ENA;EL;GL;gNAmax;gKMax];
     %    p.b = 'string '; name of HHGates function
     %    p.c = 'string '; name of input voltage
```

We extract this information as usual, assigning the variable **Pulse** to the input voltage function name.

Listing 12.60: Extracting information from the p structure

```
EK          = p.a(1);
ENA         = p.a(2);
% leakage current: run rest.m to find appropriate g_L value
% default E_leak = -50.0
5 E_leak      = p.a(3);
g_leak      = p.a(4);
% max conductance for NA: default is 120
g_NA_bar    = p.a(5);
% max conductance for K: default is 36.0
10 g_K_bar     = p.a(6);
gatename    = p.b;
Pulse       = p.c;
```

We then evaluate the incoming voltage at the time **t** using **feval(Pulse,t)** and add it to the current voltage **y(1)**. Then we send this voltage into the $\alpha-\beta$ evaluation function **GatesNAK** as we have done in the previous code.

Listing 12.61: Evaluating the gates functions

```
currentV = y(1)+feval(Pulse,t);
alpha = GatesNAK(gatename,currentV);
```

Finally, we remove the external current **IE2** as we no longer use it as a source of depolarization. The full code is given below.

Listing 12.62: Using incoming depolarizations: HHdynamicsVin

```
function f = HHdynamicsVin(p,t,y)
%       Standard Hodgkin - Huxley Model
%       voltage           mV
%       current           na
5 %     time              ms
%       concentration mM
%       conductance       micro Siemens
%       capacitance       nF
%
10 %    arguments
%       p   parameter vector
%       t   time
%       y   state vector
%
15 %    =================================
%       y vector assignments
%       =================================
%       y(1) = V
%       y(2) = m_NA
20 %     y(3) = h_NA
%       y(4) = m_K
%
%       parameter assignments
%       p comes in as a data structure.
```

```
25  %
    %      p.a(1) = EK, the potassium battery voltage
    %      p.a(2) = ENA is the sodium battery voltage
    %      p.a(3) = EL, the leakage voltage
    %      p.a(4) = GL, the leakage conductance
30  %      p.a(5) = gNAmax, maximum sodium conductance
    %      p.a(6) = gKmax, maximum potassium conductance
    %
    %      p(1,2) is the string data
    %      To use
35  %      p.a = [EK;ENA;EL;GL;gNAmax;gKMax];
    %      p.b = 'string'; name of HHGates function
    %      p.c = 'string'; name of input voltage

    EK          = p.a(1);
40  ENA         = p.a(2);
    % leakage current: run rest.m to find appropriate g_L value
    % default E_leak = −50.0
    E_leak      = p.a(3);
    g_leak      = p.a(4);
45  % max conductance for NA: default is 120
    g_NA_bar    = p.a(5);
    % max conductance for K: default is 36.0
    g_K_bar     = p.a(6);
    gatename    = p.b;
50  Pulse       = p.c;

    f = zeros(4,1);

    %      ==========================================
55  %      f vector assignments
    %      ==========================================
    %      f(1) = V dynamics
    %      f(2) = m_NA dynamics
    %      f(3) = h_NA dynamics
60  %      f(4) = m_K dynamics
    %
    currentV = y(1)+feval(Pulse,t);
    alpha = GatesNAK(gatename,currentV);
    %
65  % activation/inactivation parameters for NA
    % alpha(1), alpha(2) alpha and beta for mna
    % alpha(3), alpha(4) alpha and beta for hna
    % alpha(5), alpha(6) alpha and beta for mk
    %
70  m_NA_infinity   = alpha(1)/(alpha(1)+alpha(2));
    t_m_NA          = 1.0/(alpha(1)+alpha(2));
    f(2)            = (m_NA_infinity − y(2))/t_m_NA;
    %
    h_NA_infinity   = alpha(3)/(alpha(3)+alpha(4));
75  t_h_NA          = 1.0/(alpha(3)+alpha(4));
    f(3)            = (h_NA_infinity − y(3))/t_h_NA;
    I_NA            = g_NA_bar*(currentV−ENA)*y(2)^3*y(3);
    %
    m_K_infinity    = alpha(5)/(alpha(5)+alpha(6));
80  t_m_K           = 1.0/(alpha(5)+alpha(6));
    f(4)            = (m_K_infinity − y(4))/t_m_K;
    I_K             = g_K_bar*(currentV−EK)*y(4)^4;
    % I_L current
    I_leak = g_leak*(currentV−E_leak);
85  % Cell Capacitance
    C_M = 1.0;
    % full dynamics
    f(1) = (− I_NA − I_K − I_leak)/C_M;
    end
```

We then write code to allow us to automate the generation of the action potential. This is the function **SetUpHH**. As part of the automation, we provide our first attempt at determining if an action potential is generated. This is the function **ConvertToDigital**. This is a simple sigmoidal switch. The switch, $D(V)$, is defined by

$$D(V) = \frac{1}{2}(V_T + V_B) + \frac{1}{2}(V_T - V_B)\, \tanh\!\left(\frac{V - O}{G}\right)$$

and provides a transfer from the bottom value V_B to the top value V_T using a standard sigmoid with offset O and gain parameter G. The membrane voltage must be larger than the offset to allow the switch value to approach the top value of V_T. The implementation of the switch is given in the code below.

Listing 12.63: ConvertToDigital

```
function D = ConvertToDigital(V,VT,VB,Offset,Gain)
%
% V       = voltage
% VR      = Reference voltage
% VT      = Top voltage
% VB      = Bottomo voltage
% Offset  = pulse switch point
% Gain    = rate of transition
%
D = 0.5*(VT+VB) + 0.5*(VT-VB)*tanh( (V - Offset)/Gain );
end
```

In the **SetUpHH** code, we begin by setting up the incoming pulses for the excitable cell. Here **Vmax{ }** is the cell of pulse magnitudes, **x0{ }** is the cell of spatial locations for the pulses and **td0{ }** is the cell of the times at which the pulses are applied.

Listing 12.64: Calculating a dendritic pulse

```
% Calculate dendritic pulse
[S,DVplot,AHplot,Vin] = SetUpPulses(Vmax,x0,td0,scale,xs,ts,tf);
```

Note, when we calculate the effects of the incoming pulses, we return the function **Vin**. Then, we set the rest value for the excitable cell.

Listing 12.65: Setting up parameters

```
%
% Setup nerve cell parameters
%
E_L = -50;
V_R = -70;
g_NA_bar = 120;
g_K_bar = 36;
Kin = 400;
```

```
    Kout = 20.11;
 10 Nain = 50.0;
    Naout = 491;
    TF = 69;
    p = [Kin;Kout;Nain;Naout;TF];
    [g_L,m_NA0,h_NA0,m_K0,E_K,E_NA] = rest(E_L,V_R,g_NA_bar,g_K_bar,
        p,'HHGates');
```

Next, we initialize the data structures **pp** and **q** which we need for the call to the
function **SolveSimpleHH**. We then compute the response of the excitable nerve
cell which may include action potentials.

Listing 12.66: Initialize the pp and q structures

```
    pp.a = [E_K;E_NA;E_L;g_L;g_NA_bar;g_K_bar];
    pp.b = 'HHGates';
    pp.c = Vin;
    errortol = 1.0e-3;
 5  steptol = 1.0e-5;
    minstep = 1.0e-2;
    maxstep = 2.0;
    q = [errortol;steptol;minstep;maxstep];

 10 [tvals,yvals,fvals,hvals,g_NA,g_K,Vm,m_NA,h_NA,m_K] = ...
       SolveSimpleHH('HHdynamicsVin',q,pp,0,tf,[V_R;m_NA0;h_NA0;m_K0
          ],.6);
```

We plot the results of the axon response next.

Listing 12.67: Plot the axon response

```
    delt = .1;
    sizeTime = tf/delt+1;
    time = linspace(ts,tf,sizeTime);

 4
    figure;
    plot(tvals,Vm);
    APplot = gcf();
    xlabel('Time');
 9  ylabel('Voltage');
    title('Axon Response Voltage');
```

Finally, we plot the digitization of the axon response using the **ConvertTo-
Digital** function. We set this conversion using a top voltage of 40 mV and a
bottom voltage of $V_R = -70$ mV. Since the offset is also V_R and the gain is 20,
this function slowly transitions from the reference voltage to 40. Hence, all voltages
above 40 are output as 40.

Listing 12.68: The digital response

```
   figure
   VT = 40;
   VB = V_R;
   Offset = -70;
 5 Gain = 20;
   plot(tvals, ConvertToDigital(Vm,VT,VB, Offset , Gain));
   DPplot = gcf();
   xlabel('Time');
   ylabel('Digital Voltage');
10 title('Axon Digital Voltage');
```

The full code is listed below.

Listing 12.69: SetUpHH

```
   function [S,DVplot,AHplot,APplot,DPplot] = SetUpHH(Vmax,x0,td0,scale,
       xs,ts,tf)
   %
   % VMax  = cell of dendritic impulse sizes
   % x0    = cell of dendritic impulse locations
 5 % td0   = time dendritic impulse arrives
   % scale = multiplier for pulse
   % ts    = start time of plot
   % tf    = end time of plot
   %
10 % Calculate dendritic pulse
   [S,DVplot,AHplot,Vin] = SetUpPulses(Vmax,x0,td0,scale,xs,ts,tf);

   %
   % Setup nerve cell parameters
15 %
   E_L = -50;
   V_R = -70;
   g_NA_bar = 120;
   g_K_bar = 36;
20 Kin = 400;
   Kout = 20.11;
   Nain = 50.0;
   Naout = 491;
   TF = 69;
25 p = [Kin;Kout;Nain;Naout,TF];
   [g_L,m_NA0,h_NA0,m_K0,E_K,E_NA] = rest(E_L,V_R,g_NA_bar,g_K_bar,p,'
       HHGates');

   pp.a = [E_K;E_NA;E_L;g_L;g_NA_bar;g_K_bar];
   pp.b = 'HHGates';
30 pp.c = Vin;
   errortol = 1.0e-3;
   steptol = 1.0e-5;
   minstep = 1.0e-2;
   maxstep = 2.0;
35 q = [errortol;steptol;minstep;maxstep];

   [tvals,yvals,fvals,hvals,g_NA,g_K,Vm,m_NA,h_NA,m_K] = ...
       SolveSimpleHH('HHdynamicsVin',q,pp,0,tf,[V_R;m_NA0;h_NA0;m_K0],.6);

40 delt = .1;
   sizeTime = tf/delt+1;
   time = linspace(ts,tf,sizeTime);

   figure;
45 plot(tvals,Vm);
   APplot = gcf();
   xlabel('Time');
   ylabel('Voltage');
   title('Axon Response Voltage');
```

```
50
   figure
   VT = 40;
   VB = V_R;
   Offset = -70;
55 Gain = 20;
   plot(tvals, ConvertToDigital(Vm,VT,VB,Offset,Gain));
   DPplot = gcf();
   xlabel('Time');
   ylabel('Digital Voltage');
60 title('Axon Digital Voltage');

   end
```

We clean up our generated figures with the script **CleanUpFigures** given below.
Use it by simply typing **CleanUpFigures**.

Listing 12.70: Clean up figures

```
   CleanUpFigures.m
   N = length(S);
   for i=1:N
      delete(S{i});
 5    delete(DVplot{i});
   end
   delete(APplot);
   delete(AHplot);
   delete(DPplot);
```

To use this code is straightforward. We will send in three pulses of various magnitudes, locations and times using a scale of 1.0 and plot the results over 20 seconds. For convenience, we will denote these pulses using a dirac delta notation:
$V_{max}\delta(t - t_0, x - x_0)$ is the pulse of magnitude V_{max} at location x_0 applied at time t_0.

Listing 12.71: Setup for multiple pulses

```
 1 [S,DVplot,AHplot,APplot,DPplot] = ...
 > SetUpHH
      ({200;300;400},{1.5;2.0;3.0},{5.0;10.0;11.0},1.0,0,0,20);
   ans =    144.65
   ans =    43.640
   ans =    215.33
 6 ans =    30.054
   ans =    298.15
   ans =    -4.9560
```

The generated action potential is shown in Fig. 12.34 and the digitization of it is shown
in Fig. 12.35. Recall, the digitization here is performed by the function $D(V) =
-15 + 55 \tanh\left(\frac{V+70}{20}\right)$ which has a bottom threshold for $-70\,$mV and an upper
threshold of $40\,$mV. Since the axon potential never goes above about $20\,$mV, the
digital value does not stay near the top value for long.

Of course, by altering the parameters that shape the digitization sigmoid, we can
change this signal. To change the bottom threshold to $0\,$mV and the top to 1, we

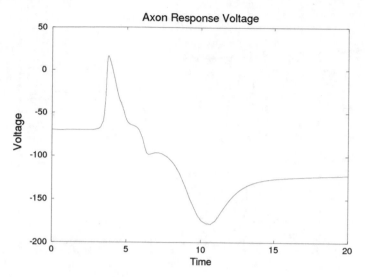

Fig. 12.34 Generated action potential for pulses $200\delta(t-5, x-1.5)$, $300\delta(t-10, x-2.0)$ and $400\delta(t-11, x-3.0)$

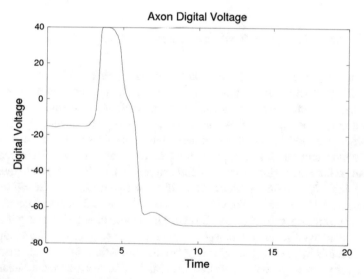

Fig. 12.35 -70 to 40 digitization of action potential for pulses $200\delta(t-5, x-1.5)$, $300\delta(t-10, x-2.0)$ and $400\delta(t-11, x-3.0)$

would set $V_B = 0$ and $V_T = 1$ and rerun the example. We also can change the offset value so that the pulse assigns a value of 0 to voltages close to V_R. In this run, we set $O = -40$; this gives a value of about 0.05 to V_R. The new switch is then defined by

$$D(V) = 0 + 1 \; \tanh\left(\frac{V+40}{20}\right)$$ and the new digitization is shown in Fig. 12.36.

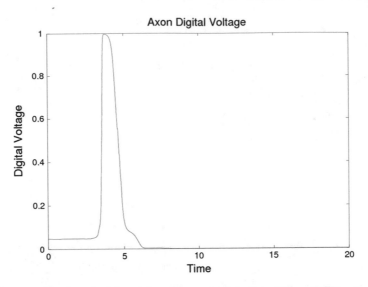

Fig. 12.36 0 to 1 digitization of action potential for pulses $200\delta(t-5, x-1.5)$, $300\delta(t-10, x-2.0)$ and $400\delta(t-11, x-3.0)$

12.12.3 Simulating Second Messenger Activity

Second messenger systems alter the axon potential generated from the inputs to the dendritic system by generating proteins which change the excitable nerve cell itself. Thus, a second messenger signal takes the existing hardware of the cell and replaces it in various ways. These systems are discussed more thoroughly in Peterson (2015) where we model generalized triggers that cause a cascade of reactions inside the cell which culminate in the production of proteins. These proteins could be additional sodium gates (increasing or decreasing g_{Na}^{Max}), potassium gates (increasing or decreasing g_{Na}^{Max}) and/or proteins that alter the hyperpolarization phase of the action potential. For example, the value of g_L could be altered leading to increased or decreased current flux across the membrane. This would change how easy it is for an action potential to be generated for a given voltage input sequence to the dendrite. We can simulate this by altering the value of g_L in the simulation. In the code **SetUpHH**, we return the value of g_L that sets up zero current flow across the membrane initially. Add this code to the function after the call to **rest**. This will print out the **rest** determined g_L value, and the changed value we get by adding 2.5.

Listing 12.72: Simulating second messenger activity

```
g_L
g_L = g_L + 2.5
```

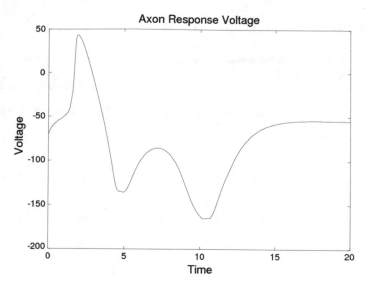

Fig. 12.37 Generated action potential for pulses $200\delta(t - 5, x - 1.5)$, $300\delta(t - 10, x - 2.0)$ and $400\delta(t - 11, x - 3.0)$ with increased g_L

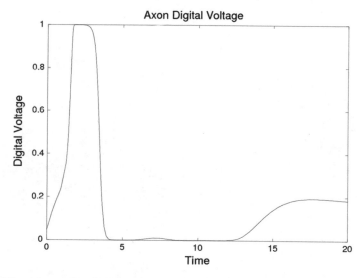

Fig. 12.38 -70 to 40 digitization of action potential for pulses $200\delta(t - 5, x - 1.5)$, $300\delta(t - 10, x - 2.0)$ and $400\delta(t - 11, x - 3.0)$ with increased g_L

The new action potential, Fig. 12.37, is significantly different from the first as the increased leakage current allows for the depolarization effects to be more pronounced.

The digitized potential now has a much higher hyperpolarization phase and a wider digitized action potential. This is shown in Fig. 12.38.

References

A. Hodgkin, The components of membrane conductance in the giant axon of Loligo. J. Physiol. (Lond.) **116**, 473–496 (1952)

A. Hodgkin, The ionic basis of electrical activity in nerve and muscle. Biol. Rev. **26**, 339–409 (1954)

A. Hodgkin, A. Huxley, Currents carried by sodium and potassium ions through the membrane of the giant axon of Loligo. J. Physiol. (Lond.) **116**, 449–472 (1952)

D. Johnston, S. Miao-Sin Wu, *Foundations of Cellular Neurophysiology* (MIT Press, Cambridge, 1995)

J. Peterson, *BioInformation Processing: A Primer on Computational Cognitive Science*. Springer Series on Cognitive Science and Technology (Springer Science+Business Media Singapore Pte Ltd., Singapore, 2015 in press)

T. Weiss, *Cellular Biophysics: Volume 1, Transport* (MIT Press, Cambridge, 1996a)

T. Weiss, *Cellular Biophysics: Volume 2, Electrical Properties* (MIT Press, Cambridge, 1996b)

Part V
Summing It All Up

Chapter 13
Final Thoughts

In this book, we have learned a lot about how to model an excitable neuron and we have seen how many approximations we must make to find models we can solve mathematically. It is important to realize that we always make error when we abstract from reality a tentative map of how variables of interest to us relate. Our choice for constructing these relationship maps here is the language of mathematics and we supplement that with another expressive language known as computer modeling. Our choice of language for that has been MatLab/Octave but that is merely convenient. Much of this material we have taught to a mixture of undergraduates and graduate students who have either taken one or two courses based on Peterson (2015a, b) or have a traditional mathematics degree where they have learned that material— without the modeling and computer work though —in other courses. On most college campuses in today's world, few students know other computer languages and so these books were written for the common denominator language MatLab which is readily available.

However, it is easy enough to do all of the computer modeling in Python or other languages. We encourage you to start exploring other programming viewpoints. In the next volume, we begin the process of setting up our tools for building network of computational nodes which interact and although all of the code is still within the MatLab/Octave family, it is clear we could perhaps do better if we started learning how to program in other choices. To start you off with Python, we recommend working through Langtangen (2012). This is a big book and will take a lot of time to process, but being able to model ideas from computational cognitive processing in both MatLab/Octave and Python will help you grow as a modeler. Another thing to do is to start learning about programming paradigms when the value of a variable can not be changed. This completely alters the way you program but it is very helpful when you try to model things using many cores on a laptop. It is easy to see why having a variable not be changeable is good in the case that you models using many processes that can access that variable. If each process can change the value, it is very easy to get into situations where the value of the variable loses meaning as it is not well defined. A good choice to begin studying this approach is to start reading about Erlang in Hébert (2013), Armstrong (2013) and Logan et al. (2011).

© Springer Science+Business Media Singapore 2016

J.K. Peterson, *Calculus for Cognitive Scientists*, Cognitive Science and Technology, DOI 10.1007/978-981-287-880-9_13

For computational speed, programming in C or C++ is very helpful and in Chap. 14 we mention books that can help you with learning those languages.

And of course, continue to learn about cognitive systems.

References

J. Armstrong, *Programming Erlang Second Edition: Software for a Concurrent World* (The Pragmatic Bookshelf, Dallas, 2013)

F. Hébert, *Learn You Some Erlang for Great Good* (No Starch Press, San Francisco, 2013)

H. Langtangen, *A Primer of Scientific Programming with Python* (Springer, New York, 2012)

M. Logan, E. Merritt, R. Carlsson, *Erlang and OTP in Actions* (Manning, Stamford, 2011)

J. Peterson, *Calculus for Cognitive Scientists: Derivatives, Integration and Modeling*, Springer Series on Cognitive Science and Technology (Science+Business Media Singapore Pte Ltd., Singapore, 2015a in press)

J. Peterson, *Calculus for Cognitive Scientists: Higher Order Models and Their Analysis*, Springer Series on Cognitive Science and Technology (Springer Science+Business Media Singapore Pte Ltd., Singapore, 2015b in press)

Part VI
Advise to the Beginner

Chapter 14
Background Reading

We have written these notes to help you if want to learn some of the basic principles of chemistry, biology and neuroscience that underlie the study of biological information processing. Perhaps you also wish to develop software models of cognition or learn better the material that underlies detailed mathematical models of neurons and other biological cells. We have been inspired by many attempts by people in disparate fields to find meaning and order in the vast compilations of knowledge that they must assimilate. Like them, we have done a fair bit of reading and study to prepare ourselves for necessary abstractions we need to make in our journey. We have learned a lot from various studies of theoretical biology and computation and so forth. You will need to make this journey too, so to help you, here are some specific comments about the sources we have used to learn from.

14.1 Biology and Physiology

It is clear that the regulation of information processing depends on many agents. These can be hormones and multiple neurotransmitters all interacting on vastly different time scales in mutually dependent ways. It is important to gain an overview understanding of these areas. The references below were how we have gotten started in this process:

- Animal physiology in Eckert et al. (1998). You will need some basic reference on animal physiology, just to see how all things fit together. This is a good source. This is also intended for browsing when you need to look something up. It has less technical detail than some of the other books in this section and because of that, it has more of an overview character that is quite useful at times.

© Springer Science+Business Media Singapore 2016
J.K. Peterson, *Calculus for Cognitive Scientists*, Cognitive Science
and Technology, DOI 10.1007/978-981-287-880-9_14

- General Neurobiology in Hall (1992). This is a standard introductory textbook to this area. It has all the requisite detail. We have read a lot of this—much of it gives needed background flavor and a useful feel for the subject.
- Ion Channels in Hille (1992a). This is an acknowledged source of material on the structure of the channels through which neurotransmitters (and other things) flow in and out of membranes. There are times when this book was a useful adjunct to our learning process. In *"Classical Biophysics of the Squid Giant Axon"* (Hille 1992b), we see a particular chapter devoted to the biophysics of the squid axonal fiber.
- The book **Foundations of Cellular Neurophysiology** by Johnston and Wu (1995a), is a detailed reference to cellular physiology. In *"Hodgkin and Huxley's Analysis of the Squid Giant Axon"* (Johnston and Wu 1995b), we note the particular chapter devoted to the Hodgkin and Huxley analysis of the giant squid axon.
- Cellular BioPhysics in Weiss (1996a, b). In the chapter *"The Hodgkin–Huxley Model"* Weiss (1996c), we reference a particular treatment of the Hodgkin–Huxley models.
- Essential Psychopharmacology in Stahl (2000).
- The chapter *"Multiple Channels and Calcium Dynamics"*, in **Methods of Neuronal Modeling** edited by Koch and Segev (Yamada et al. 1987) is a reference to a very detailed mathematical model for the generation of the output pulse of a neuron that does include second messenger effects. Although quite old, it is still very useful to read.
- Neuropharmacology in Stone (1995).
- Endrocrinology in Hadley (1996).
- Immunology in Austyn and Wood (1993).

14.2   Theoretical Biology

- The connections between genes, developmental biology and evolution from Raff and Kaufman (1983). This is the older version of his ideas, but it is very useful to a beginner to read. A more mature version of his ideas appears in Raff (1996). This is a more mature version of Raff's ideas on the connections between genes, development and evolution using more contemporary evidence.
- Neuromodulation and its importance in information processing in Katz (1999).
- The neurobiology of memory in Dudai (1989). This is a useful book that was our first introduction to theories of how nervous systems handle information processing. It is very speculative and for that reason, extremely interesting. There is much food for thought on software design buried in here!
- A theoretical approach to cell signaling, embryonic development and much more is presented in Gerhart and Kirschner (1997a). This book has influenced our

thinking enormously about generic triggers that initiate information exchange in cellular pathways. Specific chapters include *Regulatory Linkage* (Gerhart and Kirschner 1997b, Chap. 3) and *Conditionality and Compartmentalization* (Gerhart and Kirschner 1997c, Chap. 6).

14.3 Software

Out of the many books that are available for self-study in all of the areas above, some have proved to be invaluable, while others have been much less helpful. The following annotated list consists of the real gems. To learn to program effectively in an object oriented way in Python, it is helpful to know how to program in a procedural language such as C. Then, learning how to program objects within the constraints of the class syntax of C++ is very useful. This is the route we took in learning how to program in an object oriented way. The final step is to learn how to use a scripting glue language such as Python to build application software. Finally, don't be put off by the publication date of these resources! Many resources are timeless.

C++

The following books need to be on your shelf. Lippman will get you started, but you'll also need Deitel and Deitel and Olshevsky and Ponomarev for nuance.

1. C++ Primer (Lippman 1991).
   This book is the most basic resource for this area. While very complete, it has shortcomings; for example, it's discussion of call by reference is very unclear and its treatment of dynamic binding in its chapters on OOD is also murky. Nevertheless, it is a good basic introduction. It's biggest problem for us is that all of its examples are so simple (yes, even the *zoo* class is just *too* simple to give us much insight)
2. C++: How to Program (Deitel and Deitel 1994).
   We have found this book to be of great value. It intermingles excellent C++ coverage with ongoing object oriented design (OOD) material. It is full of practical advice on *software engineering* aspects of OOD design.
3. The Revolutionary Guide to OOP Using C++ (Olshevsky and Ponomarev 1994).
   This book has a wonderful discussion of call by reference and equally good material on dynamic binding.
4. Compiler Design (Wilhem and Maurer 1995).
   This book has already been mentioned in the text as the source of technical information on how an object-oriented compiler is built. This is an essential resource.

Object Oriented Programming and Design

1. Object-Oriented Analysis and Design with Applications (Booch 1994).
   This is a classic reference to one method of handling large scale OOD. As the

number of objects in your design grows there is a combinatorial explosion in the number of interaction pathways. The Booch method gives a popular software engineering tool. This is best to read on a surface level, for impressions and ideas.

2. Designing Object-Oriented C^{++} Applications Using the Booch Method (Martin 1995).

   If you decide to use the Booch method, this book is full of practical advice. It has many code examples, but it has a very heavy reliance on *templates*. This is a C^{++} language feature we have been avoiding because it complicates the architectural details. Hence, the translation of Martin's code fragments into useful insight is sometimes difficult, but nonetheless, there is much meat here.

3. Design Patterns for Object-Oriented Software Development (Pree 1995).

   The design of classes and objects is very much an art form. To some extent, like all crafts, you learn by doing. As you get more skilled, you realize how little of the real knowledge of how to write good classes is written down! This book is full of hard won real-world wisdom that comes out of actually being in the programming trenches. It is best to surface read and sample.

4. Taming C^{++}: Pattern Classes and Persistence for Large Projects (Soukup 1994).

   We have similar comments for this book. Since our proposed *neural objects* OOD project will be a rather massive undertaking, useful insight into the large scale OOD is most welcome!

5. Design Patterns: Elements of Reusable Object-Oriented Software (Gamma et al. 1995).

   As you program, you realize that many classes are essential building blocks of many disparate applications. This wonderful book brings together a large number of already worked out OOD solutions to common problems. It is extremely important to look at this book carefully. All of the different classes are presented in code sketches (not easy to follow, but well worth the effort!).

Neural Simulation Software: To model these things, we can use

1. the **Genesis** modeling language as discussed in **The Book of Genesis: Exploring Realistic Neural Models with the GEneral NEural SImulation System**, by Bower and Beeman (1998)
2. home grown code written in C++
3. home grown code written in MatLab.

References

J. Austyn, K. Wood, *Principles of Cellular and Molecular Immunology* (Oxford University Press, Oxford, 1993)

G. Booch, *Object-Oriented Analysis and Design with Applications*, 2nd edn. (Benjamin/Cummings Publishing Company Inc, Redwood City, 1994)

J. Bower, D. Beeman, *The Book of Genesis: Exploring Realistic Neural Models with the GEneral NEural SImulation System*, 2nd edn. (Springer TELOS, New York, 1998)

H. Deitel, P. Deitel, *C^{++}: How to Program* (Prentice Hal, Upper Saddle River, 1994)

Y. Dudai, *The Neurobiology of Memory: Concepts, Findings, Trends* (Oxford University Press, Oxford, 1989)

R. Eckert, D. Randall, G. Augustine, *Animal Physiology: Mechanisms and Adaptations*, 3rd edn. (W. H. freeman and Company, New York, 1998)

E. Gamma, R. Helm, R. Johnson, J. Vlissides, *Design Patterns: Elements of Reusable Object - Oriented Software* (Addison - Wesley Publishing Company, Reading, 1995)

J. Gerhart, M. Kirschner, *Cells, Embryos and Evolution: Towards a Cellular and Developmental Understanding of Phenotypic Variation and Evolutionary Adaptability* (Blackwell Science, Oxford, 1997a)

J. Gerhart, M. Kirschner, Regulatory linkage, *Cells, Embryos, and Evolution: Towards a Cellular and Developmental Understanding of Phenotypic Variation and Evolutionary Adaptability* (Blackwell Science, Oxford, 1997b), pp. 90–145

J. Gerhart, M. Kirschner, Conditionality and compartmentalization, *Cells, Embryos, and Evolution: Towards a Cellular and Developmental Understanding of Phenotypic Variation and Evolutionary Adaptability* (Blackwell Science, Oxford, 1997c), pp. 238–295

M. Hadley, *Endocrinology* (Prentice Hall, Upper Saddle River, 1996)

Z. Hall, *An Introduction to Molecular Neurobiology* (Sinauer Associates Inc., Sunderland, 1992)

B. Hille, *Ionic Channels of Excitable Membranes* (Sinauer Associates Inc, Sunderland, 1992a)

B. Hille, Classical biophysics of the squid giant axon, *Ionic Channels of Excitable Membranes* (Sinauer Associates Inc, Sunderland, 1992b), pp. 23–58

D. Johnston, S. Miao-Sin Wu, *Foundations of Cellular Neurophysiology* (MIT Press, Cambridge, 1995a)

D. Johnston, S. Miao-Sin Wu, Hodgkin and Huxley's Analysis of the Squid Giant Axon, *Foundations of Cellular Neurophysiology* (MIT Press, Cambridge, 1995b), pp. 143–182

P. Katz (ed.), *Beyond Neurotransmission: Neuromodulation and its Importance for Information Processing* (Oxford University Press, Oxford, 1999)

S. Lippman, C^{++} *Primer*, 2nd edn. (Addison–Wesley Publishing Company, Reading, 1991)

R. Martin, *Designing Object – Oriented C^{++} Applications Using the Booch Method* (Prentice Hall, Upper Saddle River, 1995)

V. Olshevsky, A. Ponomarev, *The Revolutionary Guide to OOP Using C^{++}* (WROX Publishers, Birmingham, 1994)

W. Pree, *Design Patterns for Object-Oriented Software Development*, ACM Press Books (Addison-Wesley Publishing Company, Reading, 1995)

R. Raff, T. Kaufman, *Embryos, Genes, and Evolution* (Macmillan Publishing Co., Inc, New York, 1983)

R. Raff, *The Shape of Life: Genes, Development, and the Evolution of the Animal Form* (The University of Chicago Press, Chicago, 1996)

J. Soukup, *Taming C^{++}: Pattern Classes and Persistence for Large Projects* (Addison–Wesley Publishing Company, Reading, 1994)

S. Stahl, *Essential Psychopharmacology: Neuroscientific Basis and Practical Applications*, 2nd edn. (Cambridge University Press, Cambridge, 2000)

T. Stone, *Neuropharmacology* (W. H. Freeman, Oxford, 1995)

T. Weiss, *Cellular Biophysics: Transport*, vol. 1 (MIT Press, Cambridge, 1996a)

T. Weiss, *Cellular Biophysics: Electrical Properties*, vol. 2 (MIT Press, Cambridge, 1996b)

T. Weiss, The Hodgkin - Huxley model, *Cellular Biophysics: Electrical Properties*, vol. 2 (MIT Press, Cambridge, 1996c), pp. 163–292

R. Wilhem, D. Maurer, *Compiler Design* (Addison-Wesley Publishing Company, Reading, 1995)

W. Yamada, C. Koch, P. Adams, Multiple channels and calcium dynamics, in *Methods of Neuronal Modeling*, ed. by C. Koch, I. Segev (MIT Press, Cambridge, 1987), pp. 97–134

Glossary

A

Amino acid   An α amino acid consists of the following things: an amide group NH_2, a carbonyl group $COOH$, a hydrogen atom H and a distinctive residue R. These groups are all attached to a central carbon atom which is called the α carbon. There are many common residues. An amino acid occurs in two different 3D forms. To keep it simple, look at this simple representation

$$R + y \text{ axis}$$
$$\uparrow$$
$$H \qquad \leftarrow C_\alpha \rightarrow \qquad NH_2 + x \text{ axis}$$
$$\downarrow$$
$$COOH$$

The R, H, NH_2 and $COOH$ are in the xy plane and the C_α carbon is along the positive z axis above the side groups. The NH_2 is on the positive x axis and the R is on the positive y axis. This is the L form as if you take your **right hand**, line up the fingers along the NH_2 line and rotate your fingers **left** towards the residue R. Note your thumb points out of the page towards the positive z axis location of C_α. Also, it is easy to visualize by just imaging grabbing the C_α and pulling it up out of the page that the other groups lie in. The other form is called the R form and looks like this:

$$H$$
$$\uparrow$$
$$COOH \qquad \leftarrow C_\alpha \rightarrow \qquad NH_2 + x \text{ axis}$$
$$\downarrow$$
$$R - y \text{ axis}$$

The R, NH_2, H and $COOH$ are in the xy plane and the C_α carbon is along the negative z axis below the side groups. Here the NH_2 is on the positive x axis but the R is on the negative y axis. This is the R form as if you take your **right hand**, line up the fingers along the NH_2 line and rotate your fingers **right** towards the

© Springer Science+Business Media Singapore 2016
J.K. Peterson, *Calculus for Cognitive Scientists*, Cognitive Science
and Technology, DOI 10.1007/978-981-287-880-9

residue R. Hence, we pull the C_α down below the page determined by the other groups here. Now there are a total of twenty amino acids which we show below in a table, p. 72.

Amino acid	Abbreviation	Amino acid	Abbreviation
Glycine	G, Gly	Methionine	M, Met
Alanine	A, Ala	Serine	S, Ser
Valine	V, Val	Lysine	K, Lys
Leucine	L, Leu	Threonine	T, Thr
Isoleucine	I, Ile	Arginine	R, Arg
Proline	P, Pro	Histidine	H, His
Phenylalanine	F, Phe	Aspartate	D, Asp
Tyrosine	Y, Tyr	Glutamate	E, Glu
Tryptophan	W, Trp	Asparagine	N, Asn
Cysteine	C, Cys	Glutamine	G, Gln

B

Ball stick   The salient features of the ball stick model are

- Axonal and dendritic fibers are modeled as two concentric membrane cylinders.
- The axon carries action potentials which propagate without change along the fiber once they are generated. Thus if an axon makes 100 synaptic contacts, we assume that the depolarizations of each presynaptic membrane are the same.
- Each synaptic contact on the dendritic tree generates a time and space localized depolarization of the postsynaptic membrane which is attenuated in space as the pulse travels along the fiber from the injection site and which decrease in magnitude the longer the time is since the pulse was generated.
- The effect of a synaptic contact is very dependent on the position along the dendritic fiber that the contact is made–in particular, how far was the contact from the axon hillock (i.e., in our model, how far from the soma)? Contacts made in essentially the same space locality have a high probability of reinforcing each other and thereby possibly generating a depolarization high enough to trigger an action potential.
- The effect of a synaptic contact is very dependent on the time at which the contact is made. Contacts made in essentially the same time frame have a high probability of reinforcing each other and thereby possibly generating a depolarization high enough to trigger an action potential.

We extend the simple dendritic cable model to what is called the ball and stick neuron model by using isopotential sphere to model the soma and coupling it to a single dendritic fiber input line. We model the soma as a simple parallel resistance/-capacitance network and the dendrite as a finite length cable. In Fig. 11.2, you see the terms I_0, the input current at the soma/dendrite junction starting at $\tau = 0$; I_D, the portion of the input current that enters the dendrite (effectively determined by the input conductance to the finite cable, G_D); I_S, the portion of the input current that enters the soma (effectively determined by the soma conductance G_S); and C_S, the soma membrane capacitance. We assume that the electrical properties of the soma and dendrite membrane are the same; this implies that the fundamental time and space constants of the soma and dendrite are given by the same constant (we will use our standard notation τ_M and λ_C as usual). It is possible to show that with a reasonable zero-rate left end cap condition the appropriate boundary condition at $\lambda = 0$ is given by

$$\rho \frac{\partial \hat{v}_m}{\partial \lambda}(0, \tau) = \tanh(L) \left[\hat{v}_m(0, \tau) + \frac{\partial \hat{v}_m}{\partial \tau}(0, \tau) \right],$$

where we introduce the fundamental ratio $\rho = \frac{G_D}{G_S}$, the ratio of the dendritic conductance to soma conductance. The full system to solve is therefore:

$$\frac{\partial^2 \hat{v}_m}{\partial \lambda^2} = \hat{v}_m + \frac{\partial \hat{v}_m}{\partial \tau}, \quad 0 \le \lambda \le L, \ \tau \ge 0.$$

$$\frac{\partial \hat{v}_m}{\partial \lambda}(L, \tau) = 0,$$

$$\rho \frac{\partial \hat{v}_m}{\partial \lambda}(0, \tau) = \tanh(L) \left[\hat{v}_m(0, \tau) + \frac{\partial \hat{v}_m}{\partial \tau}(0, \tau) \right].$$

which we solve using separation of variables, p. 362.

C

Cable model There are many variables are needed to describe what is happening inside and outside the membrane for a standard cable model. These variables include

Variable	Meaning	Units
V_m	Membrane potential	mV
K_m	Membrane current per length	nA/cm
K_e	Externally applied current	nA/cm
I_i	Inner current	nA
I_o	Outer current	nA
I_e	External current	nA
I_m	Membrane current	nA
V_i	Inner voltage	mV
V_o	Outer voltage	mV
r_i	Resistance inner fluid per length	μohms/cm
r_o	Resistance outer fluid per length	μohms/cm
g_m	Membrane conductance per length	μSiemens/cm
c_m	Membrane capacitance per length	nano Fahrads/cm
G_M	Membrane conductance	μSiemens/cm
C_M	Membrane capacitance	nano Fahrads/cm

The variables needed to describe what is happening inside and outside the cellular membrane and to some extent, inside the membrane are then

- V_m^0 is the rest value of the membrane potential.
- K_m^0 is the rest value of the membrane current per length density.
- K_e^0 is the rest value of the externally applied current per length density.
- I_i^0 is the rest value of the inner current.
- I_o^0 is the rest value of the outer current.
- V_i^0 is the rest value of the inner voltage.
- V_o^0 is the rest value of the outer voltage.
- r_i is the resistance of the inner fluid of the cable.
- r_o is the resistance of the outer fluid surrounding the cable.
- g_m is the membrane conductance per unit length.
- c_m is the membrane capacitance per unit length.

The membrane voltage can be shown to satisfy the first order partial differential equations

$$\frac{\partial I_i}{\partial z} = -K_m(z, t)$$

$$\frac{\partial I_o}{\partial z} = K_m(z, t) - K_e(z, t)$$

$$\frac{\partial V_i}{\partial z} = -r_i I_i(z, t)$$

$$\frac{\partial V_o}{\partial z} = -r_o I_o(z, t)$$

$$V_m = V_i - V_0$$

which we can then use to derive the standard second order partial differential equation

$$\frac{\partial^2 V_m}{\partial z^2} = (r_i + r_o)K_m - r_o K_e$$

This equation can be solved in many ways. First, we can look at cables that are infinitely long or finitely long assuming there is no dependence on time. Then, we can add time as a variable which forces us to use new tools for finding the solution. The cable equation, time dependent or independent, provides a model for the input side of a neuron. It is part of a full neuron model called the ball stick model, p. 149.

Complimentary nucleotides the purine and pyrimidine nucleotides can bond together in the following ways: A to T or T to A and C to G or G to C. We say that adenine and thymine and cytosine and guanine are complementary nucleotides. This bonding occurs because hydrogen bonds can form between the adjacent nitrogen or between adjacent nitrogen and oxygen atoms. For example, look at the T–A

bond in Fig. 4.40. Note the bases are *inside* and the sugars outside. Finally, note how the bonding is done for the cytosine and guanine components in Fig. 4.41. Now as we have said, nucleotides can link into a long chain via the phosphate bond. Each base in this chain is attracted to a complimentary base. It is energetically favorable for two chains to form: chain one and its *complement* chain 2. Each pair of complimentary nucleotides is called a *complimentary base pair*. The forces that act on the residues of the nucleotides and between the nucleotides themselves coupled with the rigid nature of the peptide bond between two nucleotides induce the two chains to form a **double helix** structure under cellular conditions which in cross-section (see Fig. 4.42) has the bases inside and the sugars outside. The double helix is called **DNA** when deoxy-ribose sugars are used on the nucleotides in our alphabet. The name **DNA** stands for *deoxy-ribose nucleic acid*. A chain structure closely related to **DNA** is what is called **RNA**, where the **R** refers to the fact that oxy-ribose sugars or simply ribose sugars are used on the nucleotides in the alphabet used to build **RNA**. In RNA, thymine is replaced by Uracil. The **RNA** alphabet is slightly different as the nucleotide Thymine, **T**, in the **DNA** alphabet is replaced by the similar nucleotide Uracil, **U**. The chemical structure of uracil is shown in Fig. 4.44 right next to the formula for thymine. Note that the only difference is that carbon $^5C'$ holds a methyl group in thymine and just a hydrogen in uracil. Despite these differences, uracil will still bond to adenine via a complimentary bond, p. 91.

E

Einstein's Relation   There is a relation between the diffusion coefficient D and the mobility μ of an ion which is called Einstein's Relation. It says

$$D = \frac{\kappa T}{q} \mu$$

where

- κ is Boltzmann's constant which is $1.38 \times 10^{-23} \frac{\text{joule}}{^\circ K}$.
- T is the temperature in degrees Kelvin.
- q is the charge of the ion c which has units of coulombs.

Further, we see that Einstein's Law says that diffusion and drift processes are additive because Ohm's Law of Drift says J_{drift} is proportional to μ which by Einstein's Law is proportional to D and hence J_{diff}, p. 101.

Excitable cell   There are specialized cells in most living creatures called neurons which are adapted for generating signals which are used for the transmission of sensory data, control of movement and cognition through mechanisms we don't fully understand. A neuron has a membrane studded with many voltage gated sodium and potassium channels. In terms of ionic permeabilities, the GHK voltage equation for the usual sodium, potassium and chlorine ions gives

$$V_0^m = 25.32 \; (mV) \; \ln \left(\frac{\mathscr{P}_K[K^+]_{out} + \mathscr{P}_{Na}[Na^+]_{out} + \mathscr{P}_{Cl}[Cl]_{in}}{\mathscr{P}_K[K^+]_{in} + \mathscr{P}_{Na}[Na^+]_{in} + \mathscr{P}_{Cl}[Cl^-]_{out}} \right)$$

which is about -60 mV at rest but which can rapidly increase to $+40$ mV upon a large shift in the sodium and potassium permeability ratio. We can also write the rest voltage in terms of conductances as

$$V_0^m = \frac{g_K}{g_K + g_{Na} + g_{Cl}} E_K + \frac{g_{Na}}{g_K + g_{Na} + g_{Cl}} E_{Na} + \frac{g_{Cl}}{g_K + g_{Na} + g_{Cl}} E_{Cl}$$

Either the conductance or the permeability model allows us to understand there is a sudden increase in voltage across the membrane in terms of either sodium to potassium permeability or conductance ratio shifts. Given the right inputs into the neuron, the potential across the membrane suddenly rises and is then followed by a sudden drop below the equilibrium voltage and then ended by a slow increase back up to the rest potential. The shape of this wave form is very characteristic and is shown in Fig. 5.10. This type of wave form is called an action potential and is a fundamental characteristic of excitable cells. It occurs because the sodium and potassium ions move through the cell in voltage dependent gates, p. 142.

F

Ficke's Law of Diffusion   is an empirical law which says the rate of change of the concentration of molecule b is proportional to the diffusion flux and is written in mathematical form as follows:

$$J_{diff} = -D \frac{\partial [b]}{\partial x}$$

where

- J_{diff} is diffusion flux which has units of $\frac{molecules}{cm^2 - second}$.
- D is the diffusion coefficient which has units of $\frac{cm^2}{second}$.
- $[b]$ is the concentration of molecule b which has units of $\frac{molecules}{cm^3}$.

The *minus* sign implies that flow is from **high** to **low** concentration; hence diffusion takes place *down* the concentration gradient. Note that D is the proportionality constant in this law, p. 100.

Fourier Series   A general trigonometric series $S(x)$ has the following form

$$S(x) = b_0 + \sum_{i=1}^{\infty} \left(a_n \sin\left(\frac{i\pi}{L}x\right) + b_n \cos\left(\frac{i\pi}{L}x\right) \right)$$

for any numbers a_n and b_n. Of course, there is no guarantee that this series will converge at any x! If we start with a function f which is continuous on the interval $[0, L]$, we can define the trigonometric series associated with f as follows

$$S(x) = \frac{1}{L} <f, \mathbf{1}>$$

$$+ \sum_{i=1}^{\infty} \left(\frac{2}{L}\left\langle f(x), \sin\left(\frac{i\pi}{L}x\right) \right\rangle \sin\left(\frac{i\pi}{L}x\right) + \frac{2}{L}\left\langle f(x), \cos\left(\frac{i\pi}{L}x\right) \right\rangle \cos\left(\frac{i\pi}{L}x\right) \right).$$

This series is called the Fourier Series for f and the coefficients in the Fourier series for f are called the *Fourier coefficients* of f. Since these coefficients are based on inner products with the normalized sin and cos functions, they are called *normalized* Fourier coefficients. The nth Fourier sin coefficient, $n \geq 1$, of f is as follows:

$$a_n(f) = \frac{2}{L} \int_0^L f(x) \sin\left(\frac{i\pi}{L}x\right) dx$$

The nth *Fourier cos coefficient, $n \geq 0$, of f* are defined similarly, p. 243:

$$b_0(f) - \frac{1}{L} \int_0^L f(x)\, dx$$

$$b_n(f) = \frac{2}{L} \int_0^L f(x) \cos\left(\frac{i\pi}{L}x\right) dx, \quad n \geq 1.$$

G

Goldman–Hodgkin–Huxley   The GHK model is based on several assumptions about how the substance c behaves in the membrane:

1. $[c^m]$ varies linearly across the membrane as discussed in the previous subsection.
2. The electric field in the membrane is constant.
3. The Nernst–Planck equation holds inside the membrane.

After much work and some approximations, we obtain another current equation

$$I_0^m = \frac{z^2 \mathscr{P} F^2 V_c^m}{RT} \left(\frac{[c]_{in} - [c]_{out} e^{-\frac{zFV_c^m}{RT}}}{1 - e^{-\frac{zFV_c^m}{RT}}} \right)$$

where $\mathscr{P}$ is the permeability of the ion which is a measure of how ions flow across the membrane. In an equilibrium situation, this current is zero and we obtain the GHK voltage equation. If the cell is permeable to say sodium, potassium and chlorine, we find the GHK currents must sum to zero:

$$I_K + I_{Na} + I_{Cl} = 0$$

Further, the associated Nernst potentials for the ions should all match because otherwise there would be current flow:

$$V_0^m = V_{K^+}^m = V_{Na^+}^m = V_{Cl^-}^m$$

where we denote this common potential by V_0^m. After manipulation, we have

$$V_0^m = \frac{RT}{F} \ln \left(\frac{\mathscr{P}_K[K^+]_{out} + \mathscr{P}_{Na}[Na^+]_{out} + \mathscr{P}_{Cl}[Cl]_{in}}{\mathscr{P}_K[K^+]_{in} + \mathscr{P}_{Na}[Na^+]_{in} + \mathscr{P}_{Cl}[Cl^-]_{out}} \right)$$

where to actually compute the GHK voltage, we would need the three ion permeabilities, p. 132.

Graham–Schmidt Orthogonalization   Graham–Schmidt orthogonalization is a tool for taking linearly independent objects and using them to construct a new set of linearly independent objects that are mutually orthogonal. This, of course, requires that the objects are from a vector space with an inner product, p. 13.

H

Hodgkin–Huxley gates   Hodgkin and Huxley modeled the sodium and potassium gates as

$$g_{Na}(V) = g_{Na}^{Max} \, \mathscr{M}_{NA}^{3}(V) \, \mathscr{H}_{NA}(V)$$
$$g_K(V) = g_K^{Max} \, \mathscr{M}_K^{4}(V)$$

where the two activation variables, $\mathscr{M}_{NA}$ and $\mathscr{M}_K$, and the one inactivation variable, $\mathscr{H}_{NA}$, all satisfy the first order kinetics

$$\tau_{\mathscr{M}_{NA}} \, \mathscr{M}_{NA}'(t) = (\mathscr{M}_{NA})_\infty - \mathscr{M}_{NA}$$
$$\tau_{\mathscr{H}_{NA}} \, \mathscr{H}_{NA}'(t) = (\mathscr{H}_{NA})_\infty - \mathscr{H}_{NA}$$
$$\tau_{\mathscr{M}_K} \, \mathscr{M}_K'(t) = (\mathscr{M}_K)_\infty - \mathscr{M}_K$$

with

$$\tau_{\mathcal{M}_{NA}} = \frac{1}{\alpha_{\mathcal{M}_{NA}} + \beta_{\mathcal{M}_{NA}}}$$

$$(\mathcal{M}_{NA})_\infty = \frac{\alpha_{\mathcal{M}_{NA}}}{\alpha_{\mathcal{M}_{NA}} + \beta_{\mathcal{M}_{NA}}}$$

$$\tau_{\mathcal{H}_{NA}} = \frac{1}{\alpha_{\mathcal{H}_{NA}} + \beta_{\mathcal{H}_{NA}}}$$

$$(\mathcal{H}_{NA})_\infty = \frac{\alpha_{\mathcal{H}_{NA}}}{\alpha_{\mathcal{H}_{NA}} + \beta_{\mathcal{H}_{NA}}}$$

$$\tau_{\mathcal{M}_K} = \frac{1}{\alpha_{\mathcal{M}_K} + \beta_{\mathcal{M}_K}}$$

$$(\mathcal{M}_K)_\infty = \frac{\alpha_{\mathcal{M}_K}}{\alpha_{\mathcal{M}_K} + \beta_{\mathcal{M}_K}}$$

Further, the coefficient functions, α and β for each variable required data fits as functions of voltage. These were determined to be

$$\alpha_{\mathcal{M}_{NA}} = -0.10 \, \frac{V + 35.0}{e^{-0.1\,(V+35.0)} - 1.0}$$

$$\beta_{\mathcal{M}_{NA}} = 4.0 \, e^{\frac{-(V+60.0)}{18.0}}$$

$$\alpha_{\mathcal{H}_{NA}} = 0.07 \, e^{-0.05\,(V+60.0)}$$

$$\beta_{\mathcal{H}_{NA}} = \frac{1.0}{(1.0 + e^{-0.1\,(V+30.0)})}$$

$$\alpha_{\mathcal{M}_K} = -\frac{0.01 * (V + 50.0)}{(e^{-0.1(V+50.0)} - 1.0)}$$

$$\beta_{\mathcal{M}_K} = 0.125 \, e^{-0.0125\,(V+60.0)}$$

Of course these data fits were obtained at a certain temperature. The model of the membrane dynamics thus consists of the following differential equations:

$$\tau_{\mathcal{M}_{NA}} \, \frac{d\mathcal{M}_{NA}}{dt} = (\mathcal{M}_{NA})_\infty - \mathcal{M}_{NA}$$

$$\tau_{\mathcal{H}_{NA}} \, \frac{d\mathcal{H}_{NA}}{dt} = (\mathcal{H}_{NA})_\infty - \mathcal{H}_{NA}$$

$$\tau_{\mathcal{M}_K} \, \frac{d\mathcal{M}_K}{dt} = (\mathcal{M}_K)_\infty - \mathcal{M}_K$$

$$\frac{dV}{dt} = \frac{I_M - I_K - I_{Na} - I_L}{C_M}$$

where the leakage current I_L is handled more simply by choosing a leakage conductance and leakage voltage that is relevant to the simulation. The model has the initial conditions

$$\mathscr{M}_{NA}(0) = (\mathscr{M}_{NA})_\infty(V_0, 0)$$
$$\mathscr{H}_{NA}(0) = (\mathscr{H}_{NA})_\infty(V_0, 0)$$
$$\mathscr{M}_K(0) = (\mathscr{M}_K)_\infty(V_0, 0)$$
$$V(0) = V_0$$

We note that at equilibrium there is no current across the membrane. Hence, the sodium and potassium currents are zero and the activation and inactivation variables should achieve their steady state values which would be m_∞, h_∞ and n_∞ computed at the equilibrium membrane potential which is here denoted by V_0, p. 410.

Hodgkin–Huxley model   The standard Hodgkin–Huxley model of an excitatory neuron consists of the equation for the total membrane current, K_M, obtained from Ohm's law

$$K_m = c_m \frac{\partial V_m}{\partial t} + K_K + K_{Na} + K_L,$$

where we have expanded the K_{ion} term to include the contributions from the sodium and potassium currents and the leakage current. The new equation for the membrane voltage is thus

$$\frac{\partial^2 V_m}{\partial z^2} = (r_i + r_o)K_m - r_o K_e$$
$$= (r_i + r_o) c_m \frac{\partial V_m}{\partial t} + (r_i + r_o) K_K + (r_i + r_o) K_{Na} + (r_i + r_o) K_L - r_o K_e$$

which can be simplified to

$$\frac{1}{r_i + r_o} \frac{\partial^2 V_m}{\partial z^2} = K_m - \frac{r_o}{r_i + r_o} K_e$$
$$= c_m \frac{\partial V_m}{\partial t} + K_K + K_{Na} + K_L - \frac{r_o}{r_i + r_o} K_e$$

where the ion current terms are modeled using Hodgkin and Huxley's model of the voltage dependent gates, p. 405.

Hydrocarbons   Hydrocarbons are molecules made up of strings of carbons. Since each carbon can make four bonds, these strings can be quite complex. Any of the four binding sites of a carbon can be filled with an arbitrarily large complex which we often called a residue and these in turn can have other residues bound to them. Hence, a hydrocarbon chain could be linear if there are no side chains or *bushy* if the chain has multiple side chains due to the attached residues. Also, five carbons can combine together into a cyclical structure called a ring which is a potent molecular component that occurs in many biologically reactive molecular assemblies, p. 71.

Hydrogen bonds   these occur when a hydrogen atom is shared between two other atoms: The atom to which the hydrogen is held more tightly is called the **hydrogen donor** and the other one which is less tightly linked is called the **hydrogen acceptor**, p. 65.

I

Infinite series   The simplest one is a series of positive terms which we write as $\sum_{n=1}^{\infty} a_n$ where each a_n is a positive number. If we add up a finite number at a time starting from the beginning, we get a sequence of what are called partial sums of the form $S_N = \sum_{n=1}^{N} a_n$, We say the series converges if this sequence of partial sums converges and as usual, this need not be true. The sequence of partial sums could blow up to infinity if we sum terms that do not decay very fast. For example, the series $\sum_{n=1}^{\infty} 1$ has partial sums $S_N = N$ which go off to infinity as we add more terms in. We can show other series such as $\sum_{n=1}^{\infty} \frac{1}{n}$ also add up to infinity although that is a bit harder. The point is that just having the nth term a_n go to zero is not enough to insure the series adds up to a finite number. In this simple case, we are always adding positive numbers, so the partial sums are always increasing. But if the terms a_n can have different algebraic signs, the behavior of these partial sums can get quite complicated. To study this behavior properly naturally enough requires a lot of mathematical analysis which we only do some of in the text. So you should feel free to read more! A general series is then of the form $\sum_{n=1}^{\infty} a_n$ where the a_n's can have differing algebraic signs. We can make it even more general, by looking at series of functions such as $\sum_{n=1}^{\infty} f_n(x)$. The idea of partial sums is still the same but we have to think about whether or not the series sums to a finite number (we say the series converges) at each x in the domain of our functions. This gets more difficult fast and requires more intellectual effort to master. When the series contains functions like $f_n(x)$, there a many new types of convergence to consider such as convergence at a given value of x but not necessarily at another value of x. This is called *pointwise convergence*. Convergence ideas on an entire interval such as $[0, 1]$ of possible x values are called *uniform convergence* notions and they imply the pointwise ideas but the implication does not go the other way. Finally, there is an idea called L_2 convergence which is also different from those. There are many details in Chap. 9 you should study, p. 235.

L

Linear partial differential equation   These are models where the variable u satisfies an equation involving the function u, its partial derivatives, second order partial derivatives and so forth. The relationship between these terms can be very nonlinear but here we will just mention a few common linear PDE which we discuss in Chap. 10.

The Wave Equation: We seek functions $\Phi(x, t)$ so that

$$\frac{\partial^2 \Phi}{\partial t^2} - c^2 \frac{\partial^2 \Phi}{\partial x^2} = 0$$

$$\Phi(x, 0) = f(x), \quad \text{for } 0 \leq x \leq L$$

$$\frac{\partial \Phi}{\partial t}(x, 0) = g(x), \quad \text{for } 0 \leq x \leq L$$

$$\Phi(0, t) = 0, \quad \text{for } 0 \leq t$$

$$\Phi(L, t) = 0, \quad \text{for } 0 \leq t$$

for some positive constant c. The solution of this equation approximates the motion of a nice string with no external forces applied. The domain here is the infinite rectangle $[0, L] \times [0, \infty)$.

Laplace's Equation: The solution $\Phi(x, y)$ of this equation is a time independent solution to a problem such as the distribution of heat on a membrane stretch over the domain given that various heat sources are applied to the boundary. Here, the domain is the finite square $[0, L] \times [0, L]$. In the problem below, three of the edges of the square are clamped to 0 and the remaining one must follow the heat profile given by the function $f(x)$.

$$\frac{\partial^2 \Phi}{\partial x^2} + \frac{\partial^2 \Phi}{\partial y^2} = 0$$

$$\frac{\partial \Phi}{\partial x}(0, y) = 0, \quad \text{for } 0 \leq y \leq L$$

$$\frac{\partial \Phi}{\partial x}(L, y) = 0, \quad \text{for } 0 \leq y \leq L$$

$$\Phi(x, L) = 0, \quad \text{for } 0 \leq x \leq L$$

$$\Phi(x, 0) = f(x), \quad \text{for } 0 \leq x \leq L$$

The Heat/Diffusion Equation: The solution of this equation, $\Phi(x, t)$, is the time dependent value of heat or temperature of a one dimensional bar which is having a heat source applied to it initially. It can also model a substance moving through a domain using diffusion with diffusion constant D as discussed in Chap. 5. The domain is again half infinite: $[0, L] \times [0, \infty)$.

$$\frac{\partial \Phi}{\partial t} - D \frac{\partial^2 \Phi}{\partial x^2} = 0$$

$$\Phi(0, t) = 0, \quad \text{for } 0 < t$$

$$\Phi(L, t) = 0, \quad \text{for } 0 < t$$

$$\Phi(x, 0) = f(x), \quad \text{for } 0 < x < L$$

This equation is very relevant to our needs. Indeed, in Peterson (2015), we derive this equation using a random walk model and learn how to interpret the diffusion constant D in terms of the space constant λ_C and the time constant τ_M. However, we will not discuss that here.

The Cable Equation: The solution $\Phi(x, t)$ of this equation for us is usually the membrane voltage in a dendrite model for an excitable neuron. Here the domain is the infinite rectangle $[0, L] \times [0, \infty)$ where the space variable represents the position on the cable. In the problem below, fluxes in the spatial direction at the endcaps of the fiber are zero and there is an imposed voltage input function over the entire cable when time is zero, p. 309.

$$\beta^2 \frac{\partial^2 \Phi}{\partial x^2} - \Phi - \alpha \frac{\partial \Phi}{\partial t} = 0, \quad \text{for } 0 \le x \le L, \ t \ge 0,$$

$$\frac{\partial \Phi}{\partial x}(0, t) = 0,$$

$$\frac{\partial \Phi}{\partial x}(L, t) = 0,$$

$$\Phi(x, 0) = f(x).$$

M

Molecular bond Molecular bonds between molecules generally involve an attraction between the negative charge on one molecule and the positive charge on another. There are many types of these bonds; for example, hydrogen bonds, disulfide bonds, Vanderwaals bonds and covalent bonds. A nice high level look at this process is to think of a given atom like carbon as having some electrons available for sharing. If carbon could add 4 electrons, it would have all of its higher locations for electron populations be completely filled which is a preferred energy state. So carbon finds other atoms it can share an electron with. A simple example is methyl CH_4 where each of the 4 hydrogens shares an electron with carbon. Another example is CO_2 where each of the two oxygens share 2 electrons with the carbon. Sharing 2 electrons gives what is called a double bond, p. 65.

N

Nernst The current form of the Nernst–Planck equation given in Eq. 5.10 describes ionic current flow driven by electro-chemical potentials (concentration gradients and electric fields). When the net current due to all of these contributions is zero, we have $I = 0$ and after suitable manipulations, we can derive the Nernst equation in terms of the inner and outer concentration of the ion c.

$$E_c = \frac{RT}{zF} \ln \frac{[c]_{out}}{[c]_{in}}$$

If we have more than one ion, say sodium, potassium and chlorine, at equilibrium, the currents would sum to zero and we would find

$$V_m = \frac{g_K}{g_K + g_{Na} + g_{Cl}} E_K + \frac{g_{Na}}{g_K + g_{Na} + g_{Cl}} E_{Na} + \frac{g_{Cl}}{g_K + g_{Na} + g_{Cl}} E_{Cl}$$

where the terms g_K and so forth are the conductances of the ions, p. 106.

Nernst Planck Under physiological conditions, ion movement across the membrane is influenced by both electric fields and concentration gradients. Let J denote the total flux, then we will assume that we can add linearly the diffusion due to the molecule c and the drift due to the ion c giving

$$J = J_{drift} + J_{diff}$$

Thus, applying Ohm's Law 5.3 and Ficke's Law 5.1, we have

$$J = -\mu z [c] \frac{\partial V}{\partial x} - D \frac{\partial [c]}{\partial x}$$

Next, we use Einstein's Relation 5.4 to replace the diffusion constant D to obtain what is called the Nernst–Planck Equation

$$J = -\mu z [c] \frac{\partial V}{\partial x} - \frac{\kappa T}{q} \mu \frac{\partial [c]}{\partial x}$$
$$= -\mu \left(z [c] \frac{\partial V}{\partial x} + \frac{\kappa T}{q} \frac{\partial [c]}{\partial x} \right)$$

We can rewrite this result by moving to units that are $\frac{moles}{cm^2-second}$ and using a known relationship between charge and moles given by Faraday's Constant F. In addition, we can convert between joules and calories using the gas constant R. We obtain a current equation

$$I = \frac{J}{N_A} zF = -\frac{\mu}{N_A} \left(z^2 F [c] \frac{\partial V}{\partial x} + z RT \frac{\partial [c]}{\partial x} \right)$$

where the symbol I denotes this current density $\frac{amps}{cm^2}$ that we obtain with this equation. The current I is the ion current that flows across the membrane per unit area due to the forces acting on the ion c, p. 103.

Nucleic acids Two important sugars are formed from cyclic hydrocarbons and are called furans and pyrans. A **pyran**'s schematic is shown in Fig. 4.30b and the **furan**'s in Fig. 4.30d. All the carbons in these ring structures are labeled and the 3D geometry of these sugars is important. A three dimensional version of the furan molecule is the $^2C'$ **endo** form shown in Fig. 4.31b. Here, $^2C'$ and $^5C'$ are out of the plane formed by $O-^1C'-^3C'-^4C'$. Another version is the one where $^3C'$ and $^5C'$ are out of the plane formed by $O-^1C'-^2C'-^4C'$ which is called the $^3C'$ **endo** form. DNA uses the $^2C'$ **endo** and RNA, the $^3C'$ **endo** form. The particular sugar of interest is **ribose** which comes in an oxygenated and non oxygenated form; **ribose** (Fig. 4.32a) and **deoxyribose** (Fig. 4.32b), respectively. There are also four special nitrogenous bases which are important. They come in two flavors: **purines** and **pyrimidines**. The **purines** have the form shown in Fig. 4.33a while the **pyrimidines** have the one shown in Fig. 4.33b. There are two purines and two pyrimidines we need to know about: the purines **adenine (A)** and **guanine (G)** and the pyrimidines **thymine (T)** and **cytosine (C)**. Figure 4.34a (Adenine is a purine with an attached amide on $^6C'$), 4.34b (Guanine is a purine with an attached oxygen on $^6C'$), 4.34d (Cytosine is a pyrimidine with an attached amide on $^4C'$), and 4.34c (Thymine is a pyrimidine with an attached oxygen on $^4C'$). These four nitrogenous bases can bond to the ribose or deoxyribose sugars to create what are called **nucleotides**. In general, a sugar plus a purine or pyrimidine nitrogenous base give us a **nucleoside**. If we add phosphate to the $^5C'$ of the sugar, we get a new molecule called a **nucleotide**. This level of detail is far more complicated and messy than we typically wish to show; hence, we generally draw this in the compact form shown in Fig. 4.37. There, we have replaced the base with a simple shaded box and simply labeled the primed carbons with the numerical ranking. In Fig. 4.38 we show how nucleotides can link up into chains: bond the $^5C'$ of the ribose on one nucleotide to the $^3C'$ of the ribose on another nucleotide with a phosphate or PO_3^- bridge. Symbolically this looks like Fig. 4.38. This chain of three nucleotides has a terminal OH on the $^5C'$ of the top sugar and a terminal OH on the $^3C'$ of the bottom sugar. We often write this even more abstractly as shown in Fig. 4.39 or just $OH-$ **Base 3** P **Base 2** P **Base 1** $P-OH$, where the P denotes a phosphate bridge. For example, for a chain with bases adenine, adenine, cytosine and guanine, we would write $OH-A-p-A-p-C-p-G-OH$ or $OHApApCpGOH$. Even this is cumbersome, so we will leave out the common phosphate bridges and terminal hydroxyl groups and simply write $AACG$. It is thus understood the left end is an OH terminated $^5C'$ and the right end an hydroxyl terminated $^3C'$, p. 85.

O

Ohm's Law of Drift   relates the electrical field due to an charged molecule, i.e. an ion, c, across a membrane to the drift of the ion across the membrane where *drift* is the amount of ions that moves across the membrane per unit area. In mathematical form

$$J_{drift} = -\partial_{el} E$$

where

- J_{drift} is the drift of the ion which has units of $\frac{\text{molecules}}{\text{cm}^2-\text{second}}$.
- ∂_{el} is electrical conductivity which has units of $\frac{\text{molecules}}{\text{volt}-\text{cm}-\text{second}}$.

From basic physics, we know an electrical field is the negative gradient of the potential so if V is the potential across the membrane and x is the variable that measures our position on the membrane, we have

$$E = -\frac{\partial V}{\partial x}$$

Let the valence of the ion c be denoted by z. It is possible to show

$$\partial_{el} = \mu z [c]$$

where dimensional analysis shows us that the proportionality constant μ, called the **mobility** of ion c, has units $\frac{\text{cm}^2}{\text{volt}-\text{second}}$. Hence, we can rewrite Ohm's Law of Drift as, p. 100.

$$J_{drift} = -\mu z[c] \frac{\partial V}{\partial x}$$

P

Peptide bond   Amino acids can link up in chains because the $COOH$ on one can bond with the NH_2 on another. The two amino acids that pair are connected by a rigid planar bond. There is a C_α^1 atom from amino acid one and another C_α^2 from amino acid two attached to this bond. The $COOH$ and NH_2 bond looks like this in block diagram form. The $COOH$ loses an OH and the NH_2 loses an H to form the bond. Think of bond as forming a rigid piece of cardboard and attached on the left is the amino acid built around C_α^1 and attached on the right is the amino acid build around C_α^2.

$$
\begin{array}{ccc}
 & O & (+y \text{ local axis}) \\
 & \uparrow & \\
C_\alpha^1 \rightarrow \leftarrow & C \rightarrow N(+x \text{ local axis}) & \rightarrow \leftarrow C_\alpha^2 \\
 & \downarrow & \\
 & H &
\end{array}
$$

Now think of C_α^1 as attached to a pencil which is plugged into the side of the peptide bond. The C_α^1 to CO bond is an axis that amino acid one is free to rotate about. Call this angle of rotation Ψ_1. We can do the same thing for the other side and talk about a rotation angle Ψ_2 for the NH to C_α^2 bond. In Fig. 4.21, R_1 is the residue or side chain for the first amino acid and R_2 is the side chain for the other.

Note amino acid one starts with an N_2H group on the left and amino acid two ends with a $COOH$ group on the right

$$C_\alpha^1 \rightarrow \Psi_1 \leftarrow \overset{\overset{O}{\uparrow}}{\underset{\underset{H}{\downarrow}}{C}} \rightarrow N(+x \text{ local axis}) \rightarrow \Psi_2 \leftarrow C_\alpha^2$$

(+y local axis)

The peptide bond allows amino acids to link into chains as we show in the next block diagram, p. 80.

$$N_2H \leftarrow C_\alpha^1 \rightarrow \Psi_1 \leftarrow \overset{\overset{O}{\uparrow}}{\underset{\underset{H}{\downarrow}}{C}} \rightarrow N(+x \text{ local axis}) \rightarrow \Psi_2 \leftarrow C_\alpha^2 \rightarrow COOH$$

(+y local axis)

Protein transcription When a protein is built, certain biological machines are used to find the appropriate place in the **DNA** double helix where a long string of nucleotides which contains the information needed to build the protein is stored. This long chain of nucleotides which encodes the information to build a protein is called a **gene**. Biological machinery *unzips* the double helix at this special point into two chains as shown in Fig. 4.45. A complimentary copy of a DNA single strand fragment is made using complimentary pairing but this time adenine pairs to uracil to create a fragment of **RNA**. This fragment of RNA serves to **transfer** information encoded in the DNA fragment to other places in the cell where the actual protein can be assembled. Hence, this RNA fragment is given a special name—**Messenger RNA or mRNA** for short. This transfer process is called **transcription**. For example, the DNA fragment $^5ACCGTTACCGT^3$ has the DNA complement $^3TGGCAATGGCA^5$ although in the cell, the complimentary RNA fragment

$$^3UGGCAAUGGCA^5$$

is produced instead. Note again that the ^{5} pairs with a ^{3} and vice versa. From what we said earlier, there are 64 different triplets that can be made from the alphabet $\{A, C, U, G\}$ and it is this mapping that is used to assemble the protein chain a little at a time. For each chain that is unzipped, a complimentary chain is attracted to it in the fashion shown by Table 4.7. This complimentary chain will however be built from the oxygenated deoxy-ribose or simply ribose nucleotides. Hence, this complimentary chain is part of a *complimentary* RNA helix. As the amino acids encoded by mRNA are built and exit from the ribosome into the fluid inside the cell, the chain of amino acids or polypeptides begins to twist and curl into its three dimensional shape based on all the forces acting on it. We can write this whole process symbolically as **DNA → mRNA → ribosome → Protein**. This is known as the **Central Dogma of Molecular Biology**. Hence to decode a particular gene stored in DNA which has been translated to its complimentary mRNA form all we need to know are which triplets are associated with which amino acids. These triplets are called **DNA Codons**. The DNA alphabet form of this mapping is given in Table 4.8. For example, the DNA sequence,

TAC|TAT|GTG|CTT|ACC|TCG|ATT

is translated into the mRNA sequence

AUG|AUA|CAC|GAA|UGG|AGC|UAA

which corresponds to the amino acid string, p. 94.

Start|Isoleucine|Histidine|Glutamic Acid|Tryptophan|Serine|Stop

R

Runge–Kutta   a method for the numerical solution of a system of differential equations. The basic idea is to approximate the true solution using first order linear approximations—i.e. tangent lines—constructed at various time and space points based on the current time and space location. The details are in the text. We typically use RK 4 which uses four such tangent line approximations to build an estimate of the true solution which has local error proportional to h^5 where h is the step size. The global error is the error made in approximating the true solution over the entire time domain of interest and it is proportional to h^4. RK4 is a nice trade off between computation and accuracy. We use 4 dynamics function evaluations at each time step but we gain h^4 accuracy. Thus, in general we can get by with a larger step size, p. 47.

Runge–Kutta–Fehlberg   This method adds an additional tangent line approximation at each time step which we use to estimate how much error we have made in approximating the true solution. As long as a true solution does exist, this is a great tool as it helps us believe the numerical solution we are constructing is indeed reasonable. We typically use RKF45 which is a base RK4 method with an additional approximation step to estimate the global error in approximating the true solution. Thus, the method is proportional to h^5 globally. The fact that we can estimate how much error we make in tracking the true solution means we can decide if the current step size is too small or too large and use that information to double or half the current step size. This allow us to start with a large step size and the method will cut that size down to a good size so we do not make too much error initially, p. 53.

S

Separation of Variables Method   A common PDE model is the general cable model which is given below is fairly abstract form.

$$\beta^2 \frac{\partial^2 \Phi}{\partial x^2} - \Phi - \alpha \frac{\partial \Phi}{\partial t} = 0, \quad \text{for } 0 \le x \le L, \ t \ge 0,$$

$$\frac{\partial \Phi}{\partial x}(0, t) = 0,$$

$$\frac{\partial \Phi}{\partial x}(L, t) = 0,$$

$$\Phi(x, 0) = f(x).$$

The domain is the usual half infinite $[0, L] \times [0, \infty)$ where the spatial part of the domain corresponds to the length of the dendritic cable in an excitable nerve cell. We won't worry too much about the details of where this model comes from as we will discuss that in another volume. The boundary conditions $u_x(0, t) = 0$ and $u_x(L, t) = 0$ are called *Neumann Boundary conditions*. The conditions $u(0, t) = 0$ and $u(L, t) = 0$ are known as *Dirichlet Boundary conditions*. One way to find the solution is to assume we can separate the variables so that we can write $\Phi(x, t) = u(x)w(t)$. We assume a solution of the form $\Phi(x, t) = u(x) \, w(t)$ and compute the needed partials. This leads to a the new equation

$$\beta^2 \frac{d^2 u}{dx^2} \, w(t) - u(x)w(t) - \alpha u(x) \frac{dw}{dt} = 0.$$

Rewriting, we find for all x and t, we must have

$$w(t) \left(\beta^2 \frac{d^2 u}{dx^2} - u(x) \right) = \alpha u(x) \frac{dw}{dt}.$$

This tells us

$$\frac{\beta^2 \frac{d^2 u}{dx^2} - u(x)}{u(x)} = \frac{\alpha \frac{dw}{dt}}{w(t)}, \quad 0 \le x \le L, \ t > 0.$$

The only way this can be true is if both the left and right hand side are equal to a constant that is usually called the **separation constant** Θ. This leads to the decoupled equations

$$\alpha \frac{dw}{dt} = \Theta \, w(t), \ t > 0,$$

$$\beta^2 \frac{d^2 u}{dx^2} = (1 + \Theta) \, u(x), \ 0 \le x \le L,$$

We also have boundary conditions.

$$\frac{du}{dx}(0) = 0$$

$$\frac{du}{dx}(L) = 0.$$

This gives us a second order ODE to solve in x and a first order ODE to solve in t. We have a lot of discussion about this in the text which you should study. In general, we find there is an infinite family of solutions that solve these coupled ODE models which we can label $u_n(x)$ and $w_n(t)$. Thus, any finite combination $\Phi_n(x, t) = \sum_{n=0}^{N} a_n u_n(x) w_n(t)$ will solve these ODE models, but we are still left with satisfying the last condition that $\Phi(x, 0) = f(x)$. We do this by finding a series solution. We can show that the data function f can be written as a series $f(x) = \sum_{n=0}^{\infty} b_n u_n(x)$ for a set of constants $\{b_0, b_1, \ldots\}$ and we can also show that the series $\Phi(x, t) = \sum_{n=0}^{\infty} a_n u_n(x) w_n(t)$ solves the last boundary condition $\Phi(x, 0) = \sum_{n=0}^{\infty} a_n u_n(x) w_n(0) = f(x)$ as long as we choose $a_n = b_n$ for all n. The idea of a series and the mathematical machinery associated with that takes a while to explain, so Chap. 9 is devoted to that, p. 229.

T

Triplet code   There are 20 amino acids and only 4 nucleotides. Hence, our alphabet here is $\{A, C, T, G\}$ The number of ways to take 3 things out of an alphabet of 4 things is 64. To see this, think of a given triplet as a set of three empty slots; there are 4 ways to fill slot 1, 4 independent ways to fill slot 2 (we know have 4×4 ways to fill the first two slots) and finally, 4 independent ways to fill slot 3. This gives a total of $4 \times 4 \times 4$ or 64 ways to fill the three slots independently. Since there are only 20 amino acids, it is clear that more than one nucleotide triplet could be mapped to a given amino acid! In a similar way, there are 64 different ways to form triplets from the RNA alphabet $\{A, C, U, G\}$. We tend to identify these two sets of triplets and the associated mapping to amino acids as it is just a matter of replacing the T in one set with an U to obtain the other set. The triplet code is shown in Table 4.8, p. 93.

V

Voltage dependent gate   A typical sodium channel as shown in Fig. 5.11 and it is a typical voltage dependent gate. When you look at the drawing of the sodium channel, you'll see it is drawn in three parts. Our idealized channel has a hinged cap which can cover the part of the gate that opens into the cell. We call this the inactivation gate. It also has a smaller flap inside the gate which can close off the throat of the channel. This is called the activation gate. As you see in the drawing, these two pieces can be in one of three positions: resting (activation gate is closed and the inactivation gate is open); open (activation gate is open and the inactivation gate is open); and closed (activation gate is closed or closed and the inactivation gate is closed). Since this is a voltage activated gate, the transition from resting to open depends on the voltage across the cell membrane. We typically use the following terminology:

- When the voltage across the membrane is above the resting membrane voltage, we say the cell is depolarized.
- When the voltage across the membrane is below the resting membrane voltage, we say the cell is hyperpolarized.

These gates transition from resting to open when the membrane depolarizes. In detail, the probability that the gate opens increases upon membrane depolarization. However, the probability that the gate transitions from open to closed is NOT voltage dependent. Hence, no matter what the membrane voltage, once a gate opens, there is a fixed probability it will close again. Hence, an action potential can be described as follows: when the cell membrane is sufficiently depolarized, there is an explosive increase in the opening of the sodium gates which causes a huge influx on sodium ions which produces a short lived rapid increase in the voltage across the membrane followed by a rapid return to the rest voltage with a typical overshoot phase which temporarily keeps the cell membrane hyperpolarized. The opening of the potassium gates lags the sodium gates and as the sodium gates begin to close, the influx of potassium ions is on the rise which brings the membrane voltage down actually below equilibrium and as the potassium gates close, there is the slow rise back up to equilibrium, p. 143.

Reference

J. Peterson, BioInformation Processing: A Primer On Computational Cognitive Science. Springer Series on Cognitive Science and Technology (Springer Science+Business Media Singapore Pte Ltd., Singapore, 2015, in press)

Index

© Springer Science+Business Media Singapore 2016
J.K. Peterson, *Calculus for Cognitive Scientists*, Cognitive Science
and Technology, DOI 10.1007/978-981-287-880-9

Printed in the United States
By Bookmasters